Bioinformatics

David Edwards • Jason Stajich • David Hansen
Editors

# Bioinformatics

## Tools and Applications

*Editors*
David Edwards
Australian Centre for Plant Functional Genomics
Institute for Molecular Biosciences
and School of Land
Crop and Food Sciences
University of Queensland
Brisbane, QLD 4072
Australia

David Hansen
Australian E-Health Research Centre
CSIRO
Qld 4027, Brisbane, Australia

Jason Stajich
Department of Plant Pathology
and Microbiology
University of California
Berkeley, CA
USA

ISBN 978-1-4614-9826-1 ISBN 978-0-387-92738-1 (eBook)
DOI 10.1007/978-0-387-92738-1
Springer New York Dordrecht Heidelberg London

© Springer Science+Business Media, LLC 2009
Softcover re-print of the Hardcover 1st edition 2009

All rights reserved. This work may not be translated or copied in whole or in part without the written permission of the publisher (Springer Science+Business Media, LLC, 233 Spring Street, New York, NY 10013, USA), except for brief excerpts in connection with reviews or scholarly analysis. Use in connection with any form of information storage and retrieval, electronic adaptation, computer software, or by similar or dissimilar methodology now known or hereafter developed is forbidden.
The use in this publication of trade names, trademarks, service marks, and similar terms, even if they are not identified as such, is not to be taken as an expression of opinion as to whether or not they are subject to proprietary rights.

Printed on acid-free paper

Springer is part of Springer Science+Business Media (www.springer.com)

# Preface

Biology has progressed tremendously in the last decade due in part to the increased automation in the generation of data from sequences to genotypes to phenotypes. Biology is now very much an information science, and bioinformatics provides the means to connect biological data to hypotheses. Within this volume, we have collated chapters describing various areas of applied bioinformatics, from the analysis of sequence, literature, and functional data to the function and evolution of organisms. The ability to process and interpret large volumes of data is essential with the application of new high throughput DNA sequencers providing an overload of sequence data. Initial chapters provide an introduction to the analysis of DNA and protein sequences, from motif detection to gene prediction and annotation, with specific chapters on DNA and protein databases as well as data visualization. Additional chapters focus on gene expression analysis from the perspective of traditional microarrays and more recent sequence-based approaches, followed by an introduction to the evolving field of phenomics, with specific chapters detailing advances in plant and microbial phenome analysis and a chapter dealing with the important issue of standards for functional genomics. Further chapters present the area of literature databases and associated mining tools which are becoming increasingly essential to interpret the vast volume of published biological information, while the final chapters present bioinformatics purely from a developer's point of view, describing the various data and databases as well as common programming languages used for bioinformatics applications. These chapters provide an introduction and motivation to further avenues for implementation. Together, this volume aims to provide a resource for biology students wanting a greater understanding of the encroaching area of bioinformatics, as well as computer scientists who are interested learning more about the field of applied bioinformatics.

Brisbane, QLD — David Edwards
Berkeley, CA — Jason E. Stajich
Brisbane, QLD — David Hansen

# Contents

# Contributors

**Jacqueline Batley** Australian Centre for Plant Functional Genomics, Centre of Excellence for Integrative Legume Research, School of Land, Crop and Food Sciences, University of Queensland, Brisbane, QLD 4072, Australia
j.batley@uq.edu.au

**John Boyle** The Institute for Systems Biology, 1441 North 34th Street, Seattle, WA 98105, USA
jboyle@systemsbiology.org

**Matthew Belgard** Centre for Comparative Genomics, Murdoch University, Perth, WA, Australia
mbellgard@ccg.murdoch.edu.au

**Scott Cain** Ontario Institute for Cancer Research, 101 College Street, Suite 800, Toronto, ON, Canada M5G0A3
cain@cshl.edu

**Helen C. Causton** MRC Clinical Sciences Centre, Imperial College London, Hammersmith Hospital Campus, Du Cane Road, London W12 0NN, UK
helen.causton@csc.mrc.ac.uk

**Stephen A. Chervitz** Affymetrix Inc., Santa Clara, CA 95051, USA
Steve_Chervitz@affymetrix.com

**Terry Clark** Australian Centre for Plant Functional Genomics, Institute for Molecular Biosciences and School of Land, Crop and Food Sciences, University of Queensland, Brisbane, QLD 4072, Australia
tclark@uq.edu.au

**Daniel Damian** Biowisdom Ltd., CB 22 7GG, Cambridge, UK
daniel.damian@biowisdom.com

**Dina Demner-Fushman** Communications Engineering Branch, Lister Hill National Center for Biomedical Communications, US National Library of Medicine, Bethesda, MD, USA
ddemner@mail.nih.gov

**Eric W. Deutsch** The Institute for Systems Biology, Seattle, WA 98105, USA
edeutsch@systemsbiology.org

**Tharram Dillon** Digital Ecosystems and Business Intelligence Institute,
Curtin University of Technology, Perth, WA, Australia
tharam.dillon@cbs.curtin.edu.au

**Chris Duran** Australian Centre for Plant Functional Genomics,
School of Land, Crop and Food Sciences,
University of Queensland, Brisbane, QLD 4072, Australia
c.duran@uq.edu.au

**Matthias Eberius** LemnaTec GmbH, Schumanstr. 1a,
52146 Wuerselen, Germany
matthias.eberius@lemnatec.com

**David Edwards** Australian Centre for Plant Functional Genomics,
Institute for Molecular Biosciences and School of land, Crop and Food Sciences,
University of Queensland, Brisbane, QLD 4072, Australia
Dave.Edwards@uq.edu.au

**Dawn Field** Natural Environmental Research Council,
Centre for Ecology and Hydrology, Oxford, OX1 3SR, UK
dfield@cch.ac.uk

**J. Lynn Fink** Skaggs School of Pharmacy and Pharmaceutical Sciences,
University of California, San Diego, CA, USA
jlfink@ucsd.edu

**Stephen S. Fong** Department of Chemical and Life Science Engineering,
Virginia Commonwealth University, P.O. Box 843028, Richmond, VA 23284, USA
ssfong@vcu.edu

**Jennifer M. Fostel** Division of Intramural Research,
National Institute of Environmental Health Sciences,
Research Triangle Park, NC 27709, USA
fostel@niehs.nih.gov

**Christopher M. Gowen** Department of Chemical and Life Science Engineering,
Virginia Commonwealth University, P.O. Box 843028, Richmond,
VA 23284, USA
gowencm@vcu.edu

**David Hansen** Australian E-Health Research Centre,
CSIRO QLD 4027, Brisbane, Australia
David.Hansen@csiro.au

**Jennifer Harrow** Wellcome Trust Sanger Institute, Morgan Building,
Wellcome Trust Genome Campus, Hinxton, Cambridgeshire CB10 1HH, UK
jla1@sanger.ac.uk

**Michael Imelfort** Australian Centre for Plant Functional Genomics, Institute for Molecular Biosciences and School of Land, Crop and Food Sciences, University of Queensland, Brisbane, QLD 4072, Australia
m.imelfort@uq.edu.au

**Laura A. Kavanaugh** Department of Molecular Genetics and Microbiology, Duke University, Durham, NC 27710, USA
laura.kavanaugh@duke.edu

**Ian Korf** UC Davis Genome Center, University of California, Davis, 451 Health Sciences Drive, Davis, CA 95616, USA
ifkorf@ucdavis.edu

**José Lima-Guerra** Keygene N.V., Agrobusiness Park 90, 6708 PW Wageningen, The Netherlands
jose.guerra@keygene.com

**William H. Majoros** Institute for Genome Sciences & Policy, Duke University, Durham, NC 27708, USA
william.majoros@duke.edu

**Lakshmi K. Matukumalli** Department of Bioinformatics and Computational Biology, George Mason University, Manassas, VA 20110, USA
Lakshmi.Matukumalli@ARS.USDA.GOV

**Sheldon McKay** Cold Spring Harbor Laboratory, 1 Bungtown Road, Cold Spring Harbor, NY 11724, USA
mckays@cshl.edu

**Alan Moses** Department of Cell & Systems Biology, University of Toronto, 25 Willcocks Street, Toronto, ON, Canada M5S 3B2
alan.moses@utoronto.ca

**Uwe Ohler** Department of Biostatistics & Bioinformatics, Institute for Genome Sciences & Policy, Duke University, Durham, NC 27708, USA
uwe.ohler@duke.edu

**Helen Parkinson** European Bioinformatics Institute, Wellcome Trust Genome Campus, Hinxton, Cambridge, UK
parkinson@ebi.ac.uk

**Philippe Rocca-Serra** European Bioinformatics Institute, Wellcome Trust Genome Campus, Hinxton, Cambridge, UK
rocca@ebi.ac.uk

**Susanna-Assunta Sansone** European Bioinformatics Institute, Wellcome Trust Genome Campus, Hinxton, Cambridge, UK
sansone@ebi.ac.uk

**Steven G. Schroeder** Bovine Functional Genomics Laboratory, US Department of Agriculture, Beltsville, MD 20705, USA
steven.schroeder@ars.usda.gov

**Amandeep S. Sidhu** Centre for Comparative Genomics, Murdoch University, Perth, WA, Australia
asidhu@ccg.murdoch.edu.au

**Saurabh Sinha** Department of Computer Science, University of Illinois, Urbana-Champaign, 201 N. Goodwin Ave, Urbana, IL 61801, USA
sinhas@uiuc.edu

**Jason Stajich** Department of Plant Pathology and Microbiology, University of California, Berkeley, CA 94720-3102, USA
jason_stajich@berkeley.edu

**Christian J. Stoeckert Jr** Department of Genetics,
Penn Center for Bioinformatics, University of Pennsylvania
School of Medicine, Philadelphia, PA 19104-6021, USA
stoeckrt@pcbi.upenn.edu

**Chris F. Taylor** European Bioinformatics Institute, Wellcome Trust Genome Campus, Hinxton, Cambridge, UK
christ@ebi.ac.uk

**Joe White** Dana-Farber Cancer Institute and Harvard School of Public Health, Harvard University, Boston, MA 02115, USA
jwhite@jimmy.harvard.edu

**Laurens Wilming** Wellcome Trust Sanger Institute, Morgan Building, Wellcome Trust Genome Campus, Hinxton, Cambridgeshire CB10 1HH, UK
lw2@sanger.ac.uk

**Sitao Wu** Center for Bioinformatics and Department of Molecular Bioscience, University of Kansas, Lawrence, KS 66047, USA
stwu@ku.edu

**Yang Zhang** Center for Bioinformatics and Department of Molecular Bioscience, University of Kansas, Lawrence, KS 66047, USA
yzhang@ku.edu

**Pierre Zweigenbaum** LIMSI-CNRS, BP 133, 91403 Orsay Cedex, France
pz@limsi.fr

# Chapter 1
# DNA Sequence Databases

**David Edwards, David Hansen, and Jason E. Stajich**

## 1.1 DNA Sequencing: An Introduction

The ability to sequence the DNA of an organism has become one of the most important tools in modern biological research. Beginning as a manual process, where DNA was sequenced a few tens or hundreds of nucleotides at a time, DNA sequencing is now performed by high throughput sequencing machines, with *billions* of bases of DNA being sequenced daily around the world. The recent development of "next generation" sequencing technology increases the throughput of sequence production many fold and reduces costs by orders of magnitude. This will eventually enable the sequencing of the whole genome of an individual for under 1,000 dollars. However, mechanisms for sharing and analysing this data, and for the efficient storage of the data, will become more critical as the amount of data being collected grows. Most importantly for biologists around the world, the analysis of this data will depend on the quality of the sequence data and annotations which are maintained in the public databases.

In this chapter we will give an overview of sequencing technology as it has changed over time, including some of the new technologies that will enable the sequencing of personal genomes. We then discuss the public DNA databases which collect, check, and publish DNA sequences from around the world. Finally we describe how to access this data.

D. Edwards (✉)
Australian Centre for Plant Functional Genomics, Institute for Molecular Biosciences and School of Land, Crop and Food Sciences, University of Queensland, Brisbane, QLD 4072, Australia
e-mail: Dave.Edwards@uq.edu.au

D. Edwards et al. (eds.), *Bioinformatics: Tools and Applications*,
DOI 10.1007/978-0-387-92738-1_1, © Springer Science+Business Media, LLC 2009

## 1.2 Sequencing Technology

Because of the cost and time consuming nature of initial DNA sequencing methods it was not considered feasible to sequence the genomes of most organisms, and DNA sequencing projects focused on single pass sequencing of expressed genes, to produce Expressed Sequence Tags (ESTs). Genes are specifically expressed in tissues in the form of messenger RNA (mRNA). A pool of mRNA is extracted from a tissue and used to produce a library of complementary DNA (cDNA) which is more stable and amenable to sequencing than the extracted RNA. Individual cDNAs from a library are then sequenced from one direction to produce the sequence tag. EST sequencing is a cost effective method for the rapid discovery of gene sequences that may be associated with development or environmental responses in the tissues from which the mRNA was extracted (Adams et al. 1991). However, mRNA species are highly redundant, with highly expressed genes demonstrating much greater abundance than genes expressed at lower levels. Similarly, only genes expressed in the tissue, growth stage and condition used to sample the mRNA are sequenced. As the number of EST sequences produced from a cDNA library increases, fewer sequences represent new genes that have not already been sampled. Thus, while EST sequencing is a valuable means to rapidly identify genes moderately or highly expressed in a tissue, the method rarely identifies genes that are expressed at lower levels, including many genes that encode regulatory proteins that are only expressed in very small quantities.

An alternative to EST sequencing is genome sequencing, aiming to sequence either the whole genome or portions of the genome. This method removes the bias associated with tissue specificity or gene expression level. However, eukaryote genomes tend to be very large, and genome sequencing only became feasible with the development of capillary-based high throughput sequencing technology. This technology has led to a move from EST sequencing to the sequencing of whole genomes, and the recent development of next generation sequencing technology will lead to genome sequencing becoming increasingly common.

The vast majority of DNA sequencing to date has been carried out using the chain-termination method developed by Frederick Sanger (Sanger et al. 1977). The modern Sanger method involves DNA synthesis from a single stranded DNA template using a mixture of deoxynucleotides spiked with fluorescent chain terminating dideoxynucleotides. A dideoxynucleotide lacks a 3′-OH group preventing continued DNA synthesis, and each of the four dideoxynucleotides, ddATP, ddGTP, ddCTP, and ddTTP are labelled with a different fluorescent marker. This reaction results in a pool of fragments differing in length by one nucleotide and terminating with a specific fluorescent nucleotide corresponding to the terminal nucleotide base (A, C, G or T). Separation of these fragments by size combined with detection of the fluorophores provides a readout of the DNA sequence in the form of a chromatogram.

The term "next generation sequencing technology" describes platforms which produce large amounts (typically millions) of short length reads (25–400 bp). One such advance is the application of pyrosequencing. The first commercially available pyrosequencing system was developed by 454 and commercialised by Roche as the

GS20, capable of sequencing over 20 million base pairs in just over 4 h. This was replaced during 2007 by the GS FLX model and more recently, by the GS FLX Titanium, capable of producing over 400 million base pairs of sequence. Three alternative ultra high throughput sequencing systems now compete with the GS FLX, Solexa technology, which is now commercialised by Illumina; the SOLiD system from Applied Biosystems (AB); and the tSMS system from Helicos Biosciences Corporation. For a summary of each of these technologies, see Table 1.1. In addition, there are at least two other systems that are in the process of being commercialized by Pacific Biosciences and VisiGen Biotechnologies.

The Roche 454 FLX system performs amplification and sequencing in a highly parallelised picoliter format. In contrast with Solexa and AB SOLiD, GS FLX Titanium read lengths average 400–500 bases per read, and with more than one million reads per run it is capable of producing over 400 Mbp of sequence with a single-read accuracy of greater than 99.5%. Emulsion PCR enables the amplification of a DNA fragment immobilized on a bead from a single fragment to 10 million identical copies, generating sufficient DNA for the subsequent sequencing reaction. Sequencing involves the sequential flow of both nucleotides and enzymes over the picoliter plate, which convert chemicals generated during nucleotide incorporation into a chemiluminescent signal that can be detected by a CCD camera. The light signal is quantified to determine the number of nucleotides incorporated during the extension of the DNA sequence.

The Solexa sequencing system, sold as the Illumina Genome Analyzer, uses reversible terminator chemistry to generate up to six thousand million bases of usable data per run. Sequencing templates are immobilized on a flow cell surface. Solid phase amplification creates clusters of up to 1,000 identical copies of each DNA molecule. Sequencing uses four proprietary fluorescently labelled nucleotides to sequence the millions of clusters on the flow cell surface. These nucleotides possess a reversible termination property, allowing each cycle of the sequencing reaction to occur simultaneously in the presence of the four nucleotides. Solexa sequencing has been developed predominantly for re-sequencing, but is also suitable for *de novo* sequencing.

The AB SOLiD System enables parallel sequencing of clonally amplified DNA fragments linked to beads. The method is based on sequential ligation with dye labelled oligonucleotides and can generate more than six gigabases of mappable data per run. The system features a two base encoding mechanism that interrogates each

**Table 1.1** Properties of the different sequencing methods

| Sequencing Machine | ABI 3730 | Roche GSFLX | Illumina Solexa | AB SOLiD | Helicos tSMS |
|---|---|---|---|---|---|
| Launched | 2000 | 2004 | 2006 | 2007 | 2008 |
| Read length, nucleotides | 800–1,100 | 250–400 | 35–70 | 25–35 | 28 |
| Reads/run | 96 | 400 K | 120 M | 170 M | 85 M |
| Throughput per run | 0.1 MB | 100 MB | 6 GB | 6 GB | 2 GB |
| Cost/GB (2007) | >$2,500 k | $84 k | $4 k | $4 k | ? |

base twice providing a form of built in error detection. The system can be used for re-sequencing and tag based applications such as gene expression and Chromatin ImmunoPrecipitation, where a large number of reads are required and where the high throughput provides greater sensitivity for the detection of lowly expressed genes.

The field of sequencing technology is advancing very rapidly, with planned improvements to current next generation systems and new systems becoming available. This will continue to place demands on data storage and interrogation and a radical rethink of the current DNA database systems may be required to manage this flood of data.

## 1.3 DNA Databases

DNA sequence data forms the core of genomics and is the foundation of much bioinformatics research. Because of this, several systems have been developed for the maintenance, annotation, and interrogation of DNA sequence information. DNA sequences are collected in many different databases and for different purposes. Often DNA sequences are the result of a collaboration effort to sequence a particular genome, such as the Human Genome Mapping Project in the 1990s or more recent efforts to sequence the genome of particular animals and plants. The data which is collected is often stored in different databases, sometimes with different annotations. For example, the DNA sequence for a particular organism might be stored in distributed databases as part of an organism specific dataset, with a subset of the data being submitted to the major international databases for distribution as part of those datasets.

The largest of the DNA sequence repositories is the International Nucleotide Sequence Database Collaboration (INSDC), made up of the DNA Data Bank of Japan (DDBJ) at The National Institute of Genetics in Mishima, Japan (Sugawara et al. 2008), GenBank at the National Center of Biotechnology Information (NCBI) in Bethesda, USA (Benson et al. 2006), and the European Molecular Biology Laboratory (EMBL) Nucleotide Sequence Database, maintained at the European Bioinformatics Institute (EBI) in the UK (Cochrane et al. 2006). Daily data exchange between these groups ensures coordinated international coverage. The Institute for Genomic Research (TIGR) based at Rockville, Maryland, USA (Lee et al. 2005) also maintains various data types including genomic sequences and annotation.

### *1.3.1 GENBANK*

GenBank® is a comprehensive database of publicly available DNA sequences for more than 300,000 named organisms, obtained through submissions from individual laboratories and batch submissions from large-scale sequencing projects.

GenBank is maintained and distributed by the National Center for Biotechnology Information (NCBI), a division of the National Library of Medicine (NLM), at the US National Institutes of Health (NIH) in Bethesda, MD. NCBI builds GenBank from several sources including the submission of sequence data from authors and from the bulk submission of expressed sequence tag (EST), genome survey sequence (GSS), whole genome shotgun (WGS) and other high-throughput data from sequencing centers. The U.S. Office of Patents and Trademarks also contributes sequences from issued patents.

The FTP release of GenBank consists of a mixture of compressed and uncompressed ASCII text files, containing sequence data and indices that cross reference author names, journal citations, gene names and keywords to individual GenBank records. For convenience, the GenBank records are partitioned into divisions according to source organism or type of sequence. Records within the same division are packaged as a set of numbered files so that records from a single division may be contained in a series of many files. The full GenBank release is offered in two formats; the GenBank "flatfile" format, and the more structured and compact Abstract Syntax Notation One (ASN.1) format used by NCBI for internal maintenance. Full releases of GenBank are made every 2 months beginning in the middle of February each year. Between full releases, daily updates are provided on the NCBI FTP site (ftp://ftp.ncbi.nih.gov/genbank/, ftp://ftp.ncbi.nih.gov/ncbi-asn1/). The Entrez system always provides access to the latest version of GenBank including the daily updates.

### 1.3.2 The Composition of GenBank

From its inception, GenBank has doubled in size about every 18 months. Contributions from WGS projects supplement the data in the traditional divisions, and the number of eukaryote genomes for which coverage and assembly are significant continues to increase as well. Database sequences are classified and can be queried using a comprehensive sequence-based taxonomy (Federhen 2003) developed by NCBI in collaboration with EMBL and DDBJ with the assistance of external advisers and curators. Detailed statistics for the current release may always be found in the GenBank release notes (ftp.ncbi.nih.gov/genbank/gbrel.txt).

### 1.3.3 EMBL

The European Bioinformatics Institute (EBI), part of the European Molecular Biology Laboratory (EMBL), is a non-profit organisation and a centre for research and services in bioinformatics. The Institute manages many databases containing biological information, including nucleic acid and protein sequences, and macromolecular structures. Data and bioinformatics services are made available freely to the scientific community.

Among the databases provided, the most widely known are the EMBL Nucleotide Sequence Database (Kanz et al. 2005), further referred to as EMBL database. The EMBL database is a typical primary (archival) database: it contains the results from sequencing experiments in laboratories together with the interpretation provided by the submitters, but with no or very limited review. The data in the EMBL database originates from large-scale genome sequencing projects, the European Patent Office, and direct submissions from individual scientists. It is important to note that the editorial rights to an entry in the EMBL database remain with the original submitter. This means that apart from the addition of cross-references, the data is not updated by EMBL database curators, unless explicitly instructed by the submitter. There is a quarterly release of the EMBL database and new and updated records are distributed on a daily basis. When EMBL entries are updated, older versions of the same entry can still be retrieved from the EMBL Sequence Version Archive (Leinonen et al. 2003).

Current statistics for the database can be found at http://www3.ebi.ac.uk/Services/DBStats/. Data in the EMBL Nucleotide Sequence Database are grouped into divisions, according to either the methodology used in their generation, such as EST (expressed sequence tag) and HTG (high throughput genome sequencing), or taxonomic origin of the sequence source, such as the PLN division for plants. This is done to create subsets of the database which separates sequences of lower quality such as EST and HTG sequences from high-quality sequences, and groups high-quality sequences into groups that reflect the main areas of interest. The advantage becomes obvious when you wish to limit your database search or sequence similarity analysis to plant entries rather than the whole database; for example, only a relatively small fraction of the total database needs to be searched when using the PLN division.

### *1.3.4 Collation and Distribution of DNA Sequence Data*

DNA sequence data is usually provided by researchers through a web application that captures the relevant sequence and associated annotation information. To maintain quality, there are processes to ensure that both the raw sequence files as well as relevant information from the sequencing machines about the quality of the sequence reads is collected. Annotation information may also be provided by the submitting scientists, and will generally include descriptions of known features and any publications related to the sequence.

While the data is stored and maintained in relational databases, distribution is in the form of a series of so called semi-structured text files. The format of a semi-structured text file is that an entry will appear as a series of lines with the first word of each line denoting what is contained on the rest of the line. In order to allow the data to be processed by a computer, a formal specification of the format is also provided in Backus–Naur form.

Some DNA databases are distributed in a XML format. In these cases the data is the same but the format of the files they are in is different. Each database entry is represented by an XML object, with each eXML object described by an XML schema or a DTD. The XML schema or document type definition (DTD) replaces the need for the Backus–Naur form description.

## *1.3.5 Data Maintained in a DNA Database*

The actual DNA sequence is sometimes not the most important part of a DNA database. Rather, it is the annotation – or known information about the sequence – which provides the real value for biological interpretation. One reason for this is that sequence similarity tools, such as the BLAST family of tools, are typically used to compare sequences of unknown function with sequences in the DNA databases. The function of the new sequence may then be inferred from the known function of like sequences. Much of the annotation is provided by the scientists submitting the sequence to the database, however there are numerous groups which apply both manual and semi automated methods of annotation to sequence databases to correct errors present during data submission and to generally improve the knowledge base of the data. Sequence annotation is covered in more detail in other chapters in this volume.

### 1.3.5.1 General Information

As well as the sequence itself, DNA databases generally maintain a host of additional information relating to the sequence. The general information relating to a DNA sequence may include:

- Unique identifiers for the sequence. Within the large international databases, EMBL, Genbank and DDBJ, entries are given a Primary Accession Number (denoted by AC in the flat file distribution). Entries also have a list of secondary Accession numbers.
- The division that the sequence belongs to (e.g. PLN, plant or WGS, whole genome shotgun).
- A version number for the sequence, to allow errors in the sequence to be corrected as new data becomes available.
- A creation date, representing when the sequence was first added to the database.
- Modification dates, representing modifications to the entry.
- A description of the sequence, including a general functional description, a list of keywords that describe the sequence, the organism from which the sequence comes and a list of terms representing the taxonomic classification of the organism.

#### 1.3.5.2 Database Cross References

With the proliferation of bioinformatics databases, the cross referencing of entries between databases is an important feature of most databases. The large DNA databases cross reference the DNA entries to many additional databases that host related information about the entry. These include:

- Protein sequence databases, such as Interpro and Uniprot, when the DNA sequence encodes a protein
- Protein structure databases, such as PDB, which again contain proteins encoded by the DNA
- GO, the gene ontology database, describing gene ontology terms that may relate to the sequence
- Pubmed for literature references relating to the sequence

#### 1.3.5.3 Feature Table

The FH lines denote the feature table header, while the actual features and corresponding qualifiers are listed on the FT lines. Features could include the source, describing the biological source of the nucleic acid sequenced; CDS (CoDing Sequence) features describing information relating to any translated protein; or mRNA describing information relating to expressed portions of the sequence. Each feature may have one or more qualifiers that provide additional information about the feature.

### *1.3.6 Model Organism Databases*

The (INSDC) database system of NCBI, EMBL and DDBJ provides a stable and persistent archive of deposited DNA sequence data. However, the task of maintaining a curated and up-to-date genome annotation for an organism requires a different type of database with more hands-on expert curation. These make up the Model Organism Databases, which include TAIR – The *Arabidopsis* Information Resource (*Arabidopsis thaliana*) (Swarbreck et al. 2008); SGD – *Saccharomyces* Genome Database (*Saccharomyces cerevisiae*) (Hong et al. 2008), FlyBase (*Drosophila melanogastor* and related species) (Drysdale 2008) and WormBase (*Caenorhabditis elegans* and related worm species) (Rogers et al. 2008). These databases capture not only the genome sequence and DNA deposited from the community and the genome sequencing project, but also information about known genetic alleles (and their sequence if known), curated literature references, and detailed curation of the genomic sequence annotation including the gene models, repetitive sequence, and other genomic features. These databases serve as clearing houses for much of the knowledge about an organism focused around its role as a model system for research.

## 1.4 Interrogating DNA Sequence Databases

The vast amount of data within DNA sequence databases provides the challenge of searching and accessing this data in a biologically relevant and rapid manner. The simplest means of accessing a sequence is to search a database with the sequence accession number, though even this may be confounded where a sequence may have multiple identifiers, not all of which are present in all databases. Other means to search sequence databases include sequence similarity or through sequence annotation and some examples of these are described below.

### *1.4.1 Sequence Similarity*

One of the most common means to search sequence databases is through sequence similarity with known sequences. The mechanisms underlying these searches are detailed in the Imelfort chapter in this volume and will not be dealt with here. Of the sequence similarity search tools available, BLAST (Altschul et al. 1990) and its various flavours is the most commonly used due to its speed. While the BLAST algorithm has been broadly adopted, the interfaces through which to conduct these searches are many and varied, with new and more complex interfaces being developed on a regular basis associated with specific species or datasets. BLAST is well documented (http://tinyurl.com/NCBIBLAST) and researchers are recommended to become familiar with the various flavours and options available to optimise sequence comparisons.

### *1.4.2 Keyword and GO Annotation*

Many of the DNA sequence databases permit the searching of sequence annotation data using keywords or structured annotation in the form of Gene Ontologies. While the annotation data and search capabilities are more advanced in the model organism databases, some key information such as sequence submitter, preliminary annotation, and taxonomy details are also maintained within the NCBI, EMBL and DDBJ databases. The introduction of gene ontologies has greatly assisted the structured searching and analysis of the increasing quantities of DNA sequence information. However, there remain concerns about the quantity of erroneous annotation, particularly from computational annotation methods. Gene annotation is addressed in more detail in other chapters in this volume.

### 1.4.3 Genome Position

Where a genome has been fully sequenced, the position of a sequence in relation to the rest of the genome is important, particularly when comparing related genomes. Comparative genomics is a growing field of research which permits the deciphering of evolutionary events and allows the translation of information from model species to non-model species for direct applications in microbiology, as well as plant and animal biology. Genome position is becoming increasingly important in the era of second generation sequencing technology, where many related genomes are sequenced or re-sequenced and compared to a reference genome. In these cases, the storage of all the new sequence information is not required, and only the positions where the new sequence information differs from the reference is of relevance. The best example of this is the human HapMap project (The International HapMap Consortium 2007). With the continued increase in DNA sequence production, there will be a move from maintaining databases of increasingly redundant DNA sequences to developing databases of genetic differences between individuals and species, leading to a fundamental change in the way we store, interrogate, and interpret DNA sequence data.

## References

Adams MD, Kelley JM, Gocayne JD, Dubnick M, Polymeropoulos MH, Xiao H et al (1991) Complementary DNA sequencing: expressed sequence tags and human genome project. Science 252:1651–1656

Altschul SF, Gish W, Miller W, Myers EW, Lipman DJ (1990) Basic local alignment search tool. J Mol Biol 215:403–410

Benson DA, Karsch-Mizrachi I, Lipman DJ, Ostell J, Wheeler DL (2006) Genbank. Nucl Acids Res 34:D16–D20

Cochrane G, Aldebert P, Althorpe N, Andersson M, Baker W, Baldwin A et al (2006) EMBL nucleotide sequence database: developments in 2005. Nucl Acids Res 34:D10–D15

Drysdale R (2008) FlyBase – A database for the Drosophila research community. Methods Mol Biol 420:45–59

Federhen, S. (2003) The taxonomy project, in The NCBI Handbook. National Center for Biotechnology Information.

Hong EL, Balakrishnan R, Dong Q, Christie KR, Park J, Binkley G et al (2008) Gene Ontology annotations at SGD: new data sources and annotation methods. Nucleic Acids Res 36:D577–D581

Kanz C, Aldebert P, Althorpe N, Baker W, Baldwin A, Bates K et al (2005) The EMBL Nucleotide Sequence Database. Nucleic Acids Res 33:D29–D33

Lee Y, Tsai J, Sunkara S, Karamycheva S, Pertea G, Sultana R et al (2005) The TIGR Gene Indices: clustering and assembling EST and known genes and integration with eukaryotic genomes. Nucl Acids Res 33:D71–D74

Leinonen R, Nardone F, Oyewole O, Redaschi N, Stoehr P (2003) The EMBL sequence version archive. Bioinformatics 19:1861–1862

Rogers A, Antoshechkin I, Bieri T, Blasiar D, Bastiani C, Canaran P et al (2008) WormBase. Nucleic Acids Res 36:D612–D617

Sanger F, Nicklen S, Coulson AR (1977) DNA sequencing with chain-terminating inhibitors. Proc Natl Acad Sci USA 74(12):5463–5467

Sugawara H, Ogasawara O, Okubo K, Gojobori T, Tateno Y (2008) DDBJ with new system and face. Nucleic Acids Res 36:D22–D24

Swarbreck S, Wilks C, Lamesch P, Berardini TZ, Garcia-Hernandez M, Foerster H et al (2008) The Arabidopsis Information Resource (TAIR): gene structure and function annotation. Nucleic Acids Res 36:D1009–D1014

The International HapMap Consortium (2007) A second generation human haplotype map of over 3.1 million SNPs. Nature 449:851–861

# Chapter 2
# Sequence Comparison Tools

**Michael Imelfort**

## 2.1 Introduction

The evolution of methods which capture genetic sequence data has inspired a parallel evolution of computational tools which can be used to analyze and compare the data. Indeed, much of the progress in modern biological research has stemmed from the application of such technology. In this chapter we provide an overview of the main classes of tools currently used for sequence comparison. For each class of tools we provide a basic overview of how they work, their history, and their current state. There have been literally hundreds of different tools produced to align, cluster, filter, or otherwise analyze sequence data and it would be impossible to list all of them in this chapter, so we supply only an overview of the tools that most readers may encounter. We apologize to researchers who feel that their particular piece of software should have been included here. The reader will notice that there is much conceptual and application overlap between tools and in many cases one tool or algorithm is used as one part of another tool's implementation. Most of the more popular sequence comparison tools are based on ideas and algorithms which can be traced back to the 1960s and 1970s when the cost of computing power first became low enough to enable wide spread development in this area. Where applicable we describe the original algorithms and then list the iterations of the idea (often by different people in different labs) noting the important changes that were included at each stage. Finally we describe the software packages currently used by today's bioinformaticians. A quick search will allow the reader to find many papers which formally compare different implementations of a particular algorithm, so while we may note that one algorithm is more efficient or accurate than another we stress that we have not performed any formal benchmarking or comparison analysis here.

The classes of tools discussed below are *sequence alignment*, including sequence homology searches and similarity scoring, *sequence filtering* methods, usually used for identifying, masking, or removing repetitive regions in sequences, and

M. Imelfort (✉)
University of Queensland, Queensland, Australia
e-mail: m.imelfort@uq.edu.au

D. Edwards et al. (eds.), *Bioinformatics: Tools and Applications*,
DOI 10.1007/978-0-387-92738-1_2, © Springer Science+Business Media, LLC 2009

*sequence assembly* and *clustering* methods. Sequence annotation tools including gene prediction and marker discovery have been covered elsewhere in this volume and are not discussed here.

## 2.2 Sequence Alignment

At the most basic level a sequence alignment is a residue by residue matching between two sequences, and algorithms which search for homologous regions between two sequences by aligning them residue by residue are arguably the most fundamental components of sequence comparison. It is also biologically relevant to consider situations where nucleotides have either been inserted into or deleted from DNA; most, but not all, sequence alignment algorithms allow the matching of a residue with a gap element or simply a gap.

Consider two sequences which are identical except that the first sequence contains one extra residue. When we view the alignment of these two sequences, the extra residue will be matched to a gap. This corresponds to an insertion event in the first sequence or a deletion event in the second. On the other hand, if we note that an insertion event has occurred in the first sequence (with respect to the second) then we know how to match that residue to a gap in the second. Thus one way to build a sequence alignment is to find a series of insertions, deletions, or replacements, collectively called *mutation events*, which will transform one sequence into the other. The number of mutation events needed to transform one sequence into the other is called the *edit distance*. As there will always be more than one series of possible mutation events which transform the first sequence into the second, it makes sense to rate each set's likelihood of occurrence. Greater confidence is placed in alignments which have a higher likelihood of occurring. Each alignment can be rated by considering both the cumulative probabilities and biological significance of each mutation event. For example, an alignment which infers a lesser amount of mutations to transform one sequence into another is almost always considered more likely to have occurred than an alignment which infers many more mutations, therefore many alignment algorithms work by minimizing the edit distance.

To resolve the issue of biological significance, information about the distribution of mutation events is used. This information is most commonly stored, in a scoring matrix. Each pair of matched residues (or residue – gap pairs) can be scored and the *similarity score* is the sum of the scores of the individual residues. Most alignment algorithms seek to produce meaningful alignments by maximizing the similarity score for two sequences. Traditionally, sequence alignment algorithms have been called *global* if they seek to optimize the overall alignment of two sequences. Often the resulting alignment can include long stretches of residues which are matched with different residues or gaps. Conversely, if the algorithm seeks to align highly conserved subsequences while ignoring any intervening unconserved regions then it is called a *local* alignment algorithm. A local alignment of two sequences can produce a number of different subsequent alignments. So far, only the case where two sequences are being compared has been described. This is called a *pairwise* alignment. The case where more than two sequences are being compared concurrently is called a *multiple* alignment.

Alignment algorithms can be broadly classified as taking a heuristic or dynamic programming approach. Generally, dynamic programming based approaches are guaranteed to produce the best alignments but are very often computationally and memory expensive, while heuristic based algorithms sacrifice guaranteed quality for speed. Note that a heuristic algorithm can produce an optimal alignment; there is just no guarantee that it will. Often dynamic programming approaches are used to *finish* or perfect alignments made using heuristics.

### *2.2.1 Substitution Matrices*

When two sequences are aligned there are often residues in one sequence which do not match residues in the other. There is usually more than one way to align two sequences, so a scoring system is needed to decide which of the possible alignments is the best. For a nucleotide alignment, a simple scoring system could award one point for every match and zero points for a mismatch or a gap. This information can be stored in a matrix called a *substitution* matrix. An example of such a matrix is shown as the first matrix in Fig. 2.1 below. This is the substitution matrix employed with good results in the original Needleman–Wunsch algorithm (Needleman and Wunsch 1970). However, this matrix was criticized as lacking in both biological relevance and mathematical rigor and there have been a number of attempts to improve on both this and some earlier methods resulting in the scoring systems used today. In 1974, Sellers introduced a metric which could be used to describe the evolutionary distance between two sequences (Sellers 1974) and this method was generalized by Waterman et al in 1976 (Waterman et al. 1976). The idea behind these scoring systems was to minimize the number of mutations needed to transform one sequence into the other while also taking into account the differing probabilities of distinct mutation events. For example, a more sophisticated scoring system could award negative scores for gaps (a *gap penalty*) and for mismatches the scores

$$
\begin{array}{c} \\ A \\ C \\ G \\ T \end{array}
\begin{array}{c} \begin{matrix} A & C & G & T \end{matrix} \\ \begin{bmatrix} 1 & 0 & 0 & 0 \\ 0 & 1 & 0 & 0 \\ 0 & 0 & 1 & 0 \\ 0 & 0 & 0 & 1 \end{bmatrix} \end{array}
\qquad
\begin{array}{c} \\ A \\ C \\ G \\ T \end{array}
\begin{array}{c} \begin{matrix} A & C & G & T \end{matrix} \\ \begin{bmatrix} 3 & -2 & -1 & -2 \\ -2 & 3 & -2 & -1 \\ -1 & -2 & 3 & -2 \\ -2 & -1 & -2 & 3 \end{bmatrix} \end{array}
$$

**Fig. 2.1** Two examples of nucleotide similarity matrices. The first matrix implements a binary scoring scheme awarding one point for a match and one for a mismatch. The second matrix introduces a more sophisticated method where biological observations such as the unequal probabilities of transitions and transversions influence the score

could differ according to whether the mismatch was a transition event or a transversion event. An example of this is the second matrix in Fig. 2.1 below.

For protein alignments, each row and column in the substitution matrix S corresponds to a particular amino acid, where each entry $S_{i,j}$ contains a value representing the probability of substituting the residue in row i for the residue in column j. The most widely used examples of such matrices are the point accepted mutation (PAM) matrices (Dayhoff 1978) and Block substitution matrices (BLOSUM) (Henikoff and Henikoff 1992). Both matrices share many similarities. They are both $20 \times 20$ matrices and in both cases identities and conservative substitutions are given high scores while unlikely replacements are given much lower scores. Both matrices are assigned numbers which identify when they should be used, for example PAM30 or BLOSUM62. However, one should use a higher numbered BLOSUM matrix when comparing more similar sequences while for PAM matrices lower numbers should be used. A more important difference is the way the matrices are built. PAM matrices are derived from an explicit model of evolution and based on observations of closely related protein sequences, while BLOSUM matrices are based directly on observations of alignments of more distantly related sequences using a much larger dataset than for PAM. As a result the BLOSUM matrices tend to produce better results than PAM matrices, particularly when aligning distantly related sequences.

Work on the PAM matrix model of protein evolution was undertaken by Dayhoff in the late 1970s (Dayhoff 1978). The main idea behind the PAM matrices is that of all possible mutations; we are going to observe only those which are accepted by natural selection. PAM1 was calculated using the observed relative frequencies of amino acids and 1,572 observed mutations in multiple alignments for 71 families of closely related proteins. Each entry in PAM1 represents the expected rates of amino acid substitution we would expect if we assume that on average only 1% of the residues in one sequence have mutated (Dayhoff 1978). By assuming that further mutations would follow the same pattern and allowing multiple substitutions at the same site, one can calculate the expected rates of substitution if we assume on average that 2% of the residues have mutated.This is the PAM2 matrix. Thus all the PAM matrices are calculated from the PAM1 matrix and are based on an explicit model of evolution based on point mutations. Matrices were calculated by Dayhoff up to PAM250.

The PAM approach performs well on closely related sequences but its performance declines for more distantly related sequences. The BLOSUM matrices were derived by Steven and Jorja Henikoff in the early 1990s to address this problem. To build a BLOSUM matrix, local alignments are made using sequences obtained from the BLOCKS database. Sequences with a similarity greater than a given cut off are combined into one sequence producing groups with a given maximum similarity. This reduces any bias caused by large numbers of highly similar sequences (Henikoff and Henikoff 1992). The value for the cut off is appended to the name of the matrix, thus the BLOSUM62 matrix is effectively made by comparing sequences with less than 62% similarity. As a result BLOSUM80 is a better matrix to use when aligning closely related sequences than BLOSUM30 which is better suited to aligning highly diverged sequences. BLOSUM62 is the default matrix used in the BLAST algorithm described below.

### 2.2.2 *Pairwise Sequence Alignment Algorithms*

At the base of many sequence comparison tools are pairwise sequence alignment algorithms. Beginning in the mid 1960s a large number of heuristic algorithms were suggested for the pairwise alignment of protein sequences. The era of modern sequence alignment techniques began in 1970 with the publication by Needleman and Wunsch of a dynamic programming method which could be used to make a global pairwise alignment of two protein sequences (Needleman and Wunsch 1970). In 1981, Smith and Waterman extended the ideas put forward by Needleman and Wunsch to create the local alignment algorithm known as the Smith–Waterman algorithm (Smith et al. 1981). Both the Needleman–Wunsch and Smith–Waterman methods belong to a class of algorithms called dynamic programming algorithms. This class of algorithms can find optimal solutions to problems but can take a long time to run, especially in complicated cases or for large data sets. These two algorithms are the most accurate pairwise alignment algorithms in existence. Nearly all of the newer local pairwise alignment algorithms use a two step approach to reduce the running time. The first stage uses heuristics to search for areas which have a high probability of producing alignments. Next, these areas are passed to a dynamic programming algorithm such as the Smith–Waterman algorithm for true alignment. The most commonly used two step approaches are FASTP/FASTA, the BLAST family of algorithms, Crossmatch/SWAT, and BLAT, although there are many others.

Higher order sequence comparison tools often employ pairwise alignment algorithms to judge similarity for use in clustering or assembly, so it is important to understand how these basic algorithms work and which sequences they are better suited to. We provide below an overview of the most common pairwise alignment algorithms.

#### 2.2.2.1 The Needleman–Wunsch Algorithm

This is a highly accurate, dynamic programming based, global pairwise alignment algorithm. It was originally developed for aligning protein sequences but can also be used to align DNA sequences. This algorithm aligns two sequences A and B with lengths m and n residues respectively, by finding a path through a two dimensional $m \times n$ array; S. As all $m \times n$ values in S must be calculated for every alignment, the work needed to align two sequences becomes intractable for large m and n. For the following example, we assume the use of the simple nucleotide similarity matrix in Fig. 2.1. First the bottom right cell $S_{m,n}$ is assigned the value 1 or 0 depending on whether the base in position m of A matches the base in position n of B. The cell diagonally above and to the left of this cell; $S_{m-1, n-1}$, is given a value of 2 if the base in position m – 1 of A matches the base in position n – 1 of B or a value of 1 otherwise. This is because a match will produce a maximum run of matches of length 2 for all bases from this point on, whereas a mismatch will produce a run of at most one match. The algorithm continues working backwards until every cell has been assigned a value. Finally the algorithm starts from the highest scoring cell in the array, and finds a path through the array which maximizes the cumulative sum of the values in the cells visited in the path. The resulting path represents a *maximally matching* global alignment.

#### 2.2.2.2 The Smith–Waterman Algorithm

In 1981 Smith and Waterman extended the ideas presented by Needleman and Wunsch to create an algorithm which is capable of finding optimal local pairwise alignments. The method uses a distance metric introduced by Sellers in 1974 which can be summarized by a matrix similar to the second example in the Fig. 2.1 (Smith et al. 1981). This algorithm uses a method similar to that of Needleman and Wunsch; first filling in all the values for an $m \times n$ matrix based on the score for a maximum length run of matches and then finding a path through the matrix. There are two main differences between the Smith–Waterman algorithm and the Needleman–Wunsch algorithm. The first is that the matrix is completed from the top left cell downwards as opposed to the backtracking done by Needleman and Wunsch. The second is that the path is built by finding the maximal valued cell in the matrix and then backtracking until a zero is found. The resulting path represents an alignment of two segments, one from each sequence. Note that while not all the bases in both sequences are aligned, there can be no other pair of segments which will produce a higher score. The algorithm was modified by Gotoh to include affine gap penalties (Gotoh 1982) and is sometimes called the Smith–Waterman–Gotoh algorithm. This algorithm is without doubt the cornerstone of modern sequence comparison.

#### 2.2.2.3 SWAT and CrossMatch

Unlike many other *fast* pairwise algorithms, SWAT does not employ first stage heuristics to speed up the Smith–Waterman algorithm. Instead, the authors of SWAT focused on speeding up the code itself by revising recursion relations and making efficient use of word-packing. This resulted in a significant reduction in the number of machine instructions executed per Smith–Waterman matrix cell. Thus they have produced a raw implementation of the Smith–Waterman–Gotoh algorithm which is about one tenth as fast as BLAST. SWAT is normally used to search query sequences against a sequence database or as an engine in other sequence comparison tools.

CrossMatch is a general-purpose sequence comparison utility based on SWAT and is used for comparing sets of DNA sequences. CrossMatch uses the same algorithm as SWAT, but allows the use of heuristics to constrain the comparison of pairs of sequences to bands of the Smith–Waterman matrix that surround one or more matching words in the sequences. This step reduces the running time for large-scale nucleotide sequence comparisons without significantly compromising sensitivity. CrossMatch and SWAT form the kernel of the Phrap assembly program and CrossMatch is used as the comparison engine in RepeatMasker. Both Phrap and RepeatMasker are described in more detail below. SWAT and CrossMatch are unpublished software; however, information can be found at: www.genome.washington.edu/UWGC/analysistools/Swat.cfm.

#### 2.2.2.4 The BLAST Family of Algorithms

BLAST (Altschul et al. 1990) and its many derivatives are arguably the most widely used pairwise local alignment algorithms. The BLAST algorithm attempts to heuristically optimize a measure of local similarity called the maximal segment pair (MSP). The MSP is defined as the highest scoring pair of identical length segments chosen from two sequences. To enable the reporting of multiple local alignments BLAST can also return other *locally* maximal segment pairs. Put simply, the speed of the BLAST algorithm is mainly due to its ability to identify and divert resources away from areas in the query sequences which have very little chance of producing high scoring alignments. Most BLAST implementations enable the user to search a pre-compiled database for high scoring segments in a set of query sequences. The database is created by running the program formatdb which produces a set of files that have been optimized for size and speed of searching. The algorithm has three distinct steps. First, using the information in the database and the query sequence, the algorithm compiles a list of high scoring words of a set length k (k-mers) from the query sequence. The database is scanned for matches to the words and where these hits occur, the algorithm tries to extend the hit to the left and right. BLAST uses a minimum score cutoff when assessing word hit quality to filter out any hits which could have occurred due to random chance. Note that the BLAST algorithm is characterized by the creation of k-mer lists for each query sequence and a linear search of the entire database for words in these lists. The BLAST algorithm has been highly successful and there are many different implementations available which have been adapted to better suit particular applications.

#### 2.2.2.5 BLAT

BLAT stands for BLAST-Like Alignment Tool and was developed by James Kent for use in the annotation and assembly of the human genome (Kent 2002). Kent was given the task of aligning many millions of mouse genomic reads against the human genome. He found that when using BLAST, the need to calculate high scoring k-mer lists for each query sequence and the linear nature of the database search proved too slow. To solve this problem, BLAT creates an indexed list of all possible non-overlapping k-mers from sequences in the database. BLAT then compiles a list of all overlapping k-mers from each query sequence and attempts to find these in the database. In the regions where multiple perfect hits occur, BLAT performs a Smith–Waterman alignment of the two sequences. This allows BLAT to maintain relatively high sensitivity, although it must be noted for example that TBLASTX can be configured to be more sensitive to distant relationships than BLAT. The reduced sensitivity is compensated for by the fact that BLAT can be run up to 50 times faster than TBLASTX (Kent 2002).

### 2.2.3 Multiple Sequence Alignment Algorithms

It is often necessary to produce an alignment of a group of three or more sequences. Examples include the comparison of the evolutionary distances between protein sequences, the evaluation of secondary structure via sequence relationship, or the identification of families of homologous genes. Efforts have been made to extend dynamic programming pairwise alignments to handle three or more sequences (Murata et al. 1985). However, the computational complexity of handling more than 4 sequences proved too much for the available computing power. Many modern multiple alignment algorithms use a method first suggested in 1987 by Feng and Doolittle called the progressive method (Feng and Doolittle 1987). The underlying assumption used in constructing the progressive method is that sequences with a high level of similarity are evolutionarily related. Given a set of sequences to be aligned, Feng and Doolittle use the Needleman–Wunsch pairwise alignment algorithm to calculate rough evolutionary distances between every pair of sequences and these are used to create a reference phylogenetic tree. Starting from the two closest branches on the tree, a pairwise alignment is made and a consensus sequence is produced which is used as a substitute for the branch. This is continued for the next closest pair of branches until all the sequences have been added and the alignment is complete.

The intermediate pairwise alignments may include two of the query sequences, one query sequence and one consensus sequence or two consensus sequences. It is important to note that the order in which sequences are added will affect the ultimate alignment and it is very difficult to repair the damage caused to the overall quality of an alignment if a less than optimal choice is made early on. However, algorithms such as MUSCLE attempt to do this. The use of a reference tree helps ensure that closely related sequences are aligned before distantly related sequences. Thus the progressive method utilizes a greedy algorithm. Feng and Doolittle stressed the point that any gaps added to the alignment in earlier stages must remain, creating the rule "once a gap, always a gap" (Feng and Doolittle 1987). This ensures that distantly related sequences cannot disturb meaningful alignments between closely related sequences, however some implementations of the progressive method do not follow this rule. Finally it is important to note that the reference tree should not be used to infer phylogenetic relationships, as there is a high probability that the tree is erroneous (in that sense). However, a new tree (or set of trees) can be made with the resulting multiple alignment and this can be used to study phylogeny. There are a number of multiple alignment algorithms based on the progressive method. The most widely used are the CLUSTAL family of algorithms, MUSCLE and T-Coffee.

#### 2.2.3.1 The CLUSTAL Family of Algorithms

The CLUSTAL family of multiple alignment algorithms includes the original program CLUSTAL as well as CLUSTAL V and CLUSTAL W. All of the CLUSTAL derivatives are based on the progressive method (Higgins and Sharp 1988; Higgins et al. 1992; Thompson 1994). The original CLUSTAL package was released as a collection of different pieces of software with each one performing one stage of a progressive alignment. CLUSTAL V was a rewrite of this system which combined all the packages into one program. CLUSTAL W is a further update to CLUSTAL V which incorporates sequence weighting, position-specific gap penalties, and weight matrix choice. For the rest of this section we describe only the features of CLUSTAL W.

Highly similar sequences will be positioned very closely on the reference tree and consequently will be added to the alignment much earlier than divergent sequences. Too many highly similar sequences in the query set can create bias in the topology of the reference tree which can lead to future alignment errors (Higgins and Sharp 1988). Sequence weighting attempts to reduce this bias by down-weighting groups of similar sequences and up-weighting divergent sequences. This feature reduces the negative impact that the topology of the reference tree can have on the final alignment (Thompson 1994). When the algorithm starts, it can use gap penalties and substitution matrices as supplied by the user. CLUSTAL W provides a choice of PAM or BLOSUM matrices with the default being BLOSUM. As the algorithm progresses, CLUSTAL W adjusts the gap penalties according to the position, content (hydrophilic or hydrophobic regions) and length of the sequences. CLUSTAL W also adjusts the weight of the substitution matrix based on the estimated evolutionary distances obtained from the reference tree. These additions to the CLUSTAL algorithm reduce the negative impact of sub-optimal parameter choices made by the user.

CLUSTAL W is the most widely used multiple sequence alignment algorithm and represents an acceptable balance between speed and accuracy. The next two algorithms are faster and more accurate respectively. The first, MUSCLE, sacrifices some accuracy for significant gains in speed, while the second, T-Coffee, makes significant gains in accuracy for a modest sacrifice in speed.

#### 2.2.3.2 MUSCLE

MUSCLE is a very fast multiple sequence alignment algorithm based on the progressive method. The algorithm is split into three phases. The first is typical of a progressive algorithm except that instead of using an alignment algorithm to generate the reference tree and evolutionary distances, MUSCLE employs the faster method of k-mer counting to judge similarity (Edgar 2004). Once the preliminary tree has been built, MUSCLE progressively adds sequences to the multiple alignment following the branching order, with closer branches being added first. At this stage, a new tree can be constructed and the progressive alignment can be returned to the user. The second phase seeks to improve the results of the first by iteratively constructing progressive alignments in the same manner as the first stage but using the most recent tree generated from the previous progressive alignment. At the end of each iteration, a new tree is made for use in the next round or phase. The third and final phase performs iterative refinement on the tree produced in the second phase. At each iteration, the tree is first separated into two pieces by removing an edge. Superfluous indels (insertions or deletions) are removed from each of the partial multiple alignments and then the tree is rejoined by re-aligning the partial multiple alignments. MUSCLE can produce multiple alignments achieving accuracy similar to CLUSTAL W but two to three orders of magnitude faster. Thus MUSCLE is suited to fast alignment of large sequence datasets.

#### 2.2.3.3 T-Coffee

Nearly all progressive based multiple alignment algorithms employ a greedy algorithm for adding sequences to the alignment. Unfortunately, errors can occur if the sequences are added in a less than ideal order. T-Coffee is an implementation of the progressive method which attempts to rectify some of the problems associated with the greedy approach to progressive alignment while minimizing speed sacrifices. To achieve this, T-Coffee first builds a library of both global and local pairwise alignments between all the query sequences. T-Coffee uses the progressive method, but in contrast to the algorithms described above, it attempts to consider the effects on every query sequence for each sequence being added. This approach seems to have worked as, on average, T-Coffee produces more accurate alignments than the competing algorithms (Notredame et al. 2000). However, this comes at the cost of increased running time, so T-Coffee may not be suited to the task of aligning large datasets.

## 2.3 Filtering, Clustering, and Assembly

This section covers the area of sequence filtering and the related areas of sequence clustering and sequence assembly. There is a great deal of overlap in the methods used for both sequence assembly and clustering. Pre-filtering reads and masking low complexity areas can improve the performance of assembly and clustering algorithms and is often a first step in many assembly/clustering pipelines.

## 2.3.1 *Filtering and Masking*

The first phase for many sequence comparison algorithms is filtering or masking regions whose presence will reduce the efficacy of tasks further down the pipeline. For example, consider the process of automated sequence annotation. One task involves querying the sequence to be annotated against a database of sequences with high confidence annotations (usually performed by making pairwise alignments). If the query sequence contains a substring which is common to many, largely unrelated or loosely related sequences, then the algorithm may return a large number of matches to sequences in the database which do not reflect meaningful annotations. These common elements are usually called *repetitive, repeats* or *low complexity sequences*. For sequence assembly, finding overlaps between reads is a fundamental task, and spurious overlaps caused by low complexity sequences can severely impede an assembly program's ability to produce accurate contigs.

Masking repetitive regions usually involves replacing all of the nucleotide bases in the repetitive region with another generic character, usually an "X" or an "N." The majority of assembly and alignment programs ignore these characters by default. In this way, results made by comparing masked sequences are usually more accurate than those where masking has not been performed. Masking can also decrease the running time of sequence comparison algorithms by reducing the number of possible alignments.

Another form of pre-filtering is sequence trimming. Often DNA sequences will begin, and possibly also end, with nucleotide bases from the vector used in the cloning stage, and for many different types of reads, the quality of the data decreases towards the end of the sequence read. An easy way to overcome these problems is to trim the ends of the sequence reads. There are a number of programs which can be used to trim sequences, however they are not discussed here. Finally, reads which contain low amounts of information can simply be removed from the data set, for example if the majority of the read consists entirely of As or Ns. When raw quality values are available, it is also common to simply discard reads whose overall quality is below a certain threshold.

#### 2.3.1.1 RepeatMasker

RepeatMasker screens DNA sequences for interspersed repeats and low complexity DNA sequences. The output of the program is a detailed annotation of the repeats present in the query sequence as well as a modified version of the query sequence where all the annotated repeats have been replaced by Ns or Xs. RepeatMasker draws information about which regions are repetitive by comparing the query sequences to a curated database of repeats. RepeatMasker uses CrossMatch for this task (Smit et al. 1996)

### 2.3.2 Sequence Clustering

With the quantity of sequence data contained in online repositories increasing at an accelerating pace, tools that can cluster related sequences into meaningful groups provide a way for researchers to efficiently sort through and make sense of this mountain of data. Many researchers are interested in clustering reads from expressed sequence tag (EST) datasets in the hope of identifying the full length genes which the ESTs represent. Another application of clustering is the identification of single nucleotide polymorphisms (SNPs). Clustering is often used to reduce redundancy in a dataset. For example, the BLOSUM substitution matrices use clustering of similar sequences as a first step to reduce the negative effect caused by including too many highly similar sequences. Clustering can also be useful as a first step in sequence assembly pipelines. Sequence assembly programs will often perform significantly better when run multiple times on sets of closely related sequences than when attempting to assemble the whole data set as one chunk. This approach can also significantly reduce the running time of the assembler. Clustering algorithms typically take as input a set of reads to be sorted and input parameters specifying the degree of similarity required for reads to be grouped together. The output is a grouping of the reads that match these criteria.

Most clustering algorithms use an agglomerative approach. At the start of the algorithm, each sequence is effectively in its own group. The algorithm successively merges groups if the similarity criteria are met, and repeats this process until no more merges are possible. These final merged groups are then returned to the user. Depending on user input or the algorithm itself, two groups will be merged when there exists a single pair of sequences (one sequence from each group) which match the similarity criteria. This is referred to as *single linkage* clustering or *transitive closure*. It is sometimes possible to raise the minimum number of pairs needed for merging to occur. If every possible pair of sequences from both groups must match the similarity criteria for merging to occur then this is called *complete linkage* clustering. Complete linkage clustering typically produces many small, high quality clusters, whereas single linkage clustering typically produces fewer, larger, lower quality clusters. Depending on the application, one approach may be more favorable than the other.

The similarity criterion for clustering is usually stated in terms of the minimum overlap and minimum percentage identity. This is sometimes augmented by limiting the maximum number of mismatches allowable. There are two main approaches available to find overlaps. The first uses information gathered from successive pairwise alignments, effectively looking at the edit distance, the number of mutation events needed to describe the distance. The second uses a k-mer counting approach, where the presence of multiple identical words is used to infer an overlap. Both methods perform well, but the k-mer counting approach has been proved to handle sequencing errors better. Furthermore, the k-mer counting approach can be implemented in linear time whereas the edit distance approach can only be as fast as the underlying alignment algorithm, which in the case of Smith–Waterman is quadratic, and slightly better for BLAST-like algorithms. Two popular clustering algorithms are WCD (Hazelhurst et al. 2008) and d2_cluster (Burke et al. 1999).

### *2.3.3 Sequence Assembly Overview*

The greatest challenge to sequencing genomes is the vast difference in scale between the size of the genomes and the lengths of the reads produced by the different sequencing methods. While there may be a 10–500-fold difference in scale between the short reads produced by next generation sequencing and modern Sanger sequencing, this still dwarfs the difference between the Sanger read length and the lengths of complete chromosomes. For example, human chromosomes vary between 47 and 245 million nucleotides in length, around 50,000–250,000 times longer than the average Sanger reads. For all technologies, the challenge is the assembly of sequence reads to produce a representation of the complete chromosomes. Whether this challenge is significantly greater for short reads is being hotly debated.

The first sequence fragment assembly algorithms were developed in the early 1980s. Early sequencing efforts focused on creating multiple overlapping alignments of the reads (typically the Sanger sequence reads) to produce a layout assembly of the data. A consensus sequence is read from the alignment and the DNA sequence is inferred from this consensus. This approach was referred to as the *overlap-layout-consensus* approach and culminated in a variety of sequence assembly applications such as CAP3 and Phrap. Previous generation DNA sequencing has produced relatively long, high quality reads which were amenable to assembly using the overlap-layout-consensus approach.

From the late 1980s to the mid 1990s research began to focus on formalizing, benchmarking, and classifying fragment assembly algorithm approaches. Three papers, (Pevzner 1987, Myers, 1995, Idury and Waterman 1995) formalized the approach of placing sequence reads or fragments in a directed graph. Myers focused on formalizing the traditional overlap-layout-repeat method while Pevzner and Idury and Waterman developed a new method for solving the assembly problem. While both methods involved the construction of graphs, they differed in that whereas the fragments in Myers graph (Myers 1995) are represented as nodes, the fragments in Pevzner's and Idury and Waterman's graphs (Pevzner 1987; Pevzner 2001; Idury and Waterman 1995) are represented as edges.

#### 2.3.3.1 The Overlap Graph Method

The overlap graph method was formalized by Myers in 1995 (Myers 1995) and is referred to here as the Myers method. In this graph, each vertex represents a read or fragment, and any two vertices are joined by an edge if the two fragments overlap (often imperfectly with some level of significance set by the user). Next, the graph is simplified by the removal of transitive edges and contained nodes

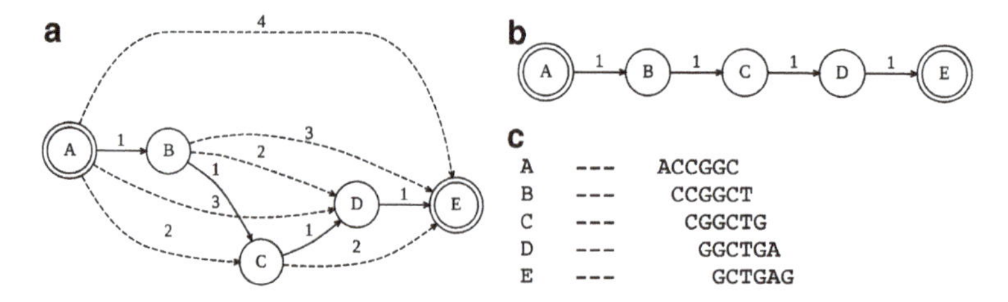


**Fig. 2.2** Removal of transitive edges from the overlap graph. (**a**) The original graph with numbers depicting offsets in the alignment of equal length reads. The dashed lines are transitive edges. (**b**) The simplified graph. (**c**) The short reads and their alignment as given by the graph yields the sequence ACCGGCTGAG

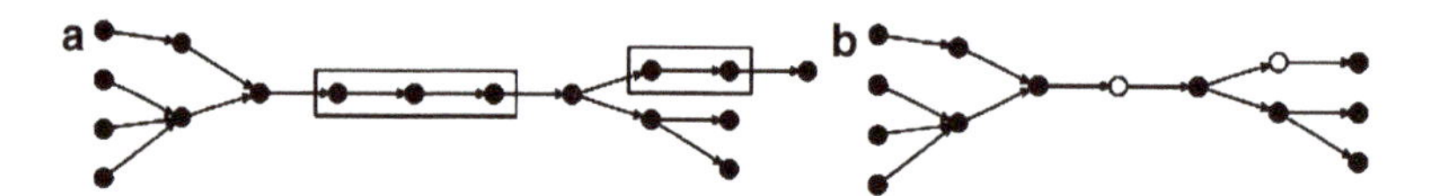


**Fig. 2.3** The collapsing of linearly connected sub graphs into single nodes greatly reduces the complexity of the overlap graph. (**a**) The original graph which contains linearly connected sub graphs. (**b**) The simplified graph with white nodes representing "chunks"

which add little or no information. The removal of transitive edges is shown in Fig. 2.2. Contained nodes occur when the graph is made from reads of different lengths and one read is completely contained within another.

Finally, chains of nodes or linearly connected sub graphs are collapsed into "chunks" which themselves are treated as single nodes in the graph. This is shown in Fig. 2.3. These graph simplification methods are very effective for reducing the computational complexity of assembly and many modern day algorithms employ these methods. Once the graph has been simplified, the Myers method finds a maximum likelihood non-cyclic (Hamiltonian) path through the graph and infers the layout of the fragments from this path.

#### 2.3.3.2 The Eulerian Path Method

Idury and Waterman proposed an algorithm which could be used to assemble data generated in sequencing by hybridization (SBH) experiments (Idury and Waterman 1995). Although the mathematics for it was developed by Pevzner in 1989 (Pevzner 1989), this is the first algorithm developed using this approach. We refer to the combined ideas of Idury and Waterman and Pevzner as the IWP method. The main application of SBH is now gene chips and not genome sequencing, but the ideas described in the IWP model can be seen in a number of sequence assembly algorithms, most notably the EULER algorithms developed by Pevzner in 2001 (Pevzner 2001). In Idury and Waterman's algorithm, sequence fragments are broken down into every possible read of some length *k* (*k* is very small, approximately 10 bases) referred to as k-tuples. The set of all k-tuples found is often referred to as the spectrum of reads (Pevzner 2001). In Idury and Waterman's model, assembled sequences are represented as paths through a de Bruijn graph where each node in the graph is a k-1 tuple. Two nodes X and Y are joined by a directed edge if there exists a read R in the spectrum where the first k-l bases of R match X and the last k-1 bases in R match Y. Thus it follows that if two edges are adjacent in the graph they will have a perfect overlap of k-1 bases. It is important to note that this model only finds perfect overlaps, while the Myers method can accept imperfect overlaps. An example of such a de Bruijn graph is shown in Fig. 2.4. Here the graph is for the sequence CAGTCGAGTTCTCTG with *k* equal to 4. Erroneous reads cause the inclusion of extra edges which can cause "tangles" in the graph. The dashed edge from TTC to TCG is due to the erroneous read TTCG being included in the spectrum. Idury and Waterman (Idury and Waterman 1995) describe a number of graph simplifications which can remove errors from the graph.

Assembly is achieved by finding an Eulerian path in the de Bruijn graph. That is a path which visits every edge exactly once. It is well known that the problem of finding Hamiltonian paths in a graph is NP hard whereas the problem of finding Eulerian paths is relatively easy. However, the theoretical advantage that the IWP method seems to have over the method of Myers has not translated into great computational or time savings. This is mainly due to the fact that heuristics have been employed to speed up the latter. The main problem with the Eulerian Path approach is that errors in real data cause the inclusion of extra edges, causing tangles. When there are too many errors in the data, the graph becomes entangled, and as a result the algorithm cannot be scaled up. An example of how erroneous reads cause graphs to become entangled is given in Fig. 2.4. In 2001, Pevzner successfully applied the method of Idury and Waterman to read sets with errors by developing an error correction algorithm which could reduce the number of errors by approximately 86% (Pevzner 2001). Pevzner introduced a number of transformations which simplify the graph and these transformations have a conceptual overlap with Myers simplifications. One transformation replaces a number of consecutive edges by one edge, in a way which mimics the collapse of linearly connected sub graphs described above. This process of edge formation/simplification is performed

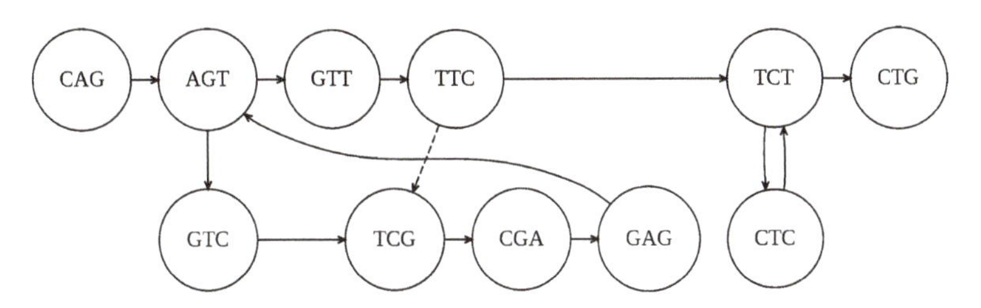


**Fig. 2.4** A de Bruijn graph for the sequence CAGTCGAGTTCTCTG. The erroneous read TTCG has been included in the spectrum causing the inclusion of the *dashed edge*. Erroneous edges cause the graph to become entangled

at the beginning of the assembly so that only the minimal number of edges possible need be processed.

The two methods described above have formed the foundation for modern assembly approaches, and all modern sequence fragment assemblers include variations of these concepts, and in some cases algorithms may borrow from both methods.

#### 2.3.3.3 Problems of Assembling Complex Genomes

One challenge of genome sequencing lies in the fact that only a small portion of the genome encodes genes, and that these genes are often surrounded by repetitive DNA which is comparably information poor. Large repeats can cause ambiguity with fragment assembly and thus pose the greatest challenge when assembling genomic data. In Fig. 2.5, we know that regions B and C are surrounded by identical repetitive regions X and that both regions lie between regions A and D, but without more information, it is impossible to know the correct ordering of B and C.

The traditional method to overcome the problems created by large repeats when assembling sequence reads is to increase the read length to such a point that every repeat is spanned by at least one read. In practice however, this is simply not possible as these repeats are frequently longer than the current Sanger read length. Modifications to the original shotgun method that attempt to overcome this problem try to increase the "effective" read length. These include using paired end sequencing, where DNA fragments of known approximate size are generated and sequenced from both ends. Information about these pairs such as average fragment size and the orientation of reads with respect to the read pair is included in the assembly process (Pevzner 2001). If the distance between the paired ends, known as the insert size, is large enough, then there is a high probability that repeats will be spanned by a pair of reads (or mates) which can remove ambiguity from the assembly. For example, if paired end data is analyzed and region B is found to have mates in regions A and C but not D, while region C has mates in regions B and D but not A, then an ordering can be inferred. This is shown in Fig. 2.6. Note that if the insert size was too large and paired ends from B reached over to region D while the paired ends of C reached over to region A there would still be doubt as to how these reads should be arranged. There is also a problem if the insert size is too small and paired reads do not reach past the repetitive region. To address these issues, a number of different size fragment libraries are often used.

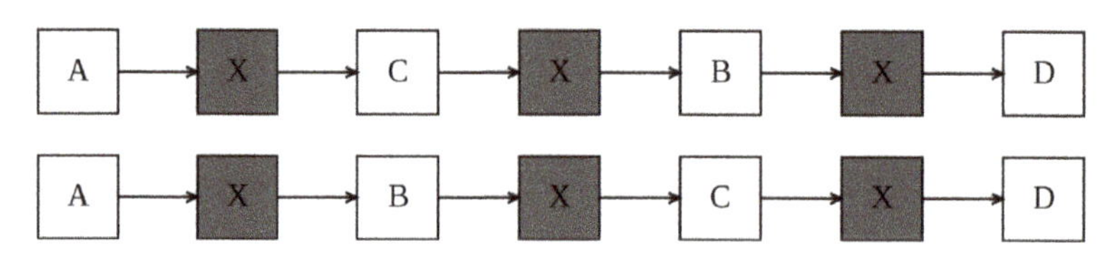


**Fig. 2.5** An example of how repeats cause ambiguity in assembly. Because both fragments B and C are surrounded by repetitive region X, there is no way to know their ordering in the assembly

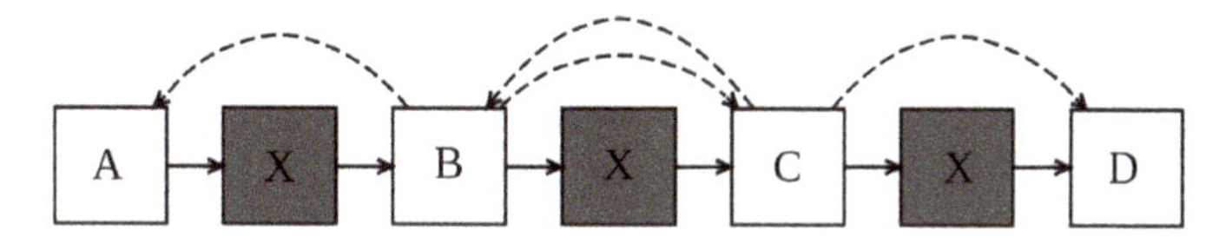


**Fig. 2.6** Resolution of ambiguities using paired end data. *Solid edges* indicate overlaps while *dashed edges* show links between reads in a region and the region(s) containing the paired mate

Errors in data further exacerbate the problem of resolving repetitive regions as it is often difficult to differentiate between reads from slightly different repetitive regions and reads from the same region that contain errors. This can cause a problem called over-collapsing, where multiple copies of a repeat will be assembled on top of each other. Although we would expect a significant increase in read depth for contigs, which are made from over collapsed regions, the read depth is frequently variable across the genome and is therefore an unreliable indicator of repeat regions. Both paired end data and various statistical methods have been applied in an attempt to solve the problem of assembling short read sequence data and these are described in more detail below.

### 2.3.4 Traditional Fragment Assembly Algorithms

For many years, the vast majority of DNA sequence data has been produced using variations of the chain termination method first introduced by Sanger in 1977 (Sanger et al. 1977). The Sanger sequence reads are typically 700–1,000 bases long and of high quality. The individual nucleotide bases in a sequence file is *called*, based on information found in a chromatogram, a *trace* file which is produced by the automatic sequencing machines. Phred is the most commonly used base calling software package (Ewing et al. 1998), and the two most commonly used programs for assembling the Sanger sequence data are Phrap and CAP3. Both programs make use of Phred generated quality scores when performing the assembly, although this data can be omitted if it is not available. Aside from being used to assemble data generated in large scale genome sequencing projects, these programs have also been used to assemble EST sequence data. Both Phrap and CAP3 use variations of a Myers-like approach to fragment assembly, though Phrap deviates from this standard template in the final consensus phase.

#### 2.3.4.1 Phrap

Phrap stands for "phragment assembly program" or "Phil's revised assembly program" and is used for assembling shotgun DNA sequence data. Unlike many other assemblers, Phrap makes use of the whole read and not just the trimmed high quality portion. Phrap can be provided with a mixture of machine generated and user supplied quality data to assist in making reliable contigs. One aspect which sets Phrap apart from many other Myers type algorithms is that Phrap returns contig sequences which are mosaics of the highest quality parts of reads rather than a consensus or majority vote.

Phrap searches for reads with matching words and then does a SWAT comparison between pairs of reads with this property. This allows Phrap to efficiently make use of the very accurate Smith–Waterman algorithm encoded in the SWAT algorithm. This first stage identifies all potential overlaps. The next stage effectively masks vector sequences. This stage also identifies near duplicate reads, reads with self matches, and reads which have more than one significant overlap in any given region. These steps help Phrap to deal with repetitive elements. Phrap then constructs contig layouts based on strong pairwise alignments using a greedy algorithm and from these layouts produces the contigs. Finally, Phrap aligns reads to the contigs identifying inconsistencies and possible misassembled sites. Phrap returns Phred-like quality scores for each base in the contig based on the consistency of the pairwise alignments at that position (http://www.phrap.org/phredphrap/phrap.html).

#### 2.3.4.2 CAP3

CAP3 is the third generation of the CAP assembly algorithm and was released in 1999 (Huang and Madan 1999). CAP3 uses a Myers-like method which makes extensive use of quality values and paired read data. The overlap stage begins by using a BLAST-like algorithm to identify areas where detailed local alignments are produced using a modified version of the Smith–Waterman algorithm which weights the substitution matrix at each position using the quality scores at the bases concerned. Where CAP3 differs from other algorithms is in the way that these overlaps are then validated. First, CAP3 identifies *good* regions in a read. A good region is a run of nucleotide bases with high quality scores and which share an overlap with a region in another read which also has high quality scores. CAP3 uses the good regions to identify which bases to trim from the ends of the reads. Once the good regions have been identified, CAP3 produces a global alignment of the reads previously identified as having local alignments and attempts to identify inconsistencies in the global alignments between good regions. There are a number of criteria each overlap must satisfy and any overlaps which do not meet all the criteria are discarded. This completes the overlap stage. CAP3 then uses a greedy algorithm to produce a layout of the reads which is validated by checking whether the paired read data (if supplied) produces any inconsistencies. Finally, the reads are aligned to the layout and a consensus produced. CAP3 also produces Phred-like quality scores which are returned to the user. In benchmarking, CAP3 generally produces better quality, shorter contigs than Phrap due to the strict methods for creating contigs (Huang and Madan 1999). However, CAP3 relies heavily on paired end data and even more so on quality values, and may not perform as well if given raw sequence data alone.

### 2.3.5 *Short Read Fragment Assembly Algorithms*

Several assemblers have been developed for short sequence reads, these include Edena, Velvet, EULER SR, SASSY and ALLPATHS. All of these algorithms borrow from the Myers or IWP models described above either implicitly or explicitly, and there are many similarities between the different algorithms in terms of their overall structure. Most algorithms are divided into up to five stages which include some or all of the following procedures: read error correction, read filtering, naïve assembly, refining of naïve assembly (using paired end data if available), and finishing. To understand what a naïve assembly is we need to define the terms *consecutive* and *linearly connected.* Two reads A and B are *consecutive* if they overlap (either the first k bases in A match the last k bases in B or vice versa) and for a graph with no transitive edges, two reads A and B are *linearly connected* if they are consecutive and there exists no read C which is consecutive with A on the same side as B or with B on the same side as A. For naïve assembly, we mean that starting with some read R we can try to extend that read (on one side) by examining the consecutive reads (on that side). If there is only one candidate, then the read can be extended in the direction of the overlap by the bases which are overhanging. This process mimics the collapsing of linearly connected sub graphs in the Myers model or edge formation in the IWP model. Thus, any string of linearly connected reads can be concatenated into one long read. For a given read there may be more than one candidate to extend with, and in this case the extension stops. Similarly, the extension stops when there are no candidates. The case where there is more than one candidate can be caused when the extension reaches the boundary of a duplicated or repetitive region in the genome or as happens much more frequently, it can be caused by errors in the data. If all possible extensions have been made for all available reads, then the resulting set of extended reads represents a naïve assembly. The accuracy of this assembly declines rapidly as both the error rate and the complexity of the organism being sequenced increase (Chaisson et al. 2004; Whiteford et al. 2005).

#### 2.3.5.1 Edena

Edena, released in 2008 (Hernandez et al. 2008), is the first short read assembly algorithm to be released which uses the traditional overlap-layout-consensus approach. Edena does not include an error correction phase before graph production, which leads to the formation of a messy sequence graph; however it does include a three step error correction phase which cleans the graph before assembly begins. The first phase of the algorithm removes duplicate reads, keeping only the original read and the number of times it has been seen. Next it uses the reads to construct an overlap graph where the reads are represented as nodes. Two nodes are joined by an edge if there is an overlap between them larger than a set minimum (defined by the user). Once this graph has been built, it contains many erroneous edges which have to be removed. First it removes transitive edges in the graph in the same manner as described by Myers (Myers 1995). Following this, all dead end paths are removed.

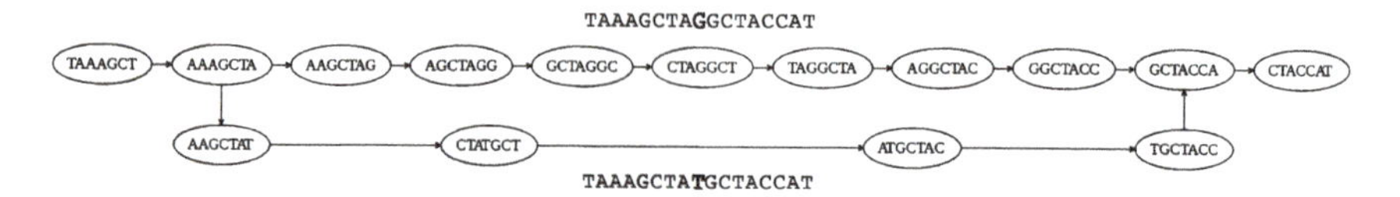


**Fig. 2.7** An example of a P-bubble most likely caused by an error. The reads making up the lower sequence will typically have a low copy number and the overlaps are very short. However, this phenomenon can also be caused by low copy number repeats

A dead end path is a series of consecutive nodes shorter than 10 reads in length which is attached to the main body of the graph on one side and to nothing on the other. These paths are caused when areas are sequenced with very low coverage, causing breaks in the sequence of consecutive reads, or when a series of errors combine to make a series of consecutive reads. Finally, the algorithm removes what are called P-bubbles. These occur when there are two regions which are identical except for a one base difference. In the case where this is caused by single nucleotide polymorphisms (SNPs) in repetitive regions, we would expect each side of the bubble to have a similar topology and copy number. Where a P-bubble is caused by an error, we would expect to see one side of the bubble with a very sparse topology and significantly lower copy number. Figure 2.7 gives such an example.

When P-bubbles are found, Edena removes the side with the lowest copy number/sparsest topology. Hernandez points out that P-bubbles may be caused by clonal polymorphisms which would account for the low coverage and sparse topology observed (Hernandez et al. 2008). However, as Edena does not take paired end information into account, the method used for eliminating P-bubbles will most certainly cause over-collapsing of low copy-number repetitive regions. Once the graph has been cleaned using the three operations described above, a naïve assembly is formed and the resulting contigs are returned to the user.

#### 2.3.5.2 Velvet

Velvet is the name given to the collection of algorithms that assemble short read data which were released by Zerbino in 2008 (Zerbino and Birney 2008). Velvet uses an IWP model to make the initial graph. Like Edena, Velvet does not include an initial error correction phase but instead uses a series of error correction algorithms to clean up the resulting graph. These algorithms work in a method analogous to the error correction phase in Edena (Hernandez et al. 2008), where tips are removed and then bubbles. In Velvet, tips are removed only if they are shorter than $2k$, where $k$ is the read length. Unlike Edena, Velvet uses an Eulerian path approach, which although highly efficient in terms of memory use, appears to further complicate the P-bubble removal step. Velvet includes an algorithm called Tour Bus which traverses the graph looking for P-bubbles, and when they are found, uses a combination of copy number and topographical information to remove the erroneous edges. Velvet then assumes that all low copy number edges that remain must be errors and removes them from the graph. Like many of the algorithms described here, Velvet does not make use of paired read information and therefore has an increased probability of over-collapsing repetitive regions.

#### 2.3.5.3 Euler SR

There have been many iterations of the original EULER algorithm developed by Pevzner in 2001 (Pevzner 2001). The latest addition to the EULER family is EULER SR which is a version of EULER optimized to handle short reads (Chaisson and Pevzner 2008). The algorithm described by Idury and Waterman did not include a step for filtering or correcting errors, however it did include a number of graph simplifications which could be used to reduce the impact of errors. Unfortunately, this method could not scale up to handle the large amounts of error present in real data. The original EULER algorithm was designed as an implementation of the Idury and Waterman algorithm, but included a novel method for error correction. A short read is broken down into a number of even shorter k-tuples which are stored in a database. In the case when the dataset contains no errors then we would expect that the k-tuples generated for a particular read R would appear a number of times in the database, as all reads overlapping with R would also contain some number of these k-tuples. Pevzner describes a read as "solid" if all of its k-tuples appear at least $n$ times (where $n$ is set by the user) or "weak" otherwise. When used with real data, if a read has been classified as weak, the algorithm tries to find the minimum number of base changes which will change its classification to strong. If that number is less than $d$ (where $d$ has been set by the user) then the changes are made, otherwise the read is discarded. Pevzner shows that this method corrects over 86% of errors with very few false positives for the dataset he analyzed (Pevzner 2001). This represents the most sophisticated and efficient approach for error correction of short reads that has been developed thus far. EULER SR builds on the original EULER algorithm and contains optimizations to make it more memory efficient, a property which is necessary for the vast amount of data produced by short read assemblers. Interestingly, in testing EULER SR, a hybrid approach was assessed where short read data was combined with longer Roche 454 read data. It was found that there was no significant improvement in assembly for the majority of reads (Pevzner 2001), which is contrary to most of the current opinion in this field. After the errors have been removed, a graph is built and a set of contigs produced by naïve assembly is returned to the user.

#### 2.3.5.4 SASSY

We are currently developing an assembly algorithm called SASSY which is based on a Myers like method that incorporates paired end data. SASSY is being developed primarily to assemble eukaryotic sequences of around 100–200 Kbp in length cloned into BACs. While SASSY shares many similarities with the software described above, there are a number of key differences. We have developed a novel iterative approach to graph construction which removes the need for some of the simplification steps typically needed for this type of implementation. We aggressively filter the data set, flagging up to approximately 90% of the reads which are set aside to be used only in later stages. With the remaining reads we construct a first round naïve assembly using only reads which have an overlap of at least k – t nucleotide bases, where *k* is the read length and *t* is very low (usually 2 or 3). The advantage of using this approach is that erroneous areas of the graph usually have a sparse topology and the number of common bases between any two reads in these areas is usually much lower than for reads in correct areas of the graph. Thus, assemblies generated in the first round represent high confidence assemblies, however their length is typically very short, with an N50 of less than 50 bases for Applied Biosystems SOLiD reads and slightly longer for Illumina Solexa reads. It should be noted that the longest contigs produced from this preliminary assembly are typically 4,000–12,000 bases long. These longer contigs are used to identify *stable* areas in the overlap graph. The next stage involves building a new overlap graph which explicitly combines the overlap data in the original graph with the paired read data. Normally this would be difficult because of the repetitive nature of the data, but by starting the graph building in stable areas, many of the problems associated with repeats are resolved. Thus we use the naïve contigs only as a guide instead of trying to extend them, which is the case for the other algorithms described in this section. Following the construction of the overlapping graph, we align all the reads flagged in the filtering stage to the assembled contigs. We examine the distribution of the insert size for the mapped reads to identify erroneous assemblies which are repaired where possible or flagged as conspicuous in the case when there is no obvious resolution. Finally, new contigs are built from the new overlap graph and these are returned to the user. SASSY is being developed to make optimal use of local topology and paired end data in order to avoid the problems of over-collapsing repetitive regions or unnecessarily breaking contigs when errors are present in the data. This software is currently still in a developmental stage; however, initial test versions promise to overcome many of the limitations inherent in current small read assembly software.

#### 2.3.5.5 ALLPATHS

ALLPATHS is another recent addition to the collection of short read assemblers based on an IWP model (Butler et al. 2008). ALLPATHS begins by correcting errors using an EULER like method, and then makes a large set of naïve assemblies which are referred to as "unipaths." At this stage, ALLPATHS leaves the model followed by EULER, and uses paired read information to sort unipaths into localized groups which can be worked on individually and in parallel. For each localized set of unipaths, ALLPATHS chooses paired reads which lie in the set, and proceeds to work out every path of consecutive unireads which could possibly be followed from the read to its mate. Once this is completed, the number of paths is trimmed down using localization information and other statistical methods until, ideally, only one path remains. This method reduces the complexity in the overall sequence graph by making local optimizations, allowing many shorter unipaths to be condensed into longer unipaths. Once the long unipath generation has been completed separately, the results from the local optimizations are stitched together to produce one long sequence graph. One limitation of this algorithm is its sensitivity to the standard deviation of the fragment length used to make the paired sequence reads. Butler notes that in some cases, a large number read-mate pairs generate over $10^3$ possible paths, and in some cases more than $10^7$ possible paths are generated, which causes ALLPATHS to return erroneous unipaths (Butler et al. 2008). The final phase incorporates both read pair information and statistics to identify erroneous assemblies, and if possible it tries to fix them. The most unique aspect of ALLPATHS is that no information is discarded at any stage in the algorithm which improves the ability to repair errors in the final phase. Unlike every other algorithm described here it returns the entire graph to the user as opposed to just the contigs.

## 2.4 Discussion

There are many branches of research into sequence comparison (more than have been covered here) with varying levels of complexity. The amount of effort being spent on solving different branches has continuously shifted as computational power has increased and the nature of the data being produced has changed. For example, multiple sequence alignment algorithms only started to receive widespread attention from the mid 1980s, almost 20 years after the merits of different phylogenetic tree making algorithms were being heavily debated, and many years after efficient algorithms had been produced for pairwise alignments. Pairwise sequence alignment has long been the base currency of sequence comparison, but graph theoretical methods; in particular k-mer distance methods and k-mer grouping/sorting have been demonstrated to be valuable for increasing the speed at which analysis can be performed. The typically long and accurate sequence reads produced using the Sanger sequencing method have been largely replaced (in terms of volume of data being produced) by next generation sequencing methods which produce copious amounts of largely error laden data, and the current focus of bioinformatics in this area has been to develop algorithms that can accurately assemble this data into long stretches of sequence. Again, graph theoretical approaches have proved valuable. Progress in the area of sequence assembly has only been feasible using computing power developed in recent years, although it should be noted that more than 40 years after the birth of comparative algorithms, the lack of ever greater computing power remains the main hindrance to progress. As an example, the original implementation of CLUSTAL was tested on a 10 MHz microcomputer (PC) with only 640K of memory, while the current iteration of the program SASSY was developed using an 8 core (1.8 GHz per core) cluster with access to 16 GB of memory and almost unlimited hard disk space. There is a clear trend that advances in computing hardware continue to spur development of ever more sophisticated comparison algorithms, allowing researchers greater insight into comparative genomics and the workings of the biological world.

## References


Altschul SF, Gish W, Miller W, Myers EW, Lipman DJ (1990) Basic local alignment search tool. J Mol Biol 215:403–410

Bentley DR (2006) Whole-genome re-sequencing. Curr Opin Genet Dev 16(6):545–552

Burke J, Davison D, Hide W (1999) d2_cluster: A validated method for clustering EST and full-length cDNA sequences. Genome Res 9:1135–1142

Butler J, MacCallum I, Kleber M, Shlyakhter IA, Belmonte MK, Lander ES et al (2008) ALLPATHS: De novo assembly of whole-genome shotgun microreads. Genome Res 18(5):810–820

Chaisson MJ, Pevzner PA (2008) Short read fragment assembly of bacterial genomes. Genome Res 18:324–330

Chaisson M, Pevzner PA, Tang HX (2004) Fragment assembly with short reads. Bioinformatics 20(13):2067–2074

Dayhoff Mo, ed., 1978, Atlas of protein Sequence and Structure, Vol 5

Dohm JC, Lottaz C, Borodina T, Himmelbauer H (2007) SHARCGS, a fast and highly accurate short-read assembly algorithm for de novo genomic sequencing. Genome Res 17:1697–1706

Edgar RC (2004) MUSCLE: multiple sequence alignment with high accuracy and high throughput. Nucleic Acids Res 32:1792–1797

Ewing B, Hillier L, Wendl MC, Green P (1998) Base-calling of automated sequencer traces using phred. 1. accuracy assessment. Genome Res 8:175–185

Feng DF, Doolittle RF (1987) Progressive sequence alignment as a prerequisite to correct phylogenetic trees. J Mol Evol 25:351–360

Gotoh O (1982) An improved algorithm for matching biological sequences. J Mol Biol 162:705–708

Hazelhurst S, Hide W, Liptak Z, Nogueira R, Starfield R (2008) An overview of the wcd EST clustering tool. Bioinformatics 24(13):1542–1546

Henikoff S, Henikoff JG (1992) Amino acid substitution matrices from protein blocks. Proc Natl Acad Sci USA 89(22):10915–10919

Hernandez D, Francois P, Farinelli L, Osteras M, Schrenzel J (2008) De novo bacterial genome sequencing: Millions of very short reads assembled on a desktop computer. Genome Res 18(5):802–809

Higgins DG, Sharp PM (1988) CLUSTAL: a package for performing multiple sequence alignment on a microcomputer. Gene 73:237–244

Higgins DG, Bleasby AJ, Fuchs R (1992) CLUSTAL V: improved software for multiple sequence alignment. Bioinformatics 8(2):189–191

Huang X, Madan A (1999) CAP3: A DNA sequence assembly program. Genome Res 9:868–877

Idury RM, Waterman MS (1995) A new algorithm for DNA sequence assembly. J Comput Biol 2:291–306

Jeck WR, Reinhardt JA, Baltrus DA, Hickenbotham MT, Magrini V, Mardis ER et al (2007) Extending assembly of short DNA sequences to handle error. Bioinformatics 23:2942–2944

Kent JW (2002) BLAT – the BLAST-like alignment tool. Genome Res 12:656–664

Murata M, Richardson JS, Sussman JL (1985) Simultaneous comparison of three protein sequences. Proc Natl Acad Sci USA 82(10):3073–3077

Myers EW (1995) Toward simplifying and accurately formulating fragment assembly. J Comput Biol 2:275–290

Needleman SB, Wunsch CD (1970) A general method applicable to the search for similarities in the amino acid sequence of two proteins. J Mol Biol 48:443–453

Notredame C, Higgins DG, Heringa J (2000) T-Coffee: A novel method for fast and accurate multiple sequence alignment. J Mol Biol 302:205–217

O'Connor M, Peifer M, Bender W (1989) Construction of large DNA segments in Escherichia coli. Science 244:1307–1312

Penzner PA (2001) Fragment assembly with double-barreled data. Bioinformatics 17:S225–S233

Pevzner PA (1989) l-tuple DNA sequencing: computer analysis. J Biomol Struct Dyn 7:63–73

Pevzner PA, Tang HX, Waterman MS (2001) An Eulerian path approach to DNA fragment assembly. Proc Natl Acad Sci USA 98(17):9748–9753

Sanger F, Nicklen S, Coulson AR (1977) DNA sequencing with chain-terminating inhibitors. Proc Natl Acad Sci USA 74(12):5463–5467

Sellers PH (1974) On the theory and computation of evolutionary distances. J Appl Math (siam) 26:787–793

Smit AFA, Hubley R, Green P RepeatMasker Open-3.0. 1996-2004. http://www.repeatmasker.org

Staden R (1979) A strategy of DNA sequencing employing computer programs. Nucleic Acids Res 6:2601–2610

Thompson JD, Higgins DG, Gibson TJ, Clustal W (1994) Improving the sensitivity of progressive multiple sequence alignment through sequence weighting, position-specific gap penalties and weight matrix choice. Nucleic Acids Res. Nov 11;22(22):4673–4680

Warren RL, Sutton GG, Jones SJM, Holt RA (2007) Assembling millions of short DNA sequences using SSAKE. Bioinformatics 23(4):500–501

Waterman MS, Smith TF, Beyer WA (1976) Some biological sequence metrics. J Adv Math 20:367–387

Wheeler DA, Srinivasan M, Egholm M, Shen Y, Chen L, McGuire A et al (2008) The complete genome of an individual by massively parallel DNA sequencing. Nature 452(7189):U872–U875

Whiteford N, Haslam N, Weber G, Prugel-Bennett A, Essex JW, Roach PL et al (2005) An analysis of the feasibility of short read sequencing. Nucleic Acids Res 33(19):e171

Zerbino DR, Birney E (2008) Velvet: algorithms for de novo short read assembly using de Bruijn graphs. Genome Res 18(5):821–829

# Chapter 3
# Genome Browsers

**Sheldon McKay and Scott Cain**

## 3.1 Introduction

The proliferation of data from genome sequencing over the past decade has brought us into an era where the volume of information available would overwhelm an individual researcher, especially one who is not computationally oriented. The need to make the bare DNA sequence, its properties, and the associated annotations more accessible is the genesis of the class of bioinformatics tools known as genome browsers. Genome browsers provide access to large amounts of sequence data via a graphical user interface. They use a visual, high-level overview of complex data in a form that can be grasped at a glance and provide the means to explore the data in increasing resolution from megabase scales down to the level of individual elements of the DNA sequence. While a user may start browsing for a particular gene, the user interface will display the area of the genome containing the gene, along with a broader context of other information available in the region of the chromosome occupied by the gene. This information is shown in "tracks," with each track showing either the genomic sequence from a particular species or a particular kind of annotation on the gene. The tracks are aligned so that the information about a particular base in the sequence is lined up and can be viewed easily. In modern browsers, the abundance of contextual information linked to a genomic region not only helps to satisfy the most directed search, but also makes available a depth of content that facilitates integration of knowledge about genes, gene expression, regulatory sequences, sequence conservation between species, and many other classes of data.

There are a number of tools now available for browsing genomes. Although there is some overlap both in terms of functionality and in types of accessible data, each offers unique features or data. In this chapter, we will give an overview of genome browsers and related software and discuss factors to be weighed in choosing the best browser for a researcher's needs.

S. McKay (✉)
Cold Spring Harbor Laboratory, 1 Bungtown Rd, Cold Spring Harbor, NY, 11724, USA
e-mail: mckays@cshl.edu

D. Edwards et al. (eds.), *Bioinformatics: Tools and Applications*,
DOI 10.1007/978-0-387-92738-1_3, © Springer Science+Business Media, LLC 2009

## 3.2 Web-based Genome Browsers

With the abundance of genome feature information available for so many organisms, an almost equally large number of web-based genome browsers have arisen to display the data. When large numbers of genomes first became available, organizations that needed to present their genomics data generally produced their own software solutions. Many of these early genome browsers have since fallen out of use as the data providers worked to standardize and reduce or redirect their development efforts. A few large, well-engineered and heavily utilized web-based genome browsing resources have grown to serve the needs of research communities. The "big three" public genome browsers are the University of California at Santa Cruz (UCSC) Genome Browser (Kent et al. 2002), Ensembl (Flicek et al. 2008) and the National Center for Biotechnology Information (NCBI) Map Viewer (Wheeler et al. 2007). Each uses a centralized model, where the web site provides access to a large public database of genome data for many species and also integrates specialized tools, such as BLAST at NCBI and Ensembl and BLAT at UCSC. The public browsers provide a valuable service to the research community by providing tools for free access to whole genome data and by supporting the complex and robust informatics infrastructure required to make the data accessible.

As these browsers are the most likely entry point to genome data for most researchers, there is a tendency to equate the user interface, the web site, with the underlying databases and infrastructure components. However, it should not be overlooked that the software driving the USCS and Ensembl browsers is open source and publicly available, and can be decoupled from the parent data sources and installed independently. Another commonly used browser, the Generic Genome Browser (GBrowse; Stein et al. 2002), takes this modularization even further. The software is specifically designed for deployment as a customized genome browser for model organisms and other genome databases and uses a generic data model that supports a variety of database types and schemas.

### *3.2.1 UCSC Genome Browser*

The Genome Browser at the UCSC (http://genome.ucsc.edu; Kent et al. 2002) provides access to genome assemblies obtained from the NCBI and external sequencing centers. The annotations available at UCSC come from many sources, including data from NCBI's RefSeq (Pruitt et al. 2007) and the Encyclopedia of DNA Elements project (ENCODE Consortium 2004). The UCSC annotation pipeline is responsible for more than half of the tracks in their browser. An example of UCSC annotations produced by this pipeline is the UCSC Genes set, which is an improved version of UCSC Known Genes (Hsu et al. 2006). These are high quality, moderately conservative gene annotations based on data from RefSeq, GenBank and Uniprot. Another prominent example of UCSC annotations is the comparative genomics tracks. These tracks show a summary sequence conservation amongst species and also include chain and net tracks (Kent et al. 2003) that highlight chromosome alignments and rearrangements. The UCSC Genome Browser for the human genome features sequence conservation data for a 28-way vertebrate genome alignment (Karolchik et al. 2008). At the time of writing, the UCSC browser supports 48 species and the human genome browser currently has 178 data tracks.

The UCSC browser features a simple user interface (Fig. 3.1). The typical entry point is a home page for each species, which provides background information about the genome build and the ability to search by either a keyword or position. The browser itself has a single window, with the detailed annotations displayed in one panel and the navigation and search functionality elsewhere on the page. Depending on the genome, a chromosome overview of the displayed region is also displayed at the top. As with most genome browsers, the orientation of the sequence is horizontal and the view can be zoomed in or out and panned right or left. Searching for genes or specific chromosome positions, using a search box and navigation buttons at the

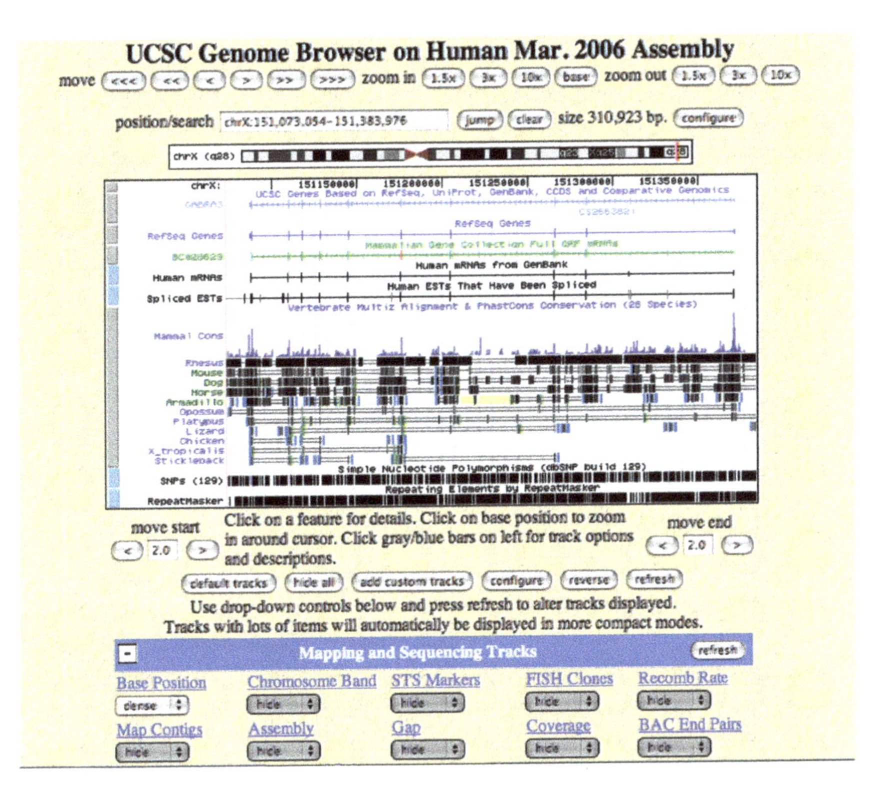


**Fig. 3.1** The UCSC Genome Browser. A portion of human chromosome X is displayed. The UCSC browser features a simple user interface with navigation controls at the *top*, a chromosome overview, if available, and a details panel. Track configuration controls are at the *bottom*. Several display options are available for each track and tracks are grouped into categories (not all are shown) that can be opened and closed by clicking the +/– buttons

top of the page, is very simple. The data are organized into linear, horizontal tracks, which contain glyphs representing individual sequence features. There can be a bewilderingly large number of tracks, but this is offset by grouping of tracks into logical categories, each of which can be opened or collapsed by the user. The browser also supports custom tracks via upload of third-party annotation files or through the table browser. Outbound data sharing via the Distributed Annotation Service protocol (DAS; Dowell et al. 2001) is also supported. After choosing a combination of tracks and display options, a user can save the session to a user account, which can be kept private or shared with colleagues. Access to the data displayed by the UCSC browser is provided in two ways. Bulk data can be downloaded from the FTP (File Transfer Protocol) site, which can be accessed via web browsers or dedicated FTP client software. Interactive data mining, custom track generation and data download are supported by the table browser interface (Fig. 3.2) (Karolchik et al. 2004).

A particular strength of the UCSC browser is the speed and stability of its user interface. UCSC is unusual amongst web-based genome browsers, in that it is written in the C programming language and uses a combination of a MySQL database and highly optimized data serialization techniques to produce a very fast, responsive browser. Other strengths of this browser include the comparative genomics resources and the well-developed support for custom tracks and third-party annotations. A minor drawback is that the user interface is spartan compared to other genome browsers. The UCSC genome browser has a loyal and enthusiastic user base, particularly amongst members of the ENCODE consortium.

The complexity of the UCSC code base and data infrastructure do not lend themselves well to off-site installation or mirroring by the casual user, but the source code and underlying data are freely available for non-commercial use. Three official mirrors are available at the Medical College of Wisconsin, Duke, and Cornell Universities (genome.ucsc.edu/mirror.html). Also, according to web access logs, there are as many as a dozen unofficial sites mirroring the UCSC data (H. Clawson personal communication).

### *3.2.2 ENSEMBL Genome Browser*

The Ensembl project's (http://www.ensembl.org; Flicek et al. 2008) genome browser has a much more nuanced, feature-rich user interface. The sequence data are derived from EMBL and other external sources. This browser is also more than just a portal to externally derived data; the Ensembl project has extensive annotation pipelines, to which a substantial part of their resources are devoted. Ensembl has its own gene prediction and annotation pipeline (Curwen et al. 2004). Some examples of recent additions to the database include genome-wide maps of protein–DNA interactions and the regulatory build, an effort to annotate all regulatory sequences (Flicek et al. 2008). The current release of Ensembl (release 52) contains 46 species, plus an additional three species in pre-release status. In the human genome browser, there are over 200 data tracks to choose from.

**Fig. 3.2** ContigView at Ensembl. A portion of human chromosome 4 is displayed. ContigView features three display panels, a chromosome view, region overview and detailed view, each at a higher zoom level. Tracks in the Detail view are configurable via a menu at the top of the panel. Navigation control buttons are available in the lower panel and rubber-band selection can be used in all three zoom levels by clicking and dragging the mouse anywhere on the panel

The Ensembl browser supports extensive keyword searching, which provides access to any textual information linked to the genome, genes, gene families, etc. An entry point to the browser is a home page for each species, which features a chromosome or karyotype view. In the species home page, clicking on a chromosome takes the user to a more detailed map view, from which more precise coordinates or search terms can be specified to enter the core of the genome browser. The Ensembl browser has three panels (with a fourth for base pair resolution in development) for different levels of resolution from chromosome to contig to base pair. The two lower panels have their own set of tracks. The top panel displays the position viewed in the chromosome context. The region overview panel provides a high-level visual summary of gene and contigs (the assembled sequences used in the genome build). The main "region in detail" panel displays a detailed view and displays tracks in a similar fashion to those at UCSC and other genome browsers. The base pair view was available in previous versions of the Ensembl browser but is still under development at the time of writing for the new Ensembl browser. It was useful for nucleotide level features such as codons, single nucleotide polymorphisms (SNPs), and restriction sites. Each of the panels can be toggled open or closed as required. A rich set of track display options is available in pull-down menus accessible from the top of the detailed view panel. An example of a feature that could be improved is the rubber-band selection. Although the selected area is represented as a red-bordered rectangle that resizes in both the vertical and horizontal dimensions, the actual selection is only the sequence coordinate range, not the particular items within the box.

Like the UCSC browser, the Ensembl browser supports uploading custom tracks, though the size of data in those tracks is limited to 5 megabytes. The Ensembl browser also makes extensive use of the Distributed Annotation System (DAS). The DAS menu allows the user to select from a number of pre-configured DAS sources or to configure their own DAS server as a source. Custom track files served from a users' own web server that follow the UCSC formatting conventions can also be provided as a data source. Although setting up a DAS server or providing custom track files on a web server is less convenient than directly uploading files, it offers the flexibility of maintaining independent remote annotations that can be updated at will and included in the browser display.

There are a variety of ways to obtain bulk data from the Ensembl site. One approach is through DAS. An Ensembl DAS server is available to export some of its more popular tracks. A second method for interactive data mining and bulk downloads is BioMart (http://www.biomart.org). BioMart provides a high-level data user interface with capabilities of complex filtering and merging operations. BioMart has its roots in Ensembl but has since grown into a generic tool that supports data mining for different genome databases, such as WormBase, HapMap, VectorBase, among others. Ensembl offers programmatic access to the live database via direct SQL queries to a MySQL server and through a well-supported Application Programming Interface (API). As with the UCSC browser, the software and raw data for Ensembl can also be downloaded from their FTP site. The Ensembl software is completely open-source for both academic and commercial users.

The Ensembl user interface is very feature-rich and does take some time to learn but documentation and training are available to get the most out of the resource. The web site administrators report an average of 250 queries made on the Ensembl database every second (Giulietta Spudich, personal communication), a good indication that this genome browser has a large following. Mirroring the Ensemble web site or setting up an independent browser with Ensembl software are encouraged and well documented. The installation of mirrors or applications that use the Ensembl infrastructure is not closely tracked but there are at least 16 known instances of external use of all or portions of the Ensembl software (data from http://www.ensembl.org/info/about/ensembl_powered.html). Some examples of websites using Ensembl software to serve their own data include Gramene (www.gramene.org) and VectorBase (http://www.vectorbase.org). Third-party genome browsers have also been written to use Ensembl data. On such example is the 3D genome browser Sockeye (Montgomery et al. 2004).

### *3.2.3 NCBI Map Viewer*

The NCBI web site is best known for GenBank and PubMed. The Map Viewer (http://www.ncbi.nlm.nih.gov/projects/mapview) is able to draw from the considerable resources of the NCBI toolkit and also from the vast stores of sequence data and annotations in GenBank. Species coverage is higher than the other browsers; the map viewer currently has 106 species. The NCBI browser is rather different from the others in terms of look and feel. For example, the display is vertically oriented (Fig. 3.3). Rather than being mapped to the reference sequence,

**Fig. 3.3** NCBI Map Viewer. A portion of human chromosome X is displayed. The chromosome ideogram and data tracks are in vertical orientation, with higher zoom level to the right. Track controls are located at the top of each track. It is possible to set track with a right-facing arrow icon as the "master map." Clicking on the non-text part of a data track will access a zoom/navigation menu

the tracks are displayed next to each other, with one of the tracks serving as a reference map. The maps are usually based on cytogenetic bands, contigs, genes, or genetic markers. The number of features displayed varies with the current zoom level and the density of features in that part of the map. The navigation features of the interface are somewhat limited and take some getting used to for users accustomed to the horizontal view typical of most genome browsers. Also, the number of tracks available for viewing is more limited than in the Ensembl or UCSC browsers.

A strength of the NCBI map viewer is its tight integration with other well-known NCBI resources, such as UniGene, LocusLink and RefSeq. Like Ensembl and the UCSC browser, sequence similarity search capabilities are also available in the form of BLAST. Unlike Ensembl and UCSC there is no mechanism for integration of third party data and the browser is neither a DAS server nor client, making integration with third party products difficult. The data that underlie the Map Viewer are publicly accessible via NCBI's web and FTP sites. Programmatic access to the data is less obvious and care must be taken not to use "screen scraping" scripts designed to harvest a lot of data from the web site interface, as NCBI will block access from computers that make excessive queries of this nature. Less obtrusive access to NCBI's data is available through tools in the BioPerl package (Stajich et al. 2002).

### 3.2.4 The Generic Genome Browser

GBrowse (http://gmod.org/wiki/GBrowse) represents a different class of genome browser in the sense that its raison d'etre is to be integrated as a user interface to third-party genome databases. In the early days of species-specific databases there was a lot of parochialism in software infrastructure development. For example, in 2000, WormBase, FlyBase, the *Saccharomyces* Genome Database (SGD), Mouse Genome Informatics (MGI), and The *Arabidopsis* Information Resource (TAIR) all used different software to display their organism's genome features. The division of efforts in the model organism community eventually gave rise to a software standardization movement, in the form of the Generic Model Organism Database project (GMOD; http://www.gmod.org). This effort has been successful in that, at the time of writing, more than a hundred sites use GBrowse including the aforementioned "core" model organism databases. GBrowse is also the principal data browser for the model organism ENCODE (modENCODE) data coordinating center (http://www.modENCODE.org) and is used by the international HapMap consortium (http://www.hapmap.org).

GBrowse offers three horizontal display panels, the chromosome overview, the region view, and the detailed view (Fig. 3.4). Data are organized into tracks and individual features are organized as glyphs with tracks potentially present at all three zoom levels (panels). The user interface offers a rich set of core functions, such as the ability to vertically re-order tracks, in-line track configuration, popup

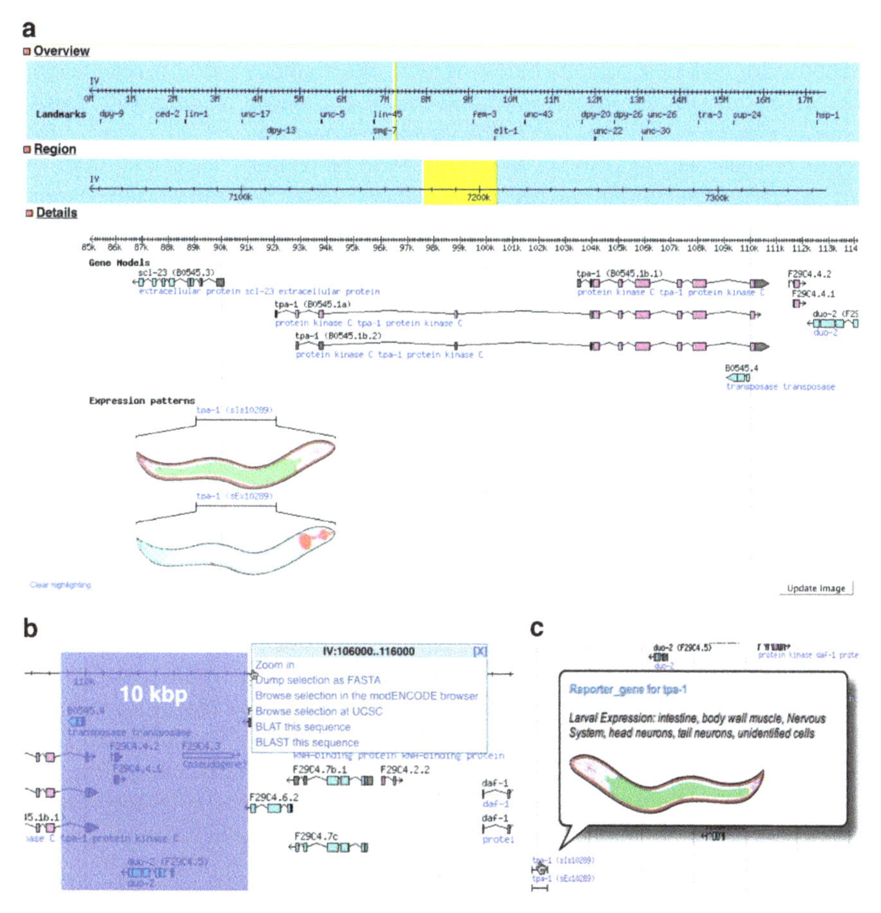


**Fig. 3.4** GBrowse: A portion of *C. elegans* chromosome IV at WormBase is displayed. (**a**) User interface. GBrowse features three display panels at increasing zoom levels. The overview, region view and details view all may display tracks. Most tracks are generally displayed in the detail view. Track display options and track sharing can be accessed by clicking icons in the title bar at the *top left* of each track. Tracks can be vertically reordered by dragging the title bar. (**b**) Rubber-band selection is available by clicking and dragging the mouse on the ruler at the top of each of the three panels. The behavior of this feature in configurable. The menu shown lists operations that can be performed on the selected sequence region. (**c**) A popup balloon tooltip. The tooltip behavior is configured via the configuration file and can be varied on a per-track basis. The example shown displays anatomical expression pattern information for promoter-GFP fusion constructs in *C. elegans*. The information displayed in the balloons is retrieved at runtime from the WormBase server via AJAX

balloon tooltips, and a convenient track sharing function called by clicking on a button. Like Ensembl, GBrowse facilitates mouse-based navigation via rubber-band selection of sequence regions in all three panels.

In order to cope with very dense quantitative data, such as microarray, the UCSC genome browser programmers developed the “wiggle” track, which uses a combination of relation database and data serialization to make dense quantitative data more tractable for real-time genome browsing. This functionality has also been adopted by GBrowse. GBrowse supports custom tracks and upload of third-party annotations in a manner similar to the UCSC genome browser. It can act as both a DAS server or client and supports an additional DAS-like protocol that allows more fine grained control of configuration and graphical rendering and allows more complex sequence features than are supported by DAS. Tracks can be shared with a click of a button to produce a popup balloon with a URL which, when provided to another GBrowse instance, will allow data to be viewed in a separate browser.

GBrowse has two main advantages over other browsers. Firstly, it is totally decoupled from the underlying data sources, ensuring that it is easy to add new data sources. Secondly, GBrowse is very easy to install and is much more configurable. A typical installation can usually be completed within a few hours and fully configured in a day or two. The look and feel of GBrowse is customizable through hooks that are accessible via the configuration files. More than 70 glyphs are currently available for displaying graphical data in the genome browser context. The style and color of a track's glyphs can be controlled dynamically by inserting Perl callbacks into the configuration file. Some examples of callback uses include changing the background color of an arrow glyph according to the strand of the feature and changing glyph types in the same track based on feature type, zoom level, or other context.

A flexible plugin architecture and API also provide the means for third-party developers to add new functionality to the browser with relative ease, without having to directly modify the software. A few default plugins come with the distribution, including support for exporting data in commonly used formats and an integrated program for designing PCR primers around features shown in the browser. GBrowse also has session management, though there are no user accounts. Through web browser "cookies" previous search sessions and configuration options and tracks are cached so the browser picks up where the user left off on the previous visit. A shareable snapshot of the current configuration can also be saved as a URL with a "bookmark this" link.

Like other browsers written in Perl, GBrowse cannot boast of the speed offered by UCSC. Recent advances in next-generation sequencing and microarray technologies have vastly increased the volume of rate of production of new data to be displayed, which strains the genome browsers' ability to retrieve and display data quickly. Recently, GBrowse developers have responded to the need for improved performance in several ways. First, image caching is part of the graphical rendering process. GBrowse remembers if a region has been drawn before and re-uses images rather than drawing them from scratch each time. Second, for dense quantitative data such as the wiggle tracks, GBrowse has adopted a data serialization strategy based on the one employed by UCSC, which has improved browser performance. As of version 2.0 (to be released in late 2009), GBrowse also uses a combination of parallel processing and uses AJAX (asynchronous JavaScript and XML) technologies to dramatically speed up graphical rendering of all types of data in the browser. By rendering each track independently, it is possible to divide the processing work across a compute farm and render tracks concurrently in the user interface.

Another aspect of this redesign is that GBrowse can display data from multiple databases simultaneously, making it possible to distribute data storage as well as data processing. For installations lacking a compute farm, GBrowse can still be run in standalone mode while enjoying the performance benefits of multiple databases and concurrent, rather than serial track rendering.

## 3.3 Standalone Annotation Browsers and Editors

While web-based genome browsers offer many advantages to users, there is a case to be made for using a standalone desktop application. From a performance perspective, responsiveness of the application is often an advantage – once the data are loaded into a standalone browser, there is no need to ask a server for more information to zoom or pan, so there is no network latency. Desktop application browsers can perform advanced operations not available in most web-based browsers, especially the ability to edit and save rather than just browse annotations. Most of the standalone browsers covered in this chapter are written in the Java programming language. An advantage of this is that web-start is often available, which makes it possible to try these applications without installing them locally. There is an abundance of standalone annotation browsers and editors available. For example, in the description of their browser Genomorama (Gans and Wolinsky 2007), the authors list 19 freely available standalone applications. Here, we will give an overview of a few of the more commonly used, open-source annotation browsers and editors.

### *3.3.1 Apollo*

Apollo (http://apollo.berkeleybop.org ; Lewis et al. 2002) is a Java application initially developed as an editor for the curators annotating the *D. melanogaster* genome for FlyBase. It has since been included in the GMOD project and several other organizations have adopted it for genome annotation, including as a tool for "community annotation," where the data providers allow interested outsiders to directly add annotations and edit gene models to their databases. It supports a wide variety of formats for importing and exporting data, including GMOD's Chado database schema, XML, GenBank, and GFF3. Apollo's main advantage as a browser is the ability for users to easily add and modify their own annotation data using a graphical tool and save changes to a remote database or a local file. However, Apollo is limited in the total amount of sequence features it can display at one time, thus limiting its utility as a genome browser for views of DNA larger than a megabase. It can be argued that this is not unreasonable, as the users of Apollo will typically be looking at a small region in detail as opposed to large scale whole genome data.

### 3.3.2 IGB

The Integrated Genome Browser (IGB, pronounced "iggbee") is another Java based stand-alone browser (http://affymetrix.com/support/developer/tools/download_igb.affx). While it can load data from a variety of flat file formats, IGB really shines when working with DAS servers. IGB retrieves data from multiple DAS1 and DAS2 servers and will intelligently retrieve an appropriate amount of data of either a whole chromosome or only the viewed region. It works well with whole genome analysis results, such as tiling array results, which is not surprising, given that development was based at the microarray company Affymetrix. Once the data are loaded, IGB scrolls and zooms smoothly, even over large sequences with many features.

IGB also has sophisticated analysis tools. Filters can be added to graphical tracks to limit upper and lower thresholds on scores and filtering the contents of one track based on the contents of another. For example, one could display only the expression results that appear where there are annotated exons. IGB can also perform a variety of other transformation operations, such as displaying the result of adding and subtracting values from tracks or performing a union or intersection of features for a set of tracks.

### 3.3.3 Artemis

Artemis (http://www.sanger.ac.uk/Software/Artemis; Berriman and Rutherford 2003) is another Java based browser developed at the Sanger Institute. Artemis features a three-panel interface that depicts the genome at different resolutions. The interface is simple and easy to master, but lacks the color and flashiness in some of the other standalone browsers. Artemis allows the user to view and edit both sequence and annotation data. Though Artemis has been around for a decade, it is still under active development and has been used to annotate many microbial genomes primarily at the Sanger Centre in Hinxton, UK. It accepts the most common sequence and feature file formats: EMBL, GENBANK, FASTA, GFF3 and the BLAST alignment format. Data can also be downloaded directly from the European Molecular Biology Laboratory (EMBL) database. Extra sequence and annotations can be integrated by loading files locally. The Artemis's display is primarily focused on curation of gene models, which is facilitated by having the display of exon or coding regions nested within the six-frame translation and stop codons. Another nice feature for eukaryotic genomes is having spliced exons visually offset into the appropriate reading frame. Although the user interface for this application is fairly basic in appearance, it should be noted that Artemis and similar applications with simple no frills interfaces are generally used by power-users or data curators who annotate whole genomes rather than casual browsers.

### *3.3.4 NCBI Genome WorkBench*

The NCBI Genome Workbench (http://www.ncbi.nlm.nih.gov/projects/gbench) is a customizable workbench of tools that allows you to organize sequence data retrieved from NCBI databases, sequence data from your own files, or both. Unlike the other annotation editors, NCBI WorkBench is written in the C++ programming language and runs as a native application on Windows, MacOS, and various Linux distributions. What this means to the end user is that installation of the program is fairly straightforward as it does not require that Java be installed.

The Genome Workbench has a richer set of features than Apollo and Artemis. It allows the user to view sequences, construct phylogenetic trees, alignments, dot-plots, and more. Also offered is an attractive zoomable graphics view that allows visual exploration of the sequence region in either horizontal or vertical orientations. A nice set of alignment analysis tools is provided and BLAST and analysis results can be saved into the project. Data can be downloaded directly from GenBank and a variety of file formats are supported for import and export of local files. It seems confusing that the Genome WorkBench will not accept GenBank files as an import format, though there are a reasonable number of alternative formats to use. The user interface could be improved by adding a progress bar or other obvious visual cues so that deployed tasks, such as BLAST, are running in the background. Of the browsers presented here, Genome WorkBench's user interface is the most visually appealing. Not all aspects are intuitive, but it is worth spending some time exploring the application and its abundant documentation.

## 3.4 Web-Based Synteny Browsers

Interest in comparative genomics has risen with the number of genome sequences due to decreasing costs of DNA sequencing. Accordingly, computational biologists have begun to address the need for visualization tools for comparative genomics. This emerging class of software is generally referred to as synteny browsers. Typically, such browsers have a look and feel similar to single genome browsers, with additional features for displaying comparative data. Generally, there needs to be a way to display two or more reference sequences as well as the relationship between the annotated features. The relationship between features can be established in several ways, including human curation, BLAST/BLAT analysis or more sophisticated comparative sequence analysis. There is considerable functional overlap amongst the various synteny browsers and it is not yet clear which will emerge as the dominant software. While a few commonly recognized file format standards have evolved or been extended for genome annotations, there is still considerable heterogeneity in the synteny browsers data. Each synteny browser has its own way of storing synteny data, and setting up a local installation is not straightforward, often involving some ad hoc scripting to convert raw comparative data or alignment data to a suitable format.

The relationship between genes and aligned regions of their respective reference sequences is a feature common to most browsers. Some browsers also provide the means to examine larger regions of aligned sequences, generally referred to as syntenic blocks, from a more zoomed-out perspective. Some synteny browsers have no predetermined upper limit in terms of the number of species shown in the browser. However, practical limitations of processing power, real-time data retrieval and available screen space impose some moderation in this regard. Here, we provide a survey of web-based Synteny browsers that are connected to the GMOD consortium.

## *3.4.1 GBrowse-Based Synteny Browsers*

### 3.4.1.1 GBrowse_syn

GBrowse_syn (http://gmod.org/wiki/GBrowse_syn) displays synteny data with a central reference genome and additional genomes displayed above and below the center panel (Fig. 3.5). Data are stored for GBrowse_syn in separate databases: each genome has its own GBrowse-compatible database (so that it can have a traditional GBrowse display in addition to GBrowse_syn), plus the alignment data are stored in a joining alignment database. This arrangement gives the system the flexibility to change the primary sequence context for the display. It also means that arbitrarily large or small blocks can be aligned in the GBrowse_syn display, so it can compare on the level of genes or larger blocks. The alignment database schema supports storage of information about insertions and deletions (indels) in the multiple sequence alignment. The indel information is then encoded in the graphical display as scaled gridlines within the shaded alignments. The configuration file for GBrowse_syn is a separate file, but similar to the GBrowse configuration file. GBrowse_syn is currently implemented at WormBase (http://dev.wormbase.org/db/seq/gbrowse_syn ).

### 3.4.1.2 SynView

SynView (http://eupathdb.org/apps/SynView; Wang et al. 2006) is a synteny browser that uses only the standard GBrowse distribution and a sophisticated GBrowse configuration file. It uses Perl callbacks to overlay a synteny view onto the GBrowse details panel (Fig. 3.6). The fact that such a display can be accomplished simply by editing the configuration file is a testament to GBrowse's flexibility (and the ingenuity of the SynView developers). There are several parts that must come together for this to work. The synteny data are stored in standard GFF3 format, and each genome that is compared to the reference is composed of two

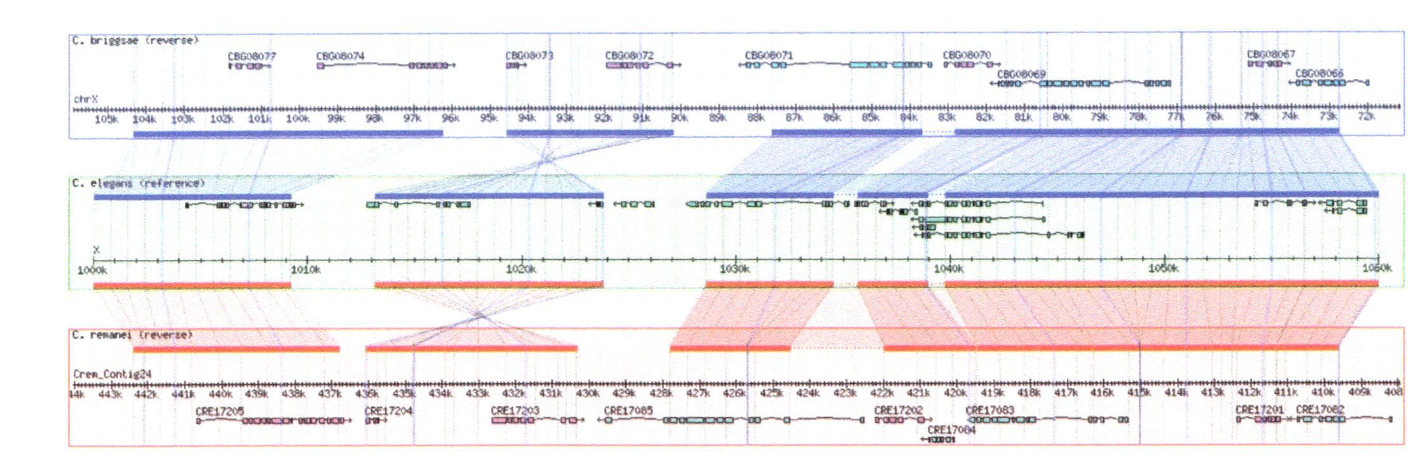


**Fig. 3.5** GBrowse_syn as implemented at WormBase (http://dev.wormbase.org/db/seq/gbrowse_syn), showing *C. elegans* as the reference sequence and *C. remanei* and *C. briggsae* as comparative sequences. Shaded polygons indicate the position and orientation of aligned region between two species. Grid-lines indicate gain or loss of sequence in *C. remanei* and *C. briggsae* relative to the *C. elegans* reference sequence. Note the inversion in the center of the panel and the changes in gene structure between the species, such as expansion or contraction of introns

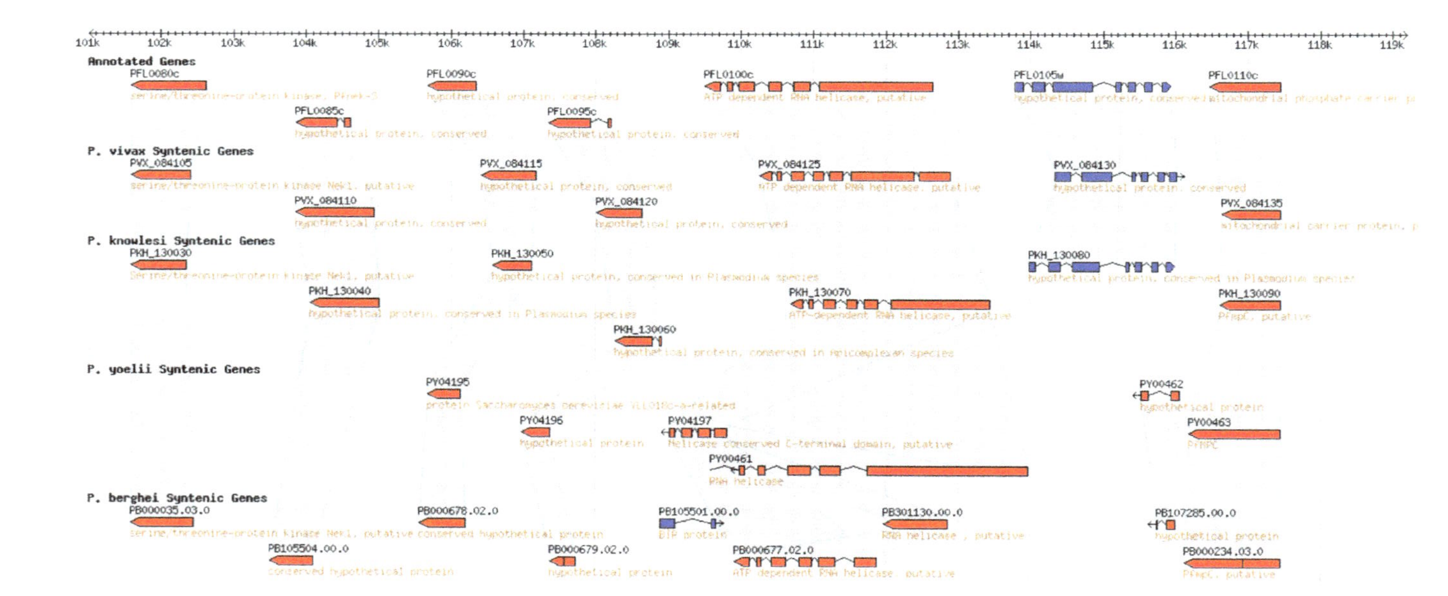


**Fig. 3.6** An example SynView, comparing several Plasmodium genomes, using shared gene homology. This data set uses *P. falciparum* as the reference genome, whose features are at the top labeled “Annotated Genes.” The other tracks appear in pairs: one with the names of the species being compared with “synteny” appended, which shows the relative base pair scale for the compared sequence, and one with the names of the compared species with “syntenic genes” appended. Shaded polygons linking intergenic sequences show syntenic relationships, which allows for relationships between one gene on one genome to correspond to multiple genes on another genome. For example, PFS0115w on the *P. falciparum* genome corresponds to three genes on the *P. vivax* genome

tracks: one to draw the syntenic genome's scale bars, and one to draw the features on the syntenic genome's features. Information on how to scale the syntenic features is stored in each feature's GFF line in its ninth column. How a syntenic gene is related to the reference gene is stored in the syntenic gene's GFF line as well. In order to draw the polygons that show the relationship, Perl callbacks that are in the GBrowse configuration file are executed to layer the graphics on the panel between the reference genes and the syntenic genes.

#### 3.4.1.3 SynBrowse

SynBrowse (http://www.synbrowse.org ) was written to take advantage of the same BioPerl (Stajich et al. 2002) infrastructure components used by GBrowse (Brendel et al. 2007). SynBrowse data are expressed as standard GFF2, using target tags to relate a region in one species to its counterpart in another species. SynBrowse also takes advantage of different "source" tags, in the second column of GFF file, to filter the view so as to display microsynteny as well as macrosynteny, depending on the zoom level of the display. For example, a single database could display synteny on the exon, gene, or multimegabase levels. The graphical display of SynBrowse is similar to GBrowse, except for the multiple detail panel linked by the alignments (Fig. 3.7)

### *3.4.2 Other Synteny Browsers*

#### 3.4.2.1 Sybil

Sybil (http://sybil.sourceforge.net) is a synteny browser that runs on a GMOD Chado database (http://gmod.org/wiki/Chado) and is populated with multiple genomes (Crabtree et al. 2007). It was initially developed at The Institute for Genomic Research (TIGR) to support their bacterial genome sequencing and annotation. The Chado database schema was designed to support data from multiple organisms, so no special modification to the schema is required. Sybil focuses on protein and gene clusters, and displays them in the genome context. In addition to synteny views, Sybil also provides a "whole genome comparison" view (Fig. 3.8), where a reference genome is displayed as a color heat map, and compared genomes use the same map so that users can see immediately what sections of the compared genomes have rearranged relative to the reference.

#### 3.4.2.2 CMap

CMap (http://gmod.org/wiki/CMap ; Faga 2007) was initially developed as a comparative mapping tool for Gramene to allow comparisons between different types of

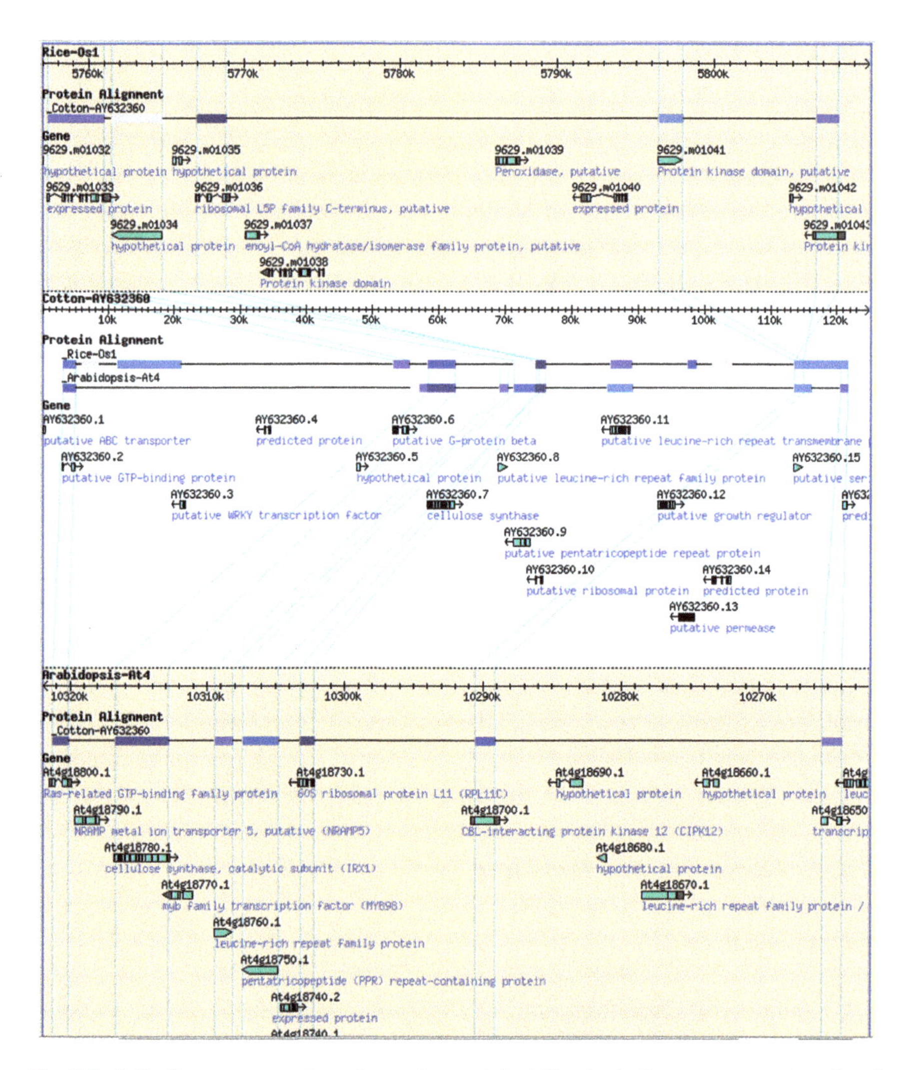


**Fig. 3.7** A SynBrowse comparison rice, cotton and *Arabidopsis thaliana* genomes using shared protein alignments. *Blue lines* drawn between genomes mark the edges of protein similarity between the compared genomes

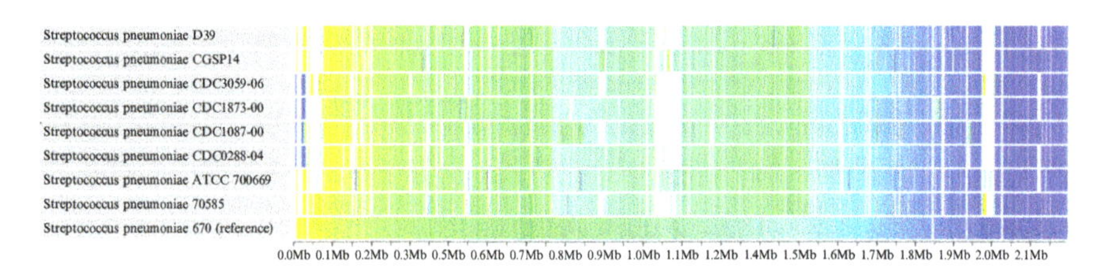


**Fig. 3.8** Using Sybil to display aligned *Streptococcus* strain genomes. The 670 strains on the bottom is the reference, and the rearranged colors on the compared genomes show how regions have rearranged in those strains. The swapped *light blue* and *green* sections in the compared genomes relative to the 670 genome shows that a rearrangement has happened in the reference relative to the other genomes. *White* sections show regions that are missing either in the reference genome or the compared genome

maps available for rice strains (physical maps, genetic maps and sequence maps). However, since CMap is organism agnostic and only "knows" about maps, it is quite easy to convert it to a synteny browser, comparing genomes of different species as maps. CMap is unique amongst synteny browsers, as it arranges the sequences vertically instead of horizontally (Fig. 3.9). Any number of genomes can be compared in this way, though side scrolling may be required just as vertical scrolling may be required with other unlimited synteny browsers.

## 3.5 How to Choose a Genome Browser

The choice of which browser to use is governed by a number of factors. In some cases, the browser of choice is a historical accident (the first browser the user happened to encounter), which is fine as long as the researcher's needs are fully met. However, it is a good idea to explore the alternatives, as relevant data or better user interface features missing from one browser of choice may be offered by other browsers.

For an end-consumer who lacks the ability or interest to build a genome browser, or whose primary interest is accessing as much information as possible about a few favorite genes, the choice may be dictated by something as simple as which browser or browsers have the species of interest. Since there is fairly substantial overlap in terms of species coverage, the choice may be guided more by user preferences for the type of interface or available features. If species choice is not limited and one prefers a fast browser with a simple interface, then UCSC would be a good choice. Nevertheless, it would still be a good idea to visit Ensembl occasionally to ensure one is getting all there is to offer for the species or genes of interest. Likewise, if one prefers a richer feature set, then Ensembl would be a good starting point. If you happen to work on an established model organism, the first browser you should visit is the model organism database (MOD) for that species, for examples see Table 3.1. Although the data will likely be mirrored to some extent on one or more of the "big three" genome browsers, the attention to detail and integration of data from all sources will be best at the MOD for your organism.

For the research group who wants to host a browser for an organism, the choice of browsers is not quite as straightforward. Preference for particular functional or aesthetic qualities of a particular browser is important but other significant factors should be considered in the decision of software platforms. For example, the size and skill-level of your informatics group will be a key factor. For smaller groups who want to have genome browser up and running in a few days, GBrowse is a good choice. If you have several skilled informatics staff available, then you might want to consider using the Ensembl or UCSC infrastructure. Integrating the latter two into an established website or genome database may be more challenging, so GBrowse could be a better choice for integration an existing infrastructure. For genome databases being built started de novo, one possibility to consider is GMODweb (O'Connor et al. 2008). GMODweb is basically a "genome database in

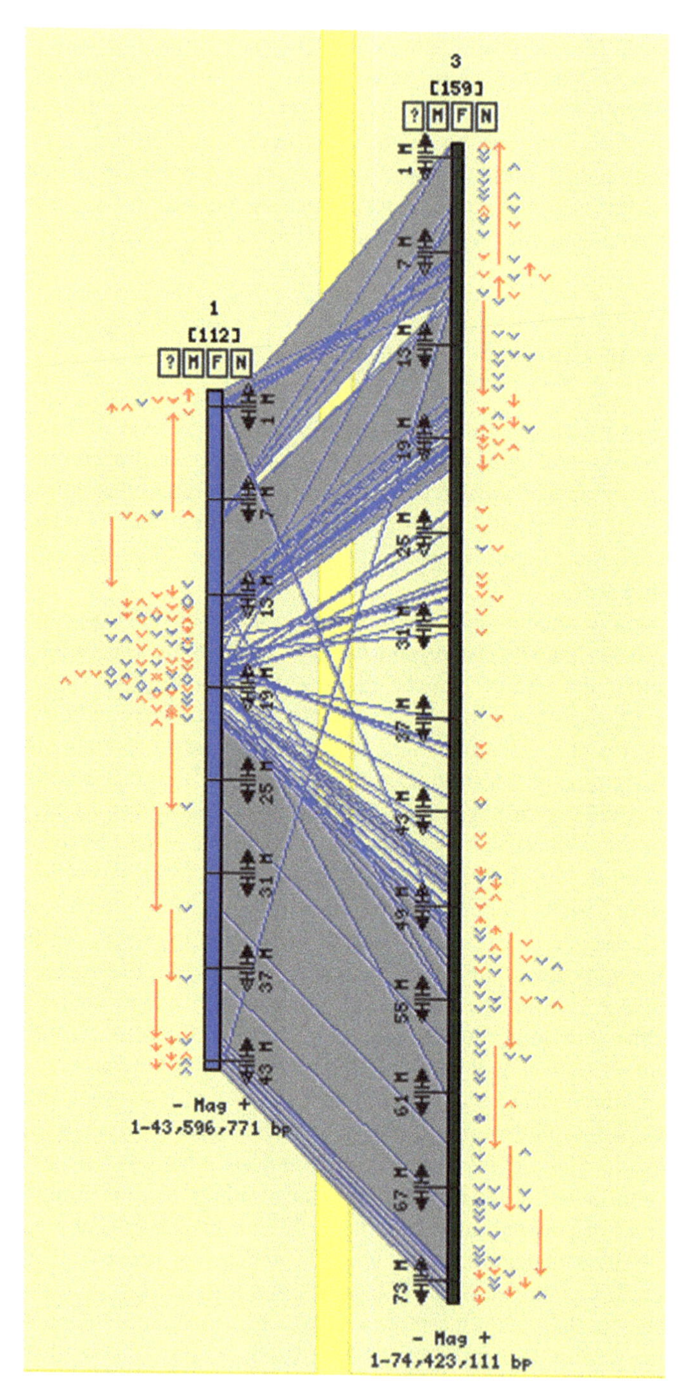


**Fig. 3.9** Using CMap as a synteny browser, showing a high-level comparison of rice chromosome 1 and sorghum chromosome 3. The *blue* shaded polygons show regions of large blocks of synteny, where the *blue lines* show individual aligned regions

**Table 3.1** Some model organism databases with genome browsers

| Organism | Web Site |
|---|---|
| *Caenorhabditis* species | wormbase.org |
| *Drosophila* species | flybase.org |
| *Mus musclulus* | www.informatics.jax.org |
| *Rattus norvegicus* | rgd.mcw.edu |
| *Saccharomyces cerevisiae* | yeastgenome.org |
| *Paramecium tetraurelia* | paramecium.cgm.cnrs-gif.fr |
| *Daphnia* species | wfleabase.org |
| Cereal species | gramene.org |
| *Arabidopsis thaliana* | www.arabidopsis.org |
| *Dictyostelium discoideum* | dictybase.org |
| *Schmidtea mediterranea* | smedgd.neuro.utah.edu |

a box" that can be set up with one's own data and a small bioinformatics staff. For example, the Paramecium MOD (Table 3.1) was constructed with GMODweb (Arnaiz et al. 2007).

Finally, if your group is actively producing or annotating genomic sequences, gene predictions, expression data, etc., a local installation of an annotation editor would be preferable to a web-based genome browser. For curating gene models based on integration of external evidence such as sequence conservation and expression data, both Apollo and Artemis are seasoned editors that have been used to annotate many genomes. The Genome WorkBench offers a very rich suite of features and integrates tightly with other major NCBI resources. There is substantial overlap in the functionality between the annotation editors covered in this chapter and the choice of which to use may depend on preference for a particular user-interface.

## 3.6 Next Generation Genome Browsers

The increasing use of JavaScript and AJAX on web sites such as Google maps™ and many others marks a trend towards enhancing web content through increasing reliance on client-side processing. Client-side processing refers to work being performed by the user's web browser rather than at the web server. What this means to the end user is that the browsing experience is an improved user interface for web-based applications. Considering the Google maps™ example, it is possible to smoothly zoom, pan, and scroll the maps without having to continuously reload the page because asynchronous queries (AJAX) are used to load the map data independent of the parent web page. Most genome browsers still use primarily server-side processing for database access and graphical rendering, which can result in perceptible delays, both from network latency and computational load imposed upon the server by large amounts of data. A few JavaScript based enhancements are already used by genome browsers such as the collapsible track sections in the UCSC browser and the rubber-band selection in the Ensembl browser. Generally, the adoption of more sophisticated JavaScript and AJAX features into genome browsers has lagged behind other classes of software but the current version of GBrowse has used JavaScript and AJAX to drive many of its user interface features, for example the popup balloons, rubber-band selection, draggable tracks, inline configuration, collapsible sections, and, in the upcoming release, parallel processing of track rendering. However, all of the web-based browsers covered in this chapter still rely heavily on server-side processing to do the heavy lifting of converting sequence annotation in the database to images that are sent out to the web browser.

A new, AJAX-based version if GBrowse (JBrowse) is currently under development (Fig. 3.10). JBrowse is a ground-up rewrite rather than an enhancement of the existing GBrowse. JBrowse uses AJAX and client-side graphical rendering to produce a smooth scrolling, Google maps™-style interface to the genome, with intuitive semantic zooming, where the glyph-types change according to the zoom level. JBrowse will offer all of the same features as the original GBrowse but will use the client's web browser to do most of the graphical rendering and display. End users will be presented with a genome browser which has a fast, smooth and intuitive user interface. More information about the AJAX GBrowse can be obtained at http://biowiki.org/view/GBrowse/WebHome.

An alternative to AJAX is Adobe Flash. Flash technology also allows rich, cross-platform web applications. An example of a Flash genome browser has been developed at The Broad Institute for a variety of genomes (Fig. 3.10). They have a parallel genome browser, synteny browser, a dot plot for genome comparison, and circular genome browser. More information can be obtained at http://www.broad.mit.edu/annotation/genome/aspergillus_group/GenomeMap.html. While this software is not yet open source, the developers are looking into releasing it after development has stabilized.

## 3.7 Online Reading and Resources

*UCSC Genome Browser*Free OpenHelix Tutorial:
http://www.openhelix.com/downloads/ucscUser's Guide:
http://genome.ucsc.edu/goldenPath/helpDownloads:
http://hgdownload.cse.ucsc.edu*Ensembl*User's Guide:
http://www.ensembl.org/info/using/index.htmlDownloads:
http://www.ensembl.org/info/downloads*NCBI Map Viewer*User's Guide:
http://www.ncbi.nlm.nih.gov/projects/mapview/static/MapViewerHelp.html*G-Browse*User's Guide:
http://gmod.org/gbrowse-cgi/tutorial/tutorial.htmlDownloads: http://gmod.org/wiki/Gbrowse#Downloads*Apollo*User's Guide:
http://apollo.berkeleybop.org/current/userguide.htmlDownloads:
http://apollo.berkeleybop.org/current/install.html*Artemis*User's Guide: http://www.sanger.ac.uk/Software/Artemis/v10/manual/Downloads:

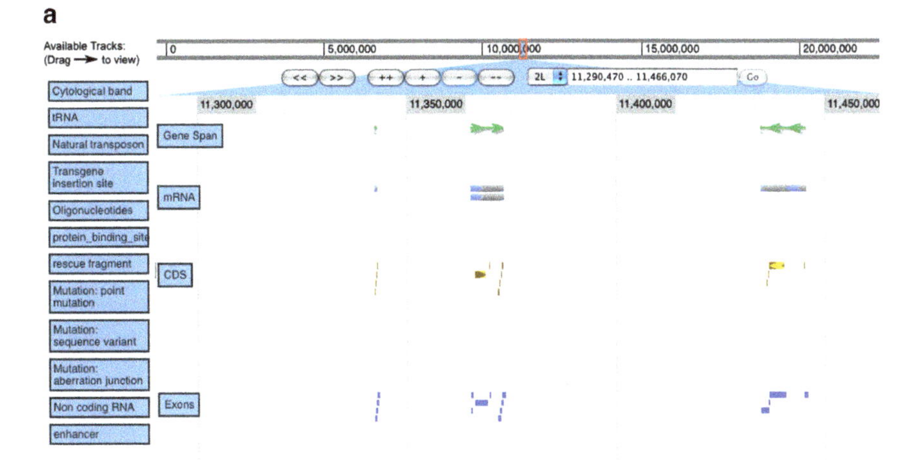


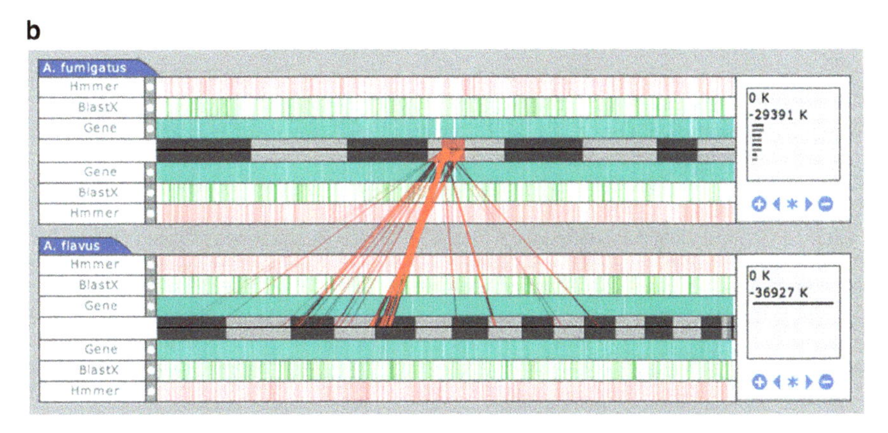


**Fig. 3.10** Next generation genome browsers. (**a**) A prototype of an AJAX GBrowse, as shown at (http://genome.biowiki.org/test/divbrowser). A portion of *Drosophila melanogaster* chromosome 3R is shown. The browser features a smooth, Google maps-style horizontal and vertical scrolling. Tracks can be added or removed from the display by simply dragging the title bar in or out of the display window. (**b**) The Broad parallel genome browser. There are two genomes displayed here, where the alternating *light* and *dark gray* stretches correspond to chromosomes in the genome. For example, the first *dark gray* section is the organism's chromosome 1, and the first *light gray* region is the organism's chromosome 2. Colored tracks above and below the *gray* tracks correspond to features in the positive and negative strands. Selecting a region results in lines to the other genome highlighting orthologous genes. Here a region of the forth chromosome *A. fumigatus* is selected and the browser shows orthologous genes on 9 chromosomes on *A. flavus*

http://www.sanger.ac.uk/Software/Artemis/v10/*IGB*User's Guide:
http://www.affymetrix.com/Auth/support/developer/tools/IGB_User_Guide.pdfDownloads:
http://www.affymetrix.com/support/developer/tools/download_igb.affx*NCBI Genome WorkBench*Tutorials:

http://www.ncbi.nlm.nih.gov/projects/gbench/tutorial.htmlDownloads:
http://www.ncbi.nlm.nih.gov/projects/gbench/download.html*Synteny Browsers*GMOD wiki:
http://gmod.org/wiki/Synteny

## References


Arnaiz O, Cain S, Cohen J, Sperling L (2007) ParameciumDB: a community resource that integrates the Paramecium tetraurelia genome sequence with genetic data. Nucleic Acids Res 35(Database issue):D439–D444

Berriman M, Rutherford K (2003) Viewing and annotating sequence data with Artemis. Brief Bioinform 4(2):124–32

Brendel V, Kurtz S, Pan X (2007) Visualization of syntenic relationships with SynBrowse. Methods Mol Biol 396:153–63

Crabtree J, Angiuoli SV, Wortman JR, White OR (2007) Sybil: methods and software for multiple genome comparison and visualization. Methods Mol Biol 408:93–108

Curwen V, Eyras E, Andrews TD, Clarke L, Mongin E, Searle SM et al (2004) The Ensembl automatic gene annotation system. Genome Res 14(5):942–50

Dowell RD, Jokerst RM, Day A, Eddy SR, Stein L (2001) The distributed annotation system. BMC Bioinformatics 2:7

ENCODE Consortium (2004) The ENCODE (ENCyclopedia Of DNA Elements) Project. Science 306(5696):636–40

Faga B (2007). Installing and configuring CMap. Curr Protoc Bioinformatics Chapter 9: Unit 9.8

Flicek P, Aken BL, Beal K, Ballester B, Caccamo M, Chen Y et al (2008) Ensembl 2008. Nucleic Acids Res 36(Database issue):D707–D714

Gans JD, Wolinsky M (2007) Genomorama: genome visualization and analysis. BMC Bioinformatics 8:204

Hsu F, Kent WJ, Clawson H, Kuhn RM, Diekhans M, Haussler D (2006) The UCSC Known Genes. Bioinformatics 22(9):1036–46

Karolchik D, Hinrichs AS, Furey TS, Roskin KM, Sugnet CW, Haussler D et al (2004) The UCSC Table Browser data retrieval tool. Nucleic Acids Res 32(Database issue):D493–6

Karolchik D, Kuhn RM, Baertsch R, Barber GP, Clawson H, Diekhans M et al (2008) The UCSC Genome Browser Database: 2008 update. Nucleic Acids Res 36(Database issue):D773–9

Kent WJ, Sugnet CW, Furey TS, Roskin KM, Pringle TH, Zahler AM et al (2002) The human genome browser at UCSC. Genome Res 12(6):996–1006

Kent WJ, Baertsch R, Hinrichs A, Miller W, Haussler D (2003) Evolution's cauldron: duplication, deletion, and rearrangement in the mouse and human genomes. Proc Natl Acad Sci USA 100(20):11484–9

Lewis SE, Searle SM, Harris N, Gibson M, Lyer V, Richter J et al (2002) Apollo: a sequence annotation editor. Genome Biol 3(12):RESEARCH0082

Montgomery SB, Astakhova T, Bilenky M, Birney E, Fu T, Hassel M et al (2004) Sockeye: a 3D environment for comparative genomics. Genome Res 14(5):956–62

Pruitt KD, Tatusova T, Maglott DR (2007) NCBI reference sequences (RefSeq): a curated non-redundant sequence database of genomes, transcripts and proteins. Nucleic Acids Res 35 (Database issue):D61–5

Stajich JE, Block D, Boulez K, Brenner SE, Chervitz SA, Dagdigian C et al (2002) The Bioperl toolkit: Perl modules for the life sciences. Genome Res 12(10):1611–8

Stein LD, Mungall C, Shu S, Caudy M, Mangone M, Day A et al (2002) The generic genome browser: a building block for a model organism system database. Genome Res 12(10):1599–610

Wang H, Su Y, Mackey AJ, Kraemer ET, Kissinger JC (2006) SynView: a GBrowse-compatible approach to visualizing comparative genome data. Bioinformatics 22(18):2308–9

Wheeler DL, Barrett T, Benson DA, Bryant SH, Canese K, Chetvernin V et al (2007) Database resources of the National Center for Biotechnology Information. Nucleic Acids Res 35(Database issue):D5–12

# Chapter 4
# Predicting Non-coding RNA Transcripts

**Laura A. Kavanaugh and Uwe Ohler**

## 4.1 Introduction

Non-coding RNAs are defined as all functional RNA transcripts other than protein encoding messenger RNAs (mRNA). Thus, they are defined more by what they are not than by what they actually are (Fig. 4.1). This unusual way of defining ncRNAs reflects a historical bias in biology. Early biological studies focused largely on prokaryotes, whose genomes are dominated by protein-coding sequence (80–95%) (Mattick 2004a; Mattick and Makunin 2006). This led to the presumption that cellular activities were carried out primarily by proteins. RNA was thought to be a passive carrier of genetic information as mRNA or as supporting molecules for the production of proteins such as transfer-RNA (tRNA) and ribosomal-RNA (rRNA).

While it certainly remains true that proteins are critical to cell function, biologists have discovered that ncRNAs exercise many key roles in the cell as well. A wide variety of functions are attributed to ncRNA, including gene regulation, chromatin remodeling, gene localization, gene modification and DNA imprinting (Leighton et al. 1995; Brannan and Bartolomei 1999; Tilghman 1999; Eddy 2001; Storz 2002; Seitz et al. 2003; Mattick and Makunin 2006; Costa 2007; Dann et al. 2007). ncRNA have also been linked to the development of cancer and associated with complex diseases (Hayashita et al. 2005; He et al. 2005; Lu et al. 2005; Sonkoly et al. 2005; Costa 2007).

Many ncRNAs may also be highly expressed, particularly in eukaryotes. In *Saccharomyces cerevisiae* ncRNA represent ~95% of RNA transcripts (Peng et al. 2003; Samanta et al. 2006). Bioinformatic approaches predict that the number of ncRNAs in bacterial genomes is in the order of hundreds, while in eukaryotic genomes it is in the order of thousands (Hershberg et al. 2003; Zhang et al. 2004; Washietl et al. 2005a, b; Huttenhofer and Vogel 2006). In humans, data suggests that the number of ncRNAs produced is comparable to the number of proteins

L.A. Kavanaugh (✉)
Department of Molecular Genetics and Microbiology, Duke University,
Durham, NC, 27710, USA

D. Edwards et al. (eds.), *Bioinformatics: Tools and Applications*,
DOI 10.1007/978-0-387-92738-1_4, © Springer Science+Business Media, LLC 2009

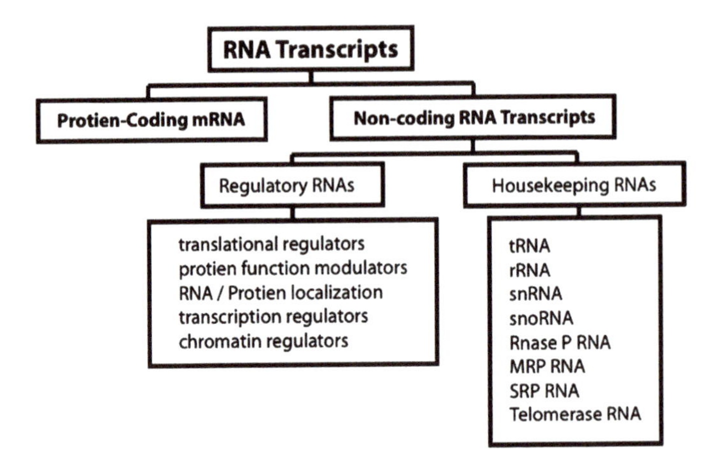


**Fig. 4.1** RNA transcripts. Many ncRNA families serve as housekeeping genes and are well characterized. Many other ncRNA families function or play a variety of regulator roles and their discovery and characterization remains an active area of research

(Kapranov et al. 2002; Cawley et al. 2004; Gardner et al. 2004; Kampa et al. 2004). This suggests that ncRNAs are likely to play a greater role in cellular function in eukaryotes than in prokaryotes (Gardner et al. 2004).

Given the prevalence and significance of ncRNAs, it is important to identify them in genomic sequence. However, when new genomes are annotated, many ncRNA families remain unannotated or under-annotated because ncRNAs are difficult to identify computationally. High-throughput experimental methods are often employed for this purpose including tiling microarrays and cDNA sequencing (Cheng et al. 2005; Hiley et al. 2005b; Kawano et al. 2005; Huttenhofer and Vogel 2006; Yin and Zhao 2007). Data from these experiments provide a genome-wide view of ncRNA transcription. These experiments provide valuable information about ncRNA expression levels and post-transcriptional processing. They often produce a sizable list of potential new ncRNA candidates. These candidates must then undergo further testing with methods like Northern hybridization, reverse transcript-PCR or rapid amplification of cDNA ends (RACE) to be confirmed as bona fide ncRNAs. Once a ncRNA is identified in this way, elucidating its function can often take years.

An intriguing discovery that has come from high-throughput methods is that much, if not most, of the eukaryotic genome is actually transcribed and may perform some function within the cell. This is contrary to the thinking that prevailed when the human genome was initially sequenced, that functional genes were considered remote islands in a sea of non-functional "junk" DNA. Recent large-scale identification of putative ncRNAs has launched a debate surrounding the question of how many of the observed transcripts represent legitimate ncRNAs and how many may result from transcriptional "noise" or experimental artifact (Babak et al.

2005; Cheng et al. 2005; Huttenhofer et al. 2005; Kapranov et al. 2005; Mattick and Makunin 2006; Mendes Soares and Valcarcel 2006; Costa 2007).

While experimental methods are essential for understanding ncRNAs, accurate computational tools for ncRNA prediction are sorely needed. Though methods for direct verification of ncRNAs are well established, most rely on an accurate prediction of the location of the ncRNAs to be useful. In addition, some ncRNAs are expressed at low levels and are difficult to detect experimentally. In many experimental systems it can be difficult to acquire enough RNA for experimental use. Experimental methods are limited in the number of samples or species they can survey, and are expensive and time consuming. Hence, coupling experimental methods with computational approaches will significantly improve our ability to identify ncRNAs.

Previous chapters have discussed protein-coding gene prediction. This chapter will explain why these methods do not work well for ncRNA gene prediction and will describe the methods that have been developed for this purpose. It will also discuss the algorithms that employ these methods to address specific ncRNA gene prediction problems. The intent is to give the reader an appreciation the field of ncRNA gene prediction and to provide guidance for selecting the best application for a specific situation.

## 4.2 Why ncRNA Prediction Is Difficult

Reasons that make computational prediction of ncRNAs difficult include (1) heterogeneity of ncRNAs, (2) lack of primary sequence features, and (3) extensive post-transcriptional processing.

### *4.2.1 Heterogeneity of ncRNAs*

To understand the challenge presented by ncRNA gene prediction, it is first necessary to gain an appreciation of the variety of ncRNAs that exist. Table 4.1 provides a list of known ncRNA families. They are involved in protein production (tRNA, rRNA), post-transcriptional modification of other RNA molecules (snoRNA), gene splicing (snRNA), control of gene expression (miRNA), chromatin structure modification (rasiRNA, siRNA), maintenance of telomere ends (telomerase), RNA turnover (RNAseP), imprinting (Eddy 2001) and chromosome silencing (X-chromosome inactivation) (Martens et al. 2004; Cheng et al. 2005; Samanta et al. 2006; Costa 2007). Some ncRNAs function in the nucleus while others are transported out of the nucleus or are encoded by the mitochondrial genome and function in the mitochondria (Cheng et al. 2005; Mendes Soares and Valcarcel 2006). There is also great variations in ncRNA length. They range from ~18 to 25 nucleotides (miRNAs), to 73–93 nucleotides (tRNA), to ~100–300 nucleotides (snoRNA), to as long as ~17,000 nucleotides (ncRNAs involved with chromosome inactivation) (Eddy 2001; Costa 2007).

**Table 4.1** Types of ncRNAs. Functional descriptions of the ncRNA discussed in this chapter are provided. Only a subset of these ncRNAs will be present in any cell and their expression is dependent on species, cell type and environmental conditions

| ncRNA | Description |
|---|---|
| miRNA (microRNA) | Small noncoding RNAs (21–25 nucleotides) that are processed from longer hairpin RNA precursors. Control expression of target genes through repressed translation or mRNA degradation. Frequently, one miRNA can target multiple mRNAs and one mRNA can be regulated by multiple miRNAs targeting different regions of the 3′ UTR (Seitz et al. 2003; Bartel 2004; Mendes Soares and Valcarcel 2006) |
| ncRNA (non-coding RNA) | All RNA transcripts that are not messenger RNA (mRNA) |
| MRP RNA | RNA component of the Ribonuclease MRP, which is an endonuclease that functions in eukaryotic pre-rRNA processing and is related to RNase P (Aspinall et al. 2007) |
| rasiRNAs (repeat-associated small interfering RNA) | Small RNA (23–27 nucleotides) typically generated from double-stranded RNA through overlapping bidirectional transcription. Encoded by repetitive elements within the genome (often transposons or retro-elements) and lead to histone and DNA modifications that induce heterochromatin formation and transcriptional repression by poorly understood mechanisms (Sontheimer and Carthew 2005; Verdel and Moazed 2005; Mendes Soares and Valcarcel 2006; Yin and Zhao 2007) |
| RNAseP | RNA component of Ribonuclese P, which is responsible for processing the 5′ end of precursor tRNAs and some rRNAs (Stark et al. 1978; Frank and Pace 1998; Eddy 2001; Storz 2002) |
| rRNA (ribosomal RNA) | RNA component of the ribosome that is present in many copies in every cell. Extensively modified, cleaved, and assembled into the ribosome (Venema and Tollervey 1999; Peng et al. 2003; Hiley et al. 2005b) |
| S/AS (sense/anti-sense transcripts) | Transcripts antisense of a coding mRNA that modulate mRNA expression by forming sense–antisense pairs. Function by affecting mRNA stability, translatability or chromatin structure (Cawley et al. 2004; Mattick 2004a, b; Katayama et al. 2005; Mendes Soares and Valcarcel 2006) |
| scaRNA (small Cajal body RNA) | A subset of H/ACA snoRNAs located in Cajal bodies (a class of small nuclear organelle) (Meier 2005; Mattick and Makunin 2006) |
| siRNA (small/short interfering RNA) | Small RNA (21–28 nucleotides) that direct cleavage of complementary mRNAs and lead to mRNA degradation and gene silencing. A natural mechanism of defense against viruses that seems to have evolved also as an important repressor of the transcriptional activity of heterochromatic regions of the genome, including centromeres and other regions with repetitive sequences and transposons (Hannon and Rossi 2004; Meister and Tuschl 2004; Bernstein and Allis 2005; Sontheimer and Carthew 2005; Zamore and Haley 2005; Mendes Soares and Valcarcel 2006; Yin and Zhao 2007) |

(continued)

**Table 4.1** (continued)

| ncRNA | Description |
|---|---|
| snoRNA (small nucleolar RNAs) | An integral part of the RNA-protein complex called snoRNAs which guides modification of a specific site within a target RNA (rRNA, tRNA, telomerase, snRNA and others). Each snoRNA guides one type of modification and belongs to one of two main classes: C/D box (which cause 2′-*O*-ribose methylation) and H/ACA box (which convert uridines to pseudouridines). They can also modulate editing or alternative splicing of pre-mRNAs (Eddy 2001; Bachellerie et al. 2002; Gesteland et al. 2006; Mendes Soares and Valcarcel 2006) |
| snRNA (small nuclear RNAs) | Important components of the spliceosome that recognize splice sites and establish an RNA scaffold that holds together the sequences involved in the splicing reaction. Some snRNAs serve in 3′-end formation of particular transcripts or other RNA-processing events (Gesteland et al. 2006; Mendes Soares and Valcarcel 2006) |
| sRNA | Term used predominantly in bacteria to mean non-coding RNA (do not function as a mRNA). Commonly found as translational regulators in bacterial cells.(Storz 2002; Gottesman 2004; Storz et al. 2004) |
| SRP RNA (signal recognition particle) | A small cytoplasmic RNA that forms the core of the signal recognition particle (SRP) required for protein translocation across membranes (Keenan et al. 2001) |
| tasiRNA (trans-acting small interfering RNA) | Small, plant-specific RNAs that regulate gene expression by guiding cleavage of target RNA. Their maturation involves miRNAs and they operate as part of an Argonaute protein complex (Hutvagner and Simard 2008) |
| telomerase RNA | An integral part of the telomerase enzyme and serves as the template for the synthesis of the chromosome ends (Chen et al. 2000; Gesteland et al. 2006) |
| tRNA (transfer RNA) | Short RNA molecules (73–93 nucleotides) that transfer a specific amino acid to a growing polypeptide chain at the ribosomal site during protein synthesis. They undergo extensive processing of their 3′ and 5′ ends, as well as covalent modifications, to achieve their mature form having a characteristic clover-leaf secondary structure (Hiley et al. 2005b; Gesteland et al. 2006; Goodenbour and Pan 2006) |
| tmRNA (transfer-messenger RNA) | Found in eubacterial genomes and named for their dual tRNA-like and mRNA-like functions. They function to liberate the mRNA from stalled ribosomes (Gillet and Felden 2001; Saguy et al. 2005) |

Another variable among ncRNAs is the degree to which structure plays a role in their function. Because ncRNAs are single stranded molecules, they can fold back onto themselves to form complex structures. The remarkable catalytic properties and versatility of ncRNAs lie in this capacity to fold into many shapes (Caprara and Nilsen 2000; Holbrook 2005; Mendes Soares and Valcarcel 2006). Some ncRNA families are highly dependent on their shape to perform their function (tRNA,

H/ACA snoRNA). Other families appear to perform their function primarily through sequence motifs. These motifs may permit complementary binding to other RNA/DNA molecules (cis or trans), assist with protein binding, and promote catalytic self-cleavage (C/D snoRNA) (Meyer 2007). Most ncRNAs probably function through a combination of structural and primary sequence components. For some ncRNAs, the functional property derives from their complementary binding?? to another transcript, and thus, no particular common features at all have been identified for the members of some families (piRNA, S/AS).

### *4.2.2 Lack of Primary Sequence Features*

Previous chapters discussed protein-coding gene prediction and described how primary sequence features are generally sufficient for predicting these genes. Protein coding genes share a common set of genomic features (splice sites, transcription factor binding motifs, polyadenylation signals, etc.), some of which are also part of non-coding transcripts, in particular those transcribed by RNA polymerase II. However, the distinctive signal stems from the protein coding content and its associated features (start codon, stop codon, large open reading frame).

Primary sequence conservation was also shown to be very useful for identifying protein-coding genes. These genes typically demonstrate a high degree of cross-species amino acid conservation over long evolutionary distances. Allowable amino acid substitutions are limited to those that retain protein shape and functional integrity, and such conformational constraints restrict nucleotide substitutions. In particular, they create a bias in codon substitution patterns at the third codon position (synonymous changes). While individual features are typically insufficient for gene prediction, they provide a reasonably good indication of where protein-coding genes are located when taken together. De novo protein coding gene predictors can typically identify more than90% of the coding bases in mammalian genomes (Jones 2006).

The situation is very different in ncRNA gene prediction. ncRNAs do not form a homogeneous class of genes processed along the same path as the central dogma once implied. In particular, they do not specify amino acid sequences and therefore necessarily lack start codons, stop codons, open reading frames, and third codon position substitution patterns. They are transcribed by multiple RNA polymerases, so they do not all share a common set of promoter motifs, splice signals or polyadenylation signals. They also typically lack significant primary sequence conservation across species (Higgs 2000; Livny and Waldor 2007). For ncRNAs that rely on structure for their catalytic activity, structural constraints exert only minimal restrictions on primary nucleotide sequence. Many different nucleotide sequences can fold into the same structure (Sect. 4.3.1 and Fig. 4.2a). For ncRNAs that function through primary sequence motifs, the brevity and degeneracy of these motifs precludes their use as generic gene-finding tools. Hence, shared primary sequence features are virtually non-existent among ncRNAs.

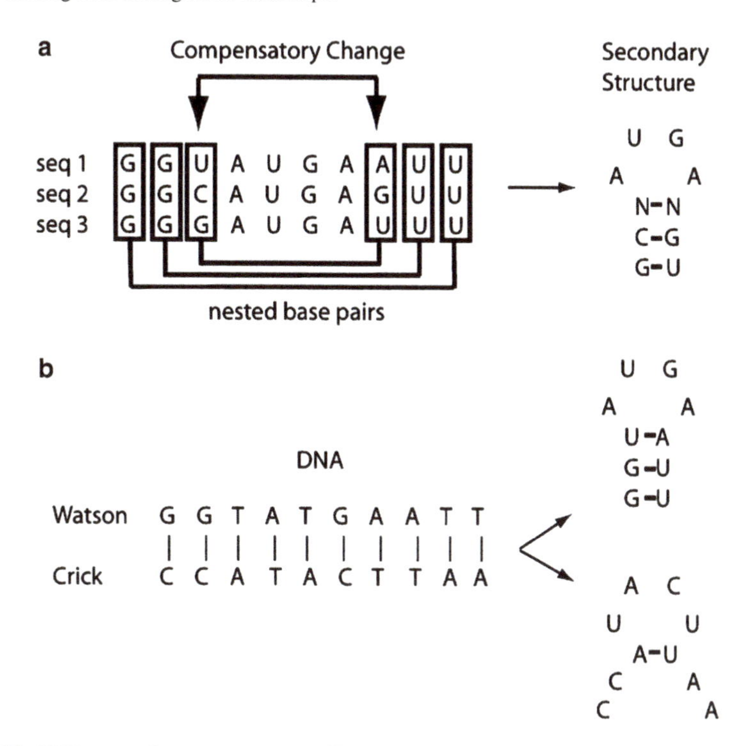


**Fig. 4.2** RNA secondary structure. (**a**) Nested bases form the ncRNA secondary structure. The structure is preserved in different sequences through compensatory changes that maintain base pairing. (**b**) The transcripts from complementary strands of DNA do not fold into the same structure

### *4.2.3 Extensive Post-transcriptional Processing*

Many ncRNAs undergo extensive post-transcriptional modification after initial transcription to reach their mature, functional form (tRNA, rRNA, snRNA, snoRNA). Modifications may include exo- and endo-nucleolytic cleavage of precursor transcripts, covalent modifications, and intron splicing (Sprinzl et al. 1998; Allmang et al. 1999; Venema and Tollervey 1999; Decatur and Fournier 2002; Hiley et al. 2005a). Many ncRNAs, including tRNA, are processed from a larger precursor transcript. MiRNAs are processed from precursor stem-loops that are themselves parts of larger primary transcripts. It is common for snoRNAs and miRNAs to be completely contained within the introns of other genes (Bachellerie et al. 2002; Seitz et al. 2003).

Post-transcriptional modifications have the potential to alter the manner in which RNA molecule folds and the complementary sequences it will match. However, our current understanding of ncRNA modifications, their role in function, and how to predict them, are still in their infancy. Consequently, ncRNA gene prediction programs do not account for many of these modifications.

## 4.3 Methods Used in ncRNA Gene Prediction

Several fundamental methods make up the core of ncRNA gene prediction. These methods form a toolkit that can be applied in slightly different ways to address a variety of problems. The myriad ncRNA algorithms are essentially variations on a theme. The following section will describe the key methods used in the field of ncRNA prediction. For a more detailed explanation of these topics, refer to the excellent discussion provided in the book by Durbin et al. (1998).

### *4.3.1 Minimum Folding Energy*

In order to understand how ncRNA structural prediction is accomplished, it is necessary to review a few basic biological principles regarding RNA. RNA is a polymer of adenine (A), cytosine (C), guanine (G) and uracil (U) nucleotides (in DNA, thymine replaces uracil). RNA is a single stranded molecule that can fold intramolecularly to form complex structures. The structure is created by base pairing between AU pairs and CG pairs (canonical base pairing) as well as GU pairs (non-canonical base pairing). While GT pairing is not typically observed in DNA, GU pairs are common in RNA structure and are almost as thermodynamically favorable as the canonical base pairs. The nested pairing of an RNA molecule's bases form its secondary structure and the same structure can be formed by many different sequences (Fig. 4.2a). The presence of GU pairing in RNA has a subtle but significant impact on structural prediction. It means that a transcribed strand of DNA and its complement will not fold into the same structure (Fig. 4.2b).

Secondary structure can be complex (Fig. 4.3a). Non-nested base pairing can also occur and form structures such as pseudoknots (Fig. 4.3b). This is referred to as the tertiary structure. While these structures are known to occur in many important RNAs, they are often ignored in practice because of the computational complexity they introduce for structure prediction (discussed later). This seems justified in many cases since the extent of pseudoknotted base pairing is typically relatively small. For example, in the well studied *E. coli* SSU rRNA molecule, only 8 of 447 base pairings are believed to be involved in a pseudoknot structure (Gutell 1993; Durbin et al. 1998). Given the allowable base pairings, it should be clear that a sequence could fold into many hypothetical structures. For an RNA of length $n$, there are approximately $1.8^n$ possible secondary structures ($3 \times 10^{25}$ structures for sequence length of 100 nucleotides) (Zuker and Sankoff 1984; Gesteland et al. 2006).

The goal of structural prediction tools is to predict the "true" (functional) ncRNA structure from among all the possibilities (assuming a structure is present). Unfortunately, predicting the functional structure of a ncRNA remains an unsolved problem in biology but biologists have developed methods to make a "best guess"

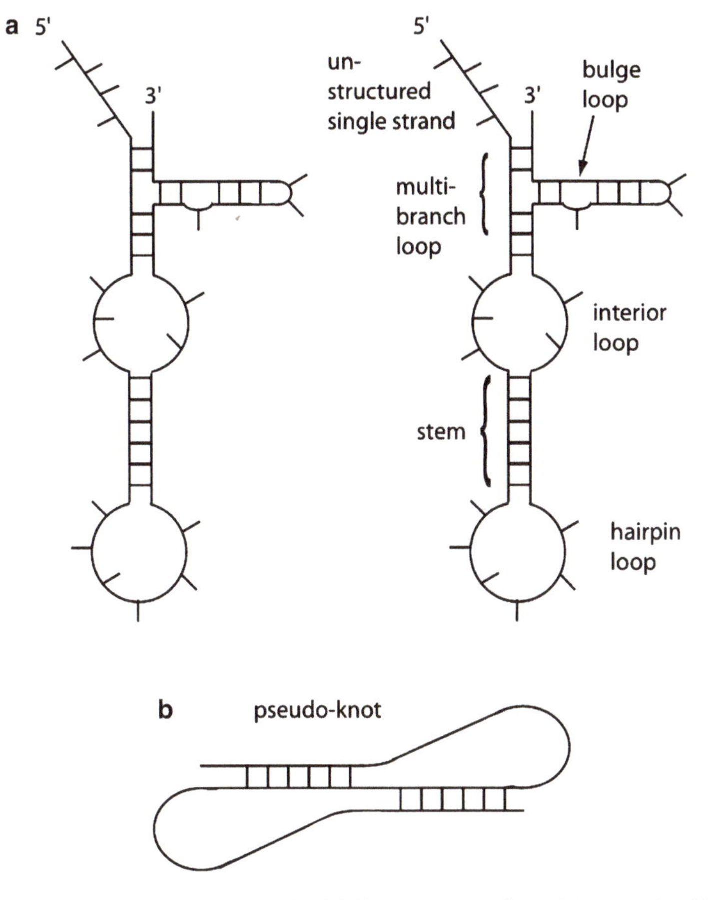


**Fig. 4.3** Complex ncRNA structures. (**a**) Multi-loop structure formed by nested pairing; (**b**) pseudoknot

at what the functional structure is likely to be. These include maximum base pairing approaches (Nussinov algorithm (Nussinov et al. 1978)), minimum folding energy approaches (Zuker algorithm (Zuker and Stiegler 1981; Zuker 1989)), and probabilistic methods that relate energies to pairing probabilities (McCaskill algorithm (McCaskill 1990)). Of these, the one used most commonly in practice is the minimum folding energy approach (MFE) (Mathews and Turner 2006). MFE approaches use experimentally derived base-pair stacking energy parameters to determine the most energetically favorable structure among all the possible folded structures.

The difficulty of determining the MFE structure grows quickly with the complexity of the RNA structure being considered. For example, the standard set of dynamic programming algorithms for MFE do not evaluate tertiary structure (pseudo-knots) and scale $O(L^3)$ in time, where $L$ is the sequence length. Dynamic programming algorithms are capable of predicting most known pseudoknot topologies scale $O(L^6)$

in time (Rivas and Eddy 1999). This means that a doubling of sequence length requires eight times as much computer time by standard methods but 64 times as much if pseudoknots are considered (Mathews and Turner 2006). There is a clear tradeoff between complexity and practicality. MFE programs must make assumptions that limit the search space of allowable structures. These assumptions are what distinguish one program from another and it is important to understand these assumptions when selecting a program to address a specific problem.

It is important to point out that a given sequence may fold into many different structures that have nearly the same MFE. Given the inaccuracies in experimental base-pair stacking energies, our incomplete understanding of thermodynamic rules, and simplification of the real physics of RNA folding, it is impossible to know which of these is the native structure (Mathews and Turner 2006). It may even be possible that the RNA molecule dynamically transitions among a series of low energy structures. This has prompted researches to augment their folding programs to also compute near-optimal structures. A variety of methods have been devised to generate selected representative low-energy structures, or even an ensemble of low energy structures (Ding and Lawrence 2003; Mathews and Turner 2006; Reeder et al. 2006). The latter allows for evaluations beyond just one structure, e.g., to assess whether particular regions are preferably open or paired under the whole ensemble of structures.

In addition, conditions in the cell may prevent an RNA molecule from actually assuming the predicted MFE structure. Binding partners, post-transcriptional modifications, folding kinetics , and other biological processing occurring during transcript synthesis can all influence the functional RNA structure (Repsilber et al. 1999; Neugebauer 2002; Gardner and Giegerich 2004; Meyer and Miklos 2004; Mathews and Turner 2006; Meyer 2007). There are currently no computational tools available that assess these effects. A perfect example of the pitfall in the minimum free folding? energy structure comes from tRNA. This is one of the best analyzed RNA families and experiments have shown that they have a functional cloverleaf structure. Yet, when 99 tRNA sequences from the Rfam database (Griffiths-Jones et al. 2003, 2005) were evaluated using a MFE approach, only 33 were predicted to have cloverleaf structure. The biological explanation for this is that tRNAs undergo many base modifications that alter their structure. Structural prediction programs are unable to account for these modifications. This example clearly shows that structural predictions should always be taken with a grain of salt (Reeder et al. 2006).

### 4.3.2 *Hidden Markov Models*

Some ncRNA families share common features on the primary sequence level. If a set of sequences belonging to the same family are available, there are several ways to search for additional family members. A naive approach would apply sequence based search methods like BLAST (Altschul et al. 1990) or FASTA (Pearson and

Lipman 1988; Pearson 2000) to look for high sequence identity in a target sequence. However, ncRNAs rarely preserve the level of sequence identity necessary for such searches (Freyhult et al. 2007). In addition, this approach can be misleading because it relies too heavily on individual sequences in the training set instead of focusing on the common features that characterize the set. It fails to take advantage of all of the available information.

When a family of homologous sequences is aligned, it is clear that some positions in the alignment are more conserved than others. Some columns of the alignment are highly conserved while others are not, and gaps and insertions are more likely to appear in some regions of the alignment than others. It makes sense to concentrate on these conserved features when looking for additional family members because these features are more likely to uncover authentic or distantly related family members. Profile-Hidden Markov Models (HMMs) are commonly used to derive a probabilistic model of sequence features in biological sequence analysis. They can be thought of as an extension of other non-probabilistic profiling approaches like position specific score matrix (PSSMs), but allow for nucleotide insertions and deletions. It should be clear that the success of the method is highly dependent on the quality of the input alignment. While HMMs can in principle be used to align primary sequences, aligning functionally equivalent regions of ncRNAs is not a trivial matter, and significant attention must be paid to this step if the profile-HMMs has to be successful.

Profile-HMMs build a model representing the consensus sequence for the family, not the sequence of any particular member. It does this by assigning position sensitive probability scores (emission probabilities) associated with observing a specific nucleotide for each column of the alignment. It also assigns position sensitive insertion and deletion probability scores (transition probabilities) for each column. One of several different approaches can be used to derive these probabilities from the alignment. Information about the evolutionary relationship of the sequences being aligned can also be incorporated to allow sequences to be weighted differently in the scoring scheme. Once these probabilities have been determined, the trained model can then be used to search a target sequence for regions similar to the training set. Determining the family of a sequence, and aligning it to the other members, often helps in drawing inferences about its function.

### 4.3.3 *Stochastic Context-Free Grammars and Covariance Models*

A beautiful mathematical method for efficiently modeling the long-distance interactions that occur in RNA structures has been developed from stochastic context-free grammars (SCFGs). This general theory for modeling strings of symbols was developed by computational linguists in an attempt to understand the structure of natural languages, but it is now being applied with great effectiveness to ncRNA structural prediction (Chomsky 1956, 1959).

SCFGs are probabilistic models that consist of symbols and production rules with associated probabilities that can capture primary sequence features as well as long-range interactions between base pairs in a RNA secondary structure. They provide a unifying framework for primary sequence-based approaches (like HMMs) and approaches that predict secondary structure. The stacking-energy derived parameters for structural prediction can be implemented as rules in the grammar to allow prediction of the MFE structure. Once the parameters of the SCFG have been determined, dynamic programming algorithms are used to derive the secondary structure that maximizes a scoring function. These algorithms make use of the fact that the overall probability of a structure can be expressed as the product of probabilities for smaller parts of the RNA structure. SCFGs do not, however, capture tertiary interactions like pseudoknots.

In analogy to profile HMMs, SCFGs with a structure that incorporates biologically meaningful patterns of insertions and deletions are known as covariance models (CMs). They can be derived if a family of known ncRNA sequences are available to form a training set. To train a CM, it is necessary to develop an accurate alignment of the available sequences. This is typically done using standard multiple sequence alignment tools like ClustalW (Chenna et al. 2003) or M-coffee (Moretti et al. 2007). The initial alignment can then be adjusted manually with the aid of a visualization program like RALEE (Griffiths-Jones 2005). The expert knowledge added in this step provides the best chance of correctly aligning functionally equivalent sequence and structural elements. The better the initial alignment, the more likely it is that the CM will capture the true footprint of compensatory changes that reveal the underlying shared structure (Gardner et al. 2004; Gardner et al. 2005).

CMs have been used extensively by the ncRNA database, RFAM, to capture the features of many known ncRNA families (Griffiths-Jones et al. 2003, 2005). The trained CM model can then be used to search a target sequence for additional family members. Each match is assigned a score that reflects the extent to which it shares the features of the training set. A cutoff threshold can be used to select the best matches.

### *4.3.4 Structural Alignment with Multiple Sequences*

Structure is typically conserved more than sequence among homologous RNA molecules that share the same function. Multiple homologous sequences can be used to derive a consensus structure that captures their common structural features. Comparative approaches are better at predicting functional RNA structure than MFE that only have information from a single sequence (Doshi et al. 2004; Gardner et al. 2004; Gesteland et al. 2006; Reeder et al. 2006).

There are three general approaches used to obtain a common structure from multiple sequences (Fig. 4.4) (Gardner et al. 2004). The best approach for a

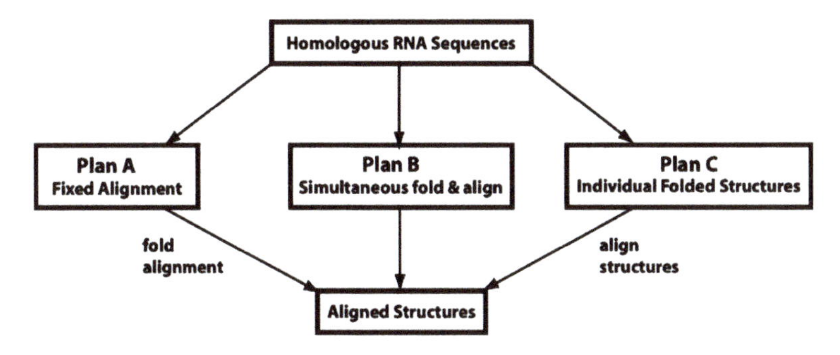


**Fig. 4.4** General approaches to structural alignment. Modified after Gardner and Giegerich

given application depends largely on the degree of sequence identity among the homologous sequences. The first approach (plan A) attempts to identify a common structure from a fixed sequence alignment. This is a good approach when sequence identity is ≥60%, allowing functionally equivalent regions to be aligned with a high degree of accuracy. The second approach (plan B) considers many different alignment possibilities and structures in order to identify the combination that yields the optimal solution. This is the best approach to take when sequence identity is less than about 50–60% and functionally equivalent regions cannot be reliably aligned based on sequence similarity alone. The third approach (plan C) folds each sequence independently and then attempts to align the resulting structures. This approach is fairly new to the field of structural alignment and holds the greatest hope for homologous sequences with little sequence identity.

To follow plan A, it is first necessary to obtain an accurate alignment of the set of sequences. Standard multiple sequence alignment tools, supplemented with expert curation, provides the greatest likelihood of accurately aligning functionally equivalent regions. Once the best possible alignment has been obtained, several methods are available to infer the consensus structure from the pattern of compensatory changes among the sequences. These methods include mutual-information measures (Chiu and Kolodziejczak 1991; Gutell et al. 1992; Gorodkin et al. 1997), combinations of MFE and covariation scores (Hofacker et al. 2002; Ruan et al. 2004a), and probabilistic models compiled from large reference data-sets (Knudsen and Hein 1999, 2003; Gardner et al. 2004).

If the pair-wise sequence identity in the set of homologous RNA sequences is below about 50–60%, plan A will not yield good results (Gardner et al. 2005). In this case, it becomes necessary to simultaneously align and fold the sequences in order to infer the common underlying structural features shared by the set (plan B). An exact solution to this simultaneous alignment and folding was presented by David Sankoff (1985). Unfortunately, the algorithm requires extreme computational resources and is not very useful in practice ($O(n^{3m})$ in time, and $O(n^{2m})$ in memory, where n is the sequence length and m is the number of sequences).

Restricted versions of this algorithm have been developed to reduce run time but this remains an active area of research.

The final approach (plan C) applies when no helpful level of sequence conservation is observed. This approach skips the sequence alignment step altogether and jumps directly to predicting the secondary structure of each homologous sequence separately. The structures, or some subset of structures, are then compared to one another to try and identify a shared structure. The crucial point in plan C is the question of whether the initial independent folding produces at least some structures that align well and hence give clues as to the underlying consensus structure, assuming one exists (Gardner et al. 2004).

## 4.4 ncRNA Gene Characterization and Prediction Algorithms

ncRNAs are a highly heterogeneous group of genes that fulfill diverse roles in the cell through diverse mechanisms. They lack a common set of primary-sequence or structural features and are rarely evolutionarily conserved at the primary-sequence level. This dearth of generic ncRNA features makes ncRNA gene prediction several orders of magnitude more difficult than protein coding gene prediction. Biologists have adapted by developing gene finders that target the unique properties of families of ncRNAs. The huge diversity in ncRNA families dictates a commensurate diversity in gene finding approaches producing a complex web of inter-related algorithms. Navigating this web can be daunting for the uninitiated. It is helpful to divide the algorithms into two broad categories: example-based approaches and de novo approaches.

Example-based approaches begin with a family of known ncRNA genes. The “family” may have only one known family member or may have many members. De novo gene predictors do not begin with a specific set of known ncRNA examples but instead search for features shared among a large number of gene families. For example, these algorithms often look for genomic regions that have well-defined structural features suggesting the presence of a ncRNA. Example-based algorithms and de novo algorithms use essentially the same methods but apply them in slightly different ways to achieve different objectives.

### *4.4.1 Example-Based Approaches*

ncRNA families are groups of ncRNA genes that perform a common function and therefore share common features. They may all have a common structure or recognize their target genes through similar patterns of sequence-complementarity. Researchers that use example-based approaches are typically asking one of two questions. Either they want to determine the features of a family because it provides insight into the function performed by the family (characterize family), or they want to use the features to search for additional family members (gene prediction). Different algorithms have been developed to perform these two different functions.

#### 4.4.1.1 Well Characterized Families

Several ncRNA gene families are well characterized. These are typically highly expressed housekeeping genes with multiple family members per genome. Family-specific features can be identified among the large training set of known examples. Examples of well-characterized families include tRNAs, tmRNAs, snoRNAs (C/D box and H/ACA box) and miRNAs. Family specific gene-finding programs have been developed for these cases and have been quite successful (Table 4.2).

Covariance models (CM) are used extensively by these programs (Sect. 4.3.3). The tRNA scan program is an example of this type of program used to predict tRNA. These ncRNAs are abundant in every cell, have a characteristic clover-leaf shape and contain a 3-basepair anti-codon sequence specifying an amino acid. The tRNAscan program is able to predict tRNAs with an accuracy of 1 false positive per 15 gigabases using a CM (Lowe and Eddy 1997). Other programs have been developed for tmRNA and snoRNA (Table 4.2) (Lowe and Eddy 1999; Omer et al. 2000; Laslett et al. 2002; Edvardsson et al. 2003; Schattner et al. 2004; Hertel et al. 2008).

MicroRNAs (miRNAs) are another family of ncRNAs that have been successfully predicted using family-specific features (Table 4.2). MiRNAs undergo significant post-transcriptional modification and it is not currently possible to computationally predict the primary transcript. Gene finders generally aim to identify the characteristic miRNA foldback structure which is excised out of the primary transcript. However, in animals, this foldback structure is relatively small (~60–70 nucleotide) and simple and, by itself, is not enough to identify miRNAs with reasonable levels of specificity. Hence, conservation patterns and additional heuristic features are often combined in a probabilistic model. Programs like MiRscan and miRRim take this approach (Lim et al. 2003; Lai 2004; Ohler et al. 2004; Terai et al. 2007). As an alternative, it is also possible to search for sequence-complementarity with potential targets to identify miRNA genes (Xie et al. 2005). This strategy has been shown to be particularly successful in plants, where the mature miRNAs target genes display longer stretches of sequence complementarity (Rhoades et al. 2002; Wang et al. 2004). Hybrid approaches, taking advantage of several different types of information, have also been developed (Jones 2006; Meyer 2007).

In addition to these family-specific programs, there are generic programs that aim to derive the parameters of a CM from any training set of known ncRNAs (Table 4.2). The INFERNAL software package, available from the RFAM database, uses a high quality alignment to derive the parameters of a CM (Eddy 2002). The program employs several techniques to extend its ability to recognize sequences more diverged than what was provided among members of the training set. It operates in $O(L\ N^{1.3})$ time and $O(N^2\log(N))$ memory where $L$ is the target sequence length and $N$ is the alignment length. The CMFINDER program trains a CM from

**Table 4.2** ncRNA Gene Prediction Programs. This table provides a list of programs associated with ncRNA gene prediction discussed in this chapter. The table should not be considered comprehensive

<table>
<tr><td rowspan="12">Known examples</td><td rowspan="2">Well Characterized Family</td><td>Family specific</td><td colspan="5">tRNA- tRNAscan; tmRNA- BRUCE; C/D box snoRNA- snoscan; C/D and H/ACA box snoRNA- SnoReport; H/ACA snoRNAs- snoGPS; miRNA- MiRscan, miRRim (Section 4.4.1.1)</td></tr>
<tr><td>Generic</td><td colspan="5">HMMer (profile-HMM); INFERNAL, CMFINDER (Covariance Model) (Section 4.4.1.1)</td></tr>
<tr><td rowspan="10">Not well characterized family</td><td rowspan="7">Single example</td><td rowspan="3">Find additional family members</td><td>PSB</td><td colspan="3">Blast, Fasta (poor results in most cases) (Section 4.3.2)</td></tr>
<tr><td>Structure known</td><td colspan="3">RSEARCH, RNAMotif (Section 4.4.1.2)</td></tr>
<tr><td>PSSB</td><td colspan="3">DYNALIGN, FOLDALIGN, CONSAN (Section 4.4.1.2)</td></tr>
<tr><td rowspan="4">Characterize family</td><td>PSB</td><td colspan="3">Insufficient information</td></tr>
<tr><td rowspan="3">SB</td><td rowspan="2">Simple Interactions</td><td>MFE structure</td><td>RNAfold, Afold, Mfold (Section 4.4.1.2)</td></tr>
<tr><td>Ensemble MFE structures</td><td>RNAsubopt, Sfold (Section 4.4.1.2)</td></tr>
<tr><td>pseudo-knots</td><td colspan="2">pknots, pknotsRG, HotKnots, (Section 4.4.1.2)</td></tr>
<tr><td rowspan="3">Multiple examples</td><td rowspan="3">Characterize family</td><td>SB</td><td colspan="3">RNAforester, MARNA, RNAcast (RNAshapes), LARA (Section 4.4.1.3)</td></tr>
<tr><td rowspan="2">PSSB</td><td>Fixed alignments</td><td colspan="2">PFOLD, RNALIFOLD, RNAlishapes, RNA-DECODER; Pseudo-knot: ILM, HXMATCH, KNetFold (Section 4.4.1.3)</td></tr>
<tr><td>Simo align & fold</td><td colspan="2">Two sequences: DYNALIGN, FOLDALIGN, CONSAN (Section 4.4.1.2)<br>Multi sequences: RNA Sampler, FoldalignM, MASTR, STEMLOC (Section 4.4.1.3)</td></tr>
<tr><td rowspan="3">De novo gene prediction</td><td>One species</td><td>PSSB</td><td colspan="5">Genome-specific methods (Section 4.4.2.1), MFE Stability (Section 4.4.2.2)</td></tr>
<tr><td rowspan="2">Multiple species</td><td rowspan="2">PSSB</td><td>Fixed alignments</td><td colspan="4">QRNA, EVOFOLD, RNAZ, AlifoldZ (Section 4.4.2.3)</td></tr>
<tr><td>Simo align & fold</td><td colspan="4">FOLDALIGN (Section 4.4.2.3)</td></tr>
</table>

a set of unaligned sequences (Yao et al. 2006). Both programs can be used to search for additional family members using the trained model. Each match found in the target sequence is assigned a score that reflects the likelihood that it is a true member of the ncRNA family.

In some cases, ncRNA families share primary-sequence features but lack significant structural features and a CM approach is inappropriate. A profile-HMM (Sect. 4.3.2) is the best approach for this case. The HMM software package is a generic profile-HMM algorithm that captures the shared sequence features among a set of homologous sequences (Eddy 1996; Meyer 2007). Once trained, the model can be used to predict additional family members in a target sequence. The program runs in $O(L\ N)$ time and $O(L\ N)$ memory for a target sequence of length $L$ that models an alignment of $N$ nucleotides.

#### 4.4.1.2 Not Well Characterized Families: Single Example

Experimental work often leads to the discovery of new ncRNA genes. It is important to characterize these genes in an attempt to gain insight into their function. They can also be used to search for additional family members. Numerous algorithms have been developed to accomplish these two goals (Table 4.2).

##### Characterize Structure

If a single example of a ncRNA has been obtained, minimum folding energy (MFE) approaches can be used to predict its structure (Sect. 4.3.1). Programs that predict the MFE structure without considering tertiary structure (pseudo-knots) include RNAfold (Hofacker 2003), Mfold (Zuker and Stiegler 1981), and Afold to name a few (Ogurtsov et al. 2006). Programs like RNAsubopt predict structures using MFE while providing information about near-optimal folding structures (Wuchty et al. 1999). Sfold is a Bayesian folding approach which identifies parts of the sequence which have a high probability of folding under an ensemble of possible structures (Ding et al. 2004). Programs capable of predicting pseudo-knot or other complex structures include pknots, pknotsRG and HotKnots (Rivas and Eddy 1999; Reeder and Giegerich 2004; Ruan et al. 2004a; Ren et al. 2005). Each of these programs makes different simplifying assumptions and provides different information about the sequence being evaluated. The most appropriate application depends on the question being asked. See Gardner and Giegerich (2004) for a comparison of some of these programs.

##### Find Additional Family Members

If the functional structure of the example is known from experimental data (crystallography/NMR), programs like RSEARCH or RNAMotif (Macke et al. 2001; Klein and Eddy 2003) can be employed to search for additional family members in a target sequence. They search for regions in the target sequence that are likely to fold into the same structure as the known example (Klein and Eddy 2003; Meyer 2007). RSEARCH uses a stochastic context-free grammar (SCFG) model and calculates a score for each match reflecting the reliability of the prediction. However, the program is computationally intensive. The RNAMotif program is highly customizable and therefore may potentially provide high sensitivity and specificity (Macke et al. 2001; Meyer 2007). The user can define a search motif, secondary and tertiary structure features, and custom scoring parameters. However, in order to supply this information, the user must have a lot of information about the search sequence – information which is often not available.

In most cases, the structure of the example ncRNA is unknown. It can still be used to search a target sequence for additional family members if alignment and folding are performed simultaneously. This general approach is based on the Sankoff algorithm (Sect. 4.3.4). An exact solution of the Sankoff algorithm is extremely computationally intensive, so programs have been developed that implement approximations of the Sankoff approach. These programs include: DYNALIGN (Mathews and Turner 2002, 2006), FOLDALIGN (Havgaard et al. 2005a, b) and CONSAN (Dowell and Eddy 2006; Reeder et al. 2006; Meyer 2007).

It is important to examine each of these programs carefully when selecting the best application for a particular problem. Both DYNALIGN and FOLDALIGN can be used to search for a shorter sequence in a long test sequence. FOLDALIGN was developed to detect local regulatory structures rather than global structures with multi-loops. DYNALIGN reduces computational complexity by placing limits on the size of internal loops in RNA structures. Neither of them explicitly models unstructured regions in the two input sequences so that the predicted results may strongly depend on the chosen sequence window (Meyer 2007). A comparison of the CONSAN, DYNALIGN and FOLDALIGN programs by Dowell and Eddy 2006 concluded that they provided comparable overall performance but have different strengths and weaknesses.

#### 4.4.1.3 Not Well Characterized Families: Multiple Examples

As discussed in Sect. 4.3.1, MFE approaches are limited in their ability to predict accurate RNA secondary structure. Better results can generally be achieved if multiple sequences are available. These approaches benefit from the combined information provided by several sequences (Doshi et al. 2004; Gardner et al. 2004; Gesteland et al. 2006; Reeder et al. 2006). A variety of approaches have been developed for application to specific circumstances (Table 4.2).

Characterize Family (Fixed Alignment)

In some cases, it is desirable to determine the shared structure of a ncRNA family for which multiple examples have been identified (Sect. 4.3.4). When primary-sequence identity among the examples is greater than about 60%, a fixed alignment approach works best. Examples of programs that determine family structure from fixed nucleotide alignments (Fig. 4.4, plan A) include PFOLD (Knudsen and Hein 2003), RNAALIFOLD (Hofacker 2007), RNAlishapes (Voss 2006) and RNA-DECODER (Pedersen et al. 2004). These programs all look for co-varying columns in the input alignment to support structural predictions (pseudo-knot free). The computational complexity of these programs is $O(L^3)$ time and $O(L^2)$ memory for an alignment of length $L$. Programs that predict pseudo-knotted structures have also been developed and include ILM (Ruan et al. 2004b), HXMATCH (Witwer et al. 2004) and KNetFold (Bindewald and Shapiro 2006). All these programs provide scoring information reflecting the reliability of the predicted consensus structure.

The programs differ in the input data they require, their underlying assumptions and their speed. For example, PFOLD and RNA-DECODER require input of an evolutionary tree relating the sequences in the alignment. RNAALIFOLD and RNA-DECODER allow a structure to be provided as an additional input constraint. RNA-DECODER can take known protein-coding regions in the input alignment explicitly into account and model un-structured regions in the input alignment. RNA-DECODER typically runs slower than PFOLD and RNAALIFOLD because it uses more complex models (Meyer 2007). See Gardner and Giegerich 2004 for a comparison of the performance of RNAalifold, PFOLD and ILM.

#### Characterize Family (Simultaneous Align and Fold)

In case of below 50–60% sequence-identity, it becomes necessary to simultaneously align and fold homologous sequences in order to reveal their common structure (Sect. 4.3.4 and Fig. 4.4, plan B). The extreme computational demands of this approach are the major obstacles to its success. Many techniques are being explored in an effort to overcome this barrier and it remains an active area of research. The breadth of these techniques makes it impossible to provide a concise summary here. One approach is to limit the comparison to only two sequences and programs that do this were discussed in the section "Find Additional Family Members." Other programs have been extended to operate on more than two sequences and examples include RNA Sampler (Xu et al. 2007), FoldalignM (Torarinsson et al. 2007), MASTR (Lindgreen et al. 2007) and STEMLOC (Holmes 2005). Additional information about Sankoff-like programs and attempts to systematically compare their performance is provided in several reviews (Gardner et al. 2004; Gardner et al. 2005; Meyer 2007). These reviews conclude that the programs vary widely in their sensitivity and specificity over different sequence lengths and homologies. Hence, it remains unclear which program will perform best for a given data set.

#### Characterize Family (Structural Alignment)

When no helpful level of sequence conservation is observed among a set of homologous ncRNAs, it is best to skip any attempts at sequence alignment and jump directly to structural comparisons (Fig. 4.4, plan C). The first step is to predict the secondary structure of each homologous sequence separately using programs like Mfold or RNAfold (Zuker and Stiegler 1981; Hofacker 2003). The resulting structures are then compared to one another to identify common structures. This is an active field of research and a variety of different structural comparison approaches have been proposed. Examples of these programs include RNAforester (Hochsmann et al. 2004), MARNA (Siebert and Backofen 2007), RNAcast (Reeder and Giegerich 2005), and LARA (Bauer et al. 2007).

RNAforester and MARNA can both be used to compute a global alignment of several unaligned input sequences whose pseudo-knot free structures are already known. They use structure and RNA sequence as input. RNAforester represents the individual structures as trees and computes a structural alignment by aligning the trees. It then calculates a score for the resulting structural alignments. In contrast, MARNA is a method that employs a progressive pair-wise alignment strategy that takes the known structures indirectly into account when calculating the global alignment. MARNA is the only one of the tree-based programs that is capable of proposing a pseudo-knot free secondary structure from input sequences whose structure is not known (Meyer 2007). The program RNAcast efficiently determines the consensus shapes of a set of sequences, and from their representative structures, a multiple alignment can be obtained with RNAforester (Reeder et al. 2006). The RNAcast program has also been integrated into a software package called RNAshapes that couples it with tools for the analysis of shape representatives and the calculation of shape probabilities (Steffen et al. 2006). See Gardner and Giegerich 2004 for a comparison of the RNAforester and MARNA programs.

## *4.4.2 De Novo ncRNA Prediction*

De novo gene predictors attempt to identify ncRNA genes in an unannotated genome(s) without the use of a training set of examples. They look for regions that have features characteristic of several ncRNA families. The challenge is distinguishing true ncRNAs from non-functional regions of the genome that contain some features of ncRNAs simply by random chance. Although it is an area of active research, many of the methods already discussed, for example-based ncRNA prediction, are used with a slight change of perspective.

### 4.4.2.1 Genome Specific Approaches

These methods exploit the unique biology of a given organism to predict ncRNAs. They were initially developed using the relatively simple genomes of prokaryotes. The small size and compact structure of prokaryotic genomes made the ncRNA prediction problem computationally tractable. Most importantly, prokaryotic genomes are well understood in terms of promoter structure, nucleotide bias and gene arrangement. Biologists were able to take advantage of this knowledge to identify features that distinguish ncRNAs from the background sequence. The approach has also been extended with some limited success to simple eukaryotes (Table 4.2).

One example of genome-specific information is di-nucleotide frequency. Though di-nucleotide frequencies vary widely among species, they tend to be relatively constant within a species (Karlin et al. 1998; Schattner 2002). Some species have evolved unusual nucleotide base composition biases in order to survive in their unique environmental niches. For example, the ncRNAs of some hyperthermophiles (*Methanococcus jannaschii*, *Pyrococcus furiosus*) can be identified by their unusually high GC nucleotide content in an otherwise AT-rich genome. The high GC content of these ncRNAs is necessary for folding and maintaining stable structures in their high temperature environment (Klein et al. 2002; Schattner 2002). However, such examples are rare and it is uncommon to find cases where a single feature is sufficient to identify ncRNAs. Most genome-specific prediction methods rely on a combination of information. They often begin by identifying intergenic regions that are unusually long, suggesting that the region contains an as-yet undetected ncRNA gene. The compact genomes of prokaryotes and some simple eukaryotes make it easy to see where unannotated genes are likely to be located. Intergenic regions can also be searched for stretches of unusually high nucleotide conservation among related species. High conservation is suggestive of a functional element that has been preserved over evolutionary time. The presence of strong promoter and terminator signals, in spatially appropriate locations, are used to further substantiate the presence of ncRNAs. Investigators have used a variety of approaches to combine these search criteria. They are often applied in a serial, heuristic manner but more sophisticated methods like neural networks have also been applied (Carter et al. 2001). The approach has proven highly successful in identifying ncRNAs in *E. coli* (Argaman et al. 2001; Eddy 2001; Wassarman et al. 2001; Lenz et al. 2004; Wilderman et al. 2004; Livny et al. 2005, 2006; Pichon and Felden 2005; Livny and Waldor 2007) and to a lesser extent *S. cerevisiae* (Olivas et al. 1997).

The development of customizable programs helps to extend this approach to new species. These programs accept genome-specific information as input (regions of conservation, intrinsic transcriptional terminators, promoter signals, secondary structure conservation, etc.) and then use it to search for new ncRNAs. When this type of information becomes available for a species, these programs can be used to search for ncRNAs in their genomes (Pichon and Felden 2003; Livny et al. 2006). Genome-specific approaches demonstrate how insight into an organism's biology can prove valuable for ncRNA prediction. This highlights the interdependent nature of experimental and computational work.

#### 4.4.2.2 MFE Stability

Many different types of ncRNAs have a significant structural component, so searching for structural features should, in theory, identify many ncRNAs. However, being able to fold a sequence into a nice structure does not mean that it is a bona fide ncRNA. Sequences that fold into structures can occur by random chance. The challenge is to distinguish real ncRNA sequences from random sequences. In the late 1980s, research in the Maizel laboratory led to the proposal that ncRNAs have energetically more stable structures than random sequences (Le et al. 1988, 1989; Chen et al. 1990). In this approach, a single sequence is folded using a MFE approach. The sequence is then randomly shuffled many times and the MFE of each of the randomly shuffled sequences is calculated. The distribution of MFEs of the random sequences is then compared to the MFE of the actual sequence. For true ncRNA, the MFE of the actual sequence will be distinct from the distribution of MFE of the random sequences.

Several researchers have investigated the general applicability of Maizel's proposal. Some have provided evidence that ncRNAs are more energetically stable than randomly shuffled sequences (Clote et al. 2005; Bonnet et al. 2004; Freyhult et al. 2005; Kavanaugh and Dietrich in press) while others have provided evidence that this is not true in general (Rivas and Eddy 2000) and the issue remains unresolved. It is certainly worth pursuing this idea further since it is the only approach currently available for potential de novo prediction of ncRNAs using a single sequence (Table 4.2). It has been shown that it is important to preserve both mono and di-nucleotide distributions when creating the randomly shuffled sequences (Workman and Krogh 1999).

#### 4.4.2.3 Comparative Approaches

It has been shown that comparative approaches produce more reliable structural predictions than MFE approaches (Doshi et al. 2004; Gardner et al. 2005; Gesteland et al. 2006; Reeder et al. 2006). Hence, it makes sense to employ these methods for de novo gene discovery. Comparative methods search for footprints of compensatory changes among multi-species alignments, suggesting the presence of a conserved secondary structure likely to be a ncRNA. Random sequences are unlikely to produce patterns of compensatory changes over evolutionary time. Computational requirements are a major obstacle in the development of de novo ncRNA gene prediction programs and necessitate simplifications in the search method (Table 4.2).

##### Fixed Sequence Alignment

One approach to reduce computational complexity is to restrict the analysis to fixed sequence alignments. Programs that take this approach include QRNA (Rivas and Eddy 2001; Rivas et al. 2001), EVOFOLD (Pedersen et al. 2006), RNAZ (Washietl et al. 2005b) and AlifoldZ (Washietl and Hofacker 2004). All of these programs take fixed sequence alignments as inputs (two in the case of QRNA and multi-sequences in the case of EVOFOLD and RNAZ). They search windows of

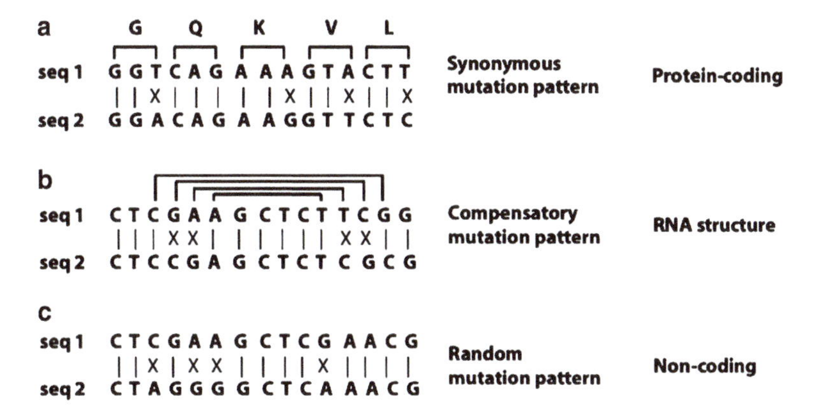


**Fig. 4.5** Patterns of mutation in different types of sequences. (**a**) A pattern of third codon position mutations (synonymous changes) indicative of protein-coding sequences. (**b**) A pattern of compensatory changes that preserve underlying structure are indicative of structural ncRNAs. (**c**) Random mutations are expected in non-coding sequence

predefined length for patterns of mutations suggestive of structural features (pseudo-knot free) (Fig. 4.5). They categorize the aligned regions as structure encoding, non-structure encoding, and (in the case of QRNA) protein-coding. A score is assigned to each prediction to reflect the degree of confidence in the classification. It remains an open question as to how the selected window size affects the performance of these programs. It is unclear as to how much difference can be tolerated between the window size and the ncRNA transcript boundaries to produce a valid prediction (Meyer 2007; Kavanaugh and Dietrich in press).

Although these programs perform fundamentally the same task, they differ significantly in their methods, assumptions and vulnerabilities. For example, EVOFOLD takes an evolutionary tree relating the aligned species as an input while the other two do not. EVOFOLD's underlying model can explicitly model non-structured regions in the input alignment, presumably making it less vulnerable to a specific window-size. The RNAz program evaluates the similarity of the encoded structures based on the similarity of their minimum free energies rather than the corresponding individual MFE structures. This may make its predictions somewhat less vulnerable to alignment errors and make it better at handling RNA structure variation. When co-evolving signals are absent in an alignment, RNAz assumes that ncRNA genes are thermodynamically more stable than expected.. The validity of this assumption remains controversial. When selecting a program for a specific task, it is important to understand their strengths and weaknesses (Meyer 2007). EvoFold, RNAz and AlifoldZ have been used to predict ncRNA genes in the human genome and have identified large sets of potential ncRNAs (Pedersen et al. 2006; Washietl et al. 2007).

Simultaneous Align and Fold

When sequence similarity is not high enough to permit an accurate fixed alignment, it becomes necessary to simultaneously align and fold the sequences (Fig. 4.4, plan B). This requires considerably more computation than fixed alignments and has been implemented in the FOLDALIGN program. In order to keep run time at bay, FOLDALIGN limits the analysis to two sequences (Gorodkin et al. 1997; Havgaard et al. 2005b). The program has been used to search for ncRNAs shared between the human and mouse genomes (Torarinsson et al. 2006). Both strands of 36,970 regions that were unalignable between the human and mouse genome were evaluated using a window length of 200 and a length difference of 15. A total of 1,800 common RNA structures were predicted.

The analysis highlights several important points about the current status of ncRNA prediction. First, such an analysis takes considerable computational resources. The FOLDALIGN analysis took ~5 months to run on a linux cluster of 70 2-GB-RAM nodes. Second, it is often unclear as to how the results of such analysis could be interpreted. It was necessary to assume a fixed window length to make the computation tractable. It is unclear as to how effective will using a fixed length be at predicting ncRNAs that range considerably in size. Finally, experimental verification of the predicted ncRNAs presents a significant challenge. The typical approach is to show that the predicted ncRNAs are expressed in the cell using genome-wide microarrays. However, even if such evidence is lacking, it is virtually impossible to say that a region of the genome is never expressed and is not a ncRNA. On the other hand, microarray studies have shown that a significant portion of the genome is expressed. It remains unclear how much of this transcription is actually functional and how much is transcriptional noise. Extensive experimentation is required to truly demonstrate that a ncRNA prediction is accurate. It is certainly not practical to perform such experiments on a genome-wide scale. Hence, the accuracy of genome-wide ncRNA predictions remain largely unverified.

## 4.5 Conclusions and Future Directions

ncRNA gene prediction is an exciting area of research where much has been accomplished but much still remains to be achieved. Accurate general-purpose ncRNA gene prediction remains elusive. It is stymied by an apparent lack of clear common primary-sequence features and the challenge of discriminating functional structures from random sequences with folding potential. Many approaches have been developed to identify subsets of ncRNA but no unified method has been identified. This has led to the creation of many different programs for specific applications. It is also important to point out that any gene predicted to be a ncRNA should be checked for the possibility that it might be a protein-coding gene instead. Potential ncRNA candidates should be evaluated for an open reading frame, a high degree of cross-species conservation, and a pattern of mutation suggestive of synonymous codon changes.

Users who apply ncRNA gene prediction programs must be wary, and should take time to determine the best algorithm for a given application. Investigators who have performed program-to-program comparisons report certain algorithm-specific eccentricities. They have also suggested that some of the most popular methods are the least accurate or are applied inappropriately. It may be wise to evaluate the output of several related algorithms before drawing specific conclusions. Another important consideration is computational requirement. Many programs come with a heavy computational cost that makes them impractical. Several reviews have been published that compare the performance of different programs and they can serve as useful guides for program selection (Gardner et al. 2004; Freyhult et al. 2005, 2007).

In the field of ncRNA gene prediction, there is a clear need for more systematic studies of the type that compare the sensitivity and specificity of different algorithms using standard data sets. Unfortunately, performing this analysis is more challenging than it may first appear. There are only a limited number of well-defined data sets that can be used for these comparisons. It is clear that more verified examples of ncRNAs must be accumulated for the field to progress. When comparing programs, it is also difficult to get an accurate estimate of a program's specificity. The fact that it is nearly impossible to prove that a predicted (ncRNA) gene is never expressed in the living organism, makes determining the degree of false positives a program may generate very difficult. Artificial data sets can be developed for this purpose but it is difficult to capture the nucleotide distribution and randomness of real biological data in artificially constructed sets (Meyer 2007).

Comparative approaches that rely on fixed sequence alignments (Fig. 4.4, plan A) present an area for further research. It is critically important that functionally equivalent regions be correctly aligned. It has been shown that structural ncRNA sequences evolve through compensatory mutations that frequently undergo transitions and rarely transversions (Gardner et al. 2004). Alignment programs that incorporated this information along with other sophisticated models of sequence divergence would probably provide better alignments for structural prediction. Window size is another area that merits further investigation. Fixed alignment programs typically assume a set window size for de novo gene discovery to reduce computational requirements (Washietl et al. 2005a; Pedersen et al. 2006). The difference in length that can be tolerated between a ncRNA transcript and the window size used by a program attempting to predict it is not clear. Current studies suggest that the tolerance is fairly small, in the order of tens of bases (Kavanaugh and Dietrich in press). Systematic studies of this type are necessary to interpret genome-wide predictions produced by thealgorithmic approach.

Structural comparison (Fig. 4.4, plan C) is another area of active research that holds significant promise. When sequence similarity is low, these methods offer the only hope for identifying shared structures. They may also avoid the pitfalls that befall approaches that operate on more closely related data sets. When sequences are very similar, the compensatory mutation signals available for structural prediction are limited. The sequences may jointly fold into structures that are not the true functional structure. More diverged sequences are unlikely to have this problem and may be more informative. Structural comparison could become a powerful approach if new developments continue to improve its speed and accuracy.

Each of the algorithms discussed in this chapter have tended to focus on a particular aspect of ncRNAs to predict or characterize these genes. They evaluate sequence similarity, folding energy, covariation signals or some heuristic combination of this information. It remains unclear how to combine all of the available data in a statistically reasonable manner. Combinatorial methods represent another direction of potential improvement in the field of ncRNA gene prediction.

While comparative methods are currently the most successful for ncRNA gene prediction, they can only find ncRNA genes that are conserved across species. Genes that are unique to a given species or that have changed their relative position in different species cannot be found by these methods. There is a need to develop tools that can identify ncRNAs without the use of comparative methods. Genome-specific methods (Sect. 4.4.2.1) and ncRNA-family specific programs (Sect. 4.4.1.1) have accomplished this goal with some success but have fairly limited application. The MFE approach (Sect. 4.4.2.2) also holds promise for achieving this goal but requires significant development to demonstrate its reliability.

All gene prediction programs, both protein-coding and ncRNA, are based on features like start codons, stop codons, open reading frames, mutational patterns, and structural cues to predict the location of genes. It is clear, however, that these are not the signals used by a living cell to transcribe genes. The cell must locate a gene and begin transcribing it before encountering any of the signals currently used by biologists to identify genes. Exactly how cells do this remains unclear. Promoter signals undoubtedly play a key role, but their variety, short length, and degenerate nature frequently prevents their use as effective guides for biologists. How, then, are cells able to use these signals effectively to transcribe their genes? What other information might the cell be using that is currently unknown to biologists? What extent does random transcription play in this process? Learning to interpret the genomic signals like what ? the cell does, is the "holy grail" of gene prediction. Both computational and experimental efforts will be required to achieve this goal.

## References

Allmang C, Kufel J, Chanfreau G, Mitchell P, Petfalski E, Tollervey D (1999) Functions of the exosome in rRNA, snoRNA and snRNA synthesis. EMBO J 18(19):5399–5410

Altschul SF, Gish W, Miller W, Myers EW, Lipman DJ (1990) Basic local alignment search tool. J Mol Biol 215(3):403–410

Argaman L, Hershberg R, Vogel J, Bejerano G, Wagner EG, Margalit H et al (2001) Novel small RNA-encoding genes in the intergenic regions of *Escherichia coli*. Curr Biol 11(12):941–950

Aspinall TV, Gordon JM, Bennett HJ, Karahalios P, Bukowski JP, Walker SC et al (2007) Interactions between subunits of *Saccharomyces cerevisiae* RNase MRP support a conserved eukaryotic RNase P/MRP architecture. Nucleic Acids Res 35(19):6439–6450

Babak T, Blencowe BJ, Hughes TR (2005) A systematic search for new mammalian noncoding RNAs indicates little conserved intergenic transcription. BMC Genomics 6:104

Bachellerie JP, Cavaille J, Huttenhofer A (2002) The expanding snoRNA world. Biochimie 84(8): 775–790

Bartel DP (2004) MicroRNAs: genomics, biogenesis, mechanism, and function. Cell 116(2): 281–297

Bauer M, Klau GW, Reinert K (2007) Accurate multiple sequence-structure alignment of RNA sequences using combinatorial optimization. BMC Bioinform 8:271

Bernstein E, Allis CD (2005) RNA meets chromatin. Genes Dev 19(14):1635–1655

Bindewald E, Shapiro BA (2006) RNA secondary structure prediction from sequence alignments using a network of k-nearest neighbor classifiers. RNA 12(3):342–352

Bonnet E, Wuyts J, Rouze P, Van de Peer Y (2004) Evidence that microRNA precursors, unlike other non-coding RNAs, have lower folding free energies than random sequences. Bioinformatics 20(17):2911–2917

Brannan CI, Bartolomei MS (1999) Mechanisms of genomic imprinting. Curr Opin Genet Dev 9(2):164–170

Caprara MG, Nilsen TW (2000) RNA: versatility in form and function. Nat Struct Biol 7(10): 831–833

Carter RJ, Dubchak I, Holbrook SR (2001) A computational approach to identify genes for functional RNAs in genomic sequences. Nucleic Acids Res 29(19):3928–3938

Cawley S, Bekiranov S, Ng HH, Kapranov P, Sekinger EA, Kampa D et al (2004) Unbiased mapping of transcription factor binding sites along human chromosomes 21 and 22 points to widespread regulation of noncoding RNAs. Cell 116(4):499–509

Chen JH, Le SY, Shapiro B, Currey KM, Maizel JV (1990) A computational procedure for assessing the significance of RNA secondary structure. Comput Appl Biosci 6(1):7–18

Chen JL, Blasco MA, Greider CW (2000) Secondary structure of vertebrate telomerase RNA. Cell 100(5):503–514

Cheng J, Kapranov P, Drenkow J, Dike S, Brubaker S, Patel S et al (2005) Transcriptional maps of 10 human chromosomes at 5-nucleotide resolution. Science 308(5725):1149–1154

Chenna R, Sugawara H, Koike T, Lopez R, Gibson TJ, Higgins DG et al (2003) Multiple sequence alignment with the Clustal series of programs. Nucleic Acids Res 31(13):3497–3500

Chiu DK, Kolodziejczak T (1991) Inferring consensus structure from nucleic acid sequences. Comput Appl Biosci 7(3):347–352

Chomsky N (1956) Three models for the description of language. IRE Transactions on Information Theory 2:113–124

Chomsky N (1959) On certain formal properties of grammers. Information and control 2(2):137–167

Clote P, Ferre F, Kranakis E, Krizanc D (2005) Structural RNA has lower folding energy than random RNA of the same dinucleotide frequency. RNA 11(5):578–591

Costa FF (2007) Non-coding RNAs: lost in translation? Gene 386(1–2):1–10

Dann CE III, Wakeman CA, Sieling CL, Baker SC, Irnov I, Winkler WC (2007) Structure and mechanism of a metal-sensing regulatory RNA. Cell 130(5):878–892

Decatur WA, Fournier MJ (2002) rRNA modifications and ribosome function. Trends Biochem Sci 27(7):344–351

Ding Y, Lawrence CE (2003) A statistical sampling algorithm for RNA secondary structure prediction. Nucleic Acids Res 31(24):7280–7301

Ding Y, Chan CY, Lawrence CE (2004) Sfold web server for statistical folding and rational design of nucleic acids. Nucleic Acids Res 32(Web Server issue):W135–W141

Doshi KJ, Cannone JJ, Cobaugh CW, Gutell RR (2004) Evaluation of the suitability of free-energy minimization using nearest-neighbor energy parameters for RNA secondary structure prediction. BMC Bioinform 5:105

Dowell RD, Eddy SR (2006) Efficient pairwise RNA structure prediction and alignment using sequence alignment constraints. BMC Bioinform 7:400

Durbin R, Eddy SR, Krogh A, Mitchison G (1998) Biological sequence analysis: probabilistic models of proteins and nucleic acids

Eddy SR (1996) Hidden Markov models. Curr Opin Struct Biol 6(3):361–365

Eddy SR (2001) Non-coding RNA genes and the modern RNA world. Nat Rev Genet 2(12):919–929

Eddy SR (2002) A memory-efficient dynamic programming algorithm for optimal alignment of a sequence to an RNA secondary structure. BMC Bioinform 3:18

Edvardsson S, Gardner PP, Poole AM, Hendy MD, Penny D, Moulton V (2003) A search for H/ACA snoRNAs in yeast using MFE secondary structure prediction. Bioinformatics 19(7):865–873

Frank DN, Pace NR (1998) Ribonuclease P: unity and diversity in a tRNA processing ribozyme. Annu Rev Biochem 67:153–180

Freyhult E, Gardner PP, Moulton V (2005) A comparison of RNA folding measures. BMC Bioinform 6:241

Freyhult EK, Bollback JP, Gardner PP (2007) Exploring genomic dark matter: a critical assessment of the performance of homology search methods on noncoding RNA. Genome Res 17(1):117–125

Gardner PP, Giegerich R (2004) A comprehensive comparison of comparative RNA structure prediction approaches. BMC Bioinform 5:140

Gardner PP, Wilm A, Washietl S (2005) A benchmark of multiple sequence alignment programs upon structural RNAs. Nucleic Acids Res 33(8):2433–2439

Gesteland RF, Cech TR, Atkins JF (2006) The RNA World. Cold Spring Harbor Laboratory Press, Cold Spring Harbor, NY

Gillet R, Felden B (2001) Emerging views on tmRNA-mediated protein tagging and ribosome rescue. Mol Microbiol 42(4):879–885

Goodenbour JM, Pan T (2006) Diversity of tRNA genes in eukaryotes. Nucleic Acids Res 34(21):6137–6146

Gorodkin J, Heyer LJ, Brunak S, Stormo GD (1997) Displaying the information contents of structural RNA alignments: the structure logos. Comput Appl Biosci 13(6):583–586

Gottesman S (2004) The small RNA regulators of *Escherichia coli*: roles and mechanisms*. Annu Rev Microbiol 58:303–328

Griffiths-Jones S (2005) RALEE–RNA ALignment editor in Emacs. Bioinformatics 21(2):257–259

Griffiths-Jones S, Bateman A, Marshall M, Khanna A, Eddy SR (2003) Rfam: an RNA family database. Nucleic Acids Res 31(1):439–441

Griffiths-Jones S, Moxon S, Marshall M, Khanna A, Eddy SR, Bateman A (2005) Rfam: annotating non-coding RNAs in complete genomes. Nucleic Acids Res 33(Database issue):D121–D124

Gutell RR (1993) Collection of small subunit (16S- and 16S-like) ribosomal RNA structures. Nucleic Acids Res 21(13):3051–3054

Gutell RR, Power A, Hertz GZ, Putz EJ, Stormo GD (1992) Identifying constraints on the higher-order structure of RNA: continued development and application of comparative sequence analysis methods. Nucleic Acids Res 20(21):5785–5795

Hannon GJ, Rossi JJ (2004) Unlocking the potential of the human genome with RNA interference. Nature 431(7006):371–378

Havgaard JH, Lyngso RB, Stormo GD, Gorodkin J (2005a) Pairwise local structural alignment of RNA sequences with sequence similarity less than 40%. Bioinformatics 21(9):1815–1824

Havgaard JH, Lyngso RB, Gorodkin J (2005a) The FOLDALIGN web server for pairwise structural RNA alignment and mutual motif search. Nucleic Acids Res 33(Web Server issue):W650–W653

Hayashita Y, Osada H, Tatematsu Y, Yamada H, Yanagisawa K, Tomida S et al (2005) A polycistronic microRNA cluster, miR-17–92, is overexpressed in human lung cancers and enhances cell proliferation. Cancer Res 65(21):9628–9632

He L, Thomson JM, Hemann MT, Hernando-Monge E, Mu D, Goodson S et al (2005) A microRNA polycistron as a potential human oncogene. Nature 435(7043):828–833

Hershberg R, Altuvia S, Margalit H (2003) A survey of small RNA-encoding genes in *Escherichia coli*. Nucleic Acids Res 31(7):1813–1820

Hertel J, Hofacker IL, Stadler PF (2008) SnoReport: computational identification of snoRNAs with unknown targets. Bioinformatics 24(2):158–164

Higgs PG (2000) RNA secondary structure: physical and computational aspects. Q Rev Biophys 33(3):199–253

Hiley SL, Babak T, Hughes TR (2005a) Global analysis of yeast RNA processing identifies new targets of RNase III and uncovers a link between tRNA 5′ end processing and tRNA splicing. Nucleic Acids Res 33(9):3048–3056

Hiley SL, Jackman J, Babak T, Trochesset M, Morris QD, Phizicky E et al (2005b) Detection and discovery of RNA modifications using microarrays. Nucleic Acids Res 33(1):e2

Hochsmann M, Voss B, Giegerich R (2004) Pure multiple RNA secondary structure alignments: a progressive profile approach. IEEE/ACM Trans Comput Biol Bioinform 1(1):53–62

Hofacker IL (2003) Vienna RNA secondary structure server. Nucleic Acids Res 31(13): 3429–3431

Hofacker IL (2007) RNA consensus structure prediction with RNAalifold. Methods Mol Biol 395:527–544

Hofacker IL, Fekete M, Stadler PF (2002) Secondary structure prediction for aligned RNA sequences. J Mol Biol 319(5):1059–1066

Holbrook SR (2005) RNA structure: the long and the short of it. Curr Opin Struct Biol 15(3):302–308

Holmes I (2005) Accelerated probabilistic inference of RNA structure evolution. BMC Bioinform 6:73

Huttenhofer A, Vogel J (2006) Experimental approaches to identify non-coding RNAs. Nucleic Acids Res 34(2):635–646

Huttenhofer A, Schattner P, Polacek N (2005) Non-coding RNAs: hope or hype? Trends Genet 21(5):289–297

Hutvagner G, Simard MJ (2008) Argonaute proteins: key players in RNA silencing. Nat Rev Mol Cell Biol 9(1):22–32

Jones SJ (2006) Prediction of genomic functional elements. Annu Rev Genomics Hum Genet 7:315–338

Kampa D, Cheng J, Kapranov P, Yamanaka M, Brubaker S, Cawley S et al (2004) Novel RNAs identified from an in-depth analysis of the transcriptome of human chromosomes 21 and 22. Genome Res 14(3):331–342

Kapranov P, Cawley SE, Drenkow J, Bekiranov S, Strausberg RL, Fodor SP et al (2002) Large-scale transcriptional activity in chromosomes 21 and 22. Science 296(5569):916–919

Kapranov P, Drenkow J, Cheng J, Long J, Helt G, Dike S et al (2005) Examples of the complex architecture of the human transcriptome revealed by RACE and high-density tiling arrays. Genome Res 15(7):987–997

Karlin S, Campbell AM, Mrazek J (1998) Comparative DNA analysis across diverse genomes. Annu Rev Genet 32:185–225

Katayama S, Tomaru Y, Kasukawa T, Waki K, Nakanishi M, Nakamura M et al (2005) Antisense transcription in the mammalian transcriptome. Science 309(5740):1564–1566

Kavanaugh LA, Dietrich FS (in press) Non-coding RNA prediction and verification in Saccharomyces cerevisiae. PLoS Genet 5(1):e1000321

Kawano M, Reynolds AA, Miranda-Rios J, Storz G (2005) Detection of 5′- and 3′-UTR-derived small RNAs and cis-encoded antisense RNAs in *Escherichia coli*. Nucleic Acids Res 33(3):1040–1050

Keenan RJ, Freymann DM, Stroud RM, Walter P (2001) The signal recognition particle. Annu Rev Biochem 70:755–775

Klein RJ, Eddy SR (2003) RSEARCH: finding homologs of single structured RNA sequences. BMC Bioinform 4:44

Klein RJ, Misulovin Z, Eddy SR (2002) Noncoding RNA genes identified in AT-rich hyperthermophiles. Proc Natl Acad Sci U S A 99(11):7542–7547
Knudsen B, Hein J (1999) RNA secondary structure prediction using stochastic context-free grammars and evolutionary history. Bioinformatics 15(6):446–454
Knudsen B, Hein J (2003) Pfold: RNA secondary structure prediction using stochastic context-free grammars. Nucleic Acids Res 31(13):3423–3428
Lai EC (2004) Predicting and validating microRNA targets. Genome Biol 5(9):115
Laslett D, Canback B, Andersson S (2002) BRUCE: a program for the detection of transfer-messenger RNA genes in nucleotide sequences. Nucleic Acids Res 30(15):3449–3453
Le SV, Chen JH, Currey KM, Maizel JV Jr (1988) A program for predicting significant RNA secondary structures. Comput Appl Biosci 4(1):153–159
Le SY, Chen JH, Maizel JV (1989) Thermodynamic stability and statistical significance of potential stem-loop structures situated at the frameshift sites of retroviruses. Nucleic Acids Res 17(15):6143–6152
Leighton PA, Ingram RS, Eggenschwiler J, Efstratiadis A, Tilghman SM (1995) Disruption of imprinting caused by deletion of the H19 gene region in mice. Nature 375(6526):34–39
Lenz DH, Mok KC, Lilley BN, Kulkarni RV, Wingreen NS, Bassler BL (2004) The small RNA chaperone Hfq and multiple small RNAs control quorum sensing in *Vibrio harveyi* and *Vibrio cholerae*. Cell 118(1):69–82
Lim LP, Lau NC, Weinstein EG, Abdelhakim A, Yekta S, Rhoades MW et al (2003) The microRNAs of *Caenorhabditis elegans*. Genes Dev 17(8):991–1008
Lindgreen S, Gardner PP, Krogh A (2007) MASTR: multiple alignment and structure prediction of non-coding RNAs using simulated annealing. Bioinformatics 23(24):3304–3311
Livny J, Waldor MK (2007) Identification of small RNAs in diverse bacterial species. Curr Opin Microbiol 10(2):96–101
Livny J, Fogel MA, Davis BM, Waldor MK (2005) sRNAPredict: an integrative computational approach to identify sRNAs in bacterial genomes. Nucleic Acids Res 33(13):4096–4105
Livny J, Brencic A, Lory S, Waldor MK (2006) Identification of 17 *Pseudomonas aeruginosa* sRNAs and prediction of sRNA-encoding genes in 10 diverse pathogens using the bioinformatic tool sRNAPredict2. Nucleic Acids Res 34(12):3484–3493
Lowe TM, Eddy SR (1997) tRNAscan-SE: a program for improved detection of transfer RNA genes in genomic sequence. Nucleic Acids Res 25(5):955–964
Lowe TM, Eddy SR (1999) A computational screen for methylation guide snoRNAs in yeast. Science 283(5405):1168–1171
Lu C, Tej SS, Luo S, Haudenschild CD, Meyers BC, Green PJ (2005) Elucidation of the small RNA component of the transcriptome. Science 309(5740):1567–1569
Macke TJ, Ecker DJ, Gutell RR, Gautheret D, Case DA, Sampath R (2001) RNAMotif, an RNA secondary structure definition and search algorithm. Nucleic Acids Res 29(22):4724–4735
Martens JA, Laprade L, Winston F (2004) Intergenic transcription is required to repress the *Saccharomyces cerevisiae* SER3 gene. Nature 429(6991):571–574
Mathews DH, Turner DH (2002) Dynalign: an algorithm for finding the secondary structure common to two RNA sequences. J Mol Biol 317(2):191–203
Mathews DH, Turner DH (2006) Prediction of RNA secondary structure by free energy minimization. Curr Opin Struct Biol 16(3):270–278
Mattick JS (2004a) RNA regulation: a new genetics? Nat Rev Genet 5(4):316–323
Mattick JS (2004b) The hidden genetic program of complex organisms. Sci Am 291(4):60–67
Mattick JS, Makunin IV (2006) Non-coding RNA. Hum Mol Genet 15(Spec No. 1):R17–R29
McCaskill JS (1990) The equilibrium partition function and base pair binding probabilities for RNA secondary structure. Biopolymers 29:1105–1119
Meier UT (2005) The many facets of H/ACA ribonucleoproteins. Chromosoma 114(1):1–14
Meister G, Tuschl T (2004) Mechanisms of gene silencing by double-stranded RNA. Nature 431(7006):343–349
Mendes Soares LM, Valcarcel J (2006) The expanding transcriptome: the genome as the 'Book of Sand'. EMBO J 25(5):923–931

Meyer IM (2007) A practical guide to the art of RNA gene prediction. Brief Bioinform 8(6):396–414

Meyer IM, Miklos I (2004) Co-transcriptional folding is encoded within RNA genes. BMC Mol Biol 5:10

Moretti S, Armougom F, Wallace IM, Higgins DG, Jongeneel CV, Notredame C (2007) The M-Coffee web server: a meta-method for computing multiple sequence alignments by combining alternative alignment methods. Nucleic Acids Res 35(Web Server issue):W645–W648

Neugebauer KM (2002) On the importance of being co-transcriptional. J Cell Sci 115(Pt 20):3865–3871

Nussinov R, Pieczenik G, Griggs JR, Kleitman DJ (1978) Algorithms for loop matchings. SIAM J Appl Math 35(1):68–82

Ogurtsov AY, Shabalina SA, Kondrashov AS, Roytberg MA (2006) Analysis of internal loops within the RNA secondary structure in almost quadratic time. Bioinformatics 22(11):1317–1324

Ohler U, Yekta S, Lim LP, Bartel DP, Burge CB (2004) Patterns of flanking sequence conservation and a characteristic upstream motif for microRNA gene identification. RNA 10(9):1309–1322

Olivas WM, Muhlrad D, Parker R (1997) Analysis of the yeast genome: identification of new non-coding and small ORF-containing RNAs. Nucleic Acids Res 25(22):4619–4625

Omer AD, Lowe TM, Russell AG, Ebhardt H, Eddy SR, Dennis PP (2000) Homologs of small nucleolar RNAs in Archaea. Science 288(5465):517–522

Pearson WR (2000) Flexible sequence similarity searching with the FASTA3 program package. Methods Mol Biol 132:185–219

Pearson WR, Lipman DJ (1988) Improved tools for biological sequence comparison. Proc Natl Acad Sci U S A 85(8):2444–2448

Pedersen JS, Meyer IM, Forsberg R, Simmonds P, Hein J (2004) A comparative method for finding and folding RNA secondary structures within protein-coding regions. Nucleic Acids Res 32(16):4925–4936

Pedersen JS, Bejerano G, Siepel A, Rosenbloom K, Lindblad-Toh K, Lander ES et al (2006) Identification and classification of conserved RNA secondary structures in the human genome. PLoS Comput Biol 2(4):e33

Peng WT, Robinson MD, Mnaimneh S, Krogan NJ, Cagney G, Morris Q et al (2003) A panoramic view of yeast noncoding RNA processing. Cell 113(7):919–933

Pichon C, Felden B (2003) Intergenic sequence inspector: searching and identifying bacterial RNAs. Bioinformatics 19(13):1707–1709

Pichon C, Felden B (2005) Small RNA genes expressed from *Staphylococcus aureus* genomic and pathogenicity islands with specific expression among pathogenic strains. Proc Natl Acad Sci U S A 102(40):14249–14254

Reeder J, Giegerich R (2004) Design, implementation and evaluation of a practical pseudoknot folding algorithm based on thermodynamics. BMC Bioinform 5:104

Reeder J, Giegerich R (2005) Consensus shapes: an alternative to the Sankoff algorithm for RNA consensus structure prediction. Bioinformatics 21(17):3516–3523

Reeder J, Hochsmann M, Rehmsmeier M, Voss B, Giegerich R (2006) Beyond Mfold: recent advances in RNA bioinformatics. J Biotechnol 124(1):41–55

Ren J, Rastegari B, Condon A, Hoos HH (2005) HotKnots: heuristic prediction of RNA secondary structures including pseudoknots. RNA 11(10):1494–1504

Repsilber D, Wiese S, Rachen M, Schroder AW, Riesner D, Steger G (1999) Formation of metastable RNA structures by sequential folding during transcription: time-resolved structural analysis of potato spindle tuber viroid (–)-stranded RNA by temperature-gradient gel electrophoresis. RNA 5(4):574–584

Rhoades MW, Reinhart BJ, Lim LP, Burge CB, Bartel B, Bartel DP (2002) Prediction of plant microRNA targets. Cell 110(4):513–520

Rivas E, Eddy SR (1999) A dynamic programming algorithm for RNA structure prediction including pseudoknots. J Mol Biol 285(5):2053–2068

Rivas E, Eddy SR (2000) Secondary structure alone is generally not statistically significant for the detection of noncoding RNAs. Bioinformatics 16(7):583–605

Rivas E, Eddy SR (2001) Noncoding RNA gene detection using comparative sequence analysis. BMC Bioinform 2(1):8

Rivas E, Klein RJ, Jones TA, Eddy SR (2001) Computational identification of noncoding RNAs in *E. coli* by comparative genomics. Curr Biol 11(17):1369–1373

Ruan J, Stormo GD, Zhang W (2004a) An iterated loop matching approach to the prediction of RNA secondary structures with pseudoknots. Bioinformatics 20(1):58–66

Ruan J, Stormo GD, Zhang W (2004b) ILM: a web server for predicting RNA secondary structures with pseudoknots. Nucleic Acids Res 32(Web Server issue):W146–W149

Saguy M, Gillet R, Metzinger L, Felden B (2005) tmRNA and associated ligands: a puzzling relationship. Biochimie 87(9–10):897–903

Samanta MP, Tongprasit W, Sethi H, Chin CS, Stolc V (2006) Global identification of noncoding RNAs in *Saccharomyces cerevisiae* by modulating an essential RNA processing pathway. Proc Natl Acad Sci U S A 103(11):4192–4197

Sankoff D (1985) Simultaneous solution of the RNA folding, alignment and protosequence problems. SIAM J Appl Math 45:810–825

Schattner P (2002) Searching for RNA genes using base-composition statistics. Nucleic Acids Res 30(9):2076–2082

Schattner P, Decatur WA, Davis CA, Ares M Jr, Fournier MJ, Lowe TM (2004) Genome-wide searching for pseudouridylation guide snoRNAs: analysis of the *Saccharomyces cerevisiae* genome. Nucleic Acids Res 32(14):4281–4296

Seitz H, Youngson N, Lin SP, Dalbert S, Paulsen M, Bachellerie JP et al (2003) Imprinted microRNA genes transcribed antisense to a reciprocally imprinted retrotransposon-like gene. Nat Genet 34(3):261–262

Siebert S, Backofen R (2007) Methods for multiple alignment and consensus structure prediction of RNAs implemented in MARNA. Methods Mol Biol 395:489–502

Sonkoly E, Bata-Csorgo Z, Pivarcsi A, Polyanka H, Kenderessy-Szabo A, Molnar G et al (2005) Identification and characterization of a novel, psoriasis susceptibility-related noncoding RNA gene, PRINS. J Biol Chem 280(25):24159–24167

Sontheimer EJ, Carthew RW (2005) Silence from within: endogenous siRNAs and miRNAs. Cell 122(1):9–12

Sprinzl M, Horn C, Brown M, Ioudovitch A, Steinberg S (1998) Compilation of tRNA sequences and sequences of tRNA genes. Nucleic Acids Res 26(1):148–153

Stark BC, Kole R, Bowman EJ, Altman S (1978) Ribonuclease P: an enzyme with an essential RNA component. Proc Natl Acad Sci U S A 75(8):3717–3721

Steffen P, Voss B, Rehmsmeier M, Reeder J, Giegerich R (2006) RNAshapes: an integrated RNA analysis package based on abstract shapes. Bioinformatics 22(4):500–503

Storz G (2002) An expanding universe of noncoding RNAs. Science 296(5571):1260–1263

Storz G, Opdyke JA, Zhang A (2004) Controlling mRNA stability and translation with small, noncoding RNAs. Curr Opin Microbiol 7(2):140–144

Terai G, Komori T, Asai K, Kin T (2007) miRRim: a novel system to find conserved miRNAs with high sensitivity and specificity. RNA 13(12):2081–2090

Tilghman SM (1999) The sins of the fathers and mothers: genomic imprinting in mammalian development. Cell 96(2):185–193

Torarinsson E, Sawera M, Havgaard JH, Fredholm M, Gorodkin J (2006) Thousands of corresponding human and mouse genomic regions unalignable in primary sequence contain common RNA structure. Genome Res 16(7):885–889

Torarinsson E, Havgaard JH, Gorodkin J (2007) Multiple structural alignment and clustering of RNA sequences. Bioinformatics 23(8):926–932

Venema J, Tollervey D (1999) Ribosome synthesis in *Saccharomyces cerevisiae*. Annu Rev Genet 33:261–311

Verdel A, Moazed D (2005) RNAi-directed assembly of heterochromatin in fission yeast. FEBS Lett 579:5872–5878

Voss B (2006) Structural analysis of aligned RNAs. Nucleic Acids Res 34(19):5471–5481

Wang XJ, Reyes JL, Chua NH, Gaasterland T (2004) Prediction and identification of *Arabidopsis thaliana* microRNAs and their mRNA targets. Genome Biol 5(9):R65
Washietl S, Hofacker IL (2004) Consensus folding of aligned sequences as a new measure for the detection of functional RNAs by comparative genomics. J Mol Biol 342(1):19–30
Washietl S, Hofacker IL, Lukasser M, Huttenhofer A, Stadler PF (2005a) Mapping of conserved RNA secondary structures predicts thousands of functional noncoding RNAs in the human genome. Nat Biotechnol 23(11):1383–1390
Washietl S, Hofacker IL, Stadler PF (2005b) Fast and reliable prediction of noncoding RNAs. Proc Natl Acad Sci U S A 102(7):2454–2459
Washietl S, Pedersen JS, Korbel JO, Stocsits C, Gruber AR, Hackermuller J et al (2007) Structured RNAs in the ENCODE selected regions of the human genome. Genome Res 17(6):852–864
Wassarman KM, Repoila F, Rosenow C, Storz G, Gottesman S (2001) Identification of novel small RNAs using comparative genomics and microarrays. Genes Dev 15(13):1637–1651
Wilderman PJ, Sowa NA, FitzGerald DJ, FitzGerald PC, Gottesman S, Ochsner UA et al (2004) Identification of tandem duplicate regulatory small RNAs in *Pseudomonas aeruginosa* involved in iron homeostasis. Proc Natl Acad Sci U S A 101(26):9792–9797
Witwer C, Hofacker IL, Stadler PF (2004) Prediction of consensus RNA secondary structures including pseudoknots. IEEE/ACM Trans Comput Biol Bioinform 1(2):66–77
Workman C, Krogh A (1999) No evidence that mRNAs have lower folding free energies than random sequences with the same dinucleotide distribution. Nucleic Acids Res 27(24):4816–4822
Wuchty S, Fontana W, Hofacker IL, Schuster P (1999) Complete suboptimal folding of RNA and the stability of secondary structures. Biopolymers 49(2):145–165
Xie X, Lu J, Kulbokas EJ, Golub TR, Mootha V, Lindblad-Toh K et al (2005) Systematic discovery of regulatory motifs in human promoters and 3′ UTRs by comparison of several mammals. Nature 434(7031):338–345
Xu X, Ji Y, Stormo GD (2007) RNA Sampler: a new sampling based algorithm for common RNA secondary structure prediction and structural alignment. Bioinformatics 23(15):1883–1891
Yao Z, Weinberg Z, Ruzzo WL (2006) CMfinder – a covariance model based RNA motif finding algorithm. Bioinformatics 22(4):445–452
Yin JQ, Zhao RC (2007) Identifying expression of new small RNAs by microarrays. Methods 43(2):123–130
Zamore PD, Haley B (2005) Ribo-gnome: the big world of small RNAs. Science 309(5740): 1519–1524
Zhang Y, Zhang Z, Ling L, Shi B, Chen R (2004) Conservation analysis of small RNA genes in *Escherichia coli*. Bioinformatics 20(5):599–603
Zuker M (1989) Computer prediction of RNA structure. Methods Enzymol 180:262–288
Zuker M, Sankoff D (1984) RNA secondary structures and their prediction. Bull Math Biol 46:591–621
Zuker M, Stiegler P (1981) Optimal computer folding of large RNA sequences using thermodynamics and auxiliary information. Nucleic Acids Res 9(1):133–148

# Chapter 5
# Gene Prediction Methods

**William H. Majoros, Ian Korf, and Uwe Ohler**

## 5.1 The Problem of Finding and Parsing Genes

Most computational gene-finding methods in current use are derived from the fields of *natural language processing* and *speech recognition*. These latter fields are concerned with parsing spoken or written language into functional components such as nouns, verbs, and phrases of various types. The parsing task is governed by a set of *syntax rules* that dictate which linguistic elements may immediately follow each other in well-formed sentences – for example,

$$subject \to verb, \text{verb} \to direct\ object, \text{etc...}$$

The problem of gene-finding is rather similar to linguistic parsing in that we wish to partition a sequence of letters into elements of biological relevance, such as *exons*, *introns*, and the *intergenic regions* separating genes. That is, we wish to not only *find* the genes, but also to predict their internal exon-intron structure so that the encoded protein(s) may be deduced. Figure 5.1 illustrates this internal structure for a typical gene.

Generalizing from this figure, we can deduce several things about the syntax of genes: (1) the *coding segment* (CDS) of the gene must begin with a *start codon* (ATG in eukaryotes); (2) it must end with a *stop codon* (TGA, TAA, or TAG in typical eukaryotes); (3) it may contain zero or more *introns*, each of which must begin with a *donor splice site* (typically GT) and must end with an *acceptor splice site* (typically AG); and (4) it will usually be flanked on the 5′ and 3′ ends by *untranslated regions* (*UTR*s).

In practice, the UTRs are generally not predicted, so that by "exon" we mean only the coding portion of an exon. Thus, the prediction of exonic structure for a CDS consists of identifying the start/stop codons and splice sites

W.H. Majoros (✉)
Institute for Genome Sciences & Policy, Duke University, Durham, NC, USA
e-mail: william.majoros@duke.edu

D. Edwards et al. (eds.), *Bioinformatics: Tools and Applications*,
DOI 10.1007/978-0-387-92738-1_5, © Springer Science+Business Media, LLC 2009

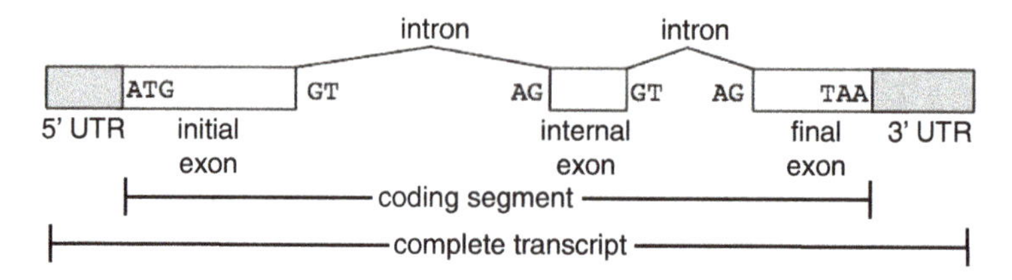


**Fig. 5.1** Eukaryotic gene structure. Canonical rules include the ATG start codon, GT-AG splice sites, and stop codon (shown here as TAA, but TAG and TGA are also typical). In the computational gene-finding field, the term "exon" typically refers to the coding parts of exons (*white boxes*) and does not include the untranslated regions (*shaded boxes*)

that mark the boundaries of coding exons. We refer to these different types of boundaries as *signals*. A *parse* of an input sequence, denoted $\phi = (s_1, s_2, \ldots, s_n)$, consists of a series of signals $s_i$ marking the boundaries of exons or introns (or intergenic regions). Denoting the four possible signals delimiting coding exons as *ATG* (for start codons), *TAG* (for any of the three stop codons: TGA, TAA, TAG), *GT* (for donor sites), and *AG* (for acceptor sites), we can encode the syntax rules for individual eukaryotic genes on the forward ("sense") strand of a chromosome as:

$$\begin{gathered} ATG \rightarrow GT \\ ATG \rightarrow TAG \\ GT \rightarrow AG \\ AG \rightarrow GT \\ AG \rightarrow TAG \end{gathered}$$

That is, if signal $s_i$ is a start codon, then $s_{i+1}$ can be either a donor site or a stop codon. Similarly, if $s_i$ is an acceptor site, then $s_{i+1}$ must be a donor site or a stop codon. Since the input sequence may contain multiple genes, we will extend our rules by introducing a *left terminus* signal, $T_L$, which denotes the 5′ end of the input sequence, and a *right terminus* signal, $T_R$, which denotes the 3′ end, with the following additional syntax rules:

$$\begin{gathered} T_L \rightarrow ATG \\ TAG \rightarrow ATG \\ TAG \rightarrow T_R \\ T_L \rightarrow T_R \end{gathered}$$

Additional rules would be necessary for genes on the reverse strand (not shown – see, e.g., Majoros 2007a). An example of a valid parse under these rules is:

$$\phi = (T_L \rightarrow ATG \rightarrow GT \rightarrow AG \rightarrow TAG \rightarrow ATG \rightarrow TAG \rightarrow T_R)$$

This parse includes two genes on the forward strand of the input sequence, the first having a single intron and the second having none. Associated with each signal in the parse would be a coordinate specifying precisely where in the input sequence the signal occurs. Given such a parse one can easily extract the putative exonic sequences for each CDS and perform translation *in silico* to arrive at the amino acid sequence for the encoded protein.

The above rules ignore an important constraint. Once the introns are spliced out of a pre-mRNA transcript (resulting in a *mature* transcript), the length of the intervening sequence between the start codon and the stop codon must be divisible by three, since the CDS must consist of an integral number of codons. We say that the stop codon is *in frame*, since it occurs in the same reading frame as all the other codons in the CDS. A related constraint is that the stop codon at the end of the CDS must be the *only* in-frame stop codon (except for the special case of *selenocysteine-bearing proteins* – see, e.g., Castellano et al. 2005).

Implicit in this formulation of the gene-parsing problem is the assumption that each gene codes for exactly one transcript, since a single parse of an input sequence provides exactly one prescribed splicing pattern for each predicted gene. This assumption is known to be false. We will revisit this issue in section 5.6 when we consider *classification-based* methods of gene finding.

Because the short nucleotide motifs for boundary signals can be expected to occur with high frequency in any typical genome, syntax rules alone do not suffice to discriminate true genes from spurious gene-like patterns, since, for example, most ATG trinucleotides on a chromosome will not function as true translation initiation sites in the living cell. We say that the syntax rules for eukaryotic genes are highly *ambiguous*, since many syntactically valid parses will not correspond to real genes. Successful gene-finding programs therefore utilize statistical measures in addition to syntax rules, in order to identify those gene parses which best match the statistical profile of a real gene in the organism of interest. These statistical profiles consist of measures such as *compositional biases* (e.g., the relative frequencies of each nucleotide, dinucleotide, trinucleotide, etc.) which are "learned" from a training set of known genes during a process called *parameter estimation*, or *training*. Because compositional biases differ from genome to genome, it is important for a gene finder to be specifically trained (when at all possible) for each genome to which it is to be applied; utilizing a gene-finder trained for one organism to predict genes in another organism can result in very poor prediction accuracy (Korf 2004). We will elaborate on training issues in the section 5.5.

Procedures utilizing properties of the DNA sequence alone in order to predict locations of genes are termed *ab initio* methods and may additionally incorporate information from homologous regions of related genomes. *Ab initio* methods utilize only *intrinsic* measures – i.e., properties of the DNA sequence itself. An alternative to *ab initio* gene finding is to incorporate expression information such as ESTs, cDNAs, or tiling arrays – termed *extrinsic* information, since it provides evidence not observable from the genomic sequence alone. We will consider all of these scenarios in the following sections.

## 5.2 Single-Genome Methods

Given only a single genomic DNA sequence, we can perform gene prediction by first identifying all the possible boundary signals, linking these together into syntactically valid gene parses, scoring all such possible parses, and then selecting the highest-scoring parse as the final prediction to be emitted by the program. This is essentially the strategy of all *ab initio* gene finders, though for efficiency reasons a full enumeration of all possible parses is either not performed or is performed using an algorithmic trick known as *dynamic programming* (see Durbin et al. 1998 for a survey of dynamic programming methods for sequence analysis). Later, we will consider two popular frameworks for evaluating gene parses – *Markov models* and *random fields*.

The process of finding putative boundary signals (i.e., start/stop codons and splice sites) is termed *signal sensing*. The simplest method of signal sensing is the common "sliding window" approach used with *position-specific weight matrices* (PWMs). An example PWM is shown in Fig. 5.2. This particular PWM consists of seven positions with each having a distribution over nucleotides which can occur at that position in the sliding window. Since this PWM is intended for finding donor splice sites, the fourth and fifth positions accept only G and T, respectively. The other five cells are nucleotide distributions which reflect compositional biases at the corresponding positions flanking true donor sites in the organism of interest.

Such a matrix can be used to identify a set of putative boundary signals as follows. At each location in the input sequence, we can superimpose the matrix onto the corresponding interval $s$ of the DNA sequence. For each such interval, we compute a signal score, $P(s|\theta)$, where $\theta$ denotes the parameters of the PWM, which were presumably estimated from known signals of the same type. $P(s|\theta)$ can be computed by multiplying the probabilities of each of the observed nucleotides falling in each position of the window where the matrix has been superimposed. The set of putative signals is obtained by retaining only those intervals for which $P(s|\theta)$ exceeds some predetermined threshold value (also determined during training). Alternatively, one can use a likelihood ratio $P(s|\theta)/P(s|\theta_{bg})$ for background model $\theta_{bg}$, in which case, a natural threshold is the value 1; one option for the background model is to simply use another PWM trained on all intervals of the target genome indiscriminately.

| | | | G | T | | |
|---|---|---|---|---|---|---|
| A : 0.28<br>C : 0.31<br>G : 0.21<br>T : 0.20 | A : 0.60<br>C : 0.13<br>G : 0.14<br>T : 0.13 | A : 0.10<br>C : 0.05<br>G : 0.75<br>T : 0.10 | A : 0.00<br>C : 0.00<br>**G : 1.00**<br>T : 0.00 | A : 0.00<br>C : 0.00<br>G : 0.00<br>**T : 1.00** | A : 0.73<br>C : 0.08<br>G : 0.10<br>T : 0.09 | A : 0.69<br>C : 0.10<br>G : 0.10<br>T : 0.11 |
| | | | consensus | | | |

**Fig. 5.2** A PWM for donor splice sites. Fixed positions flanking the GT consensus sequence are modeled using separate probability distributions over the nucleotides which may occur in those positions

Note that a PWM can have any (fixed) number of cells; in the example above, the matrix contained only two or three flanking cells on either side of the signal *consensus* (GT), presumably because for this particular organism these are the only informative positions for this type of signal. Discovering the number of informative positions flanking each signal type is a task which should be performed during model training. Given a sufficiently large set of training examples for a particular signal type, the optimal number of flanking positions can be sought via cross-validation or a similar procedure – i.e., by systematically evaluating a range of possibilities for the number of flanking positions on either side of the consensus, and choosing the configuration which gives the best cross-validation score on the training set.

The regions between signals, which comprise putative exons, introns, or intergenic regions, can be separately scored by a model called a *content sensor*. Since these regions (called *content regions*) are generally not of fixed length, PWMs are not directly applicable for this task. A type of model which is commonly used for this purpose is the *Markov chain*. An $n^{th}$*-order Markov chain* defines a probability distribution $P(x_i|x_{i-n}...x_{i-1})$, which gives the probability of the nucleotide $x_i$ at position $i$ in some putative genomic feature, conditional on the nucleotides occupying the previous $n$ positions; we call the flanking sequence $x_{i-n}...x_{i-1}$ the *context*. For some putative feature $x_1...x_m$ lying between two putative signals in the input sequence, we can evaluate $P(x_1...x_m)$ via multiplication of the terms $P(x_i|x_{i-n}...x_{i-1})$ evaluated at each position in the interval $1\leq i\leq m$; at positions $i<n$, we would of course reduce the context size so as not to take context from the preceding feature in the parse. Conditioning on preceding (or following) bases can also be performed in a PWM, resulting in a matrix with inter-column dependencies (Zhang and Marr 1993).

Note that the interval to which a content sensor is applied should not overlap with the windows for the flanking signals at either end of the content region, since for probabilistic models, this would not result in a valid joint probability when the signal and content scores are combined via multiplication to produce a score for an entire parse (see Fig. 5.3).

A demonstration of the predictive power of Markov chains is given by Fig. 5.4. The gray bars denote real coding exons; the black line plots the score computed at each position by a Markov chain trained on coding exons, normalized by scores from a background Markov chain trained on noncoding DNA. The content sensor scores can be seen to peak very strongly in real coding intervals. Comparison of the

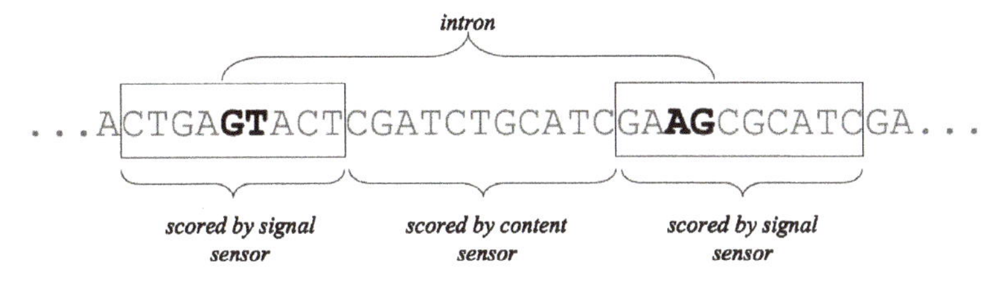


**Fig. 5.3** Evaluation windows for signal sensors and content sensors in a putative parse. By ensuring that sensor windows do not overlap, a valid joint probability for the entire parse can be obtained by multiplying together the scores for all the feature intervals comprising the parse

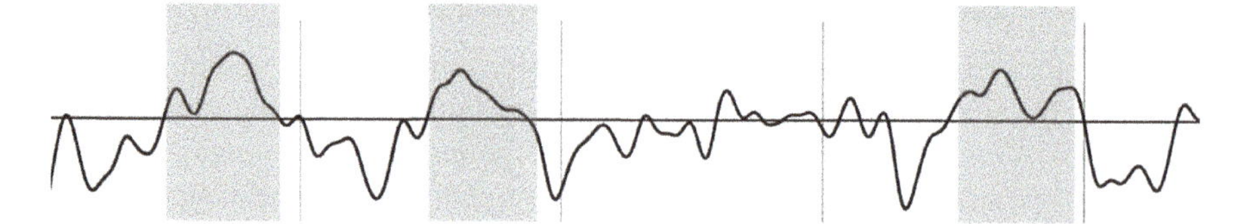

**Fig. 5.4** Likelihood ratio scores ($y$-axis) from a pair of 2nd-order Markov chains, smoothed along a ~2,000 bp sequence ($x$-axis). *Gray bars* show the true locations of coding exons. Reproduced from: Methods for Computational Gene Prediction, ©2007 W.H. Majoros. Used with permission of Cambridge University Press

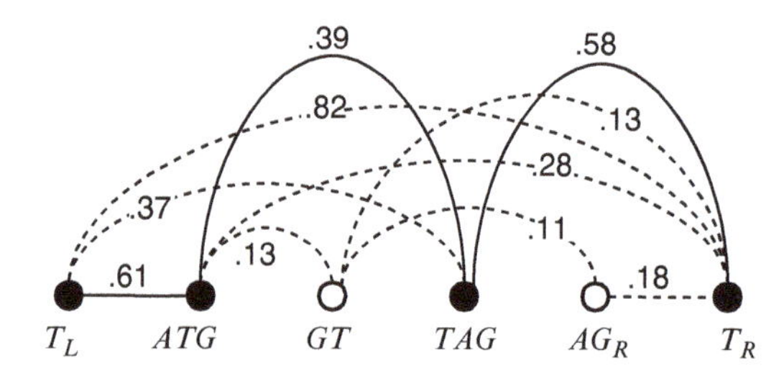


**Fig. 5.5** The shortest path (bold) through a simple trellis. Content-sensor scores are shown attached to edges in the graph; signal scores are omitted for clarity, but should be counted when computing the score of a prospective path

likelihood ratio curve $L$ with the baseline ($y=1$) shows that the regions satisfying $L>1$ provide reasonable predictions of true coding exons, though the boundaries are not precisely demarcated. More accurate prediction of exon boundaries can be accomplished by combining the scores from appropriate signal sensors with those of the content sensor. In addition, short, spurious peaks above $y=1$ can sometimes be effectively suppressed by incorporating a length distribution into the content-sensor score, as we describe shortly.

As mentioned previously, enumeration of all syntactically valid parses is (in general) computationally too expensive to perform in practice. An efficient solution to this problem can be obtained using a *shortest path* algorithm (Cormen et al. 1992). Figure 5.5 suggests how this can be done. This figure illustrates a type of graph called a *trellis*, which is a commonly-used data structure in dynamic-programming algorithms. In the case of gene prediction, we can arrange the putative signals (ordered 5′-to-3′) as a series of vertices in the graph, with edges between vertices corresponding to the syntax rules outlined above (e.g., $ATG \rightarrow GT$, etc.). We then annotate each vertex with the corresponding signal score $P(s|\theta)$ and each edge with the corresponding content score $P(x_i...x_j|\theta)$. Once we have done this, it is a simple matter to apply a shortest-path algorithm to find the highest-scoring path connecting the $T_L$ vertex to the $T_R$ vertex, which corresponds to the highest-scoring parse of the full input sequence. In the figure, we have highlighted in bold the highest scoring path, which corresponds to the optimal gene parse for the input sequence (the full sequence is not shown).

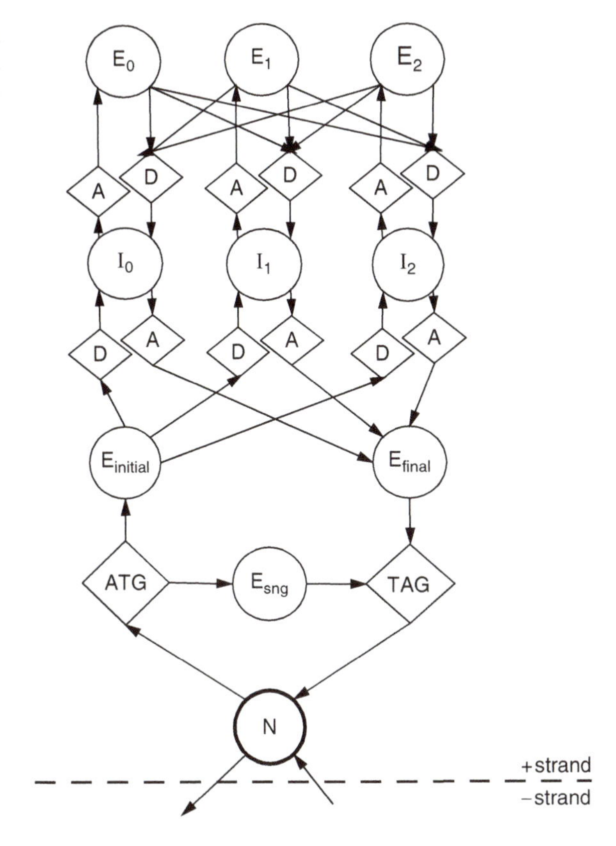


**Fig. 5.6** State-transition graph for a GHMM. *E* exon, *I* intron, *D* donor site, *A* acceptor site; integers denote phase

A popular framework for integrating signal scores and content scores in a principled way is the *generalized hidden Markov model* (GHMM); an example is illustrated in Fig. 5.6. GHMMs are *generative* models, in that their scoring framework is based on the notion that each element in the model *generates*, or *emits*, an interval of the input sequence, thereby, providing a means of scoring an input sequence by assessing the probability that the GHMM would have emitted that sequence according to some hypothesized parse. In the figure, each diamond represents a signal sensor ($D$ = donor site, $A$ = acceptor site) and each circle represents a content sensor ($E_i$ = exon beginning in reading frame $i$, $I_i$ = intron in reading frame $i$, $N$ = intergenic).

The arrows in this diagram specify the syntax rules to be enforced by the gene finder. In GHMM parlance, these are also known as *transitions*, while the circles and diamonds are known as *states*. When illustrated graphically, the result is known as a *state-transition diagram*. With each transition is associated a fixed *transition probability* (not shown in the figure). Individual states obviously produce corresponding signal scores (for the diamond states) or content scores (for the circle states); these latter two scores are known collectively as *emission probabilities*. Each content state additionally induces a distribution of *duration probabilities*, which permit the scoring of feature lengths. Any putative gene parse $\phi$ will then denote a path through the GHMM's state-transition graph, with each state having an associated set of coordinates specifying which interval of the input DNA sequence it is hypothesized to have "emitted." The score $P(\phi|S,\theta)$ of each such parse can be obtained by multiplying the transition, emission, and duration scores incurred along the corresponding path through the GHMM. The problem of identifying the highest-scoring parse with a GHMM is termed *decoding* and can be accomplished efficiently using a dynamic-programming procedure based on the shortest-path algorithm mentioned earlier (Majoros et al. 2005a).

GHMM-based gene finders have been popular for a number of years, the earliest system being the popular *GENSCAN* system (Burge and Karlin 1997) which was for years the most accurate gene-finder for the human genome; the contemporary system *GENIE* (Kulp et al. 1996) was heavily used for the annotation of the fruitfly *Drosophila melanogaster* (Adams et al. 2000). More recent GHMM-based systems which allow for end-user retraining on novel organisms include AUGUSTUS (Stanke and Waack 2003), GeneZilla (Majoros et al. 2004), GlimmerHMM (Majoros et al. 2004), and SNAP (Korf 2004).

As noted above, GHMMs are formulated on the notion of sequence *generation* – i.e., given a GHMM $\theta$ and a sequence $S$, we can compute the probability $P(S|\theta)$ that $\theta$ would, on any given invocation of the model, generate (or "emit") sequence $S$. Models of this sort are said to be *generative*, since the focus in building and training the model is on most accurately representing the process of *generating* sequences of a particular type (in our case, genes). Instead of using the model to *generate* new sequences, however, we want to use it to *parse* existing sequences – i.e., to discriminate, with the greatest possible accuracy, between genomic elements such as exons and introns. Models which focus on the optimal discrimination between classes, rather than on their generation, are called *discriminative* models. For some tasks, explicitly discriminative models have been found to produce more accurate predictions than their generative counterparts (Tong and Koller 2000; Ng and Jordan 2002).

Early attempts to derive more discriminative gene-finding models utilized generative models that were trained in a more discriminative way than usual – i.e., with the model parameters being chosen so as to maximize the expected *prediction* accuracy of the model, rather than optimizing the generative propensities of the model via maximum likelihood (e.g., Krogh 1997; Majoros and Salzberg 2004). Although these early studies showed some promise for such approaches, the preferred approach today is to employ techniques which do not model the generation of sequences at all, but instead directly model the act of parsing. By directly modeling the parsing process, such models can be more easily trained so as to maximize the expected parsing accuracy.

The most popular discriminative modeling framework for gene prediction is the *conditional random field*, or *CRF* (Lafferty et al. 2001). A CRF appears superficially much like a GHMM; indeed, the model illustrated in Fig. 5.6 could just as well have served as an example syntax model for a generalized CRF. One difference is that the scoring functions of a CRF are given as arbitrary *potential functions*, which need not evaluate to probabilities (they are automatically normalized into probabilities by the CRF framework). The ability to use arbitrary scoring functions in a CRF renders the framework very flexible, since the modeler can experiment with different scoring functions and select the set of functions which produce the most accurate predictions (e.g., via cross-validation on the training set).

Another difference is that CRFs are typically trained via *conditional maximum likelihood* (CML) rather than standard *maximum likelihood estimation* (MLE). The CML criterion – $P(\varphi|S)$ – is arguably more appropriate for training a parser than MLE – $P(S,\varphi)$ – since, during parsing, the model is used to find the parse $\varphi$ which is most probable for a given (fixed) sequence $S$ – i.e., the maximal $P(\varphi|S)$. Thus, the way in which CRFs are trained is more consistent with the way in which they are used (i.e., for parsing). Intuitively, one would expect this to translate into higher expected parsing accuracy, and this is indeed supported by recent results (e.g., Gross et al. 2008). Although GHMMs can also be discriminatively trained via CML or other non-MLE approaches, CRFs offer other advantages over GHMMs, as we will describe later.

CRFs, in their current incarnation, suffer from two possible weaknesses. The first is that they may (depending on the number of parameters in a particular CRF model) be more susceptible to the problem of *overtraining*. Overtraining occurs when the amount of training data available to train a model is insufficient to ensure a robust estimation of all parameters in the model; as a result, an overtrained model may produce highly accurate predictions on the training set, but relatively inaccurate predictions on *unseen* data, since the model is too intimately tuned to the idiosyncratic properties of the training data.

Another problem which CRFs – but not GHMMs – suffer from is that they do not explicitly model feature lengths. In the case of GHMMs, the simpler (non-generalized) hidden Markov model (HMM) can be shown to be inferior in cases where feature lengths do not follow a geometric distribution (see Fig. 5.7).

GHMMs address this problem by allowing explicit modeling of feature lengths for coding features and to a lesser extent for some noncoding features such as introns (e.g., Stanke and Waack 2003). This is easily rectified in the CRF framework, however, by instead employing a *generalized CRF* (GCRF) in which feature lengths can be explicitly modeled (Bernal et al. 2007; Majoros 2007b). An attractive feature of GCRF models is that they can additionally model longer-range dependencies than is currently feasible in non-generalized CRFs.

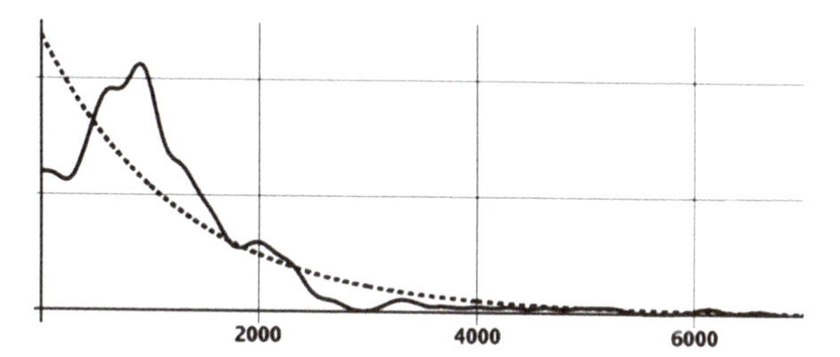


**Fig. 5.7** Observed exon length distribution (*solid line*) for *Aspergillus fumigatus*, versus a geometric distribution (*dashed line*). Reproduced from: Methods for Computational Gene Prediction, ©2007 W.H. Majoros. Used with permission of Cambridge University Press

One very significant advantage of CRFs and GCRFs is their ability to incorporate arbitrary evidence into the gene-prediction process, using either probabilistic or non-probabilistic measures. We will explore this issue further in section 5.4.

## 5.3 Multiple-Genome Methods

Whereas some *ab initio* gene finders utilize only the DNA sequence of the target genome in making their predictions, others utilize additional information in the form of observed cross-species conservation, at either the DNA or amino acid level. Such conservation, when observed between sufficiently distant species, is assumed to result from the effects of purifying selection operating (separately) on homologous regions of heterospecific genomes (Korf et al. 2001; Parra et al. 2003; Majoros et al. 2005b). Systems utilizing this information are known as *comparative* gene finders, since they compare the target genome to homologous regions of related genomes. Figure 5.8 illustrates the potential value of DNA sequence conservation evidence for gene prediction: the regions of highest conservation among human, chimpanzee, mouse, rat, and even (to a lesser degree) chicken correlate very strongly with the coding portions of exons in the example human gene.

Note from this figure that conservation between human and chimpanzee appears to offer virtually no useful evidence regarding the locations of coding exons, since the two species have only recently diverged (<6 MYA). In order to be most useful, there needs to be enough evolutionary distance between the target and "informant" genome(s) to be able to discriminate exons from the background, but not so great a distance that homologous exons have diverged too far to be recognizable. Since genes evolve at different rates, there is no ideal pair of genomes for cross-species gene finding, though the human-mouse pair (60–80 MYA) has often been used with reasonable success. One difficulty in using sequence conservation is that the amount of conserved non-coding sequence can be greater than the amount of coding sequence (Gibbs et al. 2004). Therefore, simple *phylogenetic footprinting* (as the method is sometimes called) may not identify exons with great specificity. Even so, evolutionary conservation offers an additional level of information and has the potential for finding genes overlooked by other methods (Roest Crollius et al. 2000; Siepel et al. 2007).

**Fig. 5.8** Concordance of exon structure and sequence conservation. Source: UCSC genome browser (Kent *et al.* 2002)

The modeling of sequence conservation in gene-finding systems is increasingly carried out using standard models of molecular evolution – i.e., via phylogenetic trees and substitution rate matrices such as those due to Jukes and Cantor (1969), Felsenstein (1981), and others. Because the latter models are also probabilistic, they are easily incorporated into GHMM-based systems, resulting in what are known as *phylogenetic (G)HMMs* (PhyloHMMs – Siepel and Haussler 2004) or *evolutionary (G)HMMs* (Holmes and Bruno 2001; Pedersen and Hein 2003). The phylogenetic components (one per state) of a PhyloHMM can be trained separately from the (G) HMM component and then trivially incorporated into the GHMM decoding process to allow efficient prediction. A similar approach can be taken in the GCRF framework (Majoros 2007b), though as of yet no full implementation of such an approach has been described (but see Vinson et al. 2006, for an ad hoc solution).

One shortcoming of current comparative gene finders is that the substrate for their modeling of evolutionary conservation is typically a pre-computed multi-sequence alignment produced by a general-purpose alignment program such as MULTIZ (Blanchette et al. 2004) or MAVID (Bray and Pachter 2004). Because most alignment programs do not explicitly model gene elements, alignment errors often misinform the later gene-prediction process. A counterexample is the *pair HMM*, or *PHMM*, which performs alignment and gene finding simultaneously, so that the two processes may mutually inform one another (e.g., Meyer and Durbin 2002). Unfortunately, PHMMs and their generalized cousins, *GPHMMs*, tend to be extremely computationally expensive in the absence of additional heuristics to constrain the search space of the alignment problem (i.e., the space of all possible alignments), so that simultaneous alignment and gene prediction in three or more species is likely not tractable for long sequences without liberal constraints on the search space.

Another shortcoming of many comparative gene finders is their reliance on homology information at the level of nucleotide conservation, rather than amino-acid conservation. Given a pair of orthologous genes having highly conserved functions but fairly diverged nucleotide sequences, it is reasonable to expect that the level of amino-acid similarity between the orthologous genes will be significantly higher than the level of nucleotide similarity. Indeed, amino-acid similarities have been shown to provide very strong orthology information (Majoros et al. 2005b), as suggested by Fig. 5.9. In this figure, a pair of orthologous fungal genes from

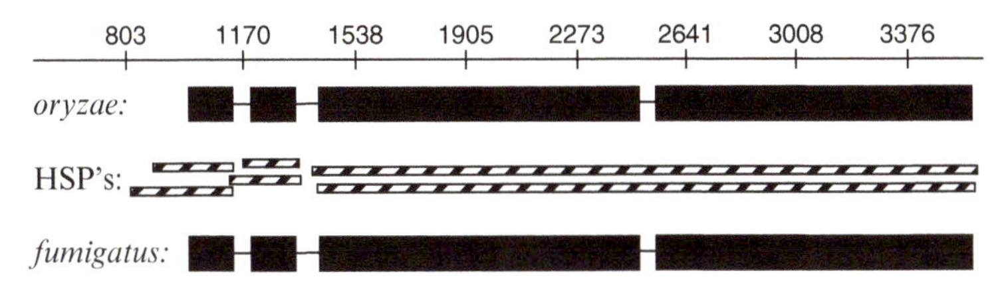


**Fig. 5.9** A pair of Aspergillus (*A. oryzae* vs. *A. fumigatus*) orthologues and their corresponding PROmer HSP's. The annotated gene structures are shown at *top* and *bottom* as *black rectangles* (exons) connected by line segments (introns). PROmer HSP's (in several frames) are shown as *thin, hatched rectangles* in the *middle* row. Reproduced from: Methods for Computational Gene Prediction, ©2007 W.H. Majoros. Used by permission of Cambridge University Press

*Aspergillus oryzae* and *A. fumigatus* is shown aligned with their amino-acid HSPs as computed by the program PROmer (Kurtz et al. 2004). As can be seen from the figure, while the boundaries of exons are not precisely demarcated in the amino-acid alignments (central track of the figure), the amino-acid HSPs provide strong indicators of individual coding regions.

The generalized pair hidden Markov model *TWAIN* (Majoros et al. 2005b) utilizes such amino-acid conservation information to predict orthologous genes in two related genomes, based on conservation scores as provided by PROmer, as well as emission, transition, and duration scores as in a standard GHMM. The result is a comparative gene-finder geared specifically toward prediction of conserved genes in pairs of moderately related genomes. A similar model was described by Alexandersson et al. (2003). A derived class of models utilizes similar computational machinery to transfer gene annotations from one genome to a related genome (Florea et al. 2005; Meyer and Durbin 2004).

Although PHMMs and GPHMMs have slipped from the spotlight in recent years, there are circumstances in which they may prove more useful than their more popular cousins, the PhyloHMMs. In cases where a genome to be annotated has one appropriately-distanced informant and few or no other related genomes available, the ability of PHMMs and GPHMMs to dynamically re-align putative orthologues provides an advantage over the "frozen alignment" approach of PhyloHMMs, which do not in practice allow for re-alignment of the informant sequences to the target genome. Thus, alignment errors introduced by the generic alignment program used to align the informants to the target genome will tend to impede PhyloHMMs more than they would a PHMM or GPHMM, with PHMMs having somewhat more flexibility to re-align informants than GPHMMs (Meyer and Durbin 2002).

Dynamic re-alignment of informants is an active research topic in the field of PhyloHMM-based gene finding (e.g., Moses et al. 2004). In the case of PHMMs and GPHMMs, the alignment of the (single) informant genome to the target genome proceeds simultaneously with the gene-prediction process, though in practice the alignment process tends to be fairly highly constrained based on heuristic preliminary alignment results, such as those based on BLAST-type programs (Altschul et al. 1990, 1997; Korf et al. 2001; Kurtz et al. 2004). Figure 5.10b illustrates a common scenario in which the optimal alignment of two genomes during gene prediction is constrained by a set of HSPs (diagonal black bars) computed via a rapid heuristic such as BLAST. Regions between successive HSPs are typically "banded" (as in *banded alignment* – Durbin et al. 1998), so that the optimal parse for the target genome may be only a *local optimum*, since it is contrained to lie within a pre-determined region of the global search space.

A related issue is that of exon splitting/merging during the evolution of orthologous genes. As suggested by Fig. 5.10a, a pair of orthologous two-exon genes derived from a common single-exon gene can mislead a PHMM-based or GPHMM-based gene finder, which may try to find the optimal parse consistent with the

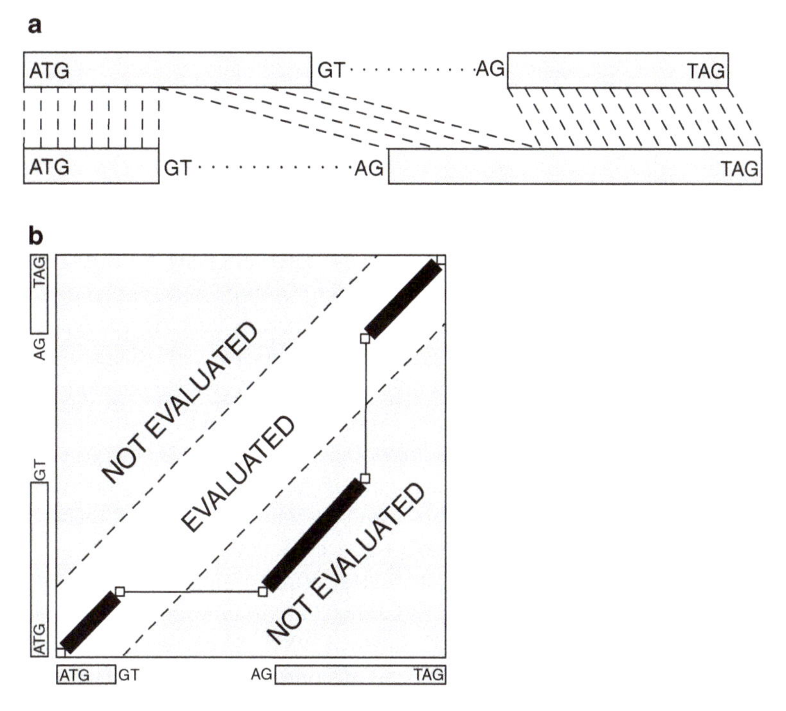


**Fig. 5.10** Intron insertion in two gene lineages. (**a**) Shows a pair of aligned orthologues in which an intron has been independently inserted in each lineage. (**b**) Shows a banded dynamic programming matrix; only the *white* region along the diagonal is evaluated. Allowing for the correct path through this matrix for the given pair of genes would require the *white* region to be expanded so as to contain a very large portion of the entire matrix, significantly reducing space efficiency of the algorithm

(erroneous) assumption that the single intron common to both genes in the orthologous pair is itself orthologous. In this hypothetical example, introns have been independently introduced in the two lineages (part (a) of the figure), at different relative locations within the genes, so that alignment of the introns will produce low homology scores. Accounting for this possibility, however, may require arbitrarily large deviations from the diagonal path through the model's dynamic programming matrix, so that decoding with a model aware of such non-orthologous intron insertions can incur large computational requirements in both space and time. In this case, aggressive banding of the matrix can result in the "correct" gene parse passing through parts of the matrix which are not evaluated, so that the correct gene parse will definitely not be found by the decoding algorithm. Addressing these issues in the context of CRFs and GCRFs seems a promising area of inquiry, though it is too early to report any progress in this area.

## 5.4 Expression-Based Methods

Apart from homology data, another useful type of external evidence is expression data, in the form of ESTs and cDNAs, which represent sequences of spliced mRNAs found in the living cell. After expressed RNAs have been mapped to locations on the genome via established programs for spliced alignment (e.g., sim4 – Florea et al. 1998; PASA – Haas et al. 2003; EXALIN – Zhang and Gish 2006; PALMA – Schulze et al. 2007), the resulting EST clusters can be used internally by a gene finder to modify its scoring system so as to weight more highly those putative coding segments supported by the existence of aligned mRNAs. Programs utilizing this type of information tend to be among the most accurate gene finders available (Guigo et al. 2006). One difficulty in the use of expression information is the effective integration of these data in a principled way. Although most expression-based gene finders have, to date, been fairly ad hoc in nature, more principled solutions are likely to emerge as modelers switch to the framework of conditional random fields, since CRFs provide a more direct means of incorporating arbitrary evidence into a predictive framework (e.g., Bernal et al. 2007; DeCaprio et al. 2007; Gross et al. 2008).

More generally, it is useful to include transcript (or even protein – see below) information when it is available and to fall back on *ab initio* methods when necessary (Hubbard et al. 2002; Allen and Salzberg 2005; Stanke et al. 2006; Wei and Brent 2006; Cantarel et al. 2008). This allows one to delineate known genes with great accuracy and to propose reasonable gene structures in the case of partial or non-existent evidence.

Expression-based methods suffer from a number of practical difficulties, especially those based on ESTs. Since ESTs are typically only 500–1,000 bp long, they frequently cover only a portion of a gene. One can assemble overlapping ESTs to produce facsimiles of complete cDNAs, but this takes a great deal of ESTs and may produce artifacts not present in a transcriptome. Newer transcript-profiling technologies, such as whole genome tiling arrays (Kampa et al. 2004; Li et al. 2007) or massively parallel sequencing (Meyers et al. 2004; Weber et al. 2007) can give a genome-wide picture of which areas are transcriptionally active, but these do not precisely delineate gene or exon boundaries. While transcript-based methods are a powerful source for gene finding, they are biased toward those genes that are abundantly and ubiquitously expressed: poorly expressed genes or those genes expressed in highly restricted developmental stages may not be represented by bulk transcript sequencing.

Another powerful source of information is the growing database of known proteins (Uniprot Consortium 2007). One can find the rough locations of exons by aligning translated nucleotide sequences to proteins from a protein database with BLAST and similar tools (e.g., Gish and States 1993; Pearson et al. 1997; Kent 2002; Li et al. 2004; Slater and Birney, 2005; Kurtz et al. 2004). Some sophisticated alignment programs even produce complete exon-intron structures (Birney et al. 2004). While protein-based alignments are very useful, they are limited to finding genes similar to known genes, and there is no guarantee that the highest scoring alignment is biologically meaningful.

## 5.5 Practical Considerations

There are a number of practical considerations which must be kept in mind when using any gene-finding software. The most important of these relate to issues of training. It is a well-known fact that a gene-finder trained for some organism *X*, but applied to the task of predicting genes in a different organism *Y*, will generally produce less accurate predictions than if the program had been trained directly on example genes from *Y*. The practice of training on one organism and then predicting genes in a different organism is called *parameter mismatching* (Korf 2004). More generally, it can be expected that a gene finder trained with larger numbers of training genes will tend to produce more accurate predictions than one trained with a smaller subset of training genes (see Fig. 5.11).

Furthermore, prediction accuracy can depend not only on the *size* of the training set, but also on the *variety* of genes in the training set. A gene-finding model is essentially a model of the *average gene*, and unfortunately, not all genes in a genome can be expected to look like the average gene, in terms of their structure or compositional biases. Training sets are also often biased toward a few well-characterized gene families, and this obviously biases the gene-finder's perception of what an "average gene" looks like. A related issue, in mammalian genomes, is that of *isochores* – regions of distinct G+C content. Early work with GHMM-based programs (Burge and Karlin 1997) showed that more accurate predictions can be obtained by including within the gene-finder several distinct models of compositional biases, one for each isochore.

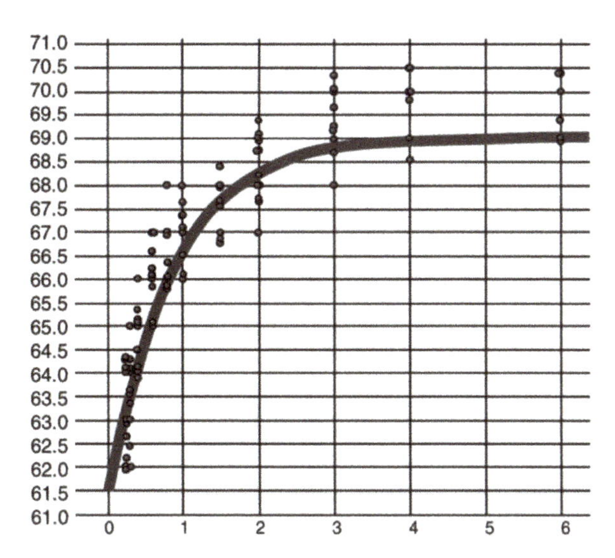


**Fig. 5.11** Gene-finding accuracy tends to increase with the size of the training set. Training set size (#genes × 1,000) are shown on the *x*-axis; whole-exon prediction accuracy on an unseen test set is shown on the *y*-axis. Reproduced with permission from (Allen, et al. 2006)

There are several options which are available when training data are very scarce. One is the practice of parameter mismatching mentioned earlier. A more promising approach is to train the gene finder on a closely related organism, use the program to predict a full complement of genes in the target organism, and then re-train the gene-finder from scratch using the first round of predicted genes as the training set for the second round of training. This approach, termed *bootstrapping* (Korf 2004), has been applied in practice with some success (Eisen et al. 2006). Unfortunately, some gene finders are not even re-trainable by end-users. There has been some success in automated *self-training*, and it may be that such methods provide a general solution to the training problem (Korf 2004; Lomsadze et al. 2005; Majoros and Salzberg 2004).

Development of novel gene-finding technologies is always limited by concerns of efficiency, in terms of both space (memory) and speed (processor cycles). The use of dynamic programming solutions is, thus, common as in other areas of bioinformatics. Dynamic programming alone is often not sufficient, however, and in practice a number of simplifying assumptions are made so as to ensure efficiency. The use of pre-computed multiple-sequence alignments in comparative gene-finders is one example. Many other assumptions are typically involved in making these systems run efficiently, or in ensuring a reasonable tradeoff between sensitivity and specificity. These include: (1) strict consensus sequences for splice sites and start/stop codons; (2) geometrically-distributed lengths for introns and other noncoding features surrounding genes; (3) that termination codons do not occur in-frame; (4) that statistical dependencies between nucleotides in a sequence do not extend over distances of greater than about 8 nucleotides; (5) that genes do not overlap; (6) that genes are not nested within other genes; (7) that the input sequence contains no sequencing errors or frameshifts; (8) that each gene has exactly one *isoform* (splice variant); (9) that functional start and stop codons are never interrupted by an intron; and (10) that the input sequence contains only As, Ts, Cs, and Gs (i.e., no *ambiguity codes* such as Y or R). Although some or all of these assumptions may be invalid for any given genome, they are typically assumed by practical software in order to ensure run-time efficiency or to avoid over-prediction of genes.

A number of heuristics have been employed in an attempt to relax some of these assumptions so as to improve the biological realism of the model (and more importantly, the predictive accuracy of the resulting software). Two examples are the modeling of non-geometric lengths for short introns (Stanke and Waack 2003) and the detection of genes coding for selenocysteine-bearing proteins (Castellano et al. 2005). Other assumptions are more difficult to circumvent in practice, though work is ongoing in this direction, as we briefly survey below.

Overlapping and nested genes are known to occur in some organisms (e.g., Normark et al. 1983; Pavesi et al. 1997; Yu et al. 2005). In many cases, the overlapping genes are on opposite strands. A few programs do allow prediction to be carried out separately on the two strands, so that such overlapping genes can be identified (e.g., *SNAP* – Korf 2004). The related issue of alternative splicing has been shown to be quite common in some organisms and is now receiving some attention (e.g., Allen et al. 2006), though the vast majority of current gene finders do not predict alternative isoforms; we comment further on this issue in the next section.

Ambiguity codes such as R (for purine) and Y (for pyrimidine) are often found in contigs resulting from low-coverage sequencing of genomes. Unfortunately, most gene finders either do not accept input containing such codes or arbitrarily change them to a valid nucleotide upon reading their input. Another common ambiguity code is N, which some assemblers will place between fragments which are believed to lie close together on the genome, but for which the intervening gap has not been properly sequenced. Genomes sequenced to low coverage often produce large numbers of fragments and contigs which the assembler may then combine with arbitrarily long runs of N's between. In order to attain high sensitivity of exon predictions in such low-coverage genomes, it is generally necessary to enable partial gene prediction for these genomes; in these cases, runs of N's may be effectively interpreted by the gene finder as the end of the contig. To our knowledge, no systematic studies have been published describing the effects of various strategies for dealing with ambiguity codes. Frameshifts are also generally not accommodated by gene-finders, though frameshifts do occur in some genome assemblies.

## 5.6 Future Directions

There are several areas worthy of future study. Two prominent issues are (1) the existence of alternative splicing, and (2) the distinctions between genes in different protein families. We will briefly consider both of these.

While it is now widely acknowledged that alternative splicing is common in many organisms, few systems at present attempt to address this issue. Instead, most gene finders predict the single, highest-scoring isoform of a putative gene. While some work has explored the sampling of suboptimal parses for prediction as alternatively-spliced isoforms (Cawley and Pachter 2003), these techniques require further investigation.

An alternative approach is to utilize classification techniques from the field of machine learning to predict individual exons, rather than predicting a global parse of the input sequence. The difference is that in a global parse the exons may not overlap and are not predicted independently of each other. A classification-based approach can be used to evaluate every possible open-reading frame to predict exons independently of each other. Predicted exons which happen to overlap may indicate alternative splicing. These approaches also require additional research, though they appear promising (e.g., Jaakkola and Haussler 1999).

Regarding the existence of protein families, it was stated earlier that current gene-finding approaches are aimed at finding the "average" gene. The success of *profile HMMs* (Durbin et al. 1998) trained on individual protein families (as opposed to general-purpose amino acid substitution matrices) for protein classification tasks (Sonnhammer et al. 1997) suggests that a similar approach may be useful for gene finding – i.e., to employ separate models for each gene family. Taken further, such an approach would permit the traditionally separate tasks of structural and functional annotation to be combined into a single computational process.

In addition to algorithmic improvements, there are also new sources of information to be utilized. Genome-wide studies of histone composition or DNAse hypersensitivity, for example, may be useful in determining the transcriptional activity of chromatin. The amount of ChIP-chip and ChIP-seq information is increasing, though these data have not yet found their way into gene-finding systems. While transcript information is utilized by a number of gene finders, there are likely to be new challenges involved in adapting these approaches to the data produced via massively parallel, short sequencing technologies.

## 5.7 Supplementary Information

Useful information about gene finding can be found at the following web site: http://www.geneprediction.org

Detailed descriptions of the models we have discussed are provided in (Durbin et al. 1998; Majoros 2007a, b).

## References


Adams MD, Celniker SE, Holt RA, 194 co-authors et al (2000) The genome sequence of *Dosophila melanogaster*. Science 287:2185–2195

Alexandersson M, Cawley S, Pachter L (2003) SLAM: cross-species gene finding and alignment with a generalized pair hidden Markov model. Genome Res 13(3):496–502

Allen JE, Salzberg SL (2005) JIGSAW: integration of multiple sources of evidence for gene prediction. Bioinformatics (Oxford, England) 21(18)):3596–3603

Allen JE, Salzberg SL (2006) A phylogenetic generalized hidden Markov model for predicting alternatively spliced exons. Algorithms Mol Biol 1:14

Altschul SF, Gish W, Miller W, Myers EW, Lipman DJ (1990) Basic local alignment search tool. J Mol Biol 215(3):403–410

Altschul SF, Madden TL, Schaffer AA, Zhang J, Zhang Z et al (1997) Gapped BLAST and PSI-BLAST: a new generation of protein database search programs. Nucleic Acids Res 25(17):3389–3402

Bernal A, Crammer K, Hatzigeorgiou A, Pereira F (2007) Global discriminative learning for higher-accuracy computational gene prediction. PLoS Comput Biol 3(3):e54

Birney E, Clamp M, Durbin R (2004) GeneWise and Genomewise. Genome Res 14(5):988–995

Blanchette M, Kent WJ, Riemer C, Elnitski L, Smit AF et al (2004) Aligning multiple genomic sequences with the threaded blockset aligner. Genome Res 14(4):708–715

Bray N, Pachter L (2004) MAVID: constrained ancestral alignment of multiple sequences. Genome Res 14(4):693–699

Burge C, Karlin S (1997) Prediction of complete gene structures in human genomic DNA. J Mol Biol 268(1):78–94

Cantarel BL, Korf I, Robb SM, Parra G, Ross E et al (2008) MAKER: an easy-to-use annotation pipeline designed for emerging model organism genomes. Genome Res 18(1):188–196

Castellano S, Lobanov AV, Chapple C, Novoselov SV, Albrecht M et al (2005) Diversity and functional plasticity of eukaryotic selenoproteins: identification and characterization of the SelJ family. Proc Natl Acad Sci USA 102(45):16188–16193

Cawley SE, Wirth AI, Speed TP (2001) Phat – a gene finding program for Plasmodium falciparum. Mol Biochem Parasitol 118(2):167–174

Cawley SL, Pachter L (2003) HMM sampling and applications to gene finding and alternative splicing. Bioinformatics (Oxford, England) 19(Suppl 2):ii36–ii41

Cormen TH, Leiserson CE, Rivest RL (1992) Introduction to algorithms. MIT, Cambridge, MA

DeCaprio D, Vinson JP, Pearson MD, Montgomery P, Doherty M et al (2007) Conrad: gene prediction using conditional random fields. Genome Res 17(9):1389–1398

Durbin R, Eddy SR, Mitchison AKG (1998) Biological sequence analysis: probabilistic models of proteins and nucleic acids. Cambridge University Press, Cambridge, p 356

Eisen JA, Coyne RS, Wu M, Wu D, Thiagarajan M, Wortman JR, Badger JH, Ren Q, Amedeo P, Jones KM, Tallon LJ, Delcher AL, Salzberg SL, Silva JC, Haas BJ, Majoros WH, Farzad M, Carlton JM, Smith RK, Garg J, Pearlman RE, Karrer KM, Sun L, Manning G, Elde NC, Turkewitz AP, Asai DJ, Wilkes DE, Wang Y, Cai H, Collins K, Stewart BA, Lee SR, Wilamowska K, Weinberg Z, Ruzzo WL, Wloga D, Gaertig J, Frankel J, Tsao CC, Gorovsky MA, Keeling PJ, Waller RF, Patron NJ, Cherry JM, Stover NA, Krieger CJ, Del Toro C, Ryder HF, Williamson SC, Barbeau RA, Hamilton EP, Orias E (2006) Macronuclear genome sequence of the ciliate *Tetrahymena thermophila*, a model eukaryote. PLoS Biol 4(9):e286

Fariselli P, Martelli PL, Casadio R (2005) A new decoding algorithm for hidden Markov models improves the prediction of the topology of all-beta membrane proteins. BMC Bioinformatics 6(Suppl 4):S12

Felsenstein J (1981) Evolutionary trees from DNA sequences: a maximum likelihood approach. J Mol Evol 17(6):368–376

Florea L, Di Francesco V, Miller J, Turner R, Yao A, Harris M, Walenz B, Mobarry C, Merkulov GV, Charlab R, Dew I, Deng Z, Istrail S, Li P, Sutton G (2005) Gene and alternative splicing annotation with AIR. Genome Res 15:54–66

Florea L, Hartzell G, Zhang Z, Rubin GM, Miller W (1998) A computer program for aligning a cDNA sequence with a genomic DNA sequence. Genome Res 8(9):967–974

Gibbs RA, Weinstock GM, Metzker ML, Muzny DM, Sodergren EJ et al (2004) Genome sequence of the Brown Norway rat yields insights into mammalian evolution. Nature 428(6982):493–521

Gish W, States DJ (1993) Identification of protein coding regions by database similarity search. Nat Genet 3:266–272

Gross SS, Do CB, Sirota M, Batzoglou S (2008) CONTRAST: a discriminative, phylogeny-free approach to multiple informant de novo gene prediction. Genome Biol 8(12):R269

Guigo R, Flicek P, Abril JF, Reymond A, Lagarde J et al (2006) EGASP: the human ENCODE Genome Annotation Assessment Project. Genome Biol 7(Suppl 1):S21–S31

Haas BJ, Delcher AL, Mount SM, Wortman JR, Smith RK Jr et al (2003) Improving the Arabidopsis genome annotation using maximal transcript alignment assemblies. Nucleic Acids Res 31(19):5654–5666

Holmes I, Bruno WJ (2001) Evolutionary HMMs: a Bayesian approach to multiple alignment. Bioinformatics 17:803–820

Hubbard T, Barker D, Birney E, Cameron G, Chen Y et al (2002) The Ensembl genome database project. Nucleic Acids Res 30(1):38–41

Jaakkola TS, Haussler D (1999) Exploiting generative models in discriminative classifiers. Adv Neural Inf Process Syst 11:487–493

JIGSAW, GeneZilla, and GlimmerHMM: puzzling out the features of human genes in the ENCODE regions. J.E. Allen, W.H. Majoros, M. Pertea, and S.L. Salzberg. Genome Biology 2006, 7(Suppl):S9

Jukes T, Cantor C (1969) Evolution of protein molecules. In: Munro H (ed) Mammalian protein metabolism. Academic, New York, NY, pp 21–132

Kall L, Krogh A, Sonnhammer EL (2005) An HMM posterior decoder for sequence feature prediction that includes homology information. Bioinformatics (Oxford, England) 21(Suppl 1):251–257

Kampa D, Cheng J, Kapranov P, Yamanaka M, Brubaker S et al (2004) Novel RNAs identified from an in-depth analysis of the transcriptome of human chromosomes 21 and 22. Genome Res 14(3):331–342

Kent WJ (2002) BLAT–the BLAST-like alignment tool. Genome Res 12(4):656–664
Kent WJ, Sugnet CW, Furey TS, Roskin KM, Pringle TH, Zahler AM, Haussler D (2002) The human genome browser at UCSC. Genome Res 12:996–1006
Korf I (2004) Gene finding in novel genomes. BMC Bioinformatics 5:59
Korf I, Flicek P, Duan D, Brent MR (2001) Integrating genomic homology into gene structure prediction. Bioinformatics (Oxford, England) 17(Suppl 1):S140–S148
Krogh A (1997) Two methods for improving performance of an HMM and their application for gene finding. Proc Int Conf Intell Syst Mol Biol 5:179–186
Kulp D, Haussler D, Reese MG, Eeckman FH (1996) A generalized hidden Markov model for the recognition of human genes in DNA. Proc Int Conf Intell Syst Mol Biol 4:134–142
Kurtz S, Phillippy A, Delcher AL, Smoot M, Shumway M, Antonescu C, Salzberg SL (2004) Versatile and open software for comparing large genomes. Genome Biol 5:R12
Lafferty J, McCallum A, Pereira F (2001) Conditional random fields: Probabilistic models for segmenting and labeling sequence data. Proc 18th International Conf on Machine Learning
Li L, Wang X, Sasidharan R, Stolc V, Deng W et al (2007) Global identification and characterization of transcriptionally active regions in the rice genome. PLoS ONE 2(3):e294
Li M, Ma B, Kisman D, Tromp J (2004) Patternhunter II: highly sensitive and fast homology search. J Bioinform Comput Biol 2(3):417–439
Lomsadze A, Ter-Hovhannisyan V, Chernoff YO, Borodovsky M (2005) Gene identification in novel eukaryotic genomes by self-training algorithm. Nucleic Acids Res 33(20):6494–6506
Lukashin AV, Borodovsky M (1998) GeneMark.hmm: new solutions for gene finding. Nucleic Acids Res 26(4):1107–1115
Majoros W (2007) Methods for Computational Gene Prediction: Cambridge University Press.
Majoros W (2007) Conditional random fields. Supplement to: methods for computational gene prediction. http://www.geneprediction.org/book/supplementary.html
Majoros WH, Salzberg SL (2004) An empirical analysis of training protocols for probabilistic gene finders. BMC Bioinformatics 5:206
Majoros WH, Pertea M, Salzberg SL (2004) TigrScan and GlimmerHMM: two open source ab initio eukaryotic gene-finders. Bioinformatics (Oxford, England) 20(16):2878–2879
Majoros WM, Pertea M, Delcher AL, Salzberg SL (2005) Efficient decoding algorithms for generalized hidden Markov model gene finders. BMC Bioinformatics 6:16
Majoros WH, Pertea M, Salzberg SL (2005) Efficient implementation of a generalized pair hidden Markov model for comparative gene finding. Bioinformatics (Oxford, England) 21(9):1782–1788
Meyer IM, Durbin R (2002) Comparative ab initio prediction of gene structures using pair HMMs. Bioinformatics (Oxford, England) 18(10):1309–1318
Meyer IM, Durbin R (2004) Gene structure conservation aids similarity based gene prediction. Nucleic Acids Res 32:776–783
Meyers BC, Vu TH, Tej SS, Ghazal H, Matvienko M et al (2004) Analysis of the transcriptional complexity of Arabidopsis thaliana by massively parallel signature sequencing. Nat Biotechnol 22(8):1006–1011
Moses AM, Chiang DY, Eisen MB (2004) Phylogenetic motif detection by expectation maximization on evolutionary mixtures. Pac Symp Biocomput 9:325–335
Ng AY, Jordan MI (2002) On discriminative vs generative classifiers: a comparison of logistic regression and naive Bayes. In: Dietterich T, Becker S, Ghahramani Z (eds.), Advances in Neural Information Processing Systems (NIPS) 14
Normark S, Bergstrom S, Edlund T, Grundstrom T, Jaurin B, Lindberg FP, Olsson O (1983) Overlapping genes. Annual Review of Genetics 17:499–525
Parra G, Blanco E, Guigo R (2000) GeneID in Drosophila. Genome Res 10(4):511–515
Parra G, Agarwal P, Abril JF, Wiehe T, Fickett JW et al (2003) Comparative gene prediction in human and mouse. Genome Res 13(1):108–117
Pavesi A, De Iaco B, Granero MI, Porati A (1997) On the informational content of overlapping genes in prokaryotic and eukaryotic viruses. Journal of Molecular Evolution 44:625–631
Pearson WR, Wood T, Zhang Z, Miller W (1997) Comparison of DNA sequences with protein sequences. Genomics 46:24–36

Pedersen JS, Hein J (2003) Gene finding with a hidden Markov model of genome structure and evolution. Bioinformatics (Oxford, England) 19(2):219–227
Reese MG, Kulp D, Tammana H, Haussler D (2000) Genie–gene finding in Drosophila melanogaster. Genome Res 10(4):529–538
Roest Crollius H, Jaillon O, Bernot A, Dasilva C, Bouneau L et al (2000) Estimate of human gene number provided by genome-wide analysis using Tetraodon nigroviridis DNA sequence. Nat Genet 25(2):235–238
Salamov AA, Solovyev VV (2000) Ab initio gene finding in Drosophila genomic DNA. Genome Res 10(4):516–522
Schulze U, Hepp B, Ong CS, Ratsch G (2007) PALMA: mRNA to genome alignments using large margin algorithms. Bioinformatics (Oxford, England) 23(15):1892–1900
Seki M, Narusaka M, Kamiya A, Ishida J, Satou M et al (2002) Functional annotation of a full-length Arabidopsis cDNA collection. Science (New York, NY) 296(5565):141–145
Siepel A, Haussler D (2004) Phylogenetic estimation of context-dependent substitution rates by maximum likelihood. Mol Biol Evol 21(3):468–488
Siepel A, Diekhans M, Brejova B, Langton L, Stevens M et al (2007) Targeted discovery of novel human exons by comparative genomics. Genome Res 17(12):1763–1773
Slater GS, Birney E (2005) Automated generation of heuristics for biological sequence comparison. BMC Bioinformatics 6:31
Sonnhammer EL, Eddy SR, Durbin R (1997) Pfam: a comprehensive database of protein domain families based on seed alignments. Proteins 28(3):405–420
Stanke M, Waack S (2003) Gene prediction with a hidden Markov model and a new intron submodel. Bioinformatics (Oxford, England) 19(Suppl 2):ii215–ii225
Stanke M, Schoffmann O, Morgenstern B, Waack S (2006) Gene prediction in eukaryotes with a generalized hidden Markov model that uses hints from external sources. BMC Bioinformatics 7:62
Strausberg RL, Feingold EA, Grouse LH, Derge JG, Klausner RD et al (2002) Generation and initial analysis of more than 15, 000 full-length human and mouse cDNA sequences. Proc Natl Acad Sci USA 99(26):16899–16903
Tong S, Koller D (2000) Restricted Bayes optimal classifiers. In: Proceedings of the Seventeenth National Conference on Artificial Intelligence, pp 658-664.
Uniprot Consortium (2007) The Universal Protein Resource (UniProt). Nucleic Acids Res 35:D193–D197
Vinson J, DeCaprio D, Luoma S, Galagan JE (2006) Gene prediction using conditional random fields (abstract). In: The Biology of Genomes, Cold Spring Harbor Laboratory, New York, May 10-14, 2006.
Weber AP, Weber KL, Carr K, Wilkerson C, Ohlrogge JB (2007) Sampling the Arabidopsis transcriptome with massively parallel pyrosequencing. Plant Physiol 144(1):32–42
Wei C, Brent MR (2006) Using ESTs to improve the accuracy of de novo gene prediction. BMC Bioinformatics 7:327
Yandell M, Bailey AM, Misra S, Shu S, Wiel C et al (2005) A computational and experimental approach to validating annotations and gene predictions in the Drosophila melanogaster genome. Proc Natl Acad Sci USA 102(5):1566–1571
Yu P, Ma D, Xu M (2005) Nested genes in the human genome. Genomics 86:414–422.
Zhang M, Gish W (2006) Improved spliced alignment from an information theoretic approach. Bioinformatics (Oxford, England) 22(1):13–20
Zhang MQ, Marr GT (1993) A weight array method for splicing signal analysis. Comput Appl Biosci 9:499–509

# Chapter 6
# Gene Annotation Methods

**Laurens Wilming and Jennifer Harrow**

## 6.1 Introduction

Gene annotation used to refer to the prediction and annotation of a coding transcript on a region of the genome, but as the complexity of the functional features on the genome increases, users require prediction of noncoding RNAs, alternatively spliced transcripts, pseudogenes, and conserved elements. Eight years after the initial draft sequence of the human genome was published, the exact number of coding genes present on this sequence is still unclear. Since new sequencing technologies have reduced the cost of sequencing and dramatically increased the speed, we can expect an enormous expansion in the amount of available genomic and transcript sequence data. To gain insight into the functional information contained within these new sequences, the features within the sequence need to be accurately annotated.

Gene annotation is not only beneficial for its identification of gene features, but it can also help improve genome sequence quality. During manual annotation, discrepancies between transcripts and genomic sequence can be identified and subsequently studied further for validation. Some discrepancies can be attributed to SNPs (Single Nucleotide Polymorphisms), DIPs (Deletion Insertion Polymorphisms) or strain or haplotype differences, while others, ranging from small insertions or deletions to debilitating substitutions, missing exons and out-of-order exon arrangements, indicate potential sequence or assembly errors. For low coverage genomic sequence (from whole genome shotgun sequencing for example), the value of annotation lies not only in the annotation itself but also in improving the genomic assembly. In this chapter we will discuss the current status of genome and gene annotation.

J. Harrow (✉)
Wellcome Trust Sanger Institute, Hinxton, Cambridgeshire, CB10 1HH, UK
e-mail: jla1@sanger.ac.uk

D. Edwards et al. (eds.), *Bioinformatics: Tools and Applications*,
DOI 10.1007/978-0-387-92738-1_6, © Springer Science+Business Media, LLC 2009

## 6.2 Automated Gene Annotation: The First Step in Annotating Genomes

Since manual annotation is labor-intensive, time consuming, and therefore expensive, it is only suitable for key reference genomes or relatively small regions of interest. As a consequence, automated annotation methods need to be employed for the vast majority of genomes. These automated methods can comprise fully de novo annotation, or include the mapping of known transcripts and protein sequences. Some automated gene-building programs, such as Ensembl, use a hybrid approach of ab initio gene predictions in combination with aligning known expressed sequences to build predicted transcripts supported by homologies to expressed genes (Hubbard et al. 2002; Flicek et al. 2008).

Gene-finding in vertebrate genome sequences is difficult due to the small proportion of sequence that actually codes for exons (generally less than 5%) and the complexity of alternative splicing. Two important aspects of a gene-finding program are the algorithm it employs and the type of information it uses. There are three types of information that algorithms can use for gene prediction: signals in the sequence such as splice sites, sequence content statistics such as codon usage, and sequence similarity to known genes/proteins. Ab initio predictors use only the first two, whereas hybrid tools such as Ensembl, incorporate similarity data to improve the quality of predictions. TWINSCAN (Korf et al. 2001), and the more recently developed NSCAN (Brown et al. 2005; Gross and Brent 2006) extend ab initio GENSCAN (Burge and Karlin 1997) predictions by including homology data from related genomes. For emerging model organisms there is also the new MAKER genome annotation pipeline (Cantarel et al. 2008): an easy to use automatic annotation system that uses gene predictions in combination with expressed sequence tag (EST) alignments to build a gene set.

Automated genome annotation systems are continually being improved and have provided a necessary service in producing a first pass annotation of draft genome sequences. They are essential for annotation of low coverage genomes (~2X) since problems experienced when predicting gene structures in draft genome assemblies (missing sequence, fragmentation, misassemblies, misplacements, small insertions/deletions/substitutions) are exacerbated. In particular, many genes will be represented only partially (or not at all) in the assembly, and many others (particularly those with large genomic extent) will be found in fragments, distributed across more than one scaffold (Hubbard et al. 2007).

In 2006, the ENCODE genome annotation assessment project (EGASP) (Guigo and Reese 2005; Guigo et al. 2006; Reese and Guigo 2006) assessed the accuracy of computational methods for predicting protein-coding genes within the ENCODE pilot regions (which encompassed 1% of the human genome sequence) (Birney et al. 2007). Eighteen groups contributed predictions from a range of programs including ab initio single genome programs (e.g., GENSCAN (Burge and Karlin 1997)), prediction methods that used expressed gene evidence (e.g., JIGSAW (Allen and Salzberg 2005)), and multigenome ab initio predictors (e.g., NSCAN (Brown et al.

2005; Gross and Brent 2006)). Predictions in a given region of the genome were assessed by comparison with a manually annotated reference gene set produced as part of the GENCODE project (Harrow et al. 2006) and only made public after the submission deadline. Results from the assessments suggested that the best automated annotation methods were able to predict at least one coding transcript correctly for over 70% of the loci. However, taking into account alternative splicing, the automated predictions only reached around 50% accuracy. Therefore, although accurate, improvements still need to be made if automatic gene prediction is to produce quality annotation approaching that from manual curators.

Genomes that are sequenced using the whole genome shotgun approach (Sundquist et al. 2007) will probably only have automated annotation. New algorithms such as GLEAN (Elsik et al. 2007), which automatically assesses the validity of gene predictions from different sources (such as Ensembl, RefSeq and Fgenesh) and takes a weighted average for a final consensus prediction, are excellent tools for this purpose. The GLEAN method of automated prediction assessment was used to create a consensus gene set for the honey bee genome and was found to be of equivalent quality to that generated by manual annotation (Elsik et al. 2007). For high-quality "finished" reference genomes, where the genomic sequence has been assembled from fully sequenced genomic clones (e.g., Bacterial Artificial Chromosomes (BACs)), manual annotation is a worthwhile investment. The human, mouse, *Arabidopsis* and *Caenorhadbditis elegans* genomes are examples of manually annotated genomes. Most of these genomes are also associated with model organism databases which serve as a hub for various types of genomic, experimental, and functional information, alongside literature curation. The only exception is for the human genome, where there is not a single database to view all information concerning *Homo sapiens,* and such data is distributed between databases.

## 6.3 Manual Annotation: The Genome Approach Versus the Transcriptome Approach

Manual gene annotation plays a significant role in annotating high quality finished genomes. However, there are two approaches to gene annotation: one transcriptome focused and one genome focused. NCBI's RefSeq collection (www.ncbi.nlm.nih.gov/RefSeq/) is a transcript centered resource of manually curated transcript sequences (Pruitt et al. 2007) and includes the entire biome with plant, bacterial, viral, vertebrate, and invertebrate sequences. The data contains predictions that are only partially supported by ESTs, cDNAs, or proteins. These have the prefix XM, while the curated transcripts have identifiers with the prefix NM. RefSeq is used internationally as a standard for gene annotation (Pruitt et al. 2007), and so when a new genome is being sequenced, researchers usually use RefSeq data to identify genes (and therefore genomic sequences) that are missing from their assembly or to identify genomic rearrangements within genes (Pruitt et al. 2007).

The genome centric approach starts with genomes, onto which transcribed sequences are mapped to determine the structure of transcripts, coding sequences (CDSs), and genes. This is how the Havana group at the Wellcome Trust Sanger Institute produces its annotation of vertebrate sequences (www.sanger.ac.uk/HGP/havana, vega.sanger.ac.uk). Currently only three vertebrate genomes – human, mouse, and zebrafish – are being fully sequenced and finished to a quality which merits manual annotation. The finished genomic sequence is analyzed using a modified Ensembl pipeline (Searle et al. 2004), so that alignments of cDNAs, ESTs, and proteins along with various ab initio predictions can be analyzed manually with the Otterlace annotation browser tool. The advantage of genomic annotation over transcriptome annotation is the improved prediction of alternatively spliced variants, as partial EST and protein evidence can be used, while cDNA annotation is limited to the availability of full-length transcripts. The genome based approach allows a far more comprehensive identification of pseudogenes (see "Pseudogene annotation" section below). The genome centric annotation approach is essential when annotating complex repetitive clusters such as histone or olfactory receptor genes. Lack of locus specific evidence for these structures means that evaluation on the transcript level would suggest a lower number of family members than would be found when examining their more complex duplicated arrangement on the genome. However, a disadvantage of the genome approach is that polymorphisms in the reference genome sequence can interfere with the annotation of coding transcripts, whereas for cDNA annotation there is usually a choice of haplotypic forms.

The CCDS (Consensus Coding Sequence) project (www.ncbi.nlm.nih.gov/CCDS/CcdsBrowse.cgi), combines the two approaches and is a collaboration between the Wellcome Trust Genome Campus institutes (Wellcome Trust Sanger Institute (www.sanger.ac.uk), the European Bioinformatics Institute (www.ebi.ac.uk)), the NCBI (RefSeq (www.ncbi.nlm.nih.gov/RefSeq/)), and the UCSC (Genome Bioinformatics Group genome.ucsc.edu)). Human and mouse RefSeq transcript models are mapped onto the genome and compared to genomic annotation from Ensembl and Vega. Where there is consensus on the prediction of the coding region, the transcript becomes a CCDS database entry. Cases of nonconsensus are discussed between the parties to attempt to reach a consensus, with the result that they are either accepted, modified and accepted, or withdrawn. The May 2008 release of the CCDS database contains over 20,000 human and 17,000 mouse CCDS transcripts, representing over 17,000 and 16,000 genes respectively. This resource provides researchers with a consistent reliable gene-set derived independently from a combination of manual and automated annotation.

An unacknowledged advantage of manual annotation is how it contributes to improved automated gene builds (such as Ensembl) because the results of manual annotation are submitted to the sequence databases and become part of the expressed sequences used to support automated gene builds. This is especially important for the annotation of genomes of species that are closely related to the manually annotated reference species, allowing the projection of the manual annotation onto the unannotated genome.

## 6.4 Pseudogene Annotation

A pseudogene is defined as a nonfunctional copy of a gene and as such pseudogenes are assumed to evolve neutrally. They are frequently considered "genomic fossils" and are often used as a calibration parameter for different models in molecular evolution (Zhang and Gerstein 2004). Pseudogenes commonly have frameshifts and/or in-frame stop codons in the coding region, rendering them nonfunctional. Pseudogenes are generated by two different mechanisms: through retrotransposition of transcribed genes back into the genome or through duplication of genomic DNA. Where the two types of pseudogenes differ is in their structure. Pseudogenes arising from retrotransposition are known as processed pseudogenes and are generated by insertion into the genome of double stranded sequence formed by reverse transcription of single stranded, processed (i.e., intron-less) mRNA (Vanin 1985). Processed pseudogenes lack introns and 5′ promoter sequences and sometimes even part of the 5′ end of the original gene. Often a genomic poly(A) tract or A-rich region is present at the 3′ end, marking the downstream insertion point. In contrast, unprocessed pseudogenes, arising from complete or partial gene duplication, have an exon structure similar to their parent gene. Duplication of DNA segments is essential for the development of complex genomes, yet exactly how this occurs is still under debate (Cooke et al. 1997; Ganfornina and Sanchez 1999).

Pseudogene identification is a major problem in computational genomics and is critical for getting an accurate view of the genomic landscape of an organism. The prevalence of pseudogenes in mammalian genomes has been problematic for gene annotation, giving rise to gene structure artifacts in automated annotation, and to some extent manual annotation. Currently, there are no consensus computational schemes for detecting different types of pseudogenes. The correct categorization of an open reading frame as part of a functional gene or a pseudogene is hampered by the fact that some processed pseudogenes still have an intact CDS and may have maintained or (re)gained function (e.g., human testis specific PGK gene (McCarrey and Thomas 1987)). Also, some unprocessed pseudogenes have a CDS very similar to their expressed versions, with only a slightly truncated 3′ end because of an in-frame stop codon close to the regular stop codon or a slightly extended or truncated 3′ end because of a frameshift close to the regular translation end. In each of these cases, automated annotation tools would consider these pseudogenes as expressed genes. Conversely, expressed pseudogenes that function as noncoding RNAs (for example Makorin1-p1 (Hirotsune et al. 2003)) may escape automatic annotation. Manual annotation allows judgment on a case-by-case basis and can also take other information, such as from publications, into account.

As part of the ENCODE pilot project, collaborators developed an approach for annotating pseudogenes (derived from protein-coding genes) that incorporated manual annotation from the GENCODE consortium alongside the union of pseudogenes predicted from four different automated methods (Zheng et al. 2007). This resulted in the identification of 201 consensus pseudogenes within the ENCODE regions, two-thirds of which were created by retrotransposition. Interestingly, when pseudogene-specific RACE was performed on these 201 pseudogenes, 20% appeared to be transcribed in one or more cell lines. This result could indicate that a number of predicted pseudogenes could have a functional role within the cell.

## 6.5 Identifying Alternative Splicing

Alternative splicing of pre-mRNA is one way higher organisms have increased proteome and transcriptome complexity from estimated gene numbers that are not that different from those of much simpler organisms such as *C. elegans* (Graveley 2001). Alternative splicing provides a versatile mechanism by which major developmental functions can be controlled through the expression of variant transcripts in a cell or tissue specific manner (Lopez 1998; Taneri et al. 2004; Yeo et al. 2004; Anderson et al. 2005). Alternative 5′ exons enable the use of different promoters for different tissues (Anderson et al. 2005), while exon skipping, alternative exons, and alternative splice sites can change the mRNA, and therefore very likely the function of the resulting protein (Taneri et al. 2004). To illustrate the amount of variation alternative splicing can introduce: the *Drosophila* Down syndrome cell adhesion molecule (*Dscam*) gene, yielding different 24-exon transcripts from a selection of 115 exons, can in theory generate over 38,000 distinct mRNA transcripts (Schmucker et al. 2000).

Alternative splicing may not only contribute to proteome diversity, but may also play a yet unappreciated regulatory role in gene expression (Stamm et al. 2005). EST coverage seems to be an influencing factor in the detection of alternative splicing. After examining seven different eukaryotic organisms with sufficient EST and mRNA coverage, Brett et al. 2002 found that alternative splicing can be detected in a large number of organisms, including invertebrates. Similarly, Hide et al. 2001 observed a correlation between EST coverage and the number of identified alternative transcripts, when they analyzed exon-skipping events in human chromosome 22. Recent experiments by Kapranov et al. (2002), Bertone et al. (2004) and Schadt et al. (2004), probing genome tiling arrays for transcribed regions and comparing the results with annotated genes, highlight many potentially transcribed regions in introns, suggesting that a considerable number of potential alternative exons and splice variants remain undiscovered.

One of the contributions of manual annotation to genome analysis is the comprehensive annotation of splice variants. Manual annotation of 1% of the human genome by the GENCODE consortium identified an average of 5.7 variants transcripts per locus, almost half of which did not appear to have a CDS. Annotated splice variants included variants based on nonhuman evidence as transcripts from other species supply a wealth of information from different developmental stages not available for humans. Genome comparison studies between mouse and human have shown that gene structures are generally conserved (Batzoglou et al. 2000; Kan et al. 2002) and that at least some alternative splicing events are conserved as well (Yeo et al. 2004).

New sequencing technologies such as Solexa, (Illumina), 454 (Roche), and SOLiD (Applied Biosystems) can generate gigabases of sequence in a single experiment, and this technology is beginning to be used to identify the transcriptomes of different organisms using a method termed mRNA-Seq (Mortazavi et al. 2008). Currently, to obtain sequence from all genes requires extraordinary depth of coverage, in the range of 5 billion bases for, for example, the yeast *Saccharomyces pombe*, equivalent to 250x the genome size (Wilhelm et al. 2008). This expression data can be used to examine alternative splicing in different tissues; for example comparisons of mRNA-Seq transcriptomes from mouse brain and muscle tissue highlighted an exon that was spliced in a specific way only in muscle (Mortazavi et al. 2008). This type of data will have a major impact on improving the annotation of tissue specific splicing events in particular and splice variation in general.

## 6.6 Comparative Genome Annotation

Comparative analysis of genomic sequence between species at different evolutionary distances is a powerful method for identifying both conserved coding and noncoding sequences (regulatory regions or noncoding genes) and for the identification of species-specific genes [42–45]. When comparing multiple genomes or transcriptomes, two classes of homologous genes are observed: orthologues, which were created after a speciation event and have similar functions in the now diverged species; and paralogues, which are the result of gene or chromosomal duplication events and typically demonstrate divergent functions. Completion of the mouse draft genome sequence provided a key informational tool for unraveling the contents of the human genome (Waterston et al. 2002). From initial comparative analysis, approximately 80% of mouse genes have a single identifiable orthologue in the human genome and less than 1% of mouse genes are without a detectable homologue in the human genome.

The choice of genomes for sequencing has now less to do with the utility of that species as a model organism than its placement on an evolutionary tree. Recently, 12 *Drosophila* species were sequenced and used to identify different "evolutionary signatures." This experiment uncovered new protein-coding genes within *Drosophila*, corrected spurious annotation and helped identify new noncoding RNA genes (Stark et al. 2007). In addition, Siepel et al. (2007) used a combination of comparative tools to identify new protein-coding exons, which had not previously been annotated in the RefSeq gene set. In total, they were able to identify and confirm by RT-PCR 160 novel gene fragments. The UCSC browser now contains a 28-way vertebrate alignment based around the human genome reference assembly, which can be used to identify additional coding exons possibly missed in automatic annotation, as well as identify new areas of the genome that may have functional importance by highlighting elements conserved across various species, (see Fig. 1) (Miller et al. 2007).

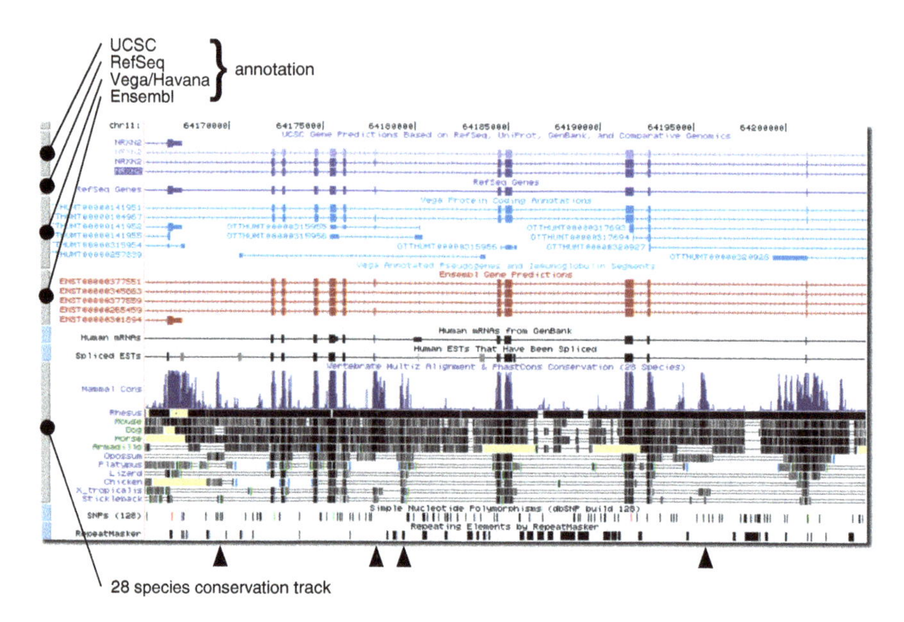


**Fig. 1** The 28 organism conservation track in the UCSC Genome Browser (human March 2006 assembly) shown for a portion of the NRXN2 gene. Note the regions of extensive conservation outside annotated exons (arrow heads)

## 6.7 Tools for Visualization and Annotation

Researchers looking for annotation and visualization tools can choose between a number of freely available open source tools that run under Unix (Linux, Mac OSX) or Windows. Alternatively, there are web-based tools where users can produce and/or display annotation. One such web-based tool is Apollo, a highly customizable Java-based desktop application with a drag-and-drop transcript building interface that is being used to annotate the *Drosophila* genome (Lewis et al. 2002; Misra and Harris 2006; Klee et al. 2007). Apollo comes with a multiple sequence alignment viewer to view protein and nucleotide alignments in detail and a graphical exon editor that allows changing of exon boundaries by simply dragging them with a cursor (see Fig. 2a). Apollo is often used in conjunction with the GMOD (Generic Model Organism Database) system to provide manual annotation. GMOD is a general system that links various data sources and analysis modules and stores the resulting biological information (Stein et al. 2002; O'Connor et al. 2008). GBrowse (Stein et al. 2002; Donlin 2007) is a popular genome viewer developed by the GMOD consortium. Many organism databases use GMOD and/or GBrowse to make their data publicly available through the web. For example TAIR (*Arabidopsis*) (Swarbreck et al. 2008), Gramene (grasses) (Liang et al. 2008), WormBase (nematode worms) (Rogers et al. 2008), SGD (budding yeast) (Christie et al. 2004), FlyBase (fruit flies) (Wilson et al. 2008), RGD (rat) (Twigger et al. 2002; Twigger et al. 2007)

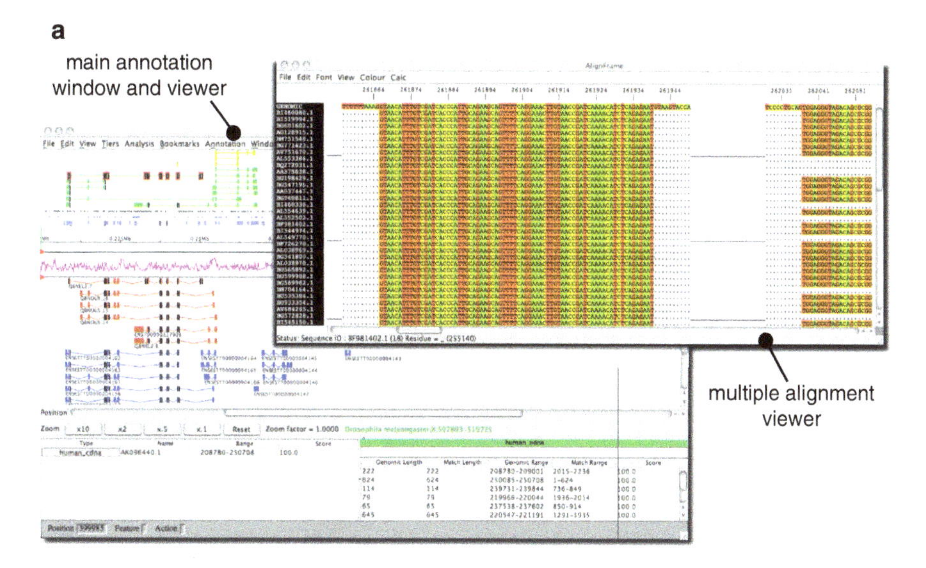


**Fig. 2a** The Apollo annotation interface. Bottom left: main window with homology features (top), annotated or imported transcripts (centre) and textual information (bottom). Top right: multiple alignment viewer showing the alignment of homology features to the genomic sequence

and MGD (mouse) (Bult et al. 2008) – all use aspects of GMOD. The Havana group at the Wellcome Trust Sanger Institute uses a custom made annotation system called Otterlace, which was developed specifically for genomic annotation but has been used successfully for transcriptome annotation as well. The software suite runs in Unix (Linux, Mac OSX) environments as a desktop application supported by central mySQL databases that store analysis and annotation in Ensembl like schemas. The software includes a transcript editor which uses copy-and-paste techniques for editing exon structures, and a graphical interface for viewing analysis and annotation. Like Apollo, Otterlace includes a multiple sequence alignment viewer in the package (see Fig. 2b).

For the annotation of small genomes such as those of pathogens, Artemis is widely used (see Fig. 2c) (Rutherford et al. 2000). Artemis is a Java application which runs on all platforms and has been used to annotate the genomes of organisms such as *Streptomyces coelicolor* (Bentley et al. 2002), *Candida albicans* (Braun et al. 2005) and *Plasmodium falciparum* (Hall et al. 2002). Where software packages like Otterlace, Apollo, and Artemis are used to both annotate and view annotation, the two processes can be separated. This allows researchers to produce annotation in their tool of choice, export it in a suitable format such as Gene Transfer Format (GTF) (mblab.wustl.edu/GTF22.html) or General Feature Format (GFF) (www.sanger.ac.uk/Software/formats/GFF/) and then display it, publicly or privately, in common genome browsers such as Ensembl or the UCSC Genome Browser using the Distributed Annotation System (DAS) [67-(Dowell et al. 2001; Finn et al. 2007).

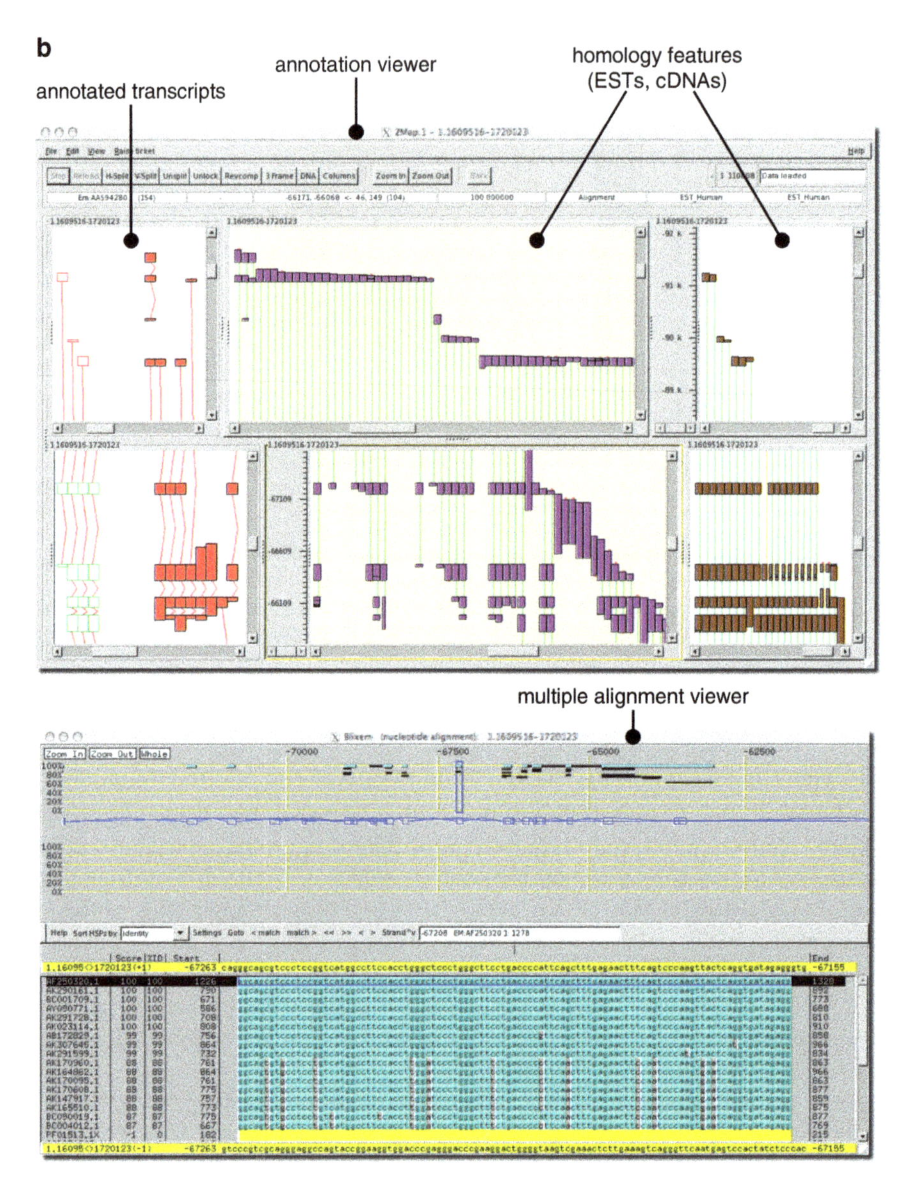


**Fig. 2b** The Otterlace annotation interface. Top: the Zmap annotation viewer with homology features and annotated transcripts. Split panels allow simultaneous viewing of different sections. Bottom: the Blixem multiple alignment viewer showing the alignment of homology features to the genomic sequence

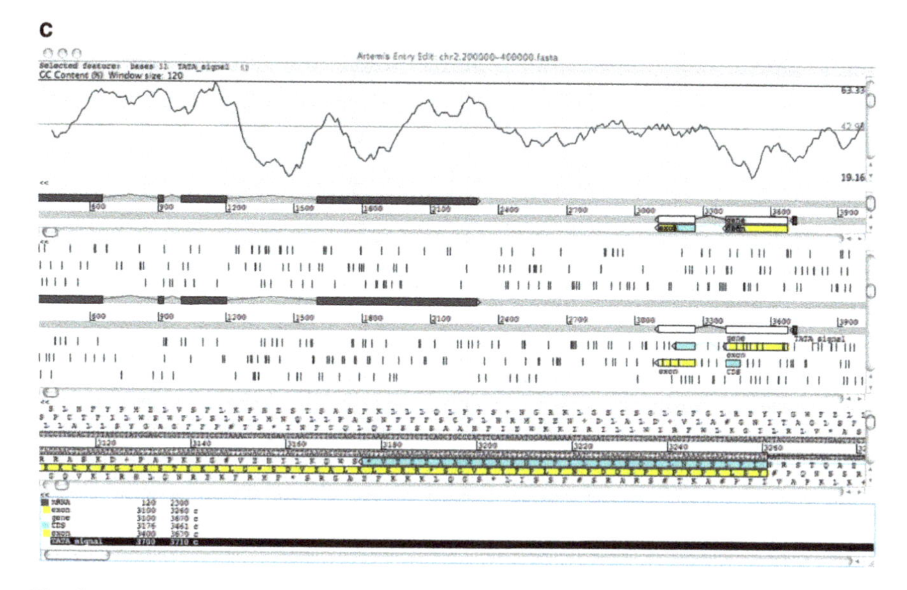

**Fig. 2c** The Artemis annotation interface showing GC content, sequence features, annotated features (colored or shaded boxes), sequence details, and textual information (bottom)

## 6.8 Gene Ontology and Community Annotation

As more genomes are sequenced and annotated, the need has arisen to unify the description of gene products within the model organism databases in the form of an agreed ontology. Ontology is a structured vocabulary, a hierarchy of terms that are precisely defined and which relate to one another in a meaningful way. The Gene Ontology (GO) Consortium (Ashburner et al. 2000) was formed to develop shared, structured vocabularies for the annotation of molecular characteristics across various organisms. These characteristics are divided into three categories and have precise definitions and relationships to one another. Terms are arranged in a directed acyclic graph allowing a term to appear in several hierarchies. Vocabulary terms are linked by "is a" and "part of" relationships, so that very general terms and very precise terms can both be represented.

Many generic databases, including Ensembl, Interpro, Entrez Gene, Mouse Genome Informatics (MGI) and UniProt, include GO terms in their annotation either manually or automatically. The assignment of GO terms/accession numbers to a gene product allows rapid comparison and cross-reference between these databases. Gene ontology is continually expanding and can represent many complex concepts, including biochemical pathways and phylogenetic associations (Thomas et al. 2007). Biological ontologies can be designed for basically any biological process, phenomenon or application, including anatomy, cell cycle, and biomedicine.

Gene Ontology annotation has also been a prominent feature in maintaining consistency in the recent FANTOM project community annotation of mouse cDNAs (Kawai et al. 2001). Community annotation was pioneered with the *Drosophila* genome annotation at Celera, where 40 scientists worked together over a two week period to complete a preliminary annotation of the *Drosophila* genome (Adams et al. 2000). Many community annotations have followed for, amongst others: human (several instances of H-invitational meetings) (Imanishi et al. 2004; Yamasaki et al. 2008); mouse (several instances of FANTOM meetings)(Kawai et al. 2001; Bono et al. 2002; Okazaki et al. 2002; Carninci et al. 2005; Maeda et al. 2006); *Xenopus tropicalis* frog; the *Bos taurus* cow genome; the *Sus scrofa* pig genome. These so-called annotation jamborees involve researchers coming together in one place and over the course of several days or weeks annotating large sets of genes to a common standard. Often participants specialize in specific families of genes or in genes related to specific diseases.

Annotation jamborees are a useful option for genome sequencing and annotation projects. However, a disadvantage of jamborees, is that they are generally one-off events, concentrating all the effort in a small amount of time, and they lack an update mechanism. For continuous community annotation it is preferable to use the Distributed Annotation System (DAS) (Dowell et al. 2001; Prlic et al. 2007). DAS allows anyone to set up a server to present analysis or annotation. Annotation can include any type of feature that can be expressed in coordinates of the relevant coordinate space. Using DAS, annotation from different sources can be shown simultaneously in a single genome browser without the need to store the data centrally. In addition, some databases have begun to incorporate Wiki pages so that users can add additional comments to entries and update information. Huss et al. (2008) have set up a Gene Wiki for the scientific community to edit. Available at en.wikipedia.org/wiki/Category:Gene_stubs, its content is based on Entrez gene data and its aim is to encourage small contributions from a large number of users. The disadvantage of this type of community annotation is that it is not peer-reviewed for quality control and not monitored except by the community in general, and therefore spurious information could be added. An alternative project at en.citizendium.org has recruited editors to monitor and review contributions and also applies a login system. This approach leads to a more trusted model for community participation in annotation projects.

## 6.9 Conclusion

With the anticipated increase in genomic and transcribed sequences the majority of future gene annotation will be produced automatically. Manual annotation will only be used for selected reference genomes and problematic regions. As demonstrated with the CCDS project, a consensus gene set, merging manual and automated procedures will be the optimal way forward for gene annotation. Regular user feedback is essential for improving annotation, however, whether this is supplied via the Wiki community approach or direct contact with the database curators is to be determined.

**Acknowledgements** This work was supported by the Wellcome Trust.

## References

Adams MD, Celniker SE, Holt RA, Evans CA, Gocayne JD, Amanatides PG et al (2000) The genome sequence of Drosophila melanogaster. Science 287(5461):2185–95
Allen JE, Salzberg SL (2005) JIGSAW: integration of multiple sources of evidence for gene prediction. Bioinformatics 21(18):3596–603
Anderson CL, Zundel MA, Werner R (2005) Variable promoter usage and alternative splicing in five mouse connexin genes. Genomics 85(2):238–44
Ashburner M, Ball CA, Blake JA, Botstein D, Butler H, Cherry JM et al (2000) Gene ontology: tool for the unification of biology. The Gene Ontology Consortium. Nat Genet 25(1):25–9
Batzoglou S, Pachter L, Mesirov JP, Berger B, Lander ES (2000) Human and mouse gene structure: comparative analysis and application to exon prediction. Genome Res 10(7):950–8
Bentley SD, Chater KF, Cerdeno-Tarraga AM, Challis GL, Thomson NR, James KD et al (2002) Complete genome sequence of the model actinomycete Streptomyces coelicolor A3(2). Nature 417(6885):141–7
Bertone P, Stolc V, Royce TE, Rozowsky JS, Urban AE, Zhu X et al (2004) Global identification of human transcribed sequences with genome tiling arrays. Science 306(5705):2242–6
Birney E, Stamatoyannopoulos JA, Dutta A, Guigo R, Gingeras TR, Margulies EH et al (2007) Identification and analysis of functional elements in 1% of the human genome by the ENCODE pilot project. Nature 447(7146):799–816
Bono H, Kasukawa T, Furuno M, Hayashizaki Y, Okazaki Y (2002) FANTOM DB: database of Functional Annotation of RIKEN Mouse cDNA Clones. Nucleic Acids Res 30(1):116–8
Braun BR, van Het Hoog M, d'Enfert C, Martchenko M, Dungan J, Kuo A et al (2005) A human-curated annotation of the Candida albicans genome. PLoS Genet 1(1):36–57
Brett D, Pospisil H, Valcarcel J, Reich J, Bork P (2002) Alternative splicing and genome complexity. Nat Genet 30(1):29–30
Brown RH, Gross SS, Brent MR (2005) Begin at the beginning: predicting genes with 5′ UTRs. Genome Res 15(5):742–7
Bult CJ, Eppig JT, Kadin JA, Richardson JE, Blake JA (2008) The Mouse Genome Database (MGD): mouse biology and model systems. Nucleic Acids Res 36(Database issue):D724–8
Burge C, Karlin S (1997) Prediction of complete gene structures in human genomic DNA. J Mol Biol 268(1):78–94
Cantarel BL, Korf I, Robb SM, Parra G, Ross E, Moore B et al (2008) MAKER: an easy-to-use annotation pipeline designed for emerging model organism genomes. Genome Res 18(1)):188–96
Carninci P, Kasukawa T, Katayama S, Gough J, Frith MC, Maeda N et al (2005) The transcriptional landscape of the mammalian genome. Science 309(5740):1559–63
Christie KR, Weng S, Balakrishnan R, Costanzo MC, Dolinski K, Dwight SS et al (2004) Saccharomyces Genome Database (SGD) provides tools to identify and analyze sequences from Saccharomyces cerevisiae and related sequences from other organisms. Nucleic Acids Res 32(Database issue):D311–4
Cooke J, Nowak MA, Boerlijst M, Maynard-Smith J (1997) Evolutionary origins and maintenance of redundant gene expression during metazoan development. Trends Genet 13(9):360–4
Donlin MJ (2007) Using the Generic Genome Browser (GBrowse). Curr Protoc Bioinformatics Chapter 9: Unit 9.9
Dowell RD, Jokerst RM, Day A, Eddy SR, Stein L (2001) The distributed annotation system. BMC Bioinformatics 2(1):7
Elsik CG, Mackey AJ, Reese JT, Milshina NV, Roos DS, Weinstock GM (2007) Creating a honey bee consensus gene set. Genome Biol 8(1):R13
Finn RD, Stalker JW, Jackson DK, Kulesha E, Clements J, Pettett R (2007) ProServer: a simple, extensible Perl DAS server. Bioinformatics 23(12):1568–70
Flicek P, Aken BL, Beal K, Ballester B, Caccamo M, Chen Y et al (2008) Ensembl 2008. Nucleic Acids Res 36(Database issue):D707–14

Ganfornina MD, Sanchez D (1999) Generation of evolutionary novelty by functional shift. Bioessays 21(5):432–9

Graveley BR (2001) Alternative splicing: increasing diversity in the proteomic world. Trends Genet 17(2):100–7

Gross SS, Brent MR (2006) Using multiple alignments to improve gene prediction. J Comput Biol 13(2):379–93

Guigo R, Reese MG (2005) EGASP: collaboration through competition to find human genes. Nat Methods 2(8):575–7

Guigo R, Flicek P, Abril JF, Reymond A, Lagarde J, Denoeud F et al (2006) EGASP: the human ENCODE Genome Annotation Assessment Project. Genome Biol 7(Suppl 1):S21–31

Hall N, Pain A, Berriman M, Churcher C, Harris B, Harris D et al (2002) Sequence of Plasmodium falciparum chromosomes 1, 3-9 and 13. Nature 419(6906):527–31

Harrow J, Denoeud F, Frankish A, Reymond A, Chen CK, Chrast J et al (2006) GENCODE: producing a reference annotation for ENCODE. Genome Biol 7(Suppl 1):S41–9

Hide WA, Babenko VN, van Heusden PA, Seoighe C, Kelso JF (2001) The contribution of exon-skipping events on chromosome 22 to protein coding diversity. Genome Res 11(11):1848–53

Hirotsune S, Yoshida N, Chen A, Garrett L, Sugiyama F, Takahashi S et al (2003) An expressed pseudogene regulates the messenger-RNA stability of its homologous coding gene. Nature 423(6935):91–6

Hubbard T, Barker D, Birney E, Cameron G, Chen Y, Clark L et al (2002) The Ensembl genome database project. Nucleic Acids Res 30(1):38–41

Hubbard TJ, Aken BL, Beal K, Ballester B, Caccamo M, Chen Y et al (2007) Ensembl 2007. Nucleic Acids Res 35(Database issue):D610–7

Huss JW, Orozco C, Goodale J, Wu C, Batalov S, Vickers TJ et al (2008) A Gene Wiki for Community Annotation of Gene Function. PLoS Biol 6(7):e175

Imanishi T, Itoh T, Suzuki Y, O'Donovan C, Fukuchi S, Koyanagi KO et al (2004) Integrative annotation of 21,037 human genes validated by full-length cDNA clones. PLoS Biol 2(6):e162

Kan Z, States D, Gish W (2002) Selecting for functional alternative splices in ESTs. Genome Res 12(12):1837–45

Kapranov P, Cawley SE, Drenkow J, Bekiranov S, Strausberg RL, Fodor SP et al (2002) Large-scale transcriptional activity in chromosomes 21 and 22. Science 296(5569):916–9

Kawai J, Shinagawa A, Shibata K, Yoshino M, Itoh M, Ishii Y et al (2001) Functional annotation of a full-length mouse cDNA collection. Nature 409(6821):685–90

Klee K, Ernst R, Spannagl M, Mayer KF (2007) Apollo2Go: a web service adapter for the Apollo genome viewer to enable distributed genome annotation. BMC Bioinformatics 8:320

Korf I, Flicek P, Duan D, Brent MR (2001) Integrating genomic homology into gene structure prediction. Bioinformatics 17:S140–8

Lewis SE, Searle SM, Harris N, Gibson M, Lyer V, Richter J et al (2002) Apollo: a sequence annotation editor. Genome Biol 3(12):RESEARCH0082

Liang C, Jaiswal P, Hebbard C, Avraham S, Buckler ES, Casstevens T et al (2008) Gramene: a growing plant comparative genomics resource. Nucleic Acids Res 36(Database issue):D947–53

Lopez AJ (1998) Alternative splicing of pre-mRNA: developmental consequences and mechanisms of regulation. Annu Rev Genet 32:279–305

Maeda N, Kasukawa T, Oyama R, Gough J, Frith M, Engstrom PG et al (2006) Transcript annotation in FANTOM3: mouse gene catalog based on physical cDNAs. PLoS Genet 2(4):e62

McCarrey JR, Thomas K (1987) Human testis-specific PGK gene lacks introns and possesses characteristics of a processed gene. Nature 326(6112):501–5

Miller W, Rosenbloom K, Hardison RC, Hou M, Taylor J, Raney B et al (2007) 28-way vertebrate alignment and conservation track in the UCSC Genome Browser. Genome Res 17(12): 1797–808

Misra S, Harris N (2006) Using Apollo to browse and edit genome annotations. Curr Protoc Bioinformatics Chapter 9: Unit 9.5

Mortazavi A, Williams BA, McCue K, Schaeffer L, Wold B (2008) Mapping and quantifying mammalian transcriptomes by RNA-Seq. Nat Methods 5(7):621–8

O'Connor BD, Day A, Cain S, Arnaiz O, Sperling L, Stein LD (2008) GMODWeb: a web framework for the Generic Model Organism Database. Genome Biol 9(6):R102
Okazaki Y, Furuno M, Kasukawa T, Adachi J, Bono H, Kondo S et al (2002) Analysis of the mouse transcriptome based on functional annotation of 60,770 full-length cDNAs. Nature 420(6915):563–73
Prlic A, Down TA, Kulesha E, Finn RD, Kahari A, Hubbard TJ (2007) Integrating sequence and structural biology with DAS. BMC Bioinformatics 8:333
Pruitt KD, Tatusova T, Maglott DR (2007) NCBI reference sequences (RefSeq): a curated non-redundant sequence database of genomes, transcripts and proteins. Nucleic Acids Res 35(Database issue):D61–5
Reese MG, Guigo R (2006) EGASP: introduction. Genome Biol 7(Suppl 1):S11–3
Rogers A, Antoshechkin I, Bieri T, Blasiar D, Bastiani C, Canaran P et al (2008) WormBase 2007. Nucleic Acids Res 36(Database issue):D612–7
Rutherford K, Parkhill J, Crook J, Horsnell T, Rice P, Rajandream MA et al (2000) Artemis: sequence visualization and annotation. Bioinformatics 16(10):944–5
Schadt EE, Edwards SW, GuhaThakurta D, Holder D, Ying L, Svetnik V et al (2004) A comprehensive transcript index of the human genome generated using microarrays and computational approaches. Genome Biol 5(10):R73
Schmucker D, Clemens JC, Shu H, Worby CA, Xiao J, Muda M et al (2000) Drosophila Dscam is an axon guidance receptor exhibiting extraordinary molecular diversity. Cell 101(6):671–84
Searle SM, Gilbert J, Iyer V, Clamp M (2004) The otter annotation system. Genome Res 14(5):963–70
Siepel A, Diekhans M, Brejova B, Langton L, Stevens M, Comstock CL et al (2007) Targeted discovery of novel human exons by comparative genomics. Genome Res 17(12):1763–73
Stamm S, Ben-Ari S, Rafalska I, Tang Y, Zhang Z, Toiber D et al (2005) Function of alternative splicing. Gene 344:1–20
Stark A, Lin MF, Kheradpour P, Pedersen JS, Parts L, Carlson JW et al (2007) Discovery of functional elements in 12 Drosophila genomes using evolutionary signatures. Nature 450(7167):219–32
Stein LD, Mungall C, Shu S, Caudy M, Mangone M, Day A et al (2002) The generic genome browser: a building block for a model organism system database. Genome Res 12(10):1599–610
Sundquist A, Ronaghi M, Tang H, Pevzner P, Batzoglou S (2007) Whole-genome sequencing and assembly with high-throughput, short-read technologies. PLoS ONE 2(5):e484
Swarbreck D, Wilks C, Lamesch P, Berardini TZ, Garcia-Hernandez M, Foerster H et al (2008) The Arabidopsis Information Resource (TAIR): gene structure and function annotation. Nucleic Acids Res 36(Database issue):D1009–14
Taneri B, Snyder B, Novoradovsky A, Gaasterland T (2004) Alternative splicing of mouse transcription factors affects their DNA-binding domain architecture and is tissue specific. Genome Biol 5(10):R75
Thomas PD, Mi H, Lewis S (2007) Ontology annotation: mapping genomic regions to biological function. Curr Opin Chem Biol 11(1):4–11
Twigger S, Lu J, Shimoyama M, Chen D, Pasko D, Long H et al (2002) Rat Genome Database (RGD): mapping disease onto the genome. Nucleic Acids Res 30(1):125–8
Twigger SN, Shimoyama M, Bromberg S, Kwitek AE, Jacob HJ (2007) The Rat Genome Database, update 2007 – easing the path from disease to data and back again. Nucleic Acids Res 35(Database issue):D658–62
Vanin EF (1985) Processed pseudogenes: characteristics and evolution. Annu Rev Genet 19:253–72
Waterston RH, Lindblad-Toh K, Birney E, Rogers J, Abril JF, Agarwal P et al (2002) Initial sequencing and comparative analysis of the mouse genome. Nature 420(6915):520–62
Wilhelm BT, Marguerat S, Watt S, Schubert F, Wood V, Goodhead I et al (2008) Dynamic repertoire of a eukaryotic transcriptome surveyed at single-nucleotide resolution. Nature 453(7199):1239–43
Wilson RJ, Goodman JL, Strelets VB (2008) FlyBase: integration and improvements to query tools. Nucleic Acids Res 36(Database issue):D588–93
Yamasaki C, Murakami K, Fujii Y, Sato Y, Harada E, Takeda J et al (2008) The H-Invitational Database (H-InvDB), a comprehensive annotation resource for human genes and transcripts. Nucleic Acids Res 36(Database issue):D793–9

Yeo G, Holste D, Kreiman G, Burge CB (2004) Variation in alternative splicing across human tissues. Genome Biol 5(10):R74

Zhang Z, Gerstein M (2004) Large-scale analysis of pseudogenes in the human genome. Curr Opin Genet Dev 14(4):328–35

Zheng D, Frankish A, Baertsch R, Kapranov P, Reymond A, Choo SW et al (2007) Pseudogenes in the ENCODE regions: consensus annotation, analysis of transcription, and evolution. Genome Res 17(6):839–51

# Chapter 7
# Regulatory Motif Analysis

**Alan Moses and Saurabh Sinha**

## 7.1 Introduction – Pattern Recognition and Discovery in *cis*-Regulatory Informatics

The first complete genome sequences of eukaryotes revealed that much of the genetic material did not code for protein sequences (Lander et al. 2001; Venter et al. 2001). Although this noncoding DNA was once thought to be "junk" DNA, it is now appreciated that large portions of it are actively conserved over evolution (Waterston et al. 2002; Johnston and Stormo 2003), suggesting that these regions contain important functional elements.

A first hypothesis about the function of this noncoding DNA is that it is involved in the regulation of gene activity. One of the best-understood mechanisms of gene regulation is the modulation of transcriptional initiation by sequence specific DNA binding proteins (or transcription factors). These proteins recognize short sequences in noncoding DNA that fall into families or contain consensus patterns or motifs.

In general, we have little understanding of how the information in noncoding regulatory sequence specifies complex patterns of gene expression. In analogy to the genetic code that translates DNA sequence to amino acids in a protein, researchers have suggested the existence of an unknown "*cis*-regulatory code" that translates DNA sequence to patterns of gene expression (Levine and Davidson 2005).

To specify complex patterns of regulation, genes are often regulated by multiple transcription factors, and the binding sites for these factors are organized into discrete regulatory regions, often called "enhancers" or "*cis*-regulatory modules." These regulatory regions are often found in the proximal 5′ promoter regions, but they may also occur much further upstream, downstream, or in intronic regions.

A. Moses (✉)
Department of Cell & Systems Biology, University of Toronto, 25 Willcocks Street, Toronto, ON, Canada, M5S 3B2
e-mail: alanmoses@utoronto.ca

S. Sinha
Dept. of Computer Sciences , University of Illinois, Urbana-Champaign, 201 N. Goodwin Ave, Urbana, IL 61801
e-mail: sinhas@uiuc.edu

D. Edwards et al. (eds.), *Bioinformatics: Tools and Applications*,
DOI 10.1007/978-0-387-92738-1_7, © Springer Science+Business Media, LLC 2009

It is these regulatory regions that execute the *cis*-regulatory code, and systematic identification of these noncoding DNA regulatory regions and the binding sites within them is of great interest in postgenome era molecular biology; the sheer vastness of the noncoding DNA sequence to be analyzed implies that computational methods will have an important role to play.

### *7.1.1 Two Major Challenges*

The biological questions regarding *cis*-regulatory sequences can be broken into two major parts. The first can be thought of as identifying the patterns or motifs associated with each transcription factor. Given this set of patterns, the next challenge is to identify the specific positions in the noncoding DNA where the transcription factors actually bind in vivo. This is directly analogous to the two steps of a statistical clustering problem; first to identify the clusters and second to assign each datapoint to a cluster. As we shall see, sophisticated statistical methods aim to solve these simultaneously. This distinction is important because the experimental approaches to attack these problems can be quite different so that historically they were distinct problems. Here we will use the terminology that "motifs" or "consensus sequences" refer to the representations of specificity or patterns associated with transcription factors, whereas "instances," "matches," or "regulatory sequences" refer to the specific places in noncoding DNA where transcription factors are predicted or known to bind.

### *7.1.2 Overview of Regulatory Informatics*

This chapter will cover three reasonably well defined types of bioinformatic applications. The first are databases and repositories for organizing, storing, and distributing experimentally identified regulatory sequences and motifs; next are pattern matching site prediction methods that begin with known motifs or patterns and attempt to predict the regulatory sequences in noncoding DNA; and finally are de novo or ab initio motif-finding methods that attempt to discover the motifs (and perhaps matches to them simultaneously). In each section, we provide a table with some examples of software implementations. However, these tables are not intended to be comprehensive, but are rather representative of the work in the area. As regulatory bioinformatics is still rapidly developing, readers should refer to recent reviews to find the latest implementations.

## 7.2 Databases and Repositories for Regulatory Sequences and Motifs

The simplest function of online databases is to store binding sites that have been characterized through biochemical and genetic experiments (Heinemeyer et al. 1998). The technically difficult aspect of these applications is to extract the experimental data from the primary biological literature. Usually this is performed by experts who read large numbers of papers and enter the results into the databases. More recently, computational text mining approaches have also been applied to extract regulatory sequences and information from the literature (Aerts et al. 2008). Several databases of curated motifs are described in Table 7.1.

### 7.2.1 Mathematical/Computational Representations of Motifs

Given a set of experimentally characterized regulatory sequences that are known to be bound by a particular factor, a first task is to identify and summarize the specificity of the transcription factor in a motif or consensus. There are two popular strategies to do this.

**Table 7.1** Databases for storing experimentally identified *cis*-regulatory sequences

| Resource | Types of data | Tools | Notes |
|---|---|---|---|
| Transfac[a] | Classification of transcription factors, experimentally proven binding sites, counts matrices | Many | Available with subscription |
| Jaspar[b] | Matrices | Logos, reverse complements, and more | Freely available, plant and animal matrices only |
| SCPD[c] | Transcription factors, characterized binding sites, counts matrices, consensus sequences | Pattern matching | Freely available, *Saccharomyces cerevisiae* only |
| REDfly[d]/Drosophila DNase I Footprint Database[e] | Transcription factor binding sites and regulatory regions (CRMs) z | Links to genome-wide alignments | Freely available, *Drosophila melanogaster* only |
| ORegAnno[f] | Regulatory regions, Transcription factor binding sites, includes evidence for each record | | Freely available, open source data and web application, integrates information from multiple databases |
| PRODORIC[g] | Transcription factor binding sites, operons, matrices, promoter architecture | Composite patterns, Genome Browser, and more | Freely available, prokaryotes only |

[a](Wingender et al. 1996), [b](Sandelin et al. 2004a), [c](Zhu and Zhang 1999), [d](Gallo et al. 2006), [e](Bergman et al. 2005), [f](Montgomery et al. 2006), [g](Münch et al. 2003)

*Matrix representation:* Here each position of the motif is treated as a multinomial distribution on the residues. This representation of motifs is used in probabilistic methods and implies an infinitely large, continuous space of motifs. Despite this, the matrix representation has several attractive features, discussed in more detail later on in the chapter:

(i) The parameters of the multinomial at each position can be readily estimated using statistical inference methods.
(ii) The multinomial distribution at each position can be used to obtain a measure of "information" contained in each position in the motif (Schneider et al. 1986).
(iii) These multinomials can be transformed into "log-odds" or weight matrices, which are a computationally convenient form to store classifiers (Stormo 2000).
(iv) Experimental and theoretical evidence suggests that this representation is related to the binding energy of the protein-DNA interactions (Berg and von Hippel 1987).

*Consensus representation:* Consensus representations of motifs are more familiar to most biologists and have also been important for computational approaches. A consensus representation of a motif may simply be the most frequent letter at each position in the motif. Alternatively, "degeneracy codes" or mismatches may be used to represent non-optimal matches. The main computational advantages of the consensus representation are:

(i) The space of motifs is discrete, so computational strategies for matching and de novo motif finding are highly efficient, and
(ii) The space of motifs is finite, so computational strategies for de novo motif-finding can aim to search exhaustively.

To illustrate the various representations of motifs, we consider a set of known binding sites (called GATA sites) from the SCPD database.

>YIR032C GATAAG
>YIR032C GGTAAG
>YIR032C GATAAG
>YJL110C GATAAT
>YKR034W GATAGA
>YKR034W GATAAC
>YKR039W GATAAG
>YKR039W GATAAC

The consensus representations for this motif might be GATAAG with one mismatch allowed or GRTARN where *R* represents *A* or *G* and *N* represents any base.

We next derive the maximum likelihood estimate (MLE) for the frequency matrix representation using this example. This example will introduce the notation and terminology that we will use later on in the chapter. We represent the sequence data at each position as a four-dimensional vector, where each dimension corresponds to one of the bases A, C, G, T.

$$\begin{array}{ccccc} & A & C & G & T \\ 1 & 0 & 0 & 8 & 0 \\ 2 & 7 & 0 & 1 & 0 \\ 3 & 0 & 0 & 0 & 8 \\ \vdots & 8 & 0 & 0 & 0 \\ & 7 & 0 & 1 & 0 \\ w & 1 & 2 & 4 & 1 \end{array}$$

This is often referred to as a "counts" matrix and such matrices are provided by many databases.

The likelihood of the data is defined as the probability of the data given the model, i.e.,

$$L(X) = p(X \mid \text{model})$$

where $p(A|B)$ represents that probability of the random variable $A$ conditioned on the random variable $B$. Under the multinomial model for each position, the likelihood of the counts matrix $X$ is the product of the probability of each base, in our case

$$\begin{aligned} L(X) &= p(X \mid \text{motif}) \\ &= f_{1G}^{\ 8} \times f_{2A}^{\ 7} \times f_{2G}^{\ 1} \times f_{3T}^{\ 8} \times f_{4A}^{\ 8} \times f_{5A}^{\ 7} \times f_{5G}^{\ 1} \times f_{6A}^{\ 1} \times f_{6C}^{\ 2} \times f_{6G}^{\ 4} \times f_{6T}^{\ 1} \end{aligned}$$

where $f$ are the parameters of the multinomial at each position in the motif. This can be written compactly as

$$p(X \mid \text{motif}) = \prod_{i=1}^{w} \prod_{b \in ACGT} f_{ib}^{X_{ib}}$$

where $i$ is the position in the motif, and $b$ indexes the bases. To find maximum likelihood estimators (MLEs) of these parameters, we simply need to find the values of the parameters that maximize this function. The simplest strategy to do this is take derivatives with respect to the parameters and set them to zero. However, in this case, as in many probabilistic models, taking derivatives of the products is difficult. To get around this issue, we instead optimize the logarithm of the likelihood, such that the products become sums. Because the logarithm is monotonic, any values of the parameters that maximize the logarithm of the likelihood will also maximize the likelihood. In addition, we note that we will not accept any values of the parameters as the MLEs: we want to enforce the constraint that the probabilities at each position must sum to one, $\sum_b f_{ib} = 1$. Such constraints can be included using Lagrange multipliers. Putting

all this together gives

$$\log[L(X)] = \sum_{i=1}^{w} \sum_{b \in ACGT} X_{ib} \log f_{ib} + \lambda\left(1 - \sum_{b} f_{ib}\right)$$

as the function we wish to maximize, where $\lambda$ is the Lagrange multiplier. We now set the derivatives with respect to each of the frequency parameters to zero.

For example, using the linearity of the derivative and that $\frac{\mathrm{d}}{\mathrm{d}x}\log(x)=\frac{1}{x}$, for the parameter at position $j$, for base $c$, we have

$$\frac{\partial}{\partial f_{jc}}\log[L(X)]=\frac{X_{jc}}{f_{jc}}-\lambda=0$$

Solving this and substituting into the constraint gives $f_{jc}=\frac{X_{jc}}{\lambda}$ and $\lambda=\sum_b X_{jb}$, the total number of observations at position $j$.

Thus, we have the intuitive result that the MLE of the frequency for each base is just the number of times we observed that base ($X_{qc}$) divided by the total number of observed bases at that position. It is important to note that in our example, our estimates of the frequency at position 1 are $f_1=(0,0,1,0)$. This implies that based on our data we conclude that there is no probability of observing "*A*," "*C*," or "*T*" at this position. Given that we have only observed 8 examples of this motif, this seems a somewhat overconfident claim. Therefore, it is common practice to "soften" the MLEs by adding some "fake" or pseudo data to each position in the counts matrix. For example, if we use 1 as the psuedocount, our estimate of the frequencies at the first position becomes $f_1=(1/12,1/12,9/12,1/12)$, and reflects our uncertainty about the estimates. These pseudocounts can be justified statistically using the concept of prior probabilities, which is discussed in detail elsewhere (Durbin et al. 1998).

## 7.3 Identifying Binding Sites Given a Known Motif

Given a matrix or consensus representation of a motif, we now consider the problem of identifying new examples of binding sites.

Given a consensus representation, it is possible to say for each possible sequence of length $w$, whether it is a match to the motif or not. For example, a consensus sequence with a mismatch allowed at any position will match $1+4w$ of the $4^w$ possible sequences of length $w$. For our example of GATAAG with one mismatch, we have $\frac{1+4\times 6}{4^6}=\frac{25}{4096}=0.0061$. This means that 0.6% of 6-mers will match this motif. For the degeneracy code representation, the number of sequences that match is the product of the degeneracies at each position. For GRTARN, this is $\frac{1\times 2\times 1\times 1\times 2\times 4}{4^6}=\frac{16}{4096}=0.0039$. Although this may seem to be a few (99.6% of sequences do not match), in a random genome of 100MB, we expect ~390,000 matches by chance! This is two orders of magnitude greater than the maximal reasonable expectation for the number of GATA sites in a genome. Although real genomes are not random, matches to motifs do occur frequently by chance, swamping the number of matches that are functionally bound in the cell. The so-called "Futility Theorem" (Wasserman and Sandelin 2004) conjectures that the large number of random matches relative to functional binding sites makes identification based on pattern matching futile.

Using the matrix representation of the motif, for any sequence of length $w$, we can follow a number of explicit statistical classification strategies to decide whether a sequence is an example of the binding site. Here we use $X$ to represent a single sequence of length $w$.

One commonly used test statistic to compare two models is the likelihood ratio (not to be confused with a likelihood ratio test). In our case, we compare the likelihood that the sequence of interest, $X$, is drawn from our motif frequency matrix, to the likelihood that $X$ was drawn from a background null distribution. There are many ways to construct such a background distribution; here we consider the simplest, namely, that the background is a single multinomial.

If the sequence we are considering is GATAAG, $X = \begin{matrix} A & 0 & 1 & 0 & 1 & 1 & 0 \\ C & 0 & 0 & 0 & 0 & 0 & 0 \\ G & 1 & 0 & 0 & 0 & 0 & 1 \\ T & 0 & 0 & 1 & 0 & 0 & 0 \end{matrix}$, we can calculate the likelihood of $X$ under the models as we did for the counts matrix above. In the case of the matrix model for the motif ($f$) and a background distribution ($g$), the likelihood ratio is simply

$$S(X) = \log \frac{p(X \mid \text{motif})}{p(X \mid bg)} = \log \frac{\prod_{i=1}^{w} \prod_{b \in ACGT} f_{ib}^{X_{ib}}}{\prod_{i=1}^{w} \prod_{b \in ACGT} g_{b}^{X_{ib}}} = \sum_{i=1}^{w} \sum_{b \in ACGT} X_{ib} \log\left(\frac{f_{ib}}{g_b}\right)$$

Thus, $S(X)$ provides a quantitative measure of how similar a sequence is to the frequency matrix. When $S(X)>0$, $X$ is more similar to the motif model than the background model.

To identify new examples of the motif in a new sequence, a typical strategy is to compute the statistic, $S$, for each overlapping subsequence of length $w$ in the sequence. For computational simplicity, this is often done using a "weight" matrix in which entries are given by $M_{ib} = \log\left(\frac{f_{ib}}{g_b}\right)$, where as above, $i$ indexes the position in the motif, and $b$ indexes the nucleotide bases. To calculate $S$, one simply adds up the entries in this matrix corresponding to the observed bases. In our notation, this can be written simply as the inner product

$$S(X) = M \cdot X$$

For the example above, using $g = (0,3,0,2,0,2,0,3)$ this is

$$S(GATAAG) = \begin{bmatrix} -1.28 & -0.875 & 1.32 & -1.28 \\ 0.799 & -0.875 & -0.182 & -1.28 \\ -1.28 & -0.875 & -0.875 & 0.916 \\ 0.916 & -0.875 & -0.875 & -1.28 \\ 0.799 & -0.875 & -0.182 & -1.28 \\ -0.588 & 0.223 & 0.734 & 0.588 \end{bmatrix} \cdot \begin{bmatrix} 0 & 1 & 0 & 1 & 1 & 0 \\ 0 & 0 & 0 & 0 & 0 & 0 \\ 1 & 0 & 0 & 0 & 0 & 1 \\ 0 & 0 & 1 & 0 & 0 & 0 \end{bmatrix} = 5.485$$

**Table 7.2** Tools for matrix matching

| Tool | Purpose | Notes |
|---|---|---|
| Patser[a] | Matching known matrices to sequences | Calculates *P*-values |
| Delila-genome[b] | Matching known matrices to sequences | Information theory-based scoring |

[a](Hertz and Stormo 1999), [b](Gadiraju et al. 2003)

the maximum possible likelihood ratio score for this matrix. Some examples of implementations of matrix matching are described in Table 7.2.

### 7.3.1 Choosing a Cutoff for the Likelihood Ratio

An important question in using such a classification framework is how high a value of $S$ is needed before we can be confident that we have identified a novel example of the motif. Several approaches to this problem have been proposed. There is a finite set of possible scores to a matrix model, and the maximum and minimum score for each matrix are different. In order to standardize the scores when comparing between multiple matrix models, the likelihood ratio for a particular sequence is often transformed into normalized score that reflects how close it is to the maximum score. For example, the transformation

$$S(X) = \frac{S(X) - S_{\text{MIN}}}{S_{\text{MAX}} - S_{\text{MIN}}}$$

standardizes the scores to fall between zero and one, which can be interpreted intuitively.

We next consider three statistically motivated approaches to standardizing the scores from matrices in order to choose a cutoff. The classical statistical treatment (Staden 1989) of this problem is to treat the background distribution as a null hypothesis and consider the *P*-value or probability of having observed the score $S$ or more under the background distribution. In order to calculate *P*-values, we must add up the probabilities of all sequences $X$ that have a score greater than $S$, which means enumerating ~$4^w$ sequences. However, because the positions in the motif are treated independently, these calculations can be done recursively in computational time ~$4w$. This allows us to calculate, for each value $S$, the *P*-value under the null hypothesis (Fig. 7.1).

It is important to note that the validity of these *P*-values depends on the accuracy of the null model or background distribution. For this reason, it is often preferred to use an "empirical" null distribution in which the *P*-value is computed simply by counting the number of times a score of $S$ or more is achieved in a large sample of "null" sequences comprising genomic sequence not thought to contain real examples of the binding site.

Regardless of the method for obtaining these *P*-values, in a sequence of length $l$, we expect to test $l - w$ subsequences, and therefore can apply a multiple testing

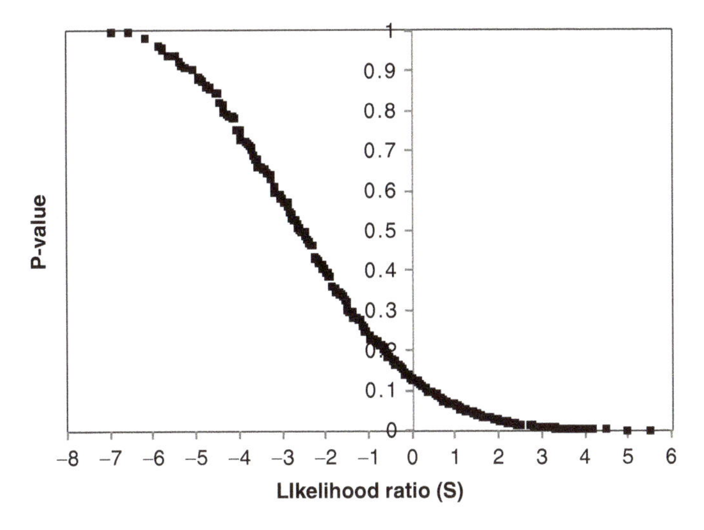


**Fig. 7.1** Exact *P*-values for the likelihood ratio score

correction to these *P*-values to control the expected false positives. For example, if we are to search a promoter of one KB of sequence, and we expect one match, we might choose the cutoff $0.05/1000 = 5\times10^{-5}$, known as the Bonferoni correction. Alternatively, we can express the confidence as an *E*-value (or expect value, which is the *P*-value multiplied by the number of tests) or using a correction for false discovery rate (Benjamini and Hochberg 1995).

A second statistical approach to choosing a threshold for classification is to note that likelihood ratio statistics such as this have the attractive property that when $S>0$, the likelihood of the sequence under the motif model is higher than that under the background model, and under the "maximum likelihood" (ML) rule for classification we should assign the data to the model that has the higher likelihood. However, this rule carries the implicit assumption that our prior expectation is that each subsequence is equally likely to be an example of the motif or the background model. In a real regulatory region, the unknown locations of binding sites might represent a small number of positions amongst thousands of basepairs of background sequences. This "prior" expectation can be incorporated in a maximum a posteriori classification rule (MAP) by using $S > \log\left(\frac{1-\pi}{\pi}\right)$, where $\pi$ is the prior probability of observing the motif.

Finally, using these priors, it is also possible to compute the posterior probability that a given position in the sequence is an example of the motif using Bayes' theorem

$$p(\text{motif} \mid X) = \frac{p(X \mid \text{motif})p(\text{motif})}{p(X \mid \text{motif})p(\text{motif}) + p(X \mid bg)p(bg)} = \frac{1}{1 + \frac{1-\pi}{\pi e^{S(X)}}}$$

This yields a number between 0 and 1 that is intuitively interpretable and can be expressed as a function of the likelihood ratio statistic $S(X)$.

Classification based on the likelihood ratio affords greater control of the false positives, as it allows us to increase the cutoff as the search becomes large, thus reducing the number of spurious matches. However, even the best possible match to the matrix will still occur by chance in about $4^w$ base-pairs. Thus, while the likelihood ratio gives a quantitative measure of how close a putative sequence is to the matrix, it does not address the large number of expected matches in random sequence – matrix matching does not escape the Futility Theorem.

### 7.3.2 Relationship to Information Theory

Given this statistical model of motifs, it is possible to ask for each frequency matrix, how strong a classifier is it. In other words, given that a sequence is a true example of the motif, how easy is it to distinguish from the random background. To measure this quantitatively, we can calculate the average or expectation of $S$ given than the sequences have come from the motif model. This average is over all possible sequences of length $w$. However, as with the $P$-value calculation above, we can use the independence of positions in the motif model to factor this sum

$$E[S(X)\,|\,motif] = \sum_{\text{all } X} S(X)p(X\,|\,\text{motif}) = \sum_{p=1}^{w} \sum_{b \in ACGT} f_{pb} \log\left(\frac{f_{pb}}{g_b}\right) = \text{I}$$

where $E[\ ]$ represents the expectation. Interestingly, this formula can also be obtained using an information theoretic approach (Schneider et al. 1986). If base 2 is used for the logarithms, $I$ is known as the "information content" in the motif and gives a value in bits (Schneider et al. 1986). Several interesting bioinformatic results can be obtained from this information theoretic perspective. For example, in the case of the uniform background distribution, the probability of observing a match to the matrix with score >0 in random sequence is given by $2^{-I}$. Furthermore, the information theoretic perspective yields an intuitive relationship between the presence of binding sites in noncoding sequence and the entropic force of random mutation. Indeed, some early "de novo" motif finding approaches (Stormo and Hartzell 1989) were motivated by the assumption that the motif in a set of binding sites would be the maximally informative motif, and this could be quantified by searching for the patterns with the most information.

The information content of a motif as defined above is also the Kullback–Leibler divergence (Kullback and Leible 1951) between the motif model and the background distribution, and can be shown to be related to the average binding energy of the transcription factor for its target binding sites (Berg and von Hippel 1987). The convergence of the statistical, information theoretic and biophysical treatments of this problem on the formula above is a great achievement in computational biology, and suggests that there are deep connections between the models that have motivated these analyses. As we shall see below, the likelihood ratio, $S(X)$ will have an important role to play in de novo motif finding as well.

## 7.4 Second Generation Regulatory Sequence Prediction Methods: Combinations and Conservation of Motifs to Improve Classification Power

A simple calculation of the $P$-values or information content for an example motif indicates that in a large genome, high-scoring matches to the motif matrix are very likely to appear even under the background null model. This is the motivation of the so-called Futility Theorem: if *bona fide* regulatory elements are rare, searching for them with motifs as described above will yield many false positives and have little power to identify functional examples of binding sites. Two major approaches have been developed to improve predictive power, and we discuss each of these in turn.

### *7.4.1 Exploiting Binding Site Clustering*

The first method is to search for combinations or clusters of transcription factor binding sites (Wasserman and Fickett 1998; Markstein and Levine 2002). Some transcription factors tend to have multiple binding sites in short regions, so as to increase the probability of binding to the DNA in that region. This results in what is sometimes called "homotypic clustering" of binding sites (Lifanov et al. 2003), i.e., an above average density of binding sites of the same factor at a locus. Moreover, transcriptional regulation is known to be combinatorial, i.e., multiple transcription factors often act in concert to regulate the activity of a target gene. Therefore, regulatory sequences may have binding sites for multiple transcription factors, a phenomenon called "heterotypic clustering." From the perspective of pattern recognition, the presence of multiple binding sites improves the signal to noise ratio.

To take advantage of the additional signal, methods (Table 7.3) have been designed to search for regions of the genome that contain multiple closely related binding sites. A simple implementation of this idea is to begin with one or more motifs, predict sites matching each motif using the method described above, and count the number of sites in a sequence window of some fixed length (Berman et al. 2002; Halfon et al. 2002). One would then scan the entire genome for windows with the largest numbers of sites and the predicted binding sites in those windows would be reported.

This simple approach has been shown to empirically add statistical power to regulatory sequence prediction. However, one potential problem with this scheme is its use of ad hoc (and usually high) thresholds on matches to motifs when the matrix representation is used. There are biological examples of regulatory sequences that function by using several weak affinity binding sites rather than one or a few strong sites (Mannervik et al. 1999). Identifying weak sites would require very low thresholds

**Table 7.3** Methods to search for clusters of binding sites

| Tool | Purpose | Notes |
|---|---|---|
| MAST[a] | Identifies matches to motif matrix | Combines *P*-values for multiple motifs |
| *cis*-analyst[b] | Identifies clusters of matrix matches | User defined sliding window and matrix cutoffs |
| Stubb[c] | Identifies clusters of matrix matches | Uses HMM; User defined sliding window |
| Cluster buster[d] | Identifies clusters of matrix matches | Uses HMM; window length automatically learned |

[a](Bailey and Gribskov 1998), [b](Berman et al. 2002), [c](Sinha et al. 2006), [d](Frith et al. 2003)

on $S(X)$ in our computational procedure (Sect. 3), leading to a large number of site predictions, including several false ones. What is needed here is a method that considers both the number and strengths of binding sites in a candidate regulatory sequence: it should accommodate the presence of weak binding sites, but more of these should be required to provide as much confidence as a smaller number of strong sites. Since the strength of binding sites cannot be captured by consensus string models, the following discussion will assume a matrix model of motifs.

One way to allow for the clustering of motifs of different strengths is to score every substring $X$ in a sequence window using the score $S(X)$ described above, and determine the sum of these scores. That is, the sequence window $Y$ is scored by $T(Y)=\sum_{i=1}^{|Y|} S(Y_i)$ where $Y_i$ is the substring at offset $i$ in $Y$. This allows us to assess the extent of homotypic clustering in $Y$, while allowing for strong as well as weak sites, and without imposing any thresholds. This scheme could be extended to work with more than one motif by simply summing over the motifs. Notice however, that adding the different $S(Y_i)$ terms amounts to multiplying probabilities of events, which is questionable since the events (different $Y_i$) are not independent. Another alternative is to use $T(Y)=\sum_{i=1}^{Y} e^{S(Y_i)}$ . This in fact is more justified statistically, as we see next. Consider a probabilistic process (Segal et al. 2003) that generates sequences of length $L_Y=|Y|$ by:

(a) Choosing, uniformly at random a number $i$ between 1 and $(L_Y-w+1)$,
(b) Sampling a site (of length $w$) from the motif,
(c) Planting this site starting at position $i$ (and ending at position $i+w-1$), and
(d) Sampling every other position (outside of $i \ldots i+w-1$) from the background frequency distribution.Denoting the random variable indicating the start position of the planted site (step c) by $i$, we have the joint probability

$$p(Y,i)=\frac{1}{L_Y-w+1}p(Y_i\,|\,\text{motif})\prod_{j\notin\{i\ldots i+w-1\}} g_{Y_j},$$

where $g_x$ is the background probability of base $x$. Summing this over all $i$ to obtain $p(Y)$, and contrasting it with the likelihood under the null model, we get the likelihood ratio as

$$\frac{p(Y \mid \text{motif})}{p(Y \mid bg)} = \frac{1}{L_Y - w + 1} \sum_i \frac{p(Y_i \mid \text{motif})}{p(Y_i \mid bg)},$$

which equals (up to a constant factor) the score $T(Y) = \sum_i e^{S(Y_i)}$ suggested above.

A more comprehensive materialization of this idea is in the form of "Hidden Markov Model" (HMM) based methods. Such methods assume a "generative model" for regulatory sequences, and compute the likelihood of the sequence under the model. The generative model is a stochastic process with states corresponding to motifs for the different transcription factors that are expected to be involved in combinatorial regulation of the genes of interest (Fig. 7.2). The process visits the states probabilistically, and emits a sample of a motif whenever it visits the state corresponding to that motif (Fig. 7.2, *red arrows*). The emitted binding site is appended to the right end of the sequence generated thus far. A "background" state (Fig. 7.2, BKG) allows for these emitted binding sites to be interspersed with randomly chosen non-binding nucleotides. At any point, the process may transition to any state with some fixed probability called the "transition probability" of that state, which is a parameter of the model. (Fig. 7.2, $p_1$, $p_2$, $p_3$, $p_b$). Different implementations take different strategies to choosing values for these parameters. The sequence of states that the process visits is called a "path" of the HMM.

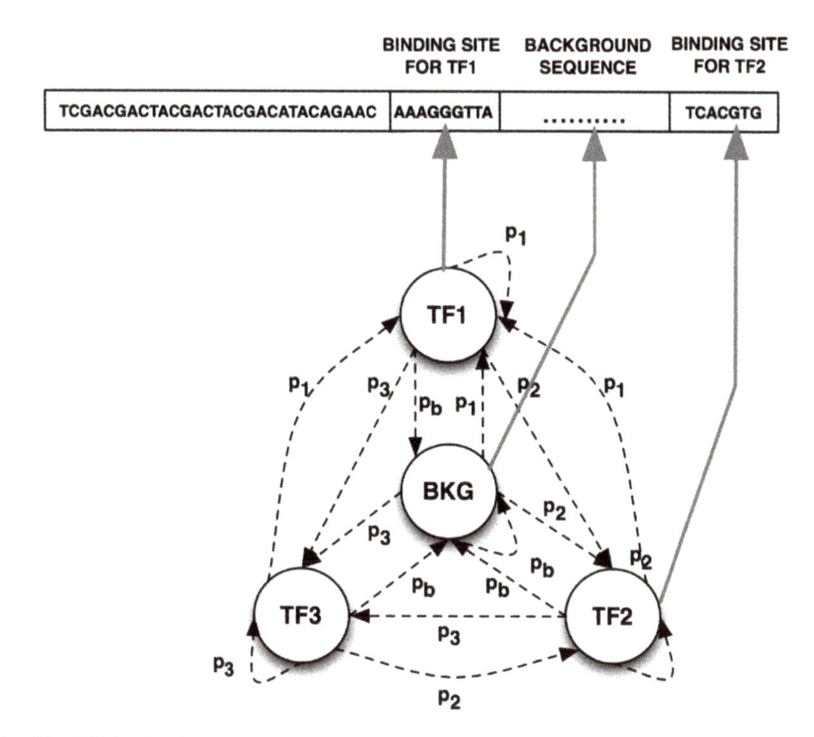


**Fig. 7.2** Hidden Markov Model for CRM discovery

Any given sequence $Y$ may be generated by many exponentially paths, and the joint probability $p(Y, \pi)$ of the sequence $Y$ and a particular path $\pi$ can be computed efficiently. The likelihood of the sequence $Y$ is then computed by summing over all possible paths, i.e., $p(Y \mid \theta) = \sum_{\pi} p(Y, \pi \mid \theta)$, where $\theta$ denotes the parameters of the HMM. This summation can be performed efficiently using the algorithmic technique of "dynamic programming." The score of sequence $Y$ is the log-ratio of this likelihood to the likelihood of $Y$ being generated by a background model (that does not use the motifs), i.e., $T(Y) = \log \frac{p(Y \mid \theta)}{p(Y \mid \theta_b)}$ where $\theta_b$ denotes the parameters of the background model, and $p(Y|\theta_b)$ is the likelihood of this model.

### 7.4.2 Evolutionary Comparisons for Regulatory Sequence Prediction

A second major class of methods to improve the predictive power in the search for regulatory sequences is the incorporation of evolutionary information. The intuition here is that mutations in functional sequences will lead to a fitness defect, and individuals carrying them will be removed from the population by natural selection. Mutations in nonfunctional sequences will have no effect on fitness, and therefore may persist in the population and become fixed through genetic drift. Thus, over long evolutionary time, functional noncoding sequences will show few changes in their sequences, while nonfunctional sequences will evolve rapidly. This is the guiding principle of comparative genomics.

In order to apply comparative methods, the first step is to identify orthologous noncoding DNA sequences. There are many ways to accomplish this. In some cases simply searching closely related genomes for similar sequences can identify the orthologous noncoding regions. More sophisticated approaches include distance or tree-based methods to rule out paralogous regions, as well as considering the homology of nearby coding regions to ensure chromosomal synteny. Once orthologous noncoding sequences have been identified, these must be aligned, preferably using a DNA multiple aligner that performs a global alignment of the shorter sequence.

The technique of identifying evolutionarily conserved sequences in alignments has been called phylogenetic footprinting, to indicate the idea that functional constraint leaves a footprint of conservation in DNA sequences. Simple approaches to phylogenetic footprinting identify regions of alignments of noncoding regions above a certain percentage identity cutoff. Such comparative methods were first combined with matrix matching approaches by requiring that the matches to the matrix fall into "conserved" regions. These approaches have been demonstrated to greatly improve the power of motif matching methods by removing large numbers of false positives.

More elegant statistical approaches to phylogenetic footprinting employ explicit probabilistic models for the evolution of noncoding DNA. Based on the hypothesis that functional sequences will evolve at a slower rate than surrounding regions, methods have been developed that explicitly compare the likelihood of each stretch of sequence under slow or fast evolutionary models. Because the sequences in the multiple alignments have evolved along an evolutionary tree, it is necessary to explicitly account for their phylogenetic relationships using an evolutionary model. This can be done by using a continuous-time Markov process to model the substitutions between the DNA bases, and a phylogenetic tree relating the sequences, in which the bases in the ancestral sequences are treated as unobserved or hidden variables. To compute the likelihood of such a model it is necessary to sum over all possible values of (or marginalize) the hidden variables (Felsenstein 1981).

Like the multinomial models for single sequences described above, probabilistic evolutionary models treat each position in the sequence independently, although rather than single bases at each position, the data are now columns in the multiple alignment. For example, for a pair wise alignment we have a tree with three nodes, the two sequences, $X$ and $Y$ and the unobserved ancestral sequence $A$. The likelihood of the pair wise alignment can be written as

$$L(X,Y) = \prod_{i=1}^{l} p(X_i, Y_i \mid R, T)$$

where the joint probability of the sequences can be written in terms of the unobserved ancestral residue as

$$p(X_i Y_i \mid R,T) = \sum_{A_i \in ACGT} p(X_i Y_i \mid A_i, R, T) p(A_i) = \sum_{A_i \in ACGT} p(X_i \mid A_i, R, T) p(Y_i \mid A_i, R, T) p(A_i)$$

where $R$ represents the transition matrix of the continuous time Markov process, $T$ represents the topology of the evolutionary tree (in this case the three nodes) and $p(A)$ are prior probabilities on the ancestral bases, and are usually assigned to be equal to the equilibrium distribution of the continuous time Markov process. An important feature of this model is that the evolution along each lineage (that leading to $X$ and that leading $Y$) is independent, conditioned on the state of the ancestor $A$. To identify conserved sequences, one can form a likelihood ratio at each position between a "background" evolutionary model, say $R_b$, and a "conserved" evolutionary model where substitutions happen at a slower rate, $R_c$.

$$U(X_i, Y_i) = \frac{p(X_i, Y_i \mid R_c, T)}{p(X_i, Y_i \mid R_b, T)}$$

Extending this approach further, it is possible to posit a "hidden" state determining whether a piece of alignment is drawn from a conserved model or from the background model, and develop an HMM to identify conserved sequences. HMMs emitting columns of alignments rather than individual residues are often referred to as "phylo-HMMs" and are increasingly used in comparative genomics.

Finally, it is possible to combine probabilistic models for sequence evolution for the specificity and evolution of the transcription factor binding sites and the

**Table 7.4** Comparative methods to identify regulatory sequences

| Tool | Purpose | Notes |
|---|---|---|
| ConSite[a] | Identifies conserved matrix matches | Pairwise analysis only |
| VISTA[b] | Identifies conserved regions, matrix matches | Popular graphical display format |
| Footprinter Package[c] | Identifies conserved regions | Uses (i) binomial distribution and (ii) parsimony-based approaches to asses conservation in windows |
| eShadow[d] | Identifies conserved regions | Uses likelihood ratio tests |
| PhastCons[e] | Identifies conserved regions | Uses phyloHMM |
| MONKEY[f] | Identifies conserved matches to matrix models | Probabilistic model of binding site evolution, computes *P*-values |

[a] (Sandelin et al. 2004b), [b] (Dubchak and Ryaboy, 2006), [c] (Blanchette and Tompa 2003), [d] (Ovcharenko et al. 2004), [e] (Siepel et al. 2005), [f] (Moses et al. 2004b)

background sequences. The critical step here is to assign the prior probabilities on the ancestral states to be the frequencies in the motif matrix. Classifiers based on these models can be constructed in much the same way as described above and have achieved much greater predictive power than approaches that match in single sequences.

Table 7.4 lists some of the implementations employing approaches utilizing comparative information.

## 7.5 De novo Motif-Finding

So far, we have assumed that the specificity of transcription factors was known, and the goal was to identify the regulatory regions or binding sites they controlled. However, in many cases, neither the sequence specificity, nor the binding sites of a transcription factor are known. Instead, the challenge is to infer the sequence specificity directly from a set of noncoding DNA sequences believed to contain binding sites. These methods rely on little biological information and are often referred to as "de novo" or "ab initio" because the computational method must identify the new motifs, starting from the beginning.

### *7.5.1 Statistical Overrepresentation of Consensus Sequence Motifs*

The first approach to the ab initio discovery of transcription factor motifs assumes that the motifs are described by their consensus sequences (Sect. 2). There are a few commonly used variants of this motif model. In the simplest model, the motif is a string over the four letter alphabet {A, C, G, T}, and binding sites are required to be exact occurrences of the string (van Helden et al. 1998). In a second variant, the binding sites are allowed to be at most one mismatch (Hamming distance 1) away from the motif sequence (Tompa 1999). A third commonly used model uses "degenerate symbols" such as "*R*" (which stands for "purine," i.e., "*A*" or "*G*") and "*Y*" (for pyrimidine, i.e., "*C*" or "*T*") in the motif alphabet, and binding sites have to be exact matches with any of the allowed nucleotides at the degenerate positions (Sinha and Tompa 2000). Typically, the motif is specified to be of a short ($<10$ bp), fixed length $k$. Each of the above motif models clearly defines a "search space" of all possible motifs; e.g., in the first variant, the search space includes all $4^k$ strings of length $k$. It also lays out a prescription to count a motif's occurrences in any given sequence. Ab initio motif discovery in this framework then amounts to finding the one (or few) motif(s) in the search space that has the greatest statistical significance, as determined by their respective counts in the given set of sequences. We will next see a simple illustration of how such statistical significance may be determined.

Suppose we are given a DNA sequence $S$ of length $L_s$, and a motif $m$. Let $N(S,m)$ denote the count of $m$ in $S$. Next, consider a random sequence $X$, also of length $L_s$, that is generated by sampling one character at a time, as per the probability distribution $\{\pi_a, \pi_c, \pi_g, \pi_t\}$. The count $N(X,m)$ is therefore a random variable defined by the generative process (the "null model"), whose probability distribution will tell us about the statistical significance of $m$. Intuitively, if it is highly unlikely that a count of $N(S,m)$ or greater is observed in a random sequence, then we should interpret the motif $m$ as being statistically overrepresented in $S$. Let us define $p_m$ as the probability that motif $m$ occurs at a specific position in $X$. In the simplest motif model of exact matches, this is given by

$$p_m = \prod_{i=1}^{k} \pi_{m_i}$$

Now, consider each of the positions $j=1$ to $j=L_s-k+1$, where the motif $m$ may occur in the random sequence $X$. The probability of occurrence of $m$ at position $j$ is given by $p_m$, for all $j$. If we further assume that these events are independent, we have $L_s-k+1$ independent and identically distributed ("i.i.d") *Bernoulli* trials with parameter $p_m$. Therefore, the number of occurrences of $m$ in $X$ follows a *Binomial* distribution with parameters $L_s-k+1$ and $p_m$. That is, the $P$-value of the observed count $N(S,m)$ is given by

$$\sum_{n \geq N(S,m)} \binom{L_s - k + 1}{n} p_m^n (1 - p_m)^{L_s - k + 1 - n}$$

This is an estimate of the statistical significance of the motif $m$ in sequence $S$ (van Helden et al. 1998). The smaller the value, the greater the significance.

In the above calculation, we made a crucial assumption that the events of motif $m$ occurring at different positions in a sequence are statistically independent. This is obviously a flawed assumption, since a motif's occurrence at a position $j$ and the next position $j+1$ (overlapping occurrences) are dependent variables: for a self-overlapping motif like "AAAAAA," occurrence at a position $j$ implies a high probability of occurrence at the very next position $j+1$, while for a motif such as

"ACGTTG," occurrence at $j$ and $j+1$ are mutually exclusive events. We therefore turn our attention to a slightly different approach to evaluating statistical significance – through the use of "*z*-scores." We shall see how the flawed independence assumption is avoided in this new approach.

We assume the same null model as above, i.e., the random sequence $X$ is generated by $L_s$ i.i.d. samples from the probability distribution $\{\pi_a, \pi_c, \pi_g, \pi_t\}$. Let $X_{mi}$ be an indicator random variable for the occurrence of motif $m$ at position $i$. That is, this variable takes the value "1" if $m$ occurs at position $i$ in $X$, and "0" otherwise. Let $X_m$ be the count of $m$ in $X$. That is,

$$X_m = \sum_{i=1}^{L} X_{mi} \text{ where } L = L_s - k + 1 \tag{7.1}$$

Note that the observed count $N(S,m)$ is the value of this random variable $X_m$ for the sequence $S$. Let $\mu_m = E(X_m)$ denote the expectation of this random variable, and $\sigma_m$ denote its standard deviation, under the i.i.d null model. Then we define the *z*-score of the motif $m$ as

$$z(S,m) = \frac{N(S,m) - \mu_m}{\sigma_m}$$

This is the number of standard deviations by which the observed count exceeds the expectation. A high value of this statistic indicates statistical significance. Our next task then is to compute the mean and standard deviation of $X_m$.

The expectation follows directly from (7.1). We note that the expectation of a sum is the sum of the expectations (the principle of "linearity of expectation"); hence we have

$$E(X_m) = E\left(\sum X_{mi}\right) = \sum E(X_{mi}) = \sum p(X_{mi} = 1) = Lp_m$$

Here, the third equality comes from the fact that the expectation of an indicator (0/1) variable is simply the probability of it being 1. Also, $p(X_{mi} = 1)$ is equal to $p_m$, as seen above. The standard deviation computation is slightly more complicated, but similar in spirit. Recalling that the variance is given by $\sigma_m^2 = E(X_m^2) - E(X_m)^2$, we need only to calculate $E(X_m^2)$, for which we have

$$E(X_m^2) = E\left(\left(\sum X_{mi}\right)^2\right) = E\left(\sum_{i,j} X_{mi} X_{mj}\right) = E\left(\sum_i X_{mi}^2 + 2\sum_i \sum_{j=i+1}^{i+k-1} X_{mi} X_{mj} + 2\sum_i \sum_{j=i+k}^{L-k+1} X_{mi} X_{mj}\right)$$

$$= E\left(\sum_i X_{mi}^2\right) + 2E\left(\sum_i \sum_{j=i+1}^{i+k-1} X_{mi} X_{mj}\right) + 2E\left(\sum_i \sum_{j=i+k}^{L-k+1} X_{mi} X_{mj}\right)$$

Note that the first term is simply $E\left(\sum_i X_{mi}\right) = E(X_m)$ since $X_{mi}$ is an indicator variable.

The second term is (twice of) the expected number of occurrences, in a sequence of length $L$, of two overlapping sites matching the motif. This may be computed by enumerating all strings of length $2k-1$ or less that have two overlapping occurrences of $m$, and adding their expectations, computed in the same way as $E(X_m)$. This term makes the variance depend on the self-overlapping structure of motif $m$. It is easy to see that among two motifs with the same $p_m$, and hence the same mean, if one has self-overlap and the other does not, the former will have the greater variance in its count. Finally, the third term amounts to (twice of)

$$E\left(\sum_i \sum_{j=i+k}^{L-k+1} X_{mi} X_{mj}\right) = \sum_i \sum_j E(X_{mi} X_{mj}) = \sum_i \sum_j p(X_{mi}=1, X_{mj}=1)$$
$$= \sum_i \sum_j p(X_{mi}=1) p(X_{mj}=1) = \sum_{i=1}^{L-2k+1} p(X_{mi}=1) \sum_{j=i+k}^{L-k+1} p(X_{mj}=1)$$
$$= \sum_{i=1}^{L-2k+1} p(X_{mi}=1)(L-2k-i+2) p_m = \frac{(L-2k+2)(L-2k+1)}{2} p_m^2$$

Here, the third equality follows from the fact that in an i.i.d. generated sequence, nonoverlapping occurrences are independent events. Note that if the null model is not i.i.d., and instead follows a higher order Markov chain (as is often the case), this independence assumption falls through, and other techniques are required to efficiently compute the third term.

The above calculations have been performed under several simplifying assumptions. In practice, the null model is often taken to be a second or third order Markov chain to capture adjacent nucleotide correlations that are present in real genomic sequences. The motif model typically allows for mismatches, so that the random variable $X_m$ must represent counts under that model. Another complication arises from the fact that motif finding is often performed on both strands of the given sequence(s). Counting occurrences of the motif on both strands leads to additional statistical dependencies that must be handled. It is possible to extend the above calculations to account for all these complications in an efficient manner (Sinha and Tompa 2000).

### 7.5.2 *De novo* Motif Finding for the Matrix Representation

The classical probabilistic formulation of the motif finding problem posits that a biological sequence is made up of short subsequences, each of which may be an instance of the motif or drawn from a random background distribution (Table 7.5). The first models used in motif-finding were designed to solve the following problem. Given a set of sequences each containing one example of an unknown motif at an unknown location, find both the motif and the locations. From this perspective motif-finding was related to multiple alignments, such that the unknown position of the binding site was the point at which the sequences could be placed into ungapped multiple alignments.

Here we treat a slightly more general, but intuitively simpler version of this problem, where there is no constraint on the input sequences or the number of

**Table 7.5** De novo motif finding methods

| Tool | Purpose | Notes |
|---|---|---|
| RSAT[a]/YMF[b] | Consensus string based moti f-finder | Word statistics with enumeration of motif space |
| MobyDick[c] | Consensus string based motif-finder | Uses a segmentation algorithm to identify optimal "words" |
| MITRA[d] | Exhaustive consensus with mismatch search | Uses suffix tree |
| Weeder[e] | Exhaustive search with statistical ranking | Best performing algorithm in a systematic comparison (Tompa et al. 2005) |
| Gibbs Motif sampler[f] | Matrix-based de novo motif finder | Original Gibbs sampler |
| MEME[g] | Matrix-based de novo motif finder | Popular EM-based method, includes several models for the distribution of motifs in the input sequences |
| Consensus[h] | Matrix-based de novo motif finder | Information based method |
| NestedMICA[i] | Matrix-based de novo motif finder | Nested sampling method; no need for initial "seeding" step |

[a](van Helden et al. 1998), [b](Sinha and Tompa 2000), [c](Bussemaker et al. 2000), [d](Eskin and Pevzner 2002), [e](Pavesi et al. 2004), [f](Lawrence et al. 1993), [g](Bailey and Elkan, 1994), [h](Stormo and Hartzell 1989), [i](Down and Hubbard 2005)

motifs in each input sequence. This implies a simple, two-component mixture model where each subsequence of length $w$ is drawn from either the motif or background multinomials. In practice, modern de novo motif finders often provide several variations on the assumptions about the distribution of motifs in the input data.

If each subsequence is considered to be independent, the likelihood of the entire sequence under the mixture model can be written as the product of all the subsequences of length $w$,

$$L(X) = \prod_{i=1}^{l-w} p(X_i \mid \text{motif})p(\text{motif}) + p(X_i \mid bg)p(bg)$$

where $i$ indexes the position of the beginning of the subsequence relative to the input sequence $X$. Above, in the case of the multinomial model for a counts matrix, it is possible to maximize the likelihood directly and obtain the parameter estimates. However, it is not possible to obtain closed form solutions for the parameter estimates by directly differentiating the likelihoods of mixture models such as the one proposed above. Two major strategies have been employed for optimization, namely sampling approaches (here we consider Gibbs sampling) and Expectation-Maximization (EM), and we discuss each in turn.

#### 7.5.2.1 Expectation-Maximization

The EM approach views the free parameters of the model as the unknown frequencies of each residue in the multinomial at each position in the motif. This means that for a DNA motif of width $w$, there are $3w$ parameters, or a likelihood surface with $3w$ dimensions. The motif-finding problem is simply the problem of estimation of these parameters by maximizing the likelihood. However, because the positions of the motifs are unknown, the EM approach is to posit the existence of unobserved (or hidden) variables that specify at each position in the input sequence data, whether a particular position is an example of a binding site or not (Lawrence and Reilly 1990).

We represent these hidden variables as a vector at each position, $Z_i=(1,0)$ if the $w$-mer starting at position $i$ is a binding site, and $Z_i=(0,1)$ if it is drawn from the background. To find parameter estimates, we assume that these hidden variables are observed, and then try to follow the maximization procedure above. Given the positions of the binding sites, we could write the "complete" likelihood:

$$L_c(X)=\prod_{i=1}^{l-w}\prod_{m\in \text{motif},bg}[p(X_i\mid m)p(m)]^{z_{im}}$$

We therefore maximize this function as above. Taking logarithms yields,

$$\log[L_c(X)]=\sum_{i=1}^{l-w}\sum_{m\in \text{motif}} z_{im}\left[\left(\sum_{p=i}^{i+w}\sum_{b\in ACGT} X_{pb}\log[p(X_{pb}\mid m)]\right)+\log\pi_m\right]$$

We can now add Lagrange multipliers for the various constraints and differentiate with respect to the parameters as above. For example, for the frequency parameters we include the constraint $\sum_b f_{pb}=1$, and obtain $\sum_{i=1}^{l-w}\frac{Z_{i0}X_{pb}}{f_{pb}}-\lambda=0$, which after rearranging and substituting into the constraint, yields

$$f_{pb}=\frac{\sum_{i=1}^{l-w}Z_{i0}X_{pb}}{\sum_{i=1}^{l-w}Z_{i0}}$$

We now recall that this derivation was done under the assumption that we actually knew the positions of the binding sites in the input, i.e., that $Z_i$ were observed. This is where the expectation step of the EM algorithm arises: we simply replace the $Z_i$ with their expectations, based on our current estimates of the parameters. The expectation of a variable that takes on only 1 or 0 is simply the probability of non-zero outcome, so the expectations of these variables can be calculated using Bayes' theorem as above.

$$Z_{i0}\to E[Z_{i0}]=p(\text{motif}\mid X_i)=\frac{1}{1+\frac{1-\pi}{\pi e^{S(X_i)}}}$$

Thus, the EM algorithm constitutes filling in these "posterior probabilities" of the unknown positions of the binding sites, recomputing the estimates of the parameters based on these, then recomputing the estimates of the hidden variables, etc., until convergence. This iterative strategy is guaranteed to increase the likelihood at each step. Once the parameter estimates have stabilized, we can be confident that we have reached a local maximum in the likelihood. It is important to note, however, that this may not represent the global maximum in the likelihood. Furthermore we have not yet addressed the issue of where to obtain the initial estimates of the parameters to begin the EM procedure. In fact, these issues must be addressed in practice with heuristics based on intuition about the problem.

#### 7.5.2.2 Gibbs Sampling

Sampling approaches posit a Markov chain whose equilibrium distribution is the posterior distribution of interest. This Markov chain starts with initial guess parameters, and then iteratively refines the guess. Once the chain has reached equilibrium, parameter estimates can be obtained by averaging over the states visited by the chain. Needless to say, the key to this procedure is how to define the transition probabilities in such chains; in other words, the rule for refining the guess. In general such approaches are called Markov-chain-Monte-Carlo or MCMC methods and have been considered elsewhere. Here we will focus on a particular type of MCMC algorithm that has been applied effectively to the de novo motif finding problem.

Gibbs Sampling is the procedure of sampling a new parameter estimate (or guess) according to the probability of the new guess conditioned on the current estimates of the remaining parameters. In our case the parameters are regarded as the unknown positions of the binding sites in the input data, and the estimates of the frequency matrix. For motif finding, therefore, at each iteration, one of the current binding site positions is selected at random to be replaced. The frequency matrix is recalculated leaving out the selected binding site and every available position in the input data is re-evaluated by computing the statistic $S(X_i)$ described above at each position.

While the derivation for the exact equations for the Gibbs Sampler is too complicated to reproduce here, it can be shown (Liu et al. 1995) that the probability required for the Gibbs Sampler is given approximately by

$$p(\text{new site at } i \mid X_i) = \frac{S(X_i)}{\sum_j S(X_j)}$$

Thus, a new binding site is then chosen by choosing randomly from the available positions with probability proportional to $S$. Interestingly, the new binding site does not necessarily improve the likelihood: the site at $p$ might have a lower likelihood by chance than the one it was replacing. However, on average this procedure will tend to sample binding sites that are near the current motif. Critically, the "tighter" or more information contained in the motif, the more likely the sampling procedure is to sample "near" it in sequence space. Thus, although the Gibbs Sampler will explore the entire space, it will be strongly biased to sample near maxima in the likelihood; the higher these maxima, the stronger this bias.

## 7.6 Second Generation Motif-Finding Methods

In addition to the computational difficulty of finding the local maxima in a high-dimensional likelihood space, several lines of evidence suggested that there is not enough information in some motifs to identify them in large regions of noncoding DNA of the eukaryotic genomes that are becoming available. The motif-finding approaches described above are general; they do not take into account specific properties of the transcription factor binding site finding problem. Recently, new motif finding methodology has been developed that used similar computational techniques and models, but included additional data about sequence specific transcription factor binding sites in the motif finding.

### *7.6.1 Associations with Functional Genomics Data*

The first methods explicitly designed to identify motifs in noncoding DNA used additional information about which genes were likely to be regulated by a transcription factor. The simplest of these cases is simply where motif finding is done on two sets of sequences, those likely to be regulated and those unlikely to be regulated. This information can be as simple as the functional classification of a gene. Searching for motifs that separate two sets of noncoding regions can be thought of as a statistical discrimination problem, and many statistical methods can be applied.

It is possible to increase the sophistication of such discriminative methods, such that the motifs are taken as explanatory variables for quantitative, possibly multivariate data. For example, genome-wide transcription factor binding data can provide a ranked list of genes that are most likely (and least likely) to be bound by a transcription factor. Motif-finding methods can exploit this information by searching for patterns that are statistically associated with these rankings. Similarly, genome-wide gene expression data can give information about which genes change expression in response to developmental changes or to the environment. If a transcription factor leads to a change in expression of transcripts in a particular condition, the expression of all (or many) of the genes containing the motif are expected to change. Therefore, motifs can be identified based on whether they can explain the variance in genome-wide gene expression data using regression and other statistical methods.

### *7.6.2 Incorporating Comparative Information into de novo Motif Finding*

Another important group of next generation de novo motif-finders are comparative or phylogenetic methods. With the availability of complete genome sequences for closely related organisms, including comparative sequence information in motif finding was a natural extension. The first methods to use comparative information did so using heuristics to encapsulate the notion that motifs should be conserved in alignments of homologous sequences. Soon after, motif-finders that incorporated explicit probabilistic models of evolution into motif finding were developed.

### 7.6.3 *Other Methods and Future Directions*

Another interesting strategy for motif finding is to use prior information about the pattern of information content in real transcription factor binding sites. This essentially attempts to reduce the number of nonbiological maxima in the likelihood function, by biasing the search away from regions of the motif space that are unlikely to represent real biological motifs (Table 7.6). In general, methods that identify additional biological features of motifs in eternal datasets, or in sequence data will be of continued interest in the short term.

**Table 7.6** De novo motif-finders that incorporate additional information

| Tool | Purpose | Notes |
|---|---|---|
| SeedSearch[a] | Consensus-based discriminative approach | Hypergeometric statistics |
| DME[b] | Identifies motifs overrepresented in one set of sequences relative to a background set | Matrix-based, enumerative discriminative approach |
| DRIM[c] | Identifies motifs in a ranked list of sequences | Consensus-based, enumerative hypergeometric statistics with corrections |
| REDUCE[d] | Identifies motifs correlated with gene expression | Uses multiple regression |
| GMEP[e] | Identifyies motifs associated with gene expression data | Uses *z*-scores |
| Footprinter[f] | Identifies conserved motifs in orthologous noncoding sequences | Parsimony based approach |
| Kellis et al.[g] | Identifies conserved motifs in genome-wide alignments | No popular implementation, computes conservation of "mini"-motifs, and combines these into larger motifs |
| EMnEM[h]/PhyME[i] | Identifies conserved motifs in orthologous noncoding sequences | Phylogenetic E–M based approach |
| PhyloGibbs[j] | Identifies conserved motifs in orthologous noncoding sequence | Phylogenetic sampling based approach. Relaxes assumption of complete conservation |
| TFEM[k] | Identifies motifs with particular information content profiles | Adds additional constraints to traditional E–M maximization |

[a](Barash et al. 2001), [b](Smith et al. 2005), [c](Eden et al. 2007), [d](Bussemaker et al. 2001), [e](Chiang et al. 2001), [f](Blanchette and Tompa 2003), [g](Kellis et al. 2004), [h](Moses et al. 2004a), [i](Sinha et al. 2004), [j](Siddharthan et al. 2005), [k](Kechris et al. 2004)

As more diverse data become available, computation systems that combine diverse data types, as well as the pattern recognition methods described in this chapter will become increasingly powerful. Indeed, recent methods have attempted to construct regulatory networks using model-based approaches to synthesize motif finding and analysis of functional genomics data (e.g., Segal et al. 2003). Computational methods will undoubtedly have an exciting role to play as we advance toward the goal of predicting gene expression from sequence (Segal et al. 2008).

## References

Aerts S, Haeussler M, van Vooren S, Griffith OL, Hulpiau P, Jones SJ et al (2008) Text-mining assisted regulatory annotation. Genome Biol 9(2):R31

Bailey TL, Elkan C (1994) Fitting a mixture model by expectation maximization to discover motifs in biopolymers. Proc Int Conf Intell Syst Mol Biol 2:28–36

Bailey TL, Gribskov M (1998) Methods and statistics for combining motif match scores. J Comput Biol 5(2):211–221

Barash Y, Bejerano G, Friedman N (2001) A simple hyper-geometric approach for discovering putative transcription factor binding sites. Proceedings of the first international workshop on algorithms in bioinformatics, Springer

Benjamini Y, Hochberg Y (1995) Controlling the false discovery rate: A practical and powerful approach to multiple testing. J Royal Stat Soc B 57(1):289–300

Berg OG, von Hippel PH (1987) Selection of DNA binding sites by regulatory proteins. Statistical-mechanical theory and application to operators and promoters. J Mol Biol 193(4):723–750

Bergman CM, Carlson JW, Celniker SE (2005) Drosophila DNase I footprint database: A systematic genome annotation of transcription factor binding sites in the fruitfly, Drosophila melanogaster. Bioinformatics 21(8):1747–1749

Berman BP, Nibu Y, Pfeiffer BD, Tomancak P, Celniker SE, Levine M et al (2002) Exploiting transcription factor binding site clustering to identify *cis*-regulatory modules involved in pattern formation in the Drosophila genome. Proc Natl Acad Sci USA 99(2):757–762

Blanchette M, Tompa M (2003) FootPrinter: A program designed for phylogenetic footprinting. Nucleic Acids Res 31(13):3840–3842

Bussemaker HJ, Li H, Siggia ED (2000) Building a dictionary for genomes: Identification of presumptive regulatory sites by statistical analysis. Proc Natl Acad Sci USA 97(18):10096–10100

Bussemaker HJ, Li H, Siggia ED (2001) Regulatory element detection using correlation with expression. Nat Genet 27(2):167–171

Chiang DY, Brown PO, Eisen MB (2001) Visualizing associations between genome sequences and gene expression data using genome-mean expression profiles. Bioinformatics 17(Suppl 1):S49–S55

Down TA, Hubbard TJ (2005) NestedMICA: Sensitive inference of over-represented motifs in nucleic acid sequence. Nucleic Acids Res 33(5):1445–1453

Dubchak I, Ryaboy DV (2006) VISTA family of computational tools for comparative analysis of DNA sequences and whole genomes. Methods Mol Biol 338:69–89

Durbin R, Eddy SR, Krogh A, Mitchison GJ (1998) Biological sequence analysis: Probalistic models of proteins and nucleic acids. Cambridge University Press, Cambridge, UK

Eden E, Lipson D, Yogev S, Yakhini Z (2007) Discovering motifs in ranked lists of DNA sequences. PLoS Comput Biol 3(3):e39

Eskin E, Pevzner PA (2002) Finding composite regulatory patterns in DNA sequences. Bioinformatics 18(Suppl 1):S354–S363

Felsenstein J (1981) Evolutionary trees from DNA sequences: A maximum likelihood approach. J Mol Evol 17(6):368–376

Frith MC, Li MC, Weng Z (2003) Cluster-Buster: Finding dense clusters of motifs in DNA sequences. Nucleic Acids Res 31(13):3666–3668
Gadiraju S, Vyhlidal CA, Leeder JS, Rogan PK (2003) Genome-wide prediction, display and refinement of binding sites with information theory-based models. BMC Bioinformatics 4:38
Gallo SM, Li L, Hu Z, Halfon MS (2006) REDfly: A regulatory element database for Drosophila. Bioinformatics 22(3):381–383
Halfon MS, Grad Y, Church GM, Michelson AM (2002) Computation-based discovery of related transcriptional regulatory modules and motifs using an experimentally validated combinatorial model. Genome Res 12(7):1019–1028
Heinemeyer T, Wingender E, Reuter I, Hermjakob H, Kel AE, Kel OV et al (1998) Databases on transcriptional regulation: TRANSFAC, TRRD and COMPEL. Nucleic Acids Res 26(1): 362–367
Hertz GZ, Stormo GD (1999) Identifying DNA and protein patterns with statistically significant alignments of multiple sequences. Bioinformatics 15(7–8):563–577
Johnston M, Stormo GD (2003) Evolution. Heirlooms in the attic. Science 302(5647):997–999
Kechris KJ, van Zwet E, Bickel PJ, Eisen MB (2004) Detecting DNA regulatory motifs by incorporating positional trends in information content. Genome Biol 5(7):R50
Kellis M, Patterson N, Birren B, Berger B, Lander ES (2004) Methods in comparative genomics: Genome correspondence, gene identification and regulatory motif discovery. J Comput Biol 11(2–3):319–355
Kullback S, Leible RA (1951) On information and sufficiency. Ann Math Stat 22(1):79–86
Lander ES, Linton LM, Birren B, Nusbaum C, Zody MC, Baldwin J et al (2001) Initial sequencing and analysis of the human genome. Nature 409(6822):860–921
Lawrence CE, Altschul SF, Boguski MS, Liu JS, Neuwald AF, Wootton JC (1993) Detecting subtle sequence signals: A Gibbs sampling strategy for multiple alignment. Science 262(5131): 208–214
Lawrence CE, Reilly AA (1990) An expectation maximization (EM) algorithm for the identification and characterization of common sites in unaligned biopolymer sequences. Proteins 7(1):41–51
Levine M, Davidson EH (2005) Gene regulatory networks for development. Proc Natl Acad Sci USA 102(14):4936–4942
Lifanov AP, Makeev VJ, Nazina AG, Papatsenko DA (2003) Homotypic regulatory clusters in Drosophila. Genome Res 13(4):579–588
Liu JS, Neuwald AF, Lawrence CE (1995) Bayesian models for multiple local sequence alignment and Gibbs sampling strategies. J Am Stat Assoc 90(432):1156–1170
Mannervik M, Nibu Y, Zhang H, Levine M (1999) Transcriptional coregulators in development. Science 284(5414):606–609
Markstein M, Levine M (2002) Decoding *cis*-regulatory DNAs in the Drosophila genome. Curr Opin Genet Dev 12(5):601–606
Montgomery SB, Griffith OL, Sleumer MC, Bergman CM, Bilenky M, Pleasance ED et al (2006) ORegAnno: An open access database and curation system for literature-derived promoters, transcription factor binding sites and regulatory variation. Bioinformatics 22(5):637–640
Moses AM, Chiang DY, Eisen MB (2004a) Phylogenetic motif detection by expectation-maximization on evolutionary mixtures. Pac Symp Biocomput:324–335
Moses AM, Chiang DY, Pollard DA, Iyer VN, Eisen MB (2004b) MONKEY: Identifying conserved transcription-factor binding sites in multiple alignments using a binding site-specific evolutionary model. Genome Biol 5(12):R98
Münch R, Hiller K, Barg H, Heldt D, Linz S, Wingender E et al (2003) PRODORIC: Prokaryotic database of gene regulation. Nucleic Acids Res 31(1):266–269
Ovcharenko I, Boffelli D, Loots GG (2004) eShadow: A tool for comparing closely related sequences. Genome Res 14(6):1191–1198
Pavesi G, Mereghetti P, Mauri G, Pesole G (2004) Weeder Web: Discovery of transcription factor binding sites in a set of sequences from co-regulated genes. Nucleic Acids Res 32(Web Server issue):W199–W203

Sandelin A, Alkema W, Engstrom P, Wasserman WW, Lenhard B (2004a) JASPAR: An open-access database for eukaryotic transcription factor binding profiles. Nucleic Acids Res 32(Database issue):D91–D94

Sandelin A, Wasserman WW, Lenhard B (2004b) ConSite: Web-based prediction of regulatory elements using cross-species comparison. Nucleic Acids Res 32(Web Server issue): W249–W252

Schneider TD, Stormo GD, Gold L, Ehrenfeucht A (1986) Information content of binding sites on nucleotide sequences. J Mol Biol 188(3):415–431

Segal E, Raveh-Sadka T, Schroeder M, Unnerstall U, Gaul U (2008) Predicting expression patterns from regulatory sequence in Drosophila segmentation. Nature 451(7178):535–540

Segal E, Yelensky R, Koller D (2003) Genome-wide discovery of transcriptional modules from DNA sequence and gene expression. Bioinformatics 19(Suppl 1):i273–i282

Siddharthan R, Siggia ED, van Nimwegen E (2005) PhyloGibbs: A Gibbs sampling motif finder that incorporates phylogeny. PLoS Comput Biol 1(7):e67

Siepel A, Bejerano G, Pedersen JS, Hinrichs AS, Hou M, Rosenbloom K et al (2005) Evolutionarily conserved elements in vertebrate, insect, worm, and yeast genomes. Genome Res 15(8):1034–1050

Sinha S, Blanchette M, Tompa M (2004) PhyME: A probabilistic algorithm for finding motifs in sets of orthologous sequences. BMC Bioinformatics 5:170

Sinha S, Liang Y, Siggia E (2006) Stubb: A program for discovery and analysis of *cis*-regulatory modules. Nucleic Acids Res 34(Web Server issue):W555–W559

Sinha S, Tompa M (2000) A statistical method for finding transcription factor binding sites. Proc Int Conf Intell Syst Mol Biol 8:344–354

Smith AD, Sumazin P, Zhang MQ (2005) Identifying tissue-selective transcription factor binding sites in vertebrate promoters. Proc Natl Acad Sci USA 102(5):1560–1565

Staden R (1989) Methods for calculating the probabilities of finding patterns in sequences. Comput Appl Biosci 5(2):89–96

Stormo GD (2000) DNA binding sites: Representation and discovery. Bioinformatics 16(1):16–23

Stormo GD, Hartzell GW III (1989) Identifying protein-binding sites from unaligned DNA fragments. Proc Natl Acad Sci USA 86(4):1183–1187

Tompa M (1999) An exact method for finding short motifs in sequences, with application to the ribosome binding site problem. Proc Int Conf Intell Syst Mol Biol:262–271

Tompa M, Li N, Bailey TL, Church GM, De Moor B, Eskin E et al (2005) Assessing computational tools for the discovery of transcription factor binding sites. Nat Biotechnol 23(1):137–144

van Helden J, Andre B, Collado-Vides J (1998) Extracting regulatory sites from the upstream region of yeast genes by computational analysis of oligonucleotide frequencies. J Mol Biol 281(5):827–842

Venter JC, Adams MD, Myers EW, Li PW, Mural RJ, Sutton GG et al (2001) The sequence of the human genome. Science 291(5507):1304–1351

Wasserman WW, Fickett JW (1998) Identification of regulatory regions which confer muscle-specific gene expression. J Mol Biol 278(1):167–181

Wasserman WW, Sandelin A (2004) Applied bioinformatics for the identification of regulatory elements. Nat Rev Genet 5(4):276–287

Waterston RH, Lindblad-Toh K, Birney E, Rogers J, Abril JF, Agarwal P et al (2002) Initial sequencing and comparative analysis of the mouse genome. Nature 420(6915):520–562

Wingender E, Dietze P, Karas H, Knuppel R (1996) TRANSFAC: A database on transcription factors and their DNA binding sites. Nucleic Acids Res 24(1):238–241

Zhu J, Zhang MQ (1999) SCPD: A promoter database of the yeast Saccharomyces cerevisiae. Bioinformatics 15(7–8):607–611

# Chapter 8
# Molecular Marker Discovery and Genetic Map Visualisation

**Chris Duran, David Edwards, and Jacqueline Batley**

## 8.1 Introduction

The bulk of variation at the nucleotide level is often not visible at the phenotypic level. However, this variation can be exploited using molecular genetic marker systems. Molecular genetic markers represent one of the most powerful tools for genome analysis and permit the association of heritable traits with underlying genomic variation. Molecular marker technology has developed rapidly over the last decade, with the development of high-throughput genotyping methods and the availability of large amounts of sequence data for automated marker discovery. Two forms of sequence based marker, Simple Sequence Repeats (SSRs), also known as microsatellites, and Single Nucleotide Polymorphisms (SNPs) are the principal markers currently applied in modern genetic analysis. This are supplemented with anonymous marker systems such as Amplified Fragment Length Polymorphisms (AFLPs; Vos et al. 1995), and Diversity Array Technology (DArT; Jaccoud et al. 2001). The reducing cost of DNA sequencing has led to the availability of large sequence data sets that enable the mining of sequence based markers, such as SSRs and SNPs, which may then be applied to diversity analysis, genetic trait mapping, association studies, and marker assisted selection.

Molecular markers have many uses in genetics, such as the detection of alleles associated with heritable disease, paternity assessment, forensics and the inference of population structure and history (Brumfield et al. 2003; Collins et al. 2004). Modern plant and animal breeding is dependent on molecular markers for trait mapping, marker assisted selection and the rapid and precise analysis of germplasm. Molecular markers can be used to select parental genotypes in breeding programs, select for traits that are difficult to measure using phenotypic assays and eliminate

J. Batley (✉)
Australian Centre for Plant Functional Genomics, ARC Centre of Excellence for Integrative Legume Research, School of Land, Crop and Food Sciences, University of Queensland, Brisbane, QLD, 4072, Australia
e-mail: j.batley@uq.edu.au

D. Edwards et al. (eds.), *Bioinformatics: Tools and Applications*,
DOI 10.1007/978-0-387-92738-1_8, © Springer Science+Business Media, LLC 2009

linkage drag in back-crossing. Furthermore, molecular markers are invaluable as a tool for genome mapping in all systems, offering the potential for generating very high density genetic maps (Rafalski 2002). Genetic linkage maps represent the order of known molecular genetic markers along a given chromosome for a given species, placing molecular genetic markers into linkage groups based on their co-segregation in a population, providing insight into genome structure and organisation. Markers can also be used for comparative mapping to identify similarities and differences between species. In comparative genomics, synteny is the preserved order of genes on chromosomes of related species which results from descent from a common ancestor. An understanding of syntenic relationships enables the transfer of information from one species to another and can assist in the reconstruction of ancestral genomes.

During the past two decades, several molecular marker technologies have been developed and applied for genome analysis. However, due to the relatively high cost associated with marker development, these methods have only been applied to a limited number of species, by a few researchers predominantly in developed countries. Even in these situations, the application of molecular markers has tended to focus on a small number of important diseases or high value traits. The increasing application of association mapping highlights the requirement to be able to identify and screen large numbers of markers, rapidly and at low cost. Bioinformatics systems that improve marker identification help to broaden the uptake of markers to more diverse species and for a greater variety of traits. In this chapter we detail the automated methods for the discovery of SSRs and SNPs and provide an overview of the diverse applications of these markers, with specific emphasis on genetic mapping and data visualisation.

### *8.1.1 SNPs*

SNPs are often considered as the ultimate form of molecular genetic marker, because a SNP represents a single nucleotide difference between two individuals at a defined location. SNPs are also the most abundant form of genetic polymorphism and may therefore provide a high density of markers near a locus of interest. There are three different categories of SNPs: transitions (C/T or G/A), transversions (C/G, A/T, C/A, or T/G) or small insertions/deletions (indels). SNPs are direct markers as the sequence information provides the exact nature of the allelic variants. Furthermore, this sequence variation can have a direct impact on the heritable phenotype. SNPs at any particular site could in principle be bi-, tri- or tetra-allelic, but in practice they are generally biallelic. This disadvantage, when compared with multiallelic markers such as SSRs, is compensated by the relative abundance of SNPs. SNPs are evolutionarily stable, not changing significantly from generation to generation and the low mutation rate of SNPs makes them excellent markers for studying complex genetic traits and as a tool for providing insight into the evolution of genomes (Syvanen 2001).

SNPs are now the dominant marker used in biomedical applications due to the availability of the human genome sequence and knowledge of allelic variation derived from the HapMap project (Altshuler et al. 2005). The ability to screen large numbers of individuals for a range of SNP variants enables the prediction of susceptibility to a wide range of diseases and opens the door to the use of personalised medicine based on the patient's genotype. There are already several companies that specialize in personal genotyping and they can predict susceptibility to a range of diseases. This ability will continue to increase as more and more associations are made between human genetic variation and heritable traits. SNPs are also used routinely in animal and crop breeding programs (Gupta et al. 2001), for genetic diversity analysis, cultivar/breed identification, phylogenetic analysis, characterisation of genetic resources and association of genetic loci with valuable traits (Rafalski 2002). The high density of SNPs makes them valuable for genome mapping, and in particular they allow the generation of ultra-high density genetic maps and haplotyping systems for genes or regions of interest, and map-based positional cloning (Batley and Edwards 2007).

### *8.1.2 SSRs*

SSRs, also known as microsatellites, are short stretches of DNA sequence occurring as tandem repeats of mono-, di-, tri-, tetra-, penta- and hexa-nucleotides. Perfect SSR repeats are without interruptions, imperfect repeats are interrupted by non-repeat nucleotides, while compound repeats are cases where two or more SSRs are found adjacent to one another. Combinations of these are also found, for example imperfect compound repeats (Weber 1990). SSRs have been found in all prokaryotic and eukaryotic genomes analysed to date and they are widely and ubiquitously distributed throughout eukaryotic genomes (Tóth et al. 2000; Katti et al. 2001). SSRs are one of the most powerful genetic markers in biology as they are highly polymorphic. The high level of polymorphism is due to mutation affecting the number of repeat units. SSRs provide several of advantages over some other molecular markers, namely that multiple SSR alleles may be detected at a single locus using a simple PCR based screen, very small quantities of DNA are required for screening, and analysis is amenable to automated allele detection and sizing (Schlötterer 2000). SSRs also demonstrate a high degree of transferability between species, as PCR primers designed to an SSR within one species frequently amplify a corresponding locus in related species. This transferability makes them suitable for genetic diversity and comparative genomic analysis. SSRs are applied to a wide range of applications, including genetic mapping, the molecular tagging of genes, genotype identification, analysis of genetic diversity, phenotype mapping and marker assisted selection (Tautz 1989; Powell et al. 1996).

Studies of the potential biological function and evolutionary relevance of SSRs provides an insight into genome structure and genomics (Subramanian et al. 2003). SSRs were initially considered to be evolutionally neutral (Awadalla and Ritland 1997), however recent evidence suggests they may play an important role in genome evolution and provide hotspots of recombination (Moxon and Wills 1999). Early suggestions that the majority of DNA was 'junk' or had no biological function are being challenged by the discovery of new functions for these sequences and various functional roles have now been attributed to SSRs. SSRs are believed to be involved in gene expression, regulation and function (Kashi et al. 1997; Gupta et al. 1994) and SSRs in non-coding regions may also be of functional significance (Mortimer et al. 2005). SSRs have been found to bind nuclear proteins and there is direct evidence that SSRs can function as transcriptional activating elements (Li et al. 2002).

## 8.2 Computational Molecular Marker Discovery Methods

Traditionally the implementation of SNPs and SSRs has been limited by the initial cost of their development. The discovery of SSR loci previously required the construction of genomic DNA libraries enriched for SSR sequences, followed by DNA sequencing of the clones and analysis of the sequence for the presence of SSRs (Edwards et al. 1996). This was both time consuming and expensive due to the large amount of specific sequencing required. SNP discovery involves finding differences between two sequences and this has traditionally been performed through PCR amplification of genes/genomic regions of interest from multiple individuals selected to represent diversity in the species or population of interest, followed by either direct sequencing of these amplicons, or cloning and sequencing the amplified products. Sequences are then aligned and any polymorphisms identified. This approach is frequently prohibitively expensive and time consuming for the identification and validation of the large number of SNPs required for most applications.

In silico methods of SNP and SSR discovery are now routinely being adopted, providing cheap and efficient marker identification. Large quantities of sequence data have been generated internationally through cDNA or genome sequencing projects and these provide a valuable resource for the mining of molecular markers. This will be further accelerated with the application of new sequencing technology. Sequence data generation is undergoing a revolution with the release of 'next generation' technologies from Roche (454), Illumina (Solexa) and Applied Biosystems (SOLiD) (Table 8.1). These technologies offer the potential to rapidly re-sequence either whole eukaryotic genomes or representative samples of genomes. While the large volume of next generation sequencing data are generally produced at the expense of sequence quality, the over sampling of genome data enables the differentiation between true SNPs and sequence error. In one of the first examples of this application, a total of 36,000 maize SNPs were identified in data from a single run of the Roche 454 GS20 DNA sequencer (Barbazuk et al. 2007). More recently, the complete genome of DNA structure pioneer, James D. Watson was re-sequenced using Roche 454 technology (Wheeler et al. 2008a), while an anonymous African male of the Yoruba people of Ibadan, who participated in the international HapMap project was completely sequenced using Illumina Solexa sequencing technology

**Table 8.1** Comparison of current DNA sequencing technologies

| Sequencing machine | ABI 3730 | Roche GSFLX | Illumina Solexa | AB SOLiD | Helicos HeliScope |
|---|---|---|---|---|---|
| Launched | 2000 | 2007 | 2006 | 2007 | 2008 |
| Read length (bp) | 800–1,100 | 250–400 | 35–50 | 25–35 | 28 |
| Reads per run | 96 | 400 K | 60 M | 85 M | 85 M |
| Throughput per run | 0.1 MB | 100 MB | 3 GB | 3 GB | 2 GB |
| Cost per GB | >$2,500 k | $84 k | $6 k | $5.8 k | ? |

(Bentley et al. 2008). Whole genome sequencing is the most robust method to identify the great variety of genetic diversity in a population and gain a greater understanding of the relationship between the inherited genome and observed heritable traits. The continued rapid advances in genome sequencing technology will lead to whole genome sequencing becoming the standard method for genetic polymorphism discovery. To date, there are over 1,700 prokaryote genome sequencing projects and over 340 eukaryote genome sequencing projects. These numbers are set to increase rapidly with the expansion of next generation sequencing technology and this data will be used for rapid, inexpensive molecular marker discovery. For a comprehensive list of the genome sequencing initiatives, see http://www.ncbi.nlm.nih.gov/genomes/static/gpstat.html.

### 8.2.1 *In Silico SNP Discovery*

The dramatic increase in the number of DNA sequences submitted to databases makes the electronic mining of SNPs possible without the need for additional specific allele sequencing. The identification of sequence polymorphisms in assembled sequence data is relatively simple; however the challenge of in silico SNP discovery is not SNP identification, but rather the ability to distinguish real polymorphisms from the often more abundant sequencing errors. High throughput sequencing remains prone to inaccuracies, Sanger sequencing produces errors as frequent as one in every one hundred base pairs, whilst some of the next generation technologies are even less accurate with errors as frequent as one in every 25 bp. These errors impede the electronic filtering of sequence data to identify potentially biologically relevant polymorphisms and several sources of sequence error need to be addressed during in silico SNP identification. The most abundant error in Sanger sequencing is incorrect base calling, towards the end of the sequence as the quality of the raw data declines. These errors are usually identified by the relatively low quality scores for these nucleotides. Further errors are due to the intrinsically high error rate of the reverse transcription and PCR processes used for the generation of cDNA libraries, and these errors are not reflected by poor sequence quality scores. A number of methods used to identify SNPs in aligned sequence data rely on sequence trace file analysis to filter out sequence errors by their dubious trace quality (Kwok et al.

1994, Marth et al. 1999, Garg et al. 1999). The major disadvantage to this approach is that the sequence trace files required are rarely available for large sequence datasets collated from a variety of sources. In cases where trace files are unavailable, two complementary approaches have been adopted to differentiate between sequence errors and true polymorphisms: (1) assessing redundancy of the polymorphism in an alignment, and (2) assessing co-segregation of SNPs to define a haplotype. These methods are employed in the following applications for in silico SNP identification (Table 8.2).

#### 8.2.1.1 SNP Discovery from Trace files

Phred is the most widely adopted software used to call bases from Sanger chromatogram data (Ewing and Green 1998; Ewing et al. 1998). The primary benefit of this software is that it provides a statistical estimate of the accuracy of calling each base, and therefore provides a primary level of confidence that a sequence difference represents true genetic variation. There are several software packages that take advantage of this feature to estimate the confidence of sequence polymorphisms within alignments.

PolyBayes

PolyBayes (Marth et al. 1999) uses a Bayesian-statistical model to find differences within assembled sequences based on the depth of coverage, the base quality values and the expected rate of polymorphic sites in the region. Base quality values are obtained by running the sequence trace files through the phred base-calling program (Ewing and Green 1998; Ewing et al. 1998), and repeats can be removed from sequences using RepeatMasker (Mallon and Strivens 1998). The output is viewed

**Table 8.2** Applications for in silico SNP and SSR discovery

| Tool | URL | Reference |
|---|---|---|
| PolyBayes | http://bioinformatics.bc.edu/marthlab/PolyBayes | Marth et al. (1999) |
| PolyPhred | http://droog.mbt.washington.edu/ | Nickerson et al. (1997) |
| SNPDetector | http://lpg.nci.nih.gov/ | Zhang et al. (2005) |
| NovoSNP | http://www.molgen.ua.ac.be/bioinfo/novosnp/ | Weckx et al. (2005) |
| AutoSNP | http://acpfg.imb.uq.edu.au | Barker et al. (2003) |
| MISA | http://pgrc.ipk-gatersleben.de/misa/ | Thiel et al. (2003) |
| SSRIT | http://www.gramene.org/db/searches/ssrtool | Temnykh et al. (2001) |
| RepeatFinder | http://www.cbcb.umd.edu/software/RepeatFinder/ | Volfovsky et al. (2001) |
| SPUTNIK | http://espressosoftware.com/pages/sputnik.jsp<br>http://cbi.labri.fr/outils/Pise/sputnik.html | Unpublished |
| TROLL | http://wsmartins.net/webtroll/troll.html | Castelo et al. (2002) |
| TRF | http://tandem.bu.edu/trf/trf.html | Benson (1999) |
| SSRPrimer | http://hornbill.cspp.latrobe.edu.au<br>http://acpfg.imb.uq.edu.au | Robinson et al. (2004); Jewell et al. (2006) |
| SSRPoly | http://acpfg.imb.uq.edu.au/ssrpoly.php | Unpublished |

through the Consed alignment viewer (Gordon et al. 1998). Recent examples of studies using PolyBayes include SNP discovery for white spruce (Pavy et al. 2006) and bird species (Sironi et al. 2006).

PolyPhred

PolyPhred (Nickerson et al. 1997) compares sequence trace files from different individuals to identify heterozygous sites. The sequence trace files are used to identify SNPs and can identify positions in the sequence where double peaks occur that are half the height of the adjacent peaks within a window. The quality of a SNP is assigned based on the spacing between peaks; the relative size of called and uncalled peaks; and the dip between peaks. PolyPhred only analyses nucleotides that have a minimum quality as determined by Phred (Ewing and Green 1998; Ewing et al. 1998). It runs on Unix, and provides output that can be viewed in Consed (Gordon et al. 1998). Recent examples of the use of PolyPhred include studies in cattle (Lee et al. 2006) and humans that have had liver transplants (Wang et al. 2007).

SNPDetector

SNPDetector (Zhang et al. 2005) uses Phred (Ewing and Green 1998; Ewing et al. 1998) to call bases and determine quality scores from trace files, and then aligns reads to a reference sequence using a Smith–Waterman algorithm. SNPs are identified where there is a sequence difference and the flanking sequence is of high quality. SNPDetector has been used to find SNPs in 454 data (Barbazuk et al. 2007) and has been included within a comprehensive SNP discovery pipeline (Matukumalli et al. 2006).

NovoSNP

NovoSNP (Weckx et al. 2005) requires both trace files and a reference sequence as input. The trace files are base-called using Phred (Ewing and Green 1998; Ewing et al. 1998) and quality clipped, then aligned to a reference sequence using BLAST (Altschul et al. 1990). A SNP confidence score is calculated for each predicted SNP. NovoSNP is written in Tcl with a graphical user interface written in Tk and runs on Linux and Windows. NovoSNP has been used in a study of genotype–phenotype correlation for human disease (Dierick et al. 2008).

#### 8.2.1.2 SNP Discovery using Redundancy Approaches

The frequency of occurrence of a polymorphism at a particular locus provides a measure of confidence in the SNP representing a true polymorphism, and is referred to as the SNP redundancy score. By examining SNPs that have a redundancy score equal than or greater than two (two or more of the aligned sequences represent the polymorphism), the vast majority of sequencing errors are removed. Although some true genetic variation is also ignored due to its presence only once within an alignment, the redundancy within the data permits the rapid identification of large numbers of SNPs without the requirement of sequence trace files. However, while redundancy based methods for SNP discovery are highly efficient, the non-random nature of sequence error may lead to certain sequence errors being repeated between runs around locations of complex DNA structure. Therefore, errors at these loci would have a relatively high SNP redundancy score and appear as confident SNPs. In order to eliminate this source of error, an additional independent SNP confidence measure is required. This can be obtained by measuring the co-segregation of SNPs defining a haplotype. True SNPs that represent divergence between homologous genes co-segregate to define a conserved haplotype, whereas sequence errors do not co-segregate with a haplotype. Thus, a co-segregation score, based on whether a SNP position contributes to defining a haplotype is a further independent measure of SNP confidence. By using the SNP score and co-segregation score together, true SNPs may be identified with reasonable confidence. Three tools currently apply the methods of redundancy and haplotype co-segregation; autoSNP (Barker et al. 2003; Batley et al. 2003), SNPServer (Savage et al. 2005) and auto SNPdb (Duran et al. 2009).

### AutoSNP

The autoSNP method (Barker et al. 2003) assembles sequences using CAP3 (Huang and Madan 1999) with the option of pre-clustering with either d2cluster (Burke et al. 1999) or TGICL (Pertea et al. 2003). Redundancy is the principle means of differentiating between sequence errors and real SNPs. While this approach ignores potential SNPs that are poorly represented in the sequence data, it offers the advantage that trace files are not required and sequences may be used directly from GenBank. AutoSNP is therefore applicable to any species for which sequence data is available. A co-segregation score is calculated based on whether multiple SNPs define a haplotype, and this is used as a second, independent measure of confidence. AutoSNP is written in Perl and is run from the Linux command line with a FASTA file of sequences as input. The output is presented as linked HTML with the index page presenting a summary of the results. AutoSNP has been applied to several species including maize (Batley et al. 2003), peach (Lazzari et al. 2005) and cattle (Corva et al. 2007). The recently developed AutoSNPdb (Duran et al. 2009) combines the SNP discovery pipeline of autoSNP with a relational database, hosting information on the polymorphisms, cultivars and gene annotations, to enable efficient mining and interrogation of the data. Users may search for SNPs within genes with specific annotation or for SNPs between defined cultivars. AutoSNPdb can integrate both Sanger and Roche 454 pyrosequencing data enabling efficient SNP discovery from next generation sequencing technologies.

SNPServer

SNPServer (Savage et al. 2005) is a real time implementation of the autoSNP method, accessed via a web server. A single FASTA sequence is pasted into the interface and similar sequences are retrieved from a nucleotide sequence database using BLAST (Altschul et al. 1990). The input sequence and matching sequences are assembled using CAP3, and SNPs are discovered using the autoSNP method. The results are presented as HTML. Alternatively, a list of FASTA sequences may be input for assembly or a preassembled ACE format file may be analysed. SNPServer has been used in studies including sea anemone (Sullivan et al. 2008) and human (Pumpernik et al. 2008). SNPServer has an advantage in being the only real time web based tool that allows users to rapidly identify novel SNPs in sequences of interest.

#### 8.2.1.3 SNP Discovery from Short Read Next Generation Sequence Data

The increased production of next or second generation sequence data provides an additional source of valuable SNP information. While the relatively long reads produced by the Roche 454 sequencers can be analyzed by currently available SNP discovery systems, the high error rates and short reads produced by the AB SOLiD and Illumina Solexa GAII require novel approaches and high sequence redundancy for efficient SNP discovery.

Mosaik

Mosaik has been developed by Michael Strömberg in the laboratory of Gabor Marth and currently accepts reads of all lengths from short read Illumina data to legacy Sanger sequences. It is written in C++ and is available from http://bioinformatics.bc.edu/marthlab/Mosaik.

MAQ

Maq (Mapping and Assembly with Quality) builds an assembly by mapping short reads produced by the Illumina Solexa platform to reference sequences. It has been produced by Heng Li and is available from http://maq.sourceforge.net. Preliminary functions are being developed to also handle AB SOLiD data

### 8.2.2 *In Silico SSR Discovery*

The availability of large quantities of sequence data makes it economical and efficient to use computational tools to mine this for SSRs. Flanking DNA sequence may then be used to design suitable forward and reverse PCR primers to assay the

SSR loci. Furthermore, when SSRs are derived from ESTs, they become gene specific and represent functional molecular markers. These features make EST–SSRs highly valuable markers for the construction and comparison of genetic maps and the association of markers with heritable traits. Several computational tools are available for the identification of SSRs in sequence data as well as for the design of PCR amplification primers. Due to redundancy in EST sequence data, and with datasets often being derived from several distinct individuals, it is now also possible to predict the polymorphism of SSRs in silico. A selection of SSR discovery tools are described below (Table 8.2).

#### 8.2.2.1 MISA

The MIcroSAtellite (MISA) tool (http://pgrc.ipk-gatersleben.de/misa/) identifies perfect, compound and interrupted SSRs. It requires a set of sequences in FASTA format and a parameter file that defines unit size and minimum repeat number of each SSR. The output includes a file containing the table of repeats found, and a summary file. MISA can also design PCR amplification primers on either side of the SSR. The tool is written in Perl and is therefore platform independent, but it requires an installation of Primer3 for the primer search (Thiel et al. 2003). MISA has been applied for SSR identification in moss (von Stackelberg et al. 2006) and coffee (Aggarwal et al. 2007).

#### 8.2.2.2 SSRIT

The tool SSRIT (Simple Sequence Repeat Identification Tool) (http://www.gramene.org/db/searches/ssrtool) uses Perl regular expressions to find perfect SSR repeats within a sequence. It can detect repeats between 2 and 10 bases in length, but eliminates mononucleotide repeats. The output is a file of SSRs in tabular format. A web based version is available that will take a single sequence, and a stand alone version is also available for download. SSRIT has been applied to rice (Temnykh et al. 2001).

#### 8.2.2.3 RepeatFinder

RepeatFinder (Volfovsky et al. 2001) (http://www.cbcb.umd.edu/software/RepeatFinder/) finds SSRs in four steps. The first step is to find all exact repeats using RepeatMatch or REPuter. The second step merges repeats together into repeat classes, for example repeats that overlap. Step three merges all of the other repeats that match those already merged, into the same classes. Finally, step four matches all repeats and classes against each other in a non-exact manner using

BLAST. The input is a genome or set of sequences, and the output is a file containing the repeat classes and number of merged repeats found in each class. RepeatFinder finds perfect, imperfect and compound repeats, and was not designed specifically to find SSRs so can find repeats of any length. It runs on Unix or Linux and has been used to identify SSRs in peanut (Jayashree et al. 2005).

#### 8.2.2.4 Sputnik

Sputnik is a commonly used SSR finder as it is fast, efficient and simple to use. It uses a recursive algorithm to search for repeats with length between 2 and 5, and finds perfect, compound and imperfect repeats. It requires sequences in FASTA format and uses a scoring system to call each SSR. The output is a file of SSRs in tabular format. Unix, Linux and windows versions of sputnik are available from http://espressosoftware.com/pages/sputnik.jsp and http://cbi.labri.fr/outils/Pise/sputnik.html (PISE enabled version). Sputnik has been applied for SSR identification in many species including Arabidopsis and barley (Cardle et al. 2000)

#### 8.2.2.5 TROLL

The SSR identification tool Tandem Repeat Occurrence Locator (TROLL) (Castelo et al. 2002) (http://wsmartins.net/webtroll/troll.html) draws a keyword tree and matches it with a technique adapted from bibliographic searches, based on the *Aho-Corasick* algorithm. It has drawbacks in that it doesn't handle very large sequences and cannot process large batches of sequences as the tree takes up large amounts of memory.

#### 8.2.2.6 Tandem Repeats Finder

Tandem Repeats Finder (TRF) (Benson, 1999) (http://tandem.bu.edu/trf/trf.html) can find very large SSR repeats, up to a length of 2,000 bp. It uses a set of statistical tests for reporting SSRs, which are based on four distributions of the pattern length, the matching probability, the indel probability and the tuple size. TRF finds perfect, imperfect and compound SSRs, and is available for Linux. TRF has been used for SSR identification in Chinese shrimp (Gao and Kong 2005) and cowpea (Chen et al. 2007).

#### 8.2.2.7 Compound Methods for SSR Discovery

The following computational SSR finders combine previous methods to produce extended output.

SSRPrimer

SSRPrimer (Robinson et al. 2004; Jewell et al. 2006) combines Sputnik and the PCR primer design software Primer3 to find SSRs and associated amplification primers. The scripts take multiple sequences in FASTA format as input and produce lists of SSRs and associated PCR primers in tabular format. This web-based tool is also available as a stand alone version for very large datasets. SSRPrimer has been applied to a wide range of species including shrimp (Perez et al. 2005), citrus (Chen et al. 2006), mint (Lindqvist et al. 2006), strawberry (Keniry et al. 2006), *Brassica* (Batley et al. 2007; Burgess et al. 2006; Hopkins et al. 2007; Ling et al. 2007), *Sclerotinia* (Winton et al. 2007) and *Eragrostis curvula* (Cervigni et al. 2008).

SSRPoly

SSRPoly (http://acpfg.imb.uq.edu.au/ssrpoly.php) is currently the only tool which is capable of identifying polymorphic SSRs from DNA sequence data. The input is a file of FASTA format sequences. SSRPoly includes a set of Perl scripts and MySQL tables that can be implemented on UNIX, Linux and Windows platforms.

## 8.3 Data Storage

Large-scale discovery projects are uncovering vast quantities of marker data. As the data size increases, the storage and logical organisation of the information becomes an important challenge. Marker databases vary between centralised repositories that integrate a variety of data for several species, to small specialised databases designed for very specific purposes. The larger repositories tend to lack detailed analytic tools, while the smaller systems may include further species specific data integration. dbSNP is becoming the default repository for SNP data, and there are a wide variety of additional marker databases specific to particular species. The most commonly used marker databases are detailed below (Table 8.3).

### *8.3.1 dbSNP*

The Single Nucleotide Polymorphism database, dbSNP (http://www.ncbi.nlm.nih.gov/projects/SNP/snp_summary.cgi), was developed by NCBI to provide a public-domain repository for simple genetic polymorphisms (Smigielski et al. 2000; Sherry et al. 1999). Although dbSNP includes data on markers such as SSRs and insertion/deletion polymorphisms, SNPs are the primary data type, comprising 97.8% of the database (Smigielski et al. 2000). Table 8.4 presents a summary of

**Table 8.3** Details of commonly used marker storage databases

| Database | URL | Reference |
|---|---|---|
| dbSNP | http://www.ncbi.nlm.nih.gov/projects/SNP/snp_summary.cgi | Smigielski et al. (2000); Sherry et al. (1999) |
| HapMap | http://www.hapmap.org/ | Stein et al. (2002) |
| IBISS | http://www.livestockgenomics.csiro.au/ibiss/ | Hawken et al. (2004) |
| MPD SNP Tools | http://www.jax.org/phenome | |
| Gramene | http://www.gramene.org/ | Jaiswal et al. (2006); Ware et al. (2002a, b) |
| GrainGenes | http://www.graingenes.org/ | Carollo et al. (2005); Matthews et al. (2003) |
| TAIR | http://www.arabidopsis.org/ | Weems et al. (2004); Rhee et al. (2003); Huala et al. (2001) |
| MaizeGDB | http://www.maizegdb.org/ | Lawrence et al. (2004) |
| AutoSNPdb | http://acpfg.imb.uq.edu.au/ | Duran et al. (2009) |

**Table 8.4** Species represented in dbSNP with greater then 1,000 submitted genetic polymorphisms

| Species | Submissions | RefSNP clusters (validated) |
|---|---|---|
| *Homo sapiens* | 55,949,131 | 14,708,752 (6,573,789) |
| *Mus musculus* | 18,645,060 | 14,380,528 (6,447,366) |
| *Gallus gallus* | 3,641,959 | 3,293,383 (3,280,002) |
| *Oryza sativa* | 5,872,081 | 5,418,373 (22,057) |
| *Canis familiaris* | 3,526,996 | 3,301,322 (217,525) |
| *Pan troglodytes* | 1,544,900 | 1,543,208 (112654) |
| *Bos taurus* | 2,233,086 | 2,223,033 (14,371) |
| *Monodelphis domestica* | 1,196,103 | 1,194,131 (0) |
| *Anopheles gambiae* | 1,368,906 | 1,131,534 (0) |
| *Apis mellifera* | 1,118,192 | 1,117,049 (16) |
| *Danio rerio* | 700,855 | 662,322 (3,091) |
| *Felis catus* | 327,037 | 327,037 (0) |
| *Plasmodium falciparum* | 185,071 | 185,071 (47) |
| *Rattus norvegicus* | 47,711 | 43,628 (1,605) |
| *Saccharum hybrid cultivar* | 42,853 | 42,853 (0) |
| *Sus scrofa* | 8,472 | 8,427 (24) |
| *Ovis aries* | 4,247 | 4,181 (66) |
| *Bos indicus* × *Bos taurus* | 2,427 | 2,484 (42) |
| *Caenorhabditis elegans* | 1,065 | 1,065 (0) |
| *Pinus pinaster* | 1,439 | 32 (0) |

species represented. Access is provided via a web interface and there are several ways to query the database. Users can search using a known SNP id or use BLAST to compare a known sequence with sequences in the database. Alternatively, dbSNP can be queried using Entrez or Locuslink. dbSNP currently hosts over 52 million refSNP clusters for 44 organisms. Of these clusters, around 16 million (30%) have been validated.

### 8.3.2 HapMap

The HapMap Consortium collates and catalogues information on human genetic polymorphisms (Gibbs et al. 2003). There are two primary methods to access the data: GBrowse (Stein et al. 2002) and Bio-Mart (Kasprzyk et al. 2004), and both methods are tailored to specific types of users. GBrowse is a genome browser and is a component of the GMOD project (http://www.gmod.org). Using the GBrowse feature of HapMap, users may browse a region of the genome or search with a specific SNP id (Fig. 8.1). Clicking the SNP location in the GBrowse viewer opens an information page, providing full details of the SNP locus. The HapMap project maintains over 3.1 million characterised human SNPs which have been genotyped in a geographically diverse selection of 270 individuals (Frazer et al. 2007).

### 8.3.3 IBISS

The Interactive Bovine in silico SNP Database (IBISS) has been created by the Commonwealth Scientific and Industrial Research Organisation of Australia (CSIRO). It is a collection of 523 448 Bovine SNPs identified from 324 031 ESTs using a custom analysis pipeline (Hawken et al. 2004). The database can be searched by keyword, accession id or by BLAST comparison with an entry sequence. Users can also browse for markers using a linked genome browser.

### 8.3.4 MPD SNP Tools

The Jackson Laboratory's Mouse Phenome Database (www.jax.org/phenome) aims to facilitate the research of human health issues through mouse models. As well as a wealth of trait information on mice, MPD also hosts a collection of over 10 million mouse SNPs (http://www.jax.org/phenome/snp.html).

### 8.3.5 Gramene

Gramene is an online comparative mapping database for rice and related grass species (Jaiswal et al. 2006, Ware et al. 2002a, b). Gramene contains information on cereal genomic and EST sequences, genetic maps, relationships between maps, details of rice mutants, and molecular genetic markers. The database uses the sequenced rice genome as its reference and annotates this genome with various data types. As well as the genome browser, Gramene also incorporates a version of the comparative map viewer, CMap. This allows users to view genetic maps and comparative genetic mapping information and provides a link between markers on genetic maps and the sequenced genome information.

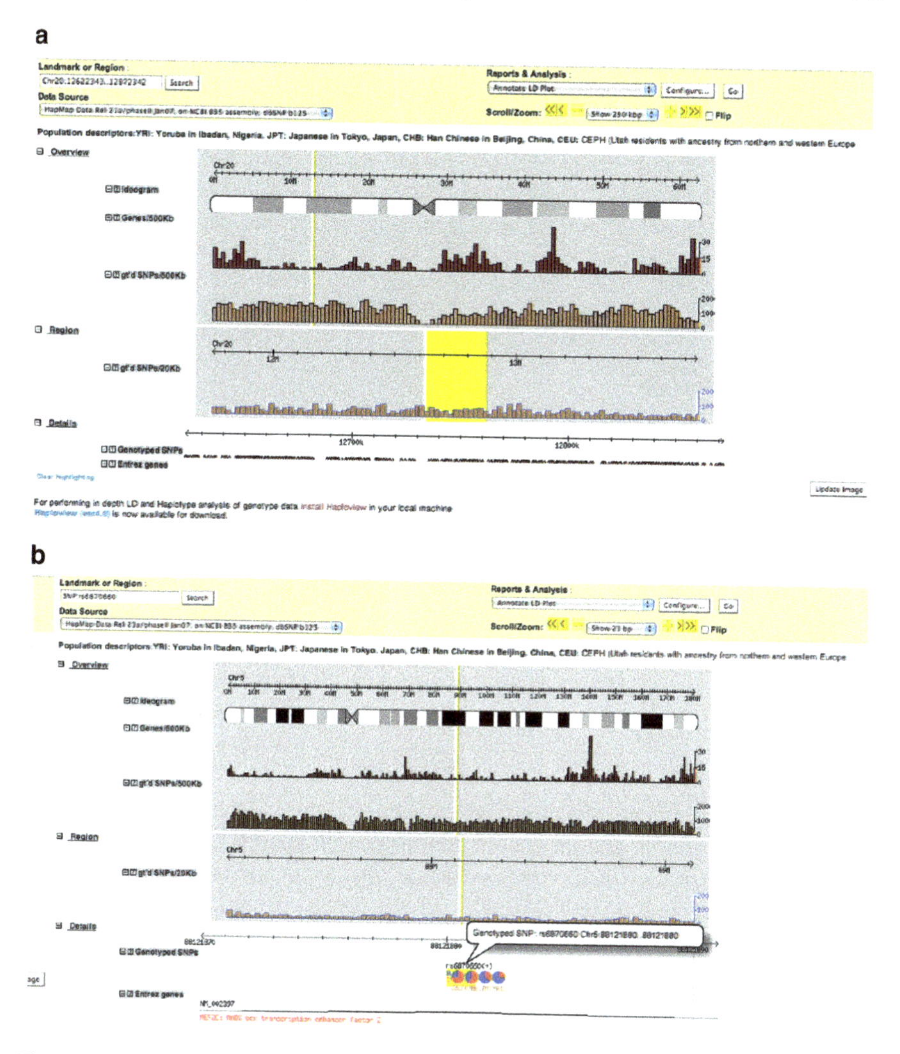


**Fig. 8.1** Examples of HapMap searches: (**a**). Chromosomal region search, showing the population of Genotyped SNPs as a custom GBrowse track (*bottom*). (**b**). SNP id search, showing the specific genome location for the SNP

### *8.3.6 GrainGenes*

GrainGenes integrates genetic data for Triticeae and Avena (Carollo et al. 2005; Matthews et al. 2003). The database includes genetic markers, map locations, alleles and key references for barley, wheat, rye, oat and related wild species. Graingenes also provides access to genetic data using CMap.

### 8.3.7 TAIR

The Arabidopsis Information Resource (TAIR) (http://www.arabidopsis.org/) provides an extensive web-based resource for the model plant *Arabidopsis thaliana* (Weems et al. 2004; Rhee et al. 2003; Huala et al. 2001) Data includes gene, marker, genetic mapping, protein sequence, gene expression and community data within a relational database.

### 8.3.8 MaizeGDB

MaizeGDB (Lawrence et al. 2004) combines information from the original MaizeDB and ZmDB (Dong et al. 2003; Gai et al. 2000) repositories with sequence data from PlantGDB (Duvick et al. 2008; Dong et al. 2004, 2005). The system maintains information on maize genomic and gene sequences, genetic markers, literature references, as well as contact information for the maize research community.

### 8.3.9 AutoSNPdb

AutoSNPdb implements the autoSNP pipeline within a relational database to enable mining for SNP and indel polymorphisms (Duran et al. 2009). A web interface enables searching and visualisation of the data, including the display of sequence alignments and SNPs (Fig. 8.2). All sequences are annotated by comparison with GenBank and UniRef90, as well as through comparison with reference genome sequences. The system allows researchers to query the results of SNP analysis to identify SNPs between specific groups of individuals or within genes of predicted function. AutoSNPdb is currently available for barley, rice and *Brassica* species and is available at: http://acpfg.imb.uq.edu.au/.

## 8.4 Data Visualisation

The effective visualisation of large amounts of data is as critical an issue as its storage. Increasing volumes of data permit researchers to draw, with increasing confidence, comparative links across the genome to phenome divide. Visualisation tools, combined with the ability to dynamically categorise data, allow the identification trends and relationships at varying tiers of resolution. Current visualisation techniques for molecular markers broadly fall into two categories: graphical map viewers and genome browsers. Map viewers display markers as a representation of a genetic

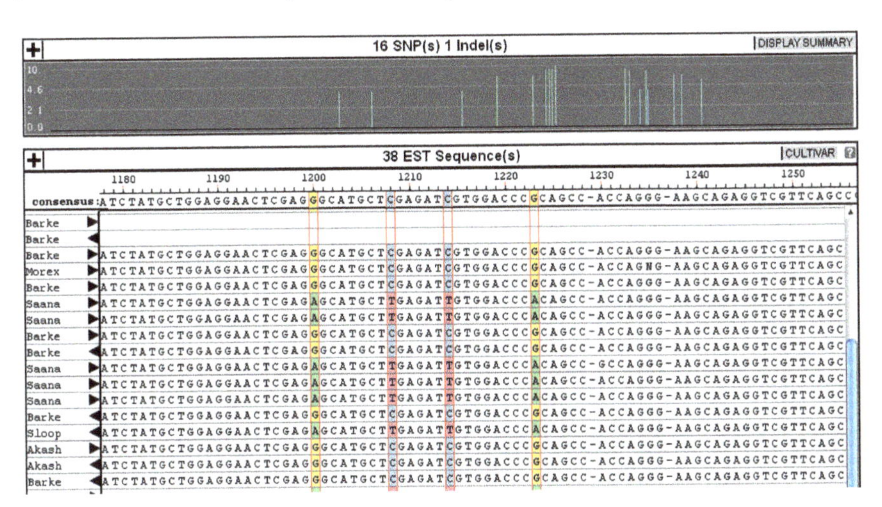


**Fig. 8.2** AutoSNPdb showing the overview of the SNPs in this assembly and the aligned sequences with the SNPs highlighted

linkage map. Genome browsers generally host a greater quantity of annotation data and may be linked to related genetic map viewers. As genome browsers are described in detail in chapter 3, details below only refer to map viewers.

### *8.4.1 Graphical Map Viewers*

The NCBI map viewer (http://www.ncbi.nih.gov/mapview) uses sets of graphically-aligned maps to visualise molecular genetic markers, genome assemblies and other annotations (Wheeler et al. 2008b). It allows users to show multiple levels of annotation in tandem for a given chromosomal segment (Fig. 8.3). As well as allowing users to view the map graphically, NCBI also provides a function to download raw mapping data in a tabular format.

CMap is a tool for viewing and comparing genetic and physical maps and has been applied successfully for the comparison of maps within and between related species (Jaiswal et al. 2006). CMap was originally developed for the Gramene project (http://www.gramene.org/CMap/) and has since been applied for the comparison of genetic maps of *Brassica* (Lim et al. 2007), sheep, cattle, pig and wallaby (Liao et al. 2007), honeybee, grasses (Jaiswal et al. 2006; Somers et al. 2004), peanut (Jesubatham and Burow 2006), Rosaceae (Jung et al. 2008) and legumes (Gonzales et al. 2005). The database specification dictates that a given species may have one or more "map sets", where a map set represents a collection of maps. The CMap database was designed with flexibility in mind, allowing the database to be used for a wide variety of mapping applications. For example, in genetic mapping studies, a map set is most likely to represent a particular

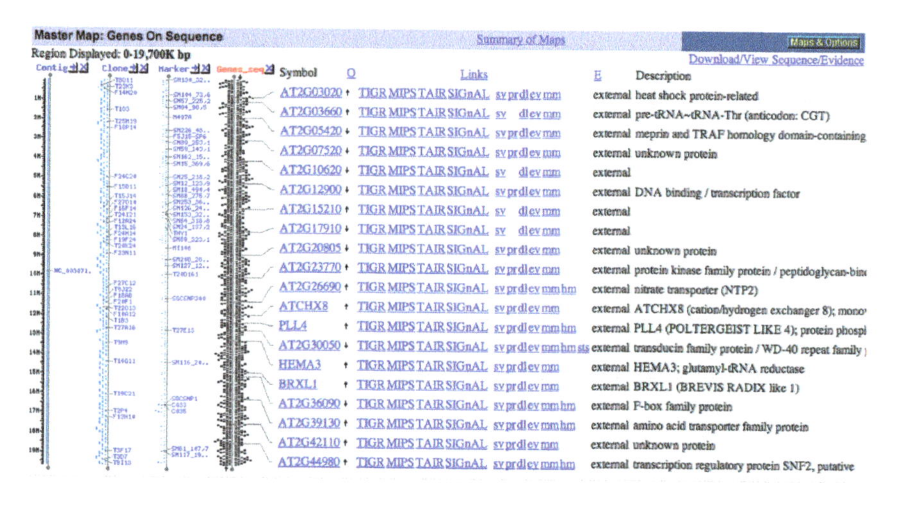


**Fig. 8.3** The NCBI map viewer displaying the overall view for *Arabidopsis thaliana* chromosome 2

organism, where the contained maps will probably be the genetic linkage groups. To generate genetic linkage maps, software such as Joinmap (Van Ooijen and Voorrips 2001), MapMaker (Lander et al. 1987) or Map Manager QTX (Manly et al. 2001) are frequently used. Once the maps are generated, they are parsed into the CMap database using a perl script provided with the CMap software package. The comparative relationships, or correspondences, between maps are generated once the maps are added to the database, using a combination of automated tools and manual curation. The automated tools included with CMap create correspondences on the basis of common feature names. If maps are generated with a consistent marker naming strategy, this can reduce the manual curation required and provide a firm basis for further curation. Curation is performed by parsing a tab-delimited file with the correspondences using a custom perl module included in CMap.

As an extension to CMap; CMap3D allows researchers to compare multiple genetic maps in three-dimensional space (Fig. 8.4). CMap3D accesses data from CMap databases, with specifications defined by the Generic Model Organism Database (GMOD) (http://www.gmod.org/CMap). The viewer is a stand-alone client written in Processing (http://www.processing.org) and is available for Windows, OSX and Linux. Information from the CMap repository enables CMap3D to generate appropriate external web links for features of a given map. By clicking on a feature in the viewing space, the user will be taken to the relevant web page. Compatibility with current CMap databases has been preserved by taking a server/client approach. The client is a user-side application that connects to servers hosting an existing CMap database known as a repository. The server is a web service that hosts a variety of scripts that communicate directly with a CMap MySQL database. The Cmap3D client first connects to a centralised repository listing server, which

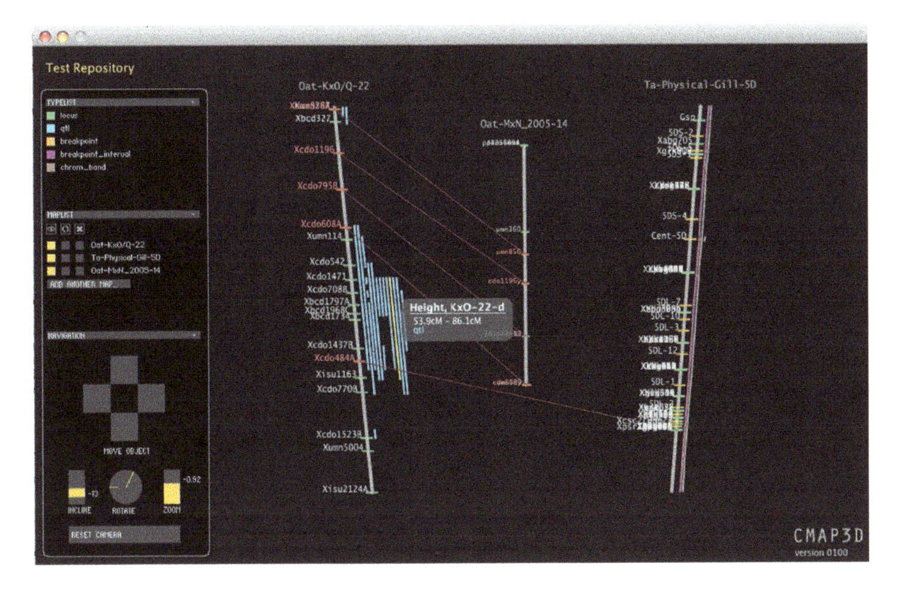


**Fig. 8.4** Screenshot of the engine view in CMap3D, showing the control panel to the left, and the viewing panel to the right. The icons above the toggles in the map controls section represent (from left to right) visibility, orientation and removal. This particular view is showing all features as visible with one of the ranged features highlighted

provides the client with a list of available repositories and their details. The client then communicates directly with the repository server to request and retrieve the required data. Data transfer uses the HTTP protocol for data transfer, to minimise institution network security conflicts.

## 8.5 Concluding Remarks

Genetic markers have played a major role in our understanding of heritable traits. In the current genomics era, molecular genetic markers are bridging the divide between these traits and increasingly available genome sequence information. Conversely, the increasing quantity of genome sequence information is a valuable source of new genetic markers. Bioinformatics tools have been developed to mine sequence data for markers and present these in a biologist friendly manner. With the expansion of next generation sequencing technologies, there will be a rapid growth in associated marker information and the use of these markers for diverse applications from crop breeding to predicting human disease risks, impacting both food production and human health for future generations.

Molecular markers have many applications in plant breeding, and the ability to detect the presence of a gene (or genes) controlling a particular desired trait has given rise to marker-assisted selection (MAS). These new technologies make it possible to speed up the breeding process. For example, a desired trait may only be observed in the mature plant, but MAS allows researchers to screen for the trait at the much earlier growth stage. Further advantages of molecular markers are that they make it possible to select simultaneously for many different plant characteristics. They can also be used to identify individual plants with a defined resistance gene without exposing the plant to the pest or pathogen in question. In order to increase throughput and decrease costs, it is necessary to eliminate bottlenecks throughout the genotyping process, as well as minimise sources of variability and human error to ensure data quality and reproducibility. These new technologies may be the way forward for the discovery and application of molecular markers and will enable the application of markers for a broader range of traits in a greater diversity of species than currently possible.

SNPs are being used for the association of genes with susceptibility to complex human diseases. For example, a study by Martin et al. (2000) found evidence of association between SNP alleles around the APOE gene to Alzheimer disease was identified. By looking at the allelic variation for a set of genes known to be associated with a particular disease, researchers may be able to assess the probability of the illness manifesting in a given individual. A greater understanding of the genetic variation underlying human disease will revolutionise drug discovery, patient treatment and lifestyle.

## References

Aggarwal RK, Hendre PS, Varshney RK, Bhat PR, Krishnakumar V, Singh L (2007) Identification, characterization and utilization of EST-derived genic microsatellite markers for genome analyses of coffee and related species. Theor Appl Genet 114:359–372

Altschul SF, Gish W, Miller W, Myers EW, Lipman DJ (1990) Basic local alignment search tool. J Mol Biol 215:403–410

Altshuler D, Brooks LD, Chakravarti A, Collins FS, Daly MJ, Donnelly P (2005) A haplotype map of the human genome. Nature 437:1299–1320

Awadalla P, Ritland K (1997) Microsatellite variation and evolution in the *Mimulus guttatus* species complex with contracting mating systems. Mol Biol Evol 14:1023–1034

Barbazuk WB, Emrich SJ, Chen HD, Li L, Schnable PS (2007) SNP discovery via 454 transcriptome sequencing. Plant J 51:910–918

Barker G, Batley J, O'Sullivan H, Edwards KJ, Edwards D (2003) Redundancy based detection of sequence polymorphisms in expressed sequence tag data using autoSNP. Bioinformatics 19:421–422

Batley J, Edwards D (2007) SNP applications in plants. In: Oraguzie NC, Rikkerink EHA, Gardiner SE, De Silva HN (eds) Association Mapping in Plants. Springer, NY, pp 95–102

Batley J, Barker G, O'Sullivan H, Edwards KJ, Edwards D (2003) Mining for single nucleotide polymorphisms and insertions/deletions in maize expressed sequence tag data. Plant Physiol 132:84–91

Batley J, Hopkins CJ, Cogan NOI, Hand M, Jewell E, Kaur J et al (2007) Identification and characterization of simple sequence repeat markers from *Brassica napus* expressed sequences. Mol Ecol Notes 7:886–889

Benson G (1999) Tandem repeats finder: a program to analyze DNA sequences. Nucleic Acids Res 27:573–580

Bentley DR, Balasubramanian S, Swerdlow HP, Smith GP, Milton J, Brown CG et al (2008) Accurate whole human genome sequencing using reversible terminator chemistry. Nature 456:53–59

Brumfield RT, Beerli P, Nickerson DA, Edwards SV (2003) The utility of single nucleotide polymorphisms in inferences of population history. Trends Ecol Evol 18:249–256

Burgess B, Mountford H, Hopkins CJ, Love C, Ling AE, Spangenberg GC et al (2006) Identification and characterization of simple sequence repeat (SSR) markers derived in silico from *Brassica oleracea* genome shotgun sequences. Mol Ecol Notes 6:1191–1194

Burke J, Davison D, Hide W (1999) d2_cluster: a validated method for clustering EST and full-length cDNA sequences. Genome Res 9:1135–1142

Cardle L, Ramsay L, Milbourne D, Macaulay M, Marshall D, Waugh R (2000) Computational and experimental characterization of physically clustered simple sequence repeats in plants. Genetics 156:847–854

Carollo V, Matthews DE, Lazo GR, Blake TK, Hummel DD, Lui N et al (2005) GrainGenes 2.0. An improved resource for the small-grains community. Plant Physiol 139:643–651

Castelo AT, Martins W, Gao GR (2002) TROLL-Tandem Repeat Occurrence Locator. Bioinformatics 18:634–636

Cervigni GDL, Paniego N, Diaz M, Selva JP, Zappacosta D, Zanazzi D et al (2008) Expressed sequence tag analysis and development of gene associated markers in a near-isogenic plant system of *Eragrostis curvula*. Plant Mol Biol 67:1–10

Chen CX, Zhou P, Choi YA, Huang S, Gmitter FG (2006) Mining and characterizing microsatellites from citrus ESTs. Theor Appl Genet 112:1248–1257

Chen XF, Laudeman TW, Rushton PJ, Spraggins TA, Timko MP (2007) CGKB: an annotation knowledge base for cowpea (*Vigna unguiculata* L.) methylation filtered genomic genespace sequences. BMC Bioinformatics 8:129

Collins A, Lau W, De la Vega FM (2004) Mapping genes for common diseases: the case for genetic (LD) maps. Hum Hered 58:2–9

Corva P, Soria L, Schor A, Villarreal E, Cenci MP, Motter M et al (2007) Association of CAPN1 and CAST gene polymorphisms with meat tenderness in *Bos taurus* beef cattle from Argentina. Genet Mol Biol 30:1064–1069

Dierick I, Baets J, Irobi J, Jacobs A, De Vriendt E, Deconinck T et al (2008) Relative contribution of mutations in genes for autosomal dominant distal hereditary motor neuropathies: a genotype–phenotype correlation study. Brain 131:1217–1227

Dong QF, Lawrence CJ, Schlueter SD, Wilkerson MD, Kurtz S, Lushbough C et al (2005) Comparative plant genomics resources at PlantGDB. Plant Physiol 139:610–618

Dong QF, Roy L, Freeling M, Walbot V, Brendel V (2003) ZmDB, an integrated database for maize genome research. Nucleic Acids Res 31:244–247

Dong QF, Schlueter SD, Brendel V (2004) PlantGDB, plant genome database and analysis tools. Nucleic Acids Res 32:D354–D359

Duran C, Appleby N, Clark T, Wood D, Imelfort M, Batley J et al (2009) AutoSNPdb: an annotated single nucleotide polymorphism database for crop plants. Nucleic Acids Res 37:D951–3

Duvick J, Fu A, Muppirala U, Sabharwal M, Wilkerson MD, Lawrence CJ et al (2008) PlantGDB: a resource for comparative plant genomics. Nucleic Acids Res 36:D959–D965

Edwards KJ, Barker JHA, Daly A, Jones C, Karp A (1996) Microsatellite libraries enriched for several microsatellite sequences in plants. Biotechniques 20:758–760

Ewing B, Green P (1998) Base-calling of automated sequencer traces using phred II. Error probabilities. Genome Res 8:186–194

Ewing B, Hillier L, Wendl MC, Green P (1998) Base-calling of automated sequencer traces using phred I. Accuracy assessment. Genome Res 8:175–185

Frazer KA, Ballinger DG, Cox DR, Hinds DA, Stuve LL, Gibbs RA et al (2007) A second generation human haplotype map of over 3.1 million SNPs. Nature 449:851–861

Gai XW, Lal S, Xing LQ, Brendel V, Walbot V (2000) Gene discovery using the maize genome database ZmDB. Nucleic Acids Res 28:94–96

Gao H, Kong J (2005) The microsatellites and minisatellites in the genome of *Fenneropenaeus chinensis*. DNA Seq 16:426–436

Garg K, Green P, Nickerson DA (1999) Identification of candidate coding region single nucleotide polymorphisms in 165 human genes using assembled expressed sequence tags. Genome Res 9:1087–1092

Gibbs RA, Belmont JW, Hardenbol P, Willis TD, Yu FL, Yang HM et al (2003) The international HapMap project. Nature 426:789–796

Gonzales MD, Archuleta E, Farmer A, Gajendran K, Grant D, Shoemaker R et al (2005) The Legume Information System (LIS): an integrated information resource for comparative legume biology. Nucleic Acids Res 33:D660–D665

Gordon D, Abajian C, Green P (1998) Consed: a graphical tool for sequence finishing. Genome Res 8:195–202

Gupta M, Chyi Y-S, Romero-Severson J, Owen JL (1994) Amplification of DNA markers from evolutionarily diverse genomes using single primers of simple-sequence re-peats. Theor Appl Genet 89:998–1006

Gupta PK, Roy JK, Prasad M (2001) Single nucleotide polymorphisms: A new paradigm for molecular marker technology and DNA polymorphism detection with emphasis on their use in plants. Curr Sci 80:524–535

Hawken RJ, Barris WC, McWilliam SM, Dalrymple BP (2004) An interactive bovine in silico SNP database (IBISS). Mamm Genome 15:819–827

Hopkins CJ, Cogan NOI, Hand M, Jewell E, Kaur J, Li X et al (2007) Sixteen new simple sequence repeat markers from *Brassica juncea* expressed sequences and their cross-species amplification. Mol Ecol Notes 7:697–700

Huala E, Dickerman AW, Garcia-Hernandez M, Weems D, Reiser L, LaFond F et al (2001) The Arabidopsis Information Resource (TAIR): a comprehensive database and web-based in-formation retrieval, analysis, and visualization system for a model plant. Nucleic Acids Res 29:102–105

Huang XQ, Madan A (1999) CAP3: a DNA sequence assembly program. Genome Res 9:868–877

Jaccoud D, Peng K, Feinstein D, Kilian A (2001) Diversity arrays: a solid state technology for sequence information independent genotyping. Nucleic Acids Res 29:e25

Jaiswal P, Ni JJ, Yap I, Ware D, Spooner W, Youens-Clark K et al (2006) Gramene: a bird's eye view of cereal genomes. Nucleic Acids Res 34:D717–D723

Jayashree B, Ferguson M, Ilut D, Doyle J, Crouch JH (2005) Analysis of genomic sequences from peanut (*Arachis hypogaea*). Electron J Biotechnol 8:3

Jesubatham AM, Burow MD (2006) PeanutMap: an online genome database for comparative molecular maps of peanut. BMC Bioinformatics 7:375

Jewell E, Robinson A, Savage D, Erwin T, Love CG, Lim GAC et al (2006) SSRPrimer and SSR Taxonomy Tree: Biome SSR discovery. Nucleic Acids Res 34:W656–W659

Jung S, Staton M, Lee T, Blenda A, Svancara R, Abbott A et al (2008) GDR (Genome Database for Rosaceae): integrated web-database for Rosaceae genomics and genetics data. Nucleic Acids Res 36:D1034–D1040

Kashi Y, King D, Soller M (1997) Simple sequence repeats as a source of quantitative genetic variation. Trends Genet 13:74–78

Kasprzyk A, Keefe D, Smedley D, London D, Spooner W, Melsopp C et al (2004) EnsMart: a generic system for fast and flexible access to biological data. Genome Res 14:160–169

Katti MV, Ranjekar PK, Gupta VS (2001) Differential distribution of simple sequence repeats in eukaryotic genome sequences. Mol Biol Evol 18:1161–1167

Keniry A, Hopkins CJ, Jewell E, Morrison B, Spangenberg GC, Edwards D et al (2006) Identification and characterization of simple sequence repeat (SSR) markers from Fragaria × ananassa expressed sequences. Mol Ecol Notes 6:319–322

Kwok PY, Carlson C, Yager TD, Ankener W, Nickerson DA (1994) Comparative analysis of human DNA variations by fluorescence-based sequencing of PCR products. Genomics 23:138–144

Lander E, Abrahamson J, Barlow A, Daly M, Lincoln S, Newburg L et al (1987) Mapmaker: a computer package for constructing genetic-linkage maps. Cytogenet Cell Genet 46:642
Lawrence CJ, Dong OF, Polacco ML, Seigfried TE, Brendel V (2004) MaizeGDB, the community database for maize genetics and genomics. Nucleic Acids Res 32:D393–D397
Lazzari B, Caprera A, Vecchietti A, Stella A, Milanesi L, Pozzi C (2005) ESTree db: a tool for peach functional genomics. BMC Bioinformatics 6(Suppl 4):S16
Lee SH, Park EW, Cho YM, Lee JW, Kim HY, Lee JH et al (2006) Confirming single nucleotide polymorphisms from expressed sequence tag datasets derived from three cattle cDNA libraries. J Biochem Mol Biol 39:183–188
Li Y-C, Korol AB, Fahima T, Beiles A, Nevo E (2002) Microsatellites: genomic distribution, putative functions and mutational mechanisms: a review. Mol Ecol 11:2453–2465
Liao W, Collins A, Hobbs M, Khatkar MS, Luo JH, Nicholas FW (2007) A comparative location database (CompLDB): map integration within and between species. Mamm Genome 18:287–299
Lim GAC, Jewell EG, Xi L, Erwin TA, Love C, Batley J et al (2007) A comparative map viewer integrating genetic maps for Brassica and Arabidopsis. BMC Plant Biol 7:40
Lindqvist C, Scheen AC, Yoo MJ, Grey P, Oppenheimer DG, Leebens-Mack JH et al (2006) An expressed sequence tag (EST) library from developing fruits of an Hawaiian endemic mint (*Stenogyne rugosa*, Lamiaceae): characterization and microsatellite markers. BMC Plant Biol 6:16
Ling AE, Kaur J, Burgess B, Hand M, Hopkins CJ, Li X et al (2007) Characterization of simple sequence repeat markers derived in silico from *Brassica rapa* bacterial artificial chromosome sequences and their application in *Brassica napus*. Mol Ecol Notes 7:273–277
Mallon AM, Strivens M (1998) DNA sequence analysis and comparative sequencing. Methods 14:160–178
Manly KF, Cudmore RH, Meer JM (2001) Map manager QTX, cross-platform software for genetic mapping, *Mamm*. Genome 12:930–932
Marth GT, Korf I, Yandell MD, Yeh RT, Gu ZJ, Zakeri H et al (1999) A general approach to single-nucleotide polymorphism discovery. Nat Genet 23:452–456
Martin ER, Lai EH, Gilbert JR, Rogala AR, Afshari AJ, Riley J et al (2000) SNPing away at complex diseases: analysis of single-nucleotide polymorphisms around *APOE* in Alzheimer disease. Am J Hum Genet 67:383–394
Matthews DE, Carollo VL, Lazo GR, Anderson OD (2003) GrainGenes, the genome database for small-grain crops. Nucleic Acids Res 31:183–186
Matukumalli LK, Grefenstette JJ, Hyten DL, Choi I-Y, Cregan PB, Van Tassell CP (2006) SNP–PHAGE – High throughput SNP discovery pipeline. BMC Bioinformatics 7:468
Mortimer J, Batley J, Love C, Logan E, Edwards D (2005) Simple Sequence Repeat (SSR) and GC distribution in the *Arabidopsis thaliana* genome. J Plant Biotechnol 7:17–25
Moxon ER, Wills C (1999) DNA microsatellites: agents of evolution. Sci Am 280:94–99
Nickerson DA, Tobe VO, Taylor SL (1997) PolyPhred: automating the detection and genotyping of single nucleotide substitutions using fluorescence-based resequencing. Nucleic Acids Res 25:2745–2751
Pavy N, Parsons LS, Paule C, MacKay J, Bousquet J (2006) Automated SNP detection from a large collection of white spruce expressed sequences: contributing factors and approaches for the categorization of SNPs. BMC Genomics 7:174
Perez F, Ortiz J, Zhinaula M, Gonzabay C, Calderon J, Volckaert F (2005) Development of EST-SSR markers by data mining in three species of shrimp: *Litopenaeus vannamei*, *Litopenaeus stylirostris*, and *Trachypenaeus birdy*. Mar Biotechnol 7:554–569
Pertea G, Huang XQ, Liang F, Antonescu V, Sultana R, Karamycheva S et al (2003) TIGR Gene Indices clustering tools (TGICL): a software system for fast clustering of large EST datasets. Bioinformatics 19:651–652
Powell W, Machray GC, Provan J (1996) Polymorphism revealed by simple sequence repeats. Trends Plant Sci 1:215–222
Pumpernik D, Oblak B, Borstnik B (2008) Replication slippage versus point mutation rates in short tandem repeats of the human genome. Mol Gen Genomics 279:53–61

Rafalski A (2002) Applications of single nucleotide polymorphisms in crop genetics. Curr Opin Plant Biol 5:94–100
Rhee SY, Beavis W, Berardini TZ, Chen GH, Dixon D, Doyle A et al (2003) The Arabidopsis Information Resource (TAIR): a model organism database providing a centralized, curated gateway to Arabidopsis biology, research materials and community. Nucleic Acids Res 31:224–228
Robinson AJ, Love CG, Batley J, Barker G, Edwards D (2004) Simple sequence repeat marker loci discovery using SSR primer. Bioinformatics 20:1475–1476
Savage D, Batley J, Erwin T, Logan E, Love CG, Lim GAC et al (2005) SNPServer: a real-time SNP discovery tool. Nucleic Acids Res 33:W493–W495
Schlötterer C (2000) Evolutionary dynamics of microsatellite DNA. Nucleic Acids Res 20:211–215
Sherry ST, Ward MH, Sirotkin K (1999) dbSNP – database for single nucleotide polymorphisms and other classes of minor genetic variation. Genome Res 9:677–679
Sironi L, Lazzari B, Ramelli P, Gorni C, Mariani P (2006) Single nucleotide polymorphism discovery in the avian Tapasin gene. Poult Sci 85:606–612
Smigielski EM, Sirotkin K, Ward M, Sherry ST (2000) dbSNP: a database of single nucleotide polymorphisms. Nucleic Acids Res 28:352–355
Somers DJ, Isaac P, Edwards K (2004) A high-density microsatellite consensus map for bread wheat (*Triticum aestivum* L.). Theor Appl Genet 109:1105–1114
Stein LD, Mungall C, Shu SQ, Caudy M, Mangone M, Day A et al (2002) The generic genome browser: a building block for a model organism system database. Genome Res 12:1599–1610
Subramanian S, Mishra RK, Singh L (2003) Genome-wide analysis of microsatellite repeats in humans: their abundance and density in specific genomic regions. Genome Biol 4:R13
Sullivan JC, Reitzel AM, Finnerty JR (2008) Upgrades to StellaBase facilitate medical and genetic studies on the starlet sea anemone, *Nematostella vectensis*. Nucleic Acids Res 36:D607–D611
Syvanen AC (2001) Genotyping single nucleotide polymorphisms. Nat Rev Genet 2:930–942
Tautz D (1989) Hypervariability of simple sequences as a general source for polymorphic DNA markers. Nucleic Acids Res 17:6463–6471
Temnykh S, DeClerck G, Lukashova A, Lipovich L, Cartinhour S, McCouch S (2001) Computational and experimental analysis of microsatellites in rice (*Oryza sativa* L.): frequency, length variation, transposon associations, and genetic marker potential. Genome Res 11:1441–1452
Thiel T, Michalek W, Varshney RK, Graner A (2003) Exploiting EST databases for the development and characterization of gene-derived SSR-markers in barley (*Hordeum vulgare* L.). Theor Appl Genet 106:411–422
Tóth G, Gáspári Z, Jurka J (2000) Microsatellites in different eukaryotic genomes: survey and analysis. Genome Res 10:967–981
Van Ooijen JW, Voorrips RE (2001) JoinMap® 3.0, software for calculation of genetic linkage maps. Plant Research International, Wageningen
Volfovsky N, Haas BJ, Salzberg SL (2001) A clustering method for repeat analysis in DNA sequences. Genome Biol 2(8):RESEARCH0027
von Stackelberg M, Rensing SA, Reski R (2006) Identification of genic moss SSR markers and a comparative analysis of twenty-four algal and plant gene indices reveal species-specific rather than group-specific characteristics of microsatellites. BMC Plant Biol 6:9
Vos P, Hogers R, Bleeker M, Reijans M, Vandelee T, Hornes M et al (1995) AFLP – a new technique for DNA fingerprinting. Nucleic Acids Res 23:4407–4414
Wang WL, Zhang GL, Wu LH, Yao MY, Jin J, Jia CK et al (2007) Efficacy of hepatitis B immunoglobulin in relation to the gene polymorphisms of human leukocyte Fc gamma receptor III (CD16) in Chinese liver trans-plant patients. Chin Med J 120:1606–1610
Ware D, Jaiswal P, Ni JJ, Pan XK, Chang K, Clark K et al (2002a) Gramene: a resource for comparative grass genomics. Nucleic Acids Res 30:103–105
Ware DH, Jaiswal PJ, Ni JJ, Yap I, Pan XK, Clark KY et al (2002b) Gramene, a tool for grass genomics. Plant Physiol 130:1606–1613

Weber JL (1990) Informativeness of human (dC–dA)n. (dG–dT) n polymorphisms. Genomics 7:524–530

Weckx S, Del-Favero J, Rademakers R, Claes L, Cruts M, De Jonghe P et al (2005) novoSNP, a novel computational tool for sequence variation discovery. Genome Res 15:436–442

Weems D, Miller N, Garcia-Hernandez M, Huala E, Rhee SY (2004) Design, implementation and maintenance of a model organism database for *Arabidopsis thaliana*. Comp Funct Genomics 5:362–369

Wheeler DA, Srinivasan M, Egholm M, Shen Y, Chen L, McGuire A et al (2008a) The complete genome of an individual by massively parallel DNA sequencing. Nature 452(7189):872–876

Wheeler DL, Barrett T, Benson DA, Bryant SH, Canese K, Chetvernin V et al (2008b) Database resources of the national center for biotechnology information. Nucleic Acids Res 36:D13–D21

Winton LM, Krohn AL, Leiner RH (2007) Microsatellite markers for Sclerotinia subarctica nom. prov., a new vegetable pathogen of the High North. Mol Ecol Notes 7:1077–1079

Zhang JH, Wheeler DA, Yakub I, Wei S, Sood R, Rowe W et al (2005) SNPdetector: a software tool for sensitive and accurate SNP detection. PLoS Comput Biol 1:395–404

# Chapter 9
# Sequence Based Gene Expression Analysis

**Lakshmi K. Matukumalli and Steven G. Schroeder**

## 9.1 Introduction

Life sciences in the twentieth century made major strides in unraveling several basic biological phenomena applicable to all living systems such as deciphering the genetic code and defining the central dogma (replication, transcription and translation), through observation and simple experimentation. However, biological research in the twenty-first century is primarily driven by high precision instrumentation for exploring the complexity of biological systems in greater detail. Very large datasets are generated from these instruments that require efficient computational tools for data mining and analysis. The definition of the term "high-throughput" has had to be redefined at regular intervals because of the exponential growth in the volume of data generated with each technological advance. For addressing the needs of modeling, simulation and visualization of large and diverse biological datasets from sequence, gene expression and proteomics datasets, "systems biology" (Hood 2003) approaches are being developed for construction of gene regulatory networks (Dojer et al. 2006; Imoto et al. 2002; Xiong 2006; Xiong et al. 2004) and for identification of key control nodes.

Among the various biomolecules present inside the cell, RNA plays a critical role in executing the inherited genetic instructions and for dynamically adapting to varying environmental conditions. The RNA component of the cell (transcriptome) is variable within different cell types in terms of both diversity and concentration of individual entities. Characterizing the transcriptome aids in gene identification, while monitoring the gene expression levels in different cell types at various time points, provides insight into the workings of the genome and biology of the organism as a whole.

L.K. Matukumalli (✉)
Department of Bioinformatics and Computational Biology, George Mason University, Manassas, VA, 20110, USA
e-mail: lmatukum@gmu.edu

D. Edwards et al. (eds.), *Bioinformatics: Tools and Applications*, 
DOI 10.1007/978-0-387-92738-1_9, © Springer Science+Business Media, LLC 2009

### *9.1.1 A Gene Atlas for Characterizing the Transcriptome*

The construction of a gene atlas is an optimal method for fully exploring the transcriptome, where gene expression is measured in diverse tissue types at distinct growth stages of an organism. This is because only small subsets of genes are expressed in each of the cell or tissue types. Gene expression is also temporal, since some genes are either switched on or off at different cell stages. A number of gene atlas projects have been implemented in plants and animals, including *Arabidopsis* (Meyers et al. 2004a, b; Peiffer et al. 2008; Weber et al. 2007), Human (Adams et al. 1995; Camargo et al. 2001; Su et al. 2004), Mouse (Mortazavi et al. 2008; Su et al. 2004), Pig (Gorodkin et al. 2007). In addition, we have developed a cattle gene atlas by deep sequencing of 100 distinct cattle tissues. (Manuscript in preparation).

## 9.2 Gene Expression by Sequencing

Advances in DNA-sequencing technology can be harnessed to explore transcriptomes in remarkable detail. Gene expression was initially performed by cloning and sequencing expressed sequence tags (ESTs) from cDNA libraries constructed from various tissue types (Adams et al. 1991; Boguski et al. 1993). EST sequencing coupled with whole genome sequence analysis helped identify 25,000–30,000 protein coding genes in humans (Boguski and Schuler 1995). Although EST sequencing provided a glimpse of gene expression in various tissues, in-depth sequencing of ESTs using Sanger sequencing remains cost prohibitive. EST sequencing remains a useful tool, but is relatively slow, provides only partial sequences that are sensitive to cloning biases, and generally cannot identify mRNAs that are expressed at low levels.

Serial Analysis of Gene Expression (SAGE) (Velculescu et al. 1995), and its variant LongSAGE (Saha et al. 2002) have been applied as a cost effective approach for deep sequencing. In this method, a short nucleotide tag, is generated from two restriction digests, the tags are concatenated into a long sequence of 20–60 tags and analyzed in one Sanger sequencing reaction (Torres et al. 2008). This method has been successfully applied for several gene expression analyses (Blackshaw et al. 2004; Hou et al. 2007). As expected, SAGE has been shown to be more sensitive in detecting low copy number transcripts as compared to EST sequencing (Sun et al. 2004) and also reproducible (Dinel et al. 2005).

## 9.3 Gene Expression by Hybridization

Global analysis of gene expression by RNA-hybridization on high-density arrays enable high-throughput profiling. Microarrays and whole genome tiling (WGT) arrays are the most common transcriptome analyses platforms. Microarrays (Su et al. 2004) are used for expression analysis of known genes whereas WGT

(Birney et al. 2007) arrays are used to capture the full complexity of the transcriptome. WGT arrays however cannot capture the splice-junction information and are associated with high costs and complexities of data analysis. Arrays that are specifically designed for detecting alternative splicing events are still immature and unable to fully address the issues on completeness and specificity (Calarco et al. 2007; Pan et al. 2004).

Microarrays rely on a continuous analog signal produced by hybridization between the templates on the arrays and the target transcripts in the sample (Velculescu and Kinzler 2007). Microarrays have played a very important role in the last decade in simultaneously monitoring the expression of thousands of molecules for large sample sizes at a reasonable cost. Large-scale application of microarrays as medical diagnostic kits for early and accurate detection of number of pathological diseases including cancer is being actively explored (Liu and Karuturi 2004).

Some limitations of the use of microarray technologies for gene expression analyses are

1. *Content*. Microarrays only allow measurement of expression of known genes that have been arrayed on the chip. Since, the knowledge of the numbers of expressed genes and their various splice isoforms is continuously expanding, it is difficult to create new arrays and analyze samples with new content.
2. *Standardization*. Microarray measurements are not easily comparable because of differences in array technologies, sample preparation methods, scanning equipment and even software. MIAME (minimum information about the microarray experiment) standards address this issue to some extent by requiring several key details to be recorded for accurate reproduction. However, the reproducibility of results between different laboratories still remains a challenge and requires validation by an independent method such as RT-PCR.
3. *Detection thresholds*. Because of the background noise and limited dynamic range (ratio of the smallest to the largest fluorescent signal) in the scanning of microarrays, it is difficult to monitor lowly expressed genes.
4. *Cross-species comparison*. Although genes are known to be conserved across closely related species as compared to nongenic regions of the genome, it is not optimal to use arrays designed in one species for another. Also, in species having high genetic diversity, the microarray results may not be reliable.
5. *Absolute vs. relative levels*. Although there is a linear relationship between the concentration of each individual RNA species and signal intensity, the intensities vary between different RNA molecules. Hence, it is difficult to quantify differences in expression of different genes within a cell.

## 9.4 Insights from the ENCODE Project

The Encyclopedia of DNA coding Elements (ENCODE) project's goal was to use high-throughput methods to identify and catalog the functional elements encoded in the human genome (Birney et al. 2007). The ENCODE regions comprise ~30 Mb

(1% of the human genome), encompassing 44 genomic regions. Approximately 15 Mb reside in 14 regions for which there is already substantial biological knowledge, whereas the other 15 Mb covers 30 regions chosen by a stratified random-sampling method. High density whole genome tiling arrays were used to identify all transcriptionally active regions (TARs) and these were validated using RACE (Wu et al., 2008b). This study provided convincing evidence that the genome is pervasively transcribed, such that the majority of its bases can be found in primary transcripts, including nonprotein-coding transcripts, and that these extensively overlap one another. Because of the increased sensitivity in measuring rare transcripts, several novel nonprotein-coding transcripts were identified, with many of these overlapping protein-coding loci and others located in regions of the genome previously thought to be transcriptionally silent. To fully capture the transcript diversity, deep sequencing was considered to be more attractive than hybridization based transcript detection methods. Sequencing the transcriptome can rapidly detect most genes at moderate sequencing depth, but identification of all genes can require very deep sequencing coverage.

## 9.5 Digital Gene Expression

Large-scale EST or SAGE based sequencing methods are called digital gene expression (DGE) measures as they provide direct transcript counts in proportion to the number of copies expressed in the cell. High-throughput DGE methods such as massively parallel signature sequencing (MPSS) or sequencing-by-synthesis (SBS) have been developed for simultaneously sequencing millions of expressed tags. The high sequencing error rate and short sequence length (~17 bases) had been a significant drawback in the initial phases of technology development (Meyers et al. 2004b), and short tag sequences do not allow for unambiguous mapping of several tags to a unique transcript or genome location.

Meyers et al sequenced 2,304,362 tags from five diverse libraries of *Arabidopsis thaliana* using MPSS to identify a total of 48,572 distinct signatures expressed at significant levels. These signatures have been compared to the existing annotation of the *A. thaliana* genomic sequence to identify matches to 17,353 and 18,361 genes with sense expression, and between 5,487 and 8,729 genes with antisense expression. An additional 6,691 MPSS signatures mapped to nonannotated regions of the genome are likely to represent novel gene content. Alternative polyadenylation was reported for more than 25% of *A. thaliana* genes transcribed in these libraries. The method was later applied for multiple species, including rice, grape and rice blast fungus (Iandolino et al. 2008; Nakano et al. 2006).

In our group, we sequenced 315 million tags from 100 distinct cattle tissues using the SBS method on the Solexa/Illumina Genome Analyzer. A total of 7.9 million distinct tag types were observed in this data, with 5.6 million tags observed only once. These singletons constituted only 1.7% of the data and a proportion of them are likely to be sequencing errors. Around 6.5 million tags could be mapped to the cattle genome, and could uniquely identify 61 % of all known RefSeq genes (17,350/28,510). From this analysis we identified 4,353 unique loci that demonstrated ubiquitous expression among all the tissues analyzed.

The technology limitations of MPSS and SBS methods described above are now being addressed by simultaneously increasing the sequence length as well as quality. Currently, several DGE platforms are available such as Genome Analyzer (Illumina) (Morin et al. 2008), SOLiD sequencing (Applied Biosystems) (Cloonan et al. 2008), 454 pyro-sequencing (Roche) (Weber et al. 2007) and polony sequencing (Kim et al. 2007). Recently, these methods have been demonstrated to be more accurate in comparison to traditional microarray methods (Marioni et al. 2008). Additionally, technology improvements for generating longer sequence reads coupled with paired end sequencing or shotgun sequencing methods such as RNA-Seq (Fig. 9.1) (Mortazavi et al. 2008) or short quantitative random RNA libraries (SQRL) (Cloonan et al. 2008) is enabling detection of full length cDNA sequences as well as all splice variants.

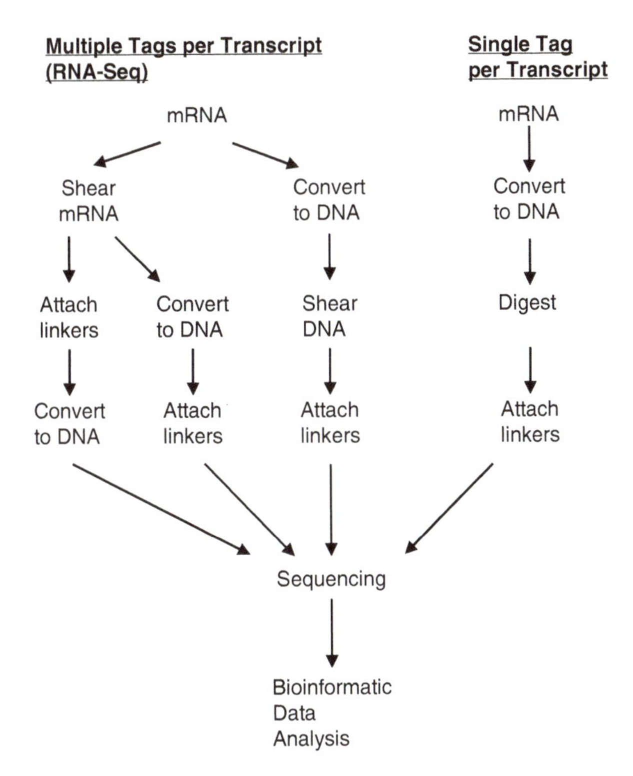


**Fig. 9.1** Simplified representation of sample processing for next generation sequencing of digital gene expression. RNA-Seq methods use mechanical or chemical shearing to generate appropriate fragment sizes for sequencing, resulting in potentially full transcript coverage. The second method uses restriction cleavage to generate fragments that more closely represent one read per transcript from the sample

Although there are several million EST sequences available at dbEST, the functional complexity of even the human transcriptome is not yet fully elucidated. Sultan et al (2008) performed RNA-Seq on human embryonic kidney and a B cell line and found that 66% of the polyadenylated transcriptome mapped to known genes and 34% to unnannotated genomic regions. From the total tags sequenced 50% mapped to unique genomic locations, 16–18% mapped to multiple regions of the genome, and 25% had no matches to the genome. From those that had unique matches to the genome, 80% corresponded to known exons. For the tags having no matches to the genome, they could attribute 14% to synthetically-computed alternative splice sites. The remaining tags are likely to be? sequencing errors or other unidentified alternative splice sites. The global survey of messenger RNA splicing events in their study revealed 94,241 splice junctions (4,096 of which were previously unidentified) and showed that exon skipping is the most prevalent form of alternative splicing. On the basis of known transcripts, they concluded that RNA-Seq can detect 25% more genes than microarrays.

Analysis of gene expression by sequencing has several advantages over microarray methods. For the development of a microarray gene expression assay, all gene sequences are required to be known prior to probe design. Microarray chip design also requires a large initial investment and long lead time. High-throughput DGE can be applied readily to any species lacking both the genome sequence and gene content. DGE methods can explore novel gene content, splice variants and rare transcripts. Hybridization techniques require the establishment of thresholds for signal-to-noise ratio and can potentially lose detection of some rare transcripts. In contrast, DGE measures have no background noise, except for the tags observed due to sequencing errors. DGE methods can also be applied to other classes of RNA transcripts such as snoRNAs (Bachellerie et al. 2002), miRNAs (Ambros 2001), piRNA (Lau et al. 2006) and siRNAs (Hamilton and Baulcombe 1999). DGE methods provide absolute tag counts (tags per million) that allow reproducible comparison of transcripts within and across various samples. Finally, for microarray experiments, SNPs and closely related genes can cause signal interference; whereas for DGE, sequencing errors and short tag lengths can create similar problems.

### *9.5.1 Choice of DGE Method: Tag Bias*

#### 9.5.1.1 Requirement for Specific Recognition Site

Some DGE methods such as MPSS and SBS require the presence of a specific restriction endonuclease recognition site within the transcript close to the poly(A) tail (Meyers et al. 2004b). Transcripts lacking the specific recognition site (or having lost the recognition site due to SNPs) are not represented in the DGE data (Silva et al. 2004).

#### 9.5.1.2 Sequencing Read Lengths

DGE methods such as 454 and LongSAGE, generate longer reads than SAGE or MPSS. Hene et al (2007) compared tag data generated from the same T cell-derived RNA sample using a LongSAGE library of 503,431 tags and a "classic" MPSS library of 1,744,173 tags, and showed that LongSAGE had 6.3-fold more genome-matching tags than MPSS. An analysis of a set of 8,132 known genes detectable by both methods, and for which there is no ambiguity about tag matching, shows that MPSS detects only half (54%) the number of transcripts identified by SAGE (3,617 vs. 1,955). Analysis of two additional MPSS libraries showed that each library sample had a different subset of transcripts, and that in combination, the three MPSS libraries (4,274,992 tags in total) detected only73% of the genes identified using SAGE. They concluded that MPSS libraries were less complex than LongSAGE libraries, revealing significant tag bias between the two methods.

#### 9.5.1.3 GC Content of the Transcript

By comparing five gene expression profiling methods: Affymetrix GeneChip, LongSAGE, LongSAGELite, "Classic" MPSS and "Signature" MPSS, it was shown that these methods are sensitive to the GC content of the transcript (Siddiqui et al. 2006). The LongSAGE method had the least bias, Signature MPSS showed a strong bias to GC rich tags, and Affymetrix data showed different bias depending on the data processing method (MAS 5.0, RMA or GC-RMA), mostly impacting genes expressed at lower levels. It was observed that despite the larger sampling of the MPSS library, SAGE identified significantly more genes (60% more RefSeq genes in a single comparison (Siddiqui et al. 2006)).

#### 9.5.1.4 Shotgun Sequencing Coupled with Long Read Lengths

Weber et al (2007) observed no tag bias in the transcripts from large-scale pyrosequencing (100–110 bp) of an *Arabidopsis* cDNA library. They performed two sequencing runs to generate 541,852 expressed sequence tags (ESTs) after quality control. Mapping of the ESTs to the *Arabidopsis* genome to The *Arabidopsis* Information Resource 7.0 cDNA models indicated detection of 17,449 transcribed gene loci. The second sequencing run only increased the number of genes identified by 10%, but increased the overall sequence coverage by 50%. Mapping of the ESTs to their predicted full-length transcripts indicated that shotgun sequencing of cDNA represented all transcripts regardless of transcript length or expression level. Over 16,000 of the ESTs that were mapped to the genome were not represented in the dbEST database, and are likely novel transcripts.

#### 9.5.1.5 5′ vs. 3′ Transcript Sequencing

Most of the transcript abundance measurement methods, including the DGE methods, rely on 3′ sequence. The cap analysis of gene expression (CAGE), method has been proposed for identification of 5′ end-specific signature sequences (Shiraki et al. 2003). In this technique, linkers are attached to the 5′ ends of full-length enriched cDNAs to introduce a recognition site for the restriction endonuclease MmeI adjacent to the 5′ ends. MmeI cleaves cDNAs 20 and 18 nucleotides away (3′) from its recognition site, generating a two-base overhang. After amplification, the sequencing tags are concatenated for high-throughput sequencing. The application of CAGE has been found to be useful in determining the transcription start sites (TSSs) along with associated gene promoters (de Hoon and Hayashizaki 2008; Kawaji et al. 2006).

#### 9.5.1.6 Availability of Whole Genome Sequence

When the genome sequence is not available, DGE methods that generate one tag per transcript such as SBS are useful for measuring expression changes between samples (Nakano et al. 2006). Shotgun transcript sequencing methods such as RNA-Seq have more utility when genome sequence is available, and can be used for measuring relative transcript abundance, full length cDNA assembly, and analysis of splice variants (Mortazavi et al. 2008).

#### 9.5.1.7 Accounting for Heterogeneous Cell Sampling

Gene expression changes are measured by sampling a homogeneous cell population. However, in complex metazoan systems it is often difficult to obtain unique cell types, hence, the gene expression measurement is often an average and is dependent on the composition of the sample. A computational deconvolution method that accounts for changes in the sizes of cell type-specific compartments can increase both the sensitivity and specificity of differential gene expression experiments performed on complex tissues (Wang et al. 2006).

### *9.5.2 Analysis of DGE Data*

DGE data from next-generation sequencing platforms consists of a large number of sequence reads. These reads, also commonly referred to as tags or signatures, present a representative sampling of the active transcriptome. A typical dataset will consist of several million reads. Each signature in a DGE data set is analyzed for quality and is compared with all others to obtain cumulative counts of all distinct tags. The unique tags are then mapped to the genome or gene models. In DGE methods such as MPSS

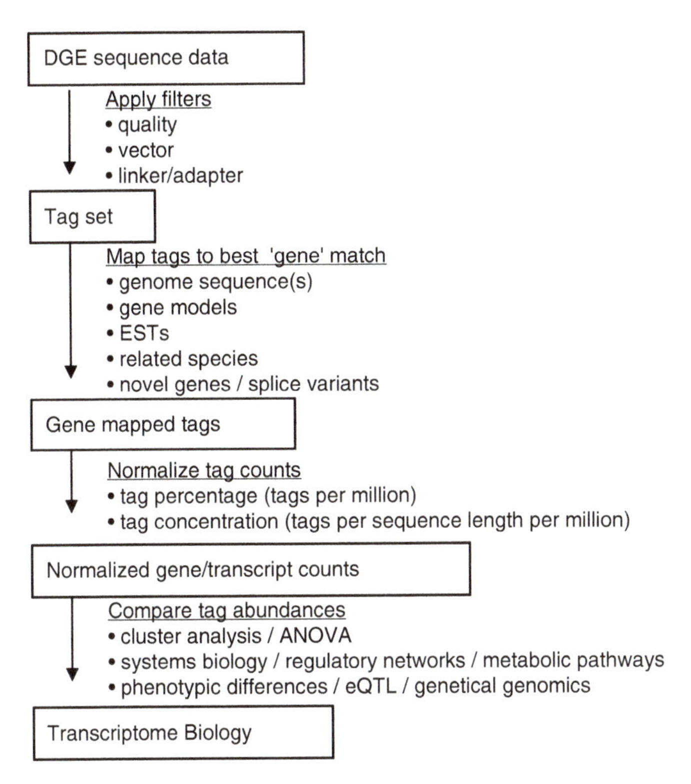


**Fig. 9.2** General flow chart of DGE data analysis. DGE sequence reads are filtered to remove low quality sequences and clustered to identify unique reads. The resulting set of reads are mapped to a combination of genomic sequences and various known genic sequences to provide identification and, in the case of RNA-seq data, to group them into defined transcripts. Reads that do not map to known exonic regions are merged into existing gene models if in appropriate proximity, or used to construct novel gene models. Mapped read abundances are used to generate normalized values for their respective assigned genes either by percentage or by concentration depending on the experiment type. Normalized expression values are then available for use in a variety of analyses to further investigate the transcriptome of the sample being studied

and SBS, where only one tag is generated per transcript, the gene expression is enumerated as transcripts per million (TPM). For other transcript shotgun DGE methods such as RNA-Seq, gene expression is measured as transcripts per kilobase per million. Using these variables, it is possible to identify the differentially expressed genes within and across the samples (Reinartz et al. 2002) (Figure 9.2).

#### 9.5.2.1 Normalization and Filtering of Tag Signatures

The first stage of processing in DGE is to reduce the set of distinct signatures to determine an "abundance" value. This consists primarily of counting signatures based on their unique sequence. Raw abundance counts from multiple sequencing runs can be merged. Abundance values are determined by normalization to yield a relative percentage for each observed signature as transcripts per million (TPM):

$$\text{abundance}_{\text{normalized}} = \frac{\sum \text{raw abundance}_{\text{signature}}}{\sum \text{raw abundance}_{\text{all signature}}} \times 10^6.$$

The normalized value is largely independent of the total number of signatures in the library allowing comparison across libraries. In most cases normalization can be performed either before or after tag mapping. In the case of RNA-Seq experiments, normalization is refined during the mapping process. In this case, where each RNA transcript is represented by multiple unique reads, a more representative normalization value is to use a read "concentration" such as reads per length of the full-length transcript per million total reads. This should account for the distribution of potential reads between long and short transcripts better than an observed read percentage value.

The set of distinct signatures can be filtered using two criteria: reliability and significance. These filters are designed to remove noisy sequences that may have resulted from erroneous processes or systematic errors such as adapters or vector sequences. The reliability filter removes signatures observed in only one of the sequencing runs across all libraries (e.g., inconsistent expression, sequencing errors), and some literature suggests filtering signatures that are below 2 tpm, as very low expression (Meyers et al. 2004a).

#### 9.5.2.2 Database Resources for Mapping Signature Sequences

Signature sequences can be annotated by mapping to databases of known genes, annotated genome sequence, expressed sequence tags (EST), and protein sequences in both sense and antisense directions. For species lacking these resources, sequences from closely related species can be used for annotation because of the relatively high sequence conservation between genes. Gene annotation methods require classification of transcript signatures into mRNA, tRNA, rRNA, miRNA and snoRNAs. Accurate gene models can be constructed from signature sequence by discovery of splice junctions, promoters and transcription start sites.

#### 9.5.2.3 Generic Software for Mapping Signature Sequences

For mapping signatures to genes there are several software packages already available such as BLAST (Altschul et al. 1990), BLAT (Kent 2002) and MUMMER (Delcher et al. 1999). However, for next-generation sequencing data, more efficient index-filtering algorithms such as ELAND (Illumina Inc), SeqMap (Jiang and Wong 2008) and MAQ (Li et al. 2008) have been developed. These programs can map tens of millions of short sequences to a genome of several billions of nucleotides and can also handle multiple substitutions and insertions/deletions to account for sequencing errors and polymorphisms. Each of the software packages has distinct features. SeqMap can be run in parallel using multiple processors, while MAQ can process paired sequences. Although these programs can generate both ungapped and gapped alignments, gaps are not explicitly evaluated for splice junctions

#### 9.5.2.4 Transcript Mapping Software with Spliced Alignments

The open source software QPALMA (De Bona et al. 2008) has been designed for accurate spliced alignments. QPALMA uses a training set of spliced reads for computing splice junctions. The software uses a large margin approach similar to support vector machines, to estimate its parameters for maximizing alignment accuracy. To facilitate mapping of massive amounts of sequencing data, this method is combined with a fast mapping pipeline based on enhanced suffix arrays. The QPALMA algorithm has been optimized and demonstrated using reads produced with the Illumina Genome Analyzer for the model plant *A. thaliana*.

#### 9.5.2.5 Software Pipeline for DGE Data Analysis

The ERANGE software package (Mortazavi et al. 2008) developed by the Wold group performs analysis of DGE data from RNA-Seq experiments. ERANGE implements the following computational steps:

(a) Read mapping to the gene/genome by either direct assignment for unique matches or identify "best" site(s) for multireads.
(b) Detection of alternative splice sites
(c) Organization of reads that cluster together into candidate exons
(d) Calculation of the prevalence of transcripts for each known or new mRNA, based on normalized counts of unique reads, spliced reads, and multireads.

A list of additional software for DGE data analysis is given in Table 9.1. For more recent updates, the next generation sequencing community website: http://seqanswers.com can be helpful.

**Table 9.1** DGE processing software

| Name | URL | Ref |
|---|---|---|
| MAQ | http://sourceforge.net/projects/maq/ | (Li et al. 2008) |
| SeqMap | http://biogibbs.stanford.edu/~jiangh/SeqMap/ | (Jiang and Wong 2008) |
| ELAND | http://www.illumina.com/pages.ilmn?ID=268 | |
| ERANGE | http://woldlab.caltech.edu/rnaseq/ | (Mortazavi et al. 2008) |
| QPALMA | http://www.fml.mpg.de/raetsch/projects/qpalma | (De Bona et al. 2008) |
| SimCluster | http://xerad.systemsbiology.net/simcluster/ | (Vencio et al. 2007) |
| MSAGE | http://bioinf.wehi.edu.au/folders/msage/ | (Robinson and Smyth 2007) |

#### 9.5.2.6 Handling Unique and Redundant Matches

Sequence reads can be divided into two primary groups, unique reads and nonunique reads. Nonunique reads, although not identical, have high sequence similarity with other reads (e.g., a family of related sequences). A significant fraction of unique signatures can be mapped unambiguously to genomic sequence and known gene models or ESTs. Some signatures can be mapped to regions outside known exons, and these can either be merged into existing gene models or classified as novel genes. Similarly, nonunique reads that are usually the result of duplicated genes and segmental genome duplication can be mapped to a "best" match location based on abundance levels (Mortazavi et al. 2008). The remaining unassigned reads may be the result of sequencing errors or noncoding RNA such as tRNA, rRNA, miRNA and snoRNAs.

#### 9.5.2.7 Examining Patterns of Gene Expression

The abundance of mapped signatures allows for investigation of expression patterns within and across the samples. Comparisons between individual samples can be made using some of the methods employed in the analyses of microarrays and SAGE data (Reimers and Carey 2006). One such method is the use of the correlation coefficient $r$ between the logarithms of the gene expression vectors of all pairs of samples, using $d=(1-r)$ as a measure of the difference between the members of a pair (Jongeneel et al. 2005).

Hierarchical clustering can also be used to display the relationship between expression profiles. This can use the same distance measure as presented above, or alternatives such as Shannon entropy (Schug et al. 2005). The given expression levels of a gene in $N$ tissues, defines the relative expression of a gene g in a tissue t as

$$p_{t|g} = \frac{w_{\mathrm{g,t}}}{\sum\limits_{1 \geq \mathrm{t} \geq N} w_{\mathrm{g,t}}},$$

where $w_{\mathrm{g,t}}$ is the expression level of the gene in the tissue.

The entropy (Shannon 1949) of a gene's expression distribution is

$$H_{\mathrm{g}} = \sum_{1 \geq \mathrm{t} \geq N} -p_{t|g} \log(p_{t|g}).$$

$H_{\mathrm{g}}$ has units of bits and ranges from zero for genes expressed in a single tissue to $\log_2(N)$ for genes expressed uniformly in all tissues considered. The maximum value of $H_{\mathrm{g}}$ depends on the number of tissues considered, so this number is reported when appropriate. With the use of relative expression, the entropy of a gene is not sensitive to absolute expression levels.

Categorical tissue specificity can be defined as

$$Q_{g|t} = H_g - \log_2(p_{t|g}).$$

The quantity $-\log_2(p_{t|g})$ also has units of bits and has a minimum of zero that occurs when a gene is expressed in a single tissue, and grows unboundedly as the relative expression level drops to zero. Thus $Q_{g|t}$ is near its minimum of zero bits when a gene is relatively highly expressed in a small number of tissues, including the tissue of interest, and becomes higher as either the number of tissues expressing the gene becomes higher, or as the relative contribution of the tissue to the gene's overall pattern becomes smaller. By itself, the term $-\log_2(p_{t|g})$ is equivalent to $p_{t|g}$. Adding the entropy term serves to favor genes that are not expressed highly in the tissue of interest, but are expressed only in a small number of other tissues. Such genes should be considered as categorically tissue-specific since their expression pattern is very restricted (Schug et al. 2005).

One software tool that can be used for clustering genes based on expression level is SimCluster (Vencio et al. 2007), which was specifically designed for the efficient clustering of large data sets – DGE data in particular. SimCluster performs calculations on the gene expression data in the simplex vector space in $d$ dimensions (count of distinct tags) that is defined as

$$S_{d-1} = \left\{\pi \mid \pi \in \mathrm{R}_{+}^{d}, \pi 1' = 1\right\}.$$

Here, $d$ represents the number of distinct tags and 1 is a vec tor of 1's. Simcluster's method can be described as the use of a Bayesian inference step to determine the expected abundance simplex vectors from the observed counts. Where $p = \mathbb{E}[\pi|x]$, the translations on the simplex space is defined as

$$p \oplus t = \frac{(p \cdot t)}{(p \cdot t)1'},$$

where $\cdot$ is the usual Hadamard product and the division is vector-evaluated. SimCluster uses an Aitchisonean distance metric that is defined as

$$\Delta(p,q) = \sqrt{\ln\left(\frac{p_{-d}/p_d}{q_{-d}/q_d}\right)\left(I + 1' \times 1\right)^{-1} \ln\left(\frac{p_{-d}/p_d}{q_{-d}/q_d}\right)'}$$

where $I$ is the identity matrix, $\times$ is the Kronecker product, $-d$ subscript is a notation for "excluding the $d$th element," and elementary operations are vector-evaluated.

Several algorithms were implemented in Simcluster that include: k-means, k-medoids and self-organizing maps (SOM) for partition clustering, PCA for inferring the number of variability sources present, and common variants of agglomerative hierarchical clustering. Currently, the Simcluster package comprises Simtree, for hierarchical clustering; Simpart, for partition clustering; Simpca for Principal Component Analysis (PCA); and several utilities such as TreeDraw, a program to draw hierarchical clustering dendrograms with customizable tree leaf colors.

Robinson and Smyth (Robinson and Smyth 2007) developed a statistical package (edgeR) that uses a negative binomial distribution for modeling the over-dispersion of DGE data and weighted maximum likelihood estimating the dispersion.

### 9.5.3 DGE Data Databases, Visualization and Presentation

Given the large volumes of data generated from these high-throughput sequencers, it is important to manage the data effectively as it can easily overwhelm the available resources. For example, most laboratories now find it cost-effective to prepare a new sequencing run than store terabytes of image data. Meyers et al. (2004a) proposed a database architecture along with web-enabled query and display interface for DGE data. Human, Mouse and Cow gene atlas projects have used either custom browsers (Su et al. 2004) or present expression information as tracks in the UCSC Genome Browser (Kent et al. 2002) or Gbrowse (Stein et al. 2002) for data queries and graphical display. All DGE datasets can be deposited and queried from GEO (Barrett et al. 2007) database at NCBI. Genome and chromosome level views of mapped expression levels provide the additional benefit of aiding in the identification of transcriptionally active regions (TARs) along their sequence span. Large-scale gene expression datasets can be used as phenotypes in association studies for the identification of genomic regions termed expression QTL (eQTL) (Damerval et al. 1994; Stranger et al. 2007; Wu et al., 2008a).

## 9.6 Summary

Digital gene expression using next generation sequencing has several advantages over the traditional EST sequencing, SAGE or microarrays. Although advances in various sequencing technologies are enabling further reduction in the sequencing costs, algorithms and software for efficient data handling are still evolving and require significant improvements to meet current and future data analysis challenges. It is also important to consider the impact of bias introduced from each of the different sequencing platforms, while designing a DGE experiment. Currently, genome wide association studies (GWAS) in humans are largely conducted using high density SNP arrays or by sequencing candidate genes. Similar to the use of whole genome sequencing (1,000 genomes project) for GWAS, expression based association analyses (eQTL) will likely gain prominence in the near future. Significant reductions in the sequencing costs is also enabling large-scale transcriptome analysis in several plant and animal species that have been so far restricted to humans and few other model organisms.

## References

Adams MD, Kelley JM, Gocayne JD, Dubnick M, Polymeropoulos MH, Xiao H et al (1991) Complementary DNA sequencing: expressed sequence tags and human genome project. Science 252:1651–1656

Adams MD, Kerlavage AR, Fleischmann RD, Fuldner RA, Bult CJ, Lee NH et al (1995) Initial assessment of human gene diversity and expression patterns based upon 83 million nucleotides of cDNA sequence. Nature 377:3–174

Altschul SF, Gish W, Miller W, Myers EW, Lipman DJ (1990) Basic local alignment search tool. J Mol Biol 215:403–410

Ambros V (2001) microRNAs: tiny regulators with great potential. Cell 107:823–826

Bachellerie JP, Cavaille J, Huttenhofer A (2002) The expanding snoRNA world. Biochimie 84:775–790

Barrett T, Troup DB, Wilhite SE, Ledoux P, Rudnev D, Evangel-ista C et al (2007) NCBI GEO: mining tens of millions of expression profiles – database and tools update. Nucleic Acids Res 35:D760–D765

Birney E, Stamatoyannopoulos JA, Dutta A, Guigo R, Gingeras TR, Margulies EH et al (2007) Identification and analysis of functional elements in 1% of the human genome by the ENCODE pilot project. Nature 447:799–816

Blackshaw S, Harpavat S, Trimarchi J, Cai L, Huang H, Kuo WP et al (2004) Genomic analysis of mouse retinal development. PLoS Biol 2:E247

Boguski MS, Schuler GD (1995) Establishing a human transcript map. Nat Genet 10:369–371

Boguski MS, Lowe TM, Tolstoshev CM (1993) dbEST – database for "expressed sequence tags". Nat Genet 4:332–333

Calarco JA, Saltzman AL, Ip JY, Blencowe BJ (2007) Technologies for the global discovery and analysis of alternative splicing. Adv Exp Med Biol 623:64–84

Camargo AA, Samaia HP, Dias-Neto E, Simao DF, Migotto IA, Briones MR et al (2001) The contribution of 700,000 ORF sequence tags to the definition of the human transcriptome. Proc Natl Acad Sci USA 98:12103–12108

Claude E, Shannon A, mathematical theory of communication. Bell System Technical Journal, 27:379–423 and 623–656, July and October 1948. http://cm.bell-labs.com/cm/ms/what/shannonday/shannon1948.pdf

Cloonan N, Forrest AR, Kolle G, Gardiner BB, Faulkner GJ, Brown MK et al (2008) Stem cell transcriptome profiling via massive-scale mRNA sequencing. Nat Methods 5:613–619

Damerval C, Maurice A, Josse JM, de Vienne D (1994) Quantitative trait loci underlying gene product variation: a novel perspective for analyzing regulation of genome expression. Genetics 137:289–301

De Bona F, Ossowski S, Schneeberger K, Ratsch G (2008) Optimal spliced alignments of short sequence reads. Bioinformatics 24:i174–i180

de Hoon, M, Hayashizaki, Y (2008) Deep cap analysis gene expression (CAGE): genome-wide identification of promoters, quantification of their expression, and network inference. Biotechniques 44:627–628, 630, 632

Delcher AL, Kasif S, Fleischmann RD, Peterson J, White O, and Salzberg SL (1999) Nucleic Acids Research 27:11, 2369–2376

Delcher AL, Salzberg SL, Phillippy AM (2003) Using MUMmer to identify similar regions in large sequence sets. Current Protocols in Bioinformatics, Chapter 10:3

Dinel S, Bolduc C, Belleau P, Boivin A, Yoshioka M, Calvo E et al (2005) Reproducibility, bioinformatic analysis and power of the SAGE method to evaluate changes in transcriptome. Nucleic Acids Res 33:e26

Dojer N, Gambin A, Mizera A, Wilczynski B, Tiuryn J (2006) Applying dynamic Bayesian networks to perturbed gene expression data. BMC Bioinform 7:249

Gorodkin J, Cirera S, Hedegaard J, Gilchrist MJ, Panitz F, Jorgensen C et al (2007) Porcine transcriptome analysis based on 97 non-normalized cDNA libraries and assembly of 1, 021, 891 expressed sequence tags. Genome Biol 8:R45

Hamilton AJ, Baulcombe DC (1999) A species of small antisense RNA in posttranscriptional gene silencing in plants. Science 286:950–952

Hene L, Sreenu VB, Vuong MT, Abidi SH, Sutton JK, Rowland-Jones SL et al (2007) Deep analysis of cellular transcriptomes – LongSAGE versus classic MPSS. BMC Genomics 8:333

Hood L (2003) Systems biology: integrating technology, biology, and computation. Mech Ageing Dev 124:9–16

Hou J, Charters AM, Lee SC, Zhao Y, Wu MK, Jones SJ et al (2007) A systematic screen for genes expressed in definitive endoderm by serial analysis of gene expression (SAGE). BMC Dev Biol 7:92

Iandolino A, Nobuta K, da Silva FG, Cook DR, Meyers BC (2008) Comparative expression profiling in grape (Vitis vinifera) berries derived from frequency analysis of ESTs and MPSS signatures. BMC Plant Biol 8:53

Imoto S, Goto T, Miyano S (2002) Estimation of genetic networks and functional structures between genes by using Bayesian networks and nonparametric regression. Pac Symp Biocomput 175–186

Jiang H, Wong WH (2008) SeqMap : mapping massive amount of oligonucleotides to the genome. Bioinformatics 24:2395–2396

Jongeneel CV, Delorenzi M, Iseli C, Zhou D, Haudenschild CD, Khrebtukova I et al (2005) An atlas of human gene expression from massively parallel signature sequencing (MPSS). Genome Res 15:1007–1014

Kawaji H, Kasukawa T, Fukuda S, Katayama S, Kai C, Kawai J et al (2006) CAGE Basic/Analysis Databases: the CAGE resource for comprehensive promoter analysis. Nucleic Acids Res 34:D632–D636

Kent WJ (2002) BLAT – the BLAST-like alignment tool. Genome Res 12:656–664

Kent WJ, Sugnet CW, Furey TS, Roskin KM, Pringle TH, Zahler AM et al (2002) The human genome browser at UCSC. Genome Res 12:996–1006

Kim JB, Porreca GJ, Song L, Greenway SC, Gorham JM, Church GM et al (2007) Polony multiplex analysis of gene expression (PMAGE) in mouse hypertrophic cardiomyopathy. Science 316:1481–1484

Lau NC, Seto AG, Kim J, Kuramochi-Miyagawa S, Nakano T, Bartel DP et al (2006) Characterization of the piRNA complex from rat testes. Science 313:363–367

Li H, Ruan J, Durbin R (2008) Mapping short DNA sequencing reads and calling variants using mapping quality scores. Genome Res 18:1851–1858

Liu ET, Karuturi KR (2004) Microarrays and clinical investigations. N Engl J Med 350:1595–1597

Marioni JC, Mason CE, Mane SM, Stephens M, Gilad Y (2008) RNA-seq: An assessment of technical reproducibility and comparison with gene expression arrays. Genome Res 18:1509–1517

Meyers BC, Lee DK, Vu TH, Tej SS, Edberg SB, Matvienko M et al (2004a) *Arabidopsis* MPSS. An online resource for quantitative expression analysis. Plant Physiol 135:801–813

Meyers BC, Tej SS, Vu TH, Haudenschild CD, Agrawal V, Edberg SB et al (2004b) The use of MPSS for whole-genome transcriptional analysis in *Arabidopsis*. Genome Res 14:1641–1653

Morin R, Bainbridge M, Fejes A, Hirst M, Krzywinski M, Pugh T (2008) Profiling the HeLa S3 transcriptome using randomly primed cDNA and massively parallel short-read sequencing. Biotechniques 45:81–94

Mortazavi A, Williams BA, McCue K, Schaeffer L, Wold B (2008) Mapping and quantifying mammalian transcriptomes by RNA-Seq. Nat Methods 5:621–628

Nakano M, Nobuta K, Vemaraju K, Tej SS, Skogen JW, Meyers BC (2006) Plant MPSS databases: signature-based transcriptional resources for analyses of mRNA and small RNA. Nucleic Acids Res 34:D731–D735

Pan Q, Shai O, Misquitta C, Zhang W, Saltzman AL, Mohammad N et al (2004) Revealing global regulatory features of mammalian alternative splicing using a quantitative microarray platform. Mol Cell 16:929–941

Peiffer JA, Kaushik S, Sakai H, Arteaga-Vazquez M, Sanchez-Leon N, Ghazal H et al (2008) A spatial dissection of the *Arabidopsis* floral transcriptome by MPSS. BMC Plant Biol 8:43
Reimers M, Carey VJ (2006) Bioconductor: an open source framework for bioinformatics and computational biology. Methods Enzymol 411:119–134
Reinartz J, Bruyns E, Lin JZ, Burcham T, Brenner S, Bowen B et al (2002) Massively parallel signature sequencing (MPSS) as a tool for in-depth quantitative gene expression profiling in all organisms. Brief Funct Genomic Proteomic 1:95–104
Robinson MD, Smyth GK (2007) Moderated statistical tests for assessing differences in tag abundance. Bioinformatics 23:2881–2887
Saha S, Sparks AB, Rago C, Akmaev V, Wang CJ, Vogelstein B et al (2002) Using the transcriptome to annotate the genome. Nat Biotechnol 20:508–512
Schug J, Schuller WP, Kappen C, Salbaum JM, Bucan M, Stoeckert CJ Jr (2005) Promoter features related to tissue specificity as measured by Shannon entropy. Genome Biol 6:R33
Shannon, C (1949) The Mathematical Theory of Communication
Shiraki T, Kondo S, Katayama S, Waki K, Kasukawa T, Kawaji H et al (2003) Cap analysis gene expression for high-throughput analysis of transcriptional starting point and identification of promoter usage. Proc Natl Acad Sci USA 100:15776–15781
Siddiqui AS, Delaney AD, Schnerch A, Griffith OL, Jones SJM, Marra MA (2006) Sequence biases in large scale gene expression profiling data. Nucleic Acids Res 34:e83
Silva AP, De Souza JE, Galante PA, Riggins GJ, de Souza SJ, Camargo AA (2004) The impact of SNPs on the interpretation of SAGE and MPSS experimental data. Nucleic Acids Res 32:6104–6110
Stein LD, Mungall C, Shu S, Caudy M, Mangone M, Day A et al (2002) The generic genome browser: a building block for a model organism system database. Genome Res 12:1599–1610
Stranger BE, Forrest MS, Dunning M, Ingle CE, Beazley C, Thorne N et al (2007) Relative impact of nucleotide and copy number variation on gene expression phenotypes. Science 315:848–853
Su AI, Wiltshire T, Batalov S, Lapp H, Ching KA, Block D et al (2004) A gene atlas of the mouse and human protein-encoding transcriptomes. Proc Natl Acad Sci USA 101:6062–6067
Sultan M, Schulz MH, Richard H, Magen A, Klingenhoff A, Scherf M et al (2008) A global view of gene activity and alternative splicing by deep sequencing of the human transcriptome. Science 321:956–960
Sun M, Zhou G, Lee S, Chen J, Shi RZ, Wang SM (2004) SAGE is far more sensitive than EST for detecting low-abundance transcripts. BMC Genomics 5:1
Torres TT, Metta M, Ottenwalder B, Schlotterer C (2008) Gene expression profiling by massively parallel sequencing. Genome Res 18:172–177
Velculescu VE, Kinzler KW (2007) Gene expression analysis goes digital. Nat Biotechnol 25:878–880
Velculescu VE, Zhang L, Vogelstein B, Kinzler KW (1995) Serial analysis of gene expression. Science 270:484–487
Vencio RZ, Varuzza L, de BPC, Brentani H, Shmulevich I. (2007) Simcluster: clustering enumeration gene expression data on the simplex space. BMC Bioinform 8:246
Wang M, Master SR, Chodosh LA (2006) Computational expression deconvolution in a complex mammalian organ. BMC Bioinform 7:328
Weber AP, Weber KL, Carr K, Wilkerson C, Ohlrogge JB (2007) Sampling the *Arabidopsis* transcriptome with massively parallel pyrosequencing. Plant Physiol 144:32–42
Wu C, Delano DL, Mitro N, Su SV, Janes J, McClurg P et al (2008a) Gene set enrichment in eQTL data identifies novel annotations and pathway regulators. PLoS Genet 4:e1000070
Wu JQ, Du J, Rozowsky J, Zhang Z, Urban AE, Euskirchen G et al (2008a) Systematic analysis of transcribed loci in ENCODE regions using RACE sequencing reveals extensive transcription in the human genome. Genome Biol 9:R3
Xiong H (2006) Non-linear tests for identifying differentially expressed genes or genetic networks. Bioinformatics 22:919–923
Xiong M, Li J, Fang X (2004) Identification of genetic networks. Genetics 166:1037–1052

# Chapter 10
# Protein Sequence Databases

Terry Clark

## 10.1 Introduction

The near exponential growth in protein sequence data is at the foundation of transformations in biological research and related technological developments. The use of protein sequence data is widespread in fields including agronomy, biochemistry, ecology, etymology, evolution, genetics, genetic engineering, genomics, molecular phylogenetics and systematics, pharmacology, and toxicology. The remarkable increase in available protein sequences will most likely continue with the proliferation of genome sequencing projects, the latter enabled by ongoing improvements in DNA sequencing technology.[1] Along with opportunities, protein sequence data bring scientifically challenging problems.

Proteins are the unbranched polymers formed by a sequence of amino acids forming the protein polymer chain. This sequence describes the order of the amino acids and consequently the covalent structure of the polypeptide chain. The vast majority of protein sequence data are represented by the sequence of amino acids. The complete covalent structure is the *primary structure* of the protein and may include covalent cross links between distinct polypeptide chains. Sequence and primary structure are used interchangeably for single chain structures without cross links (Cantor and Schimmel 1980). It is customary to represent each of the twenty standard amino acids as a single letter code in the literature and in digital collections. Many bioinformatics analyses conducted on proteins consider the sequence of single letter codes alone. This representation is useful since the amino acid sequence is fundamental to protein structure and function. Several *secondary structure* elements are also useful in characterizing proteins. Some secondary structures can persist independently of the overall protein conformation. Examples of such

T. Clark (✉)
University of Queensland, Brisbane, QLD, Australia
e-mail: tclark@uq.edu.au

[1] The number of bases in GenBank has doubled in content approximately every 18 months since 1982 (Fig. 10.1). This growth is in stride with the doubling of components of integrated circuit density at approximately 18–24 month intervals, commonly referred to as Moore's Law (Moore 1965).

D. Edwards et al. (eds.), *Bioinformatics: Tools and Applications*,
DOI 10.1007/978-0-387-92738-1_10, © Springer Science+Business Media, LLC 2009

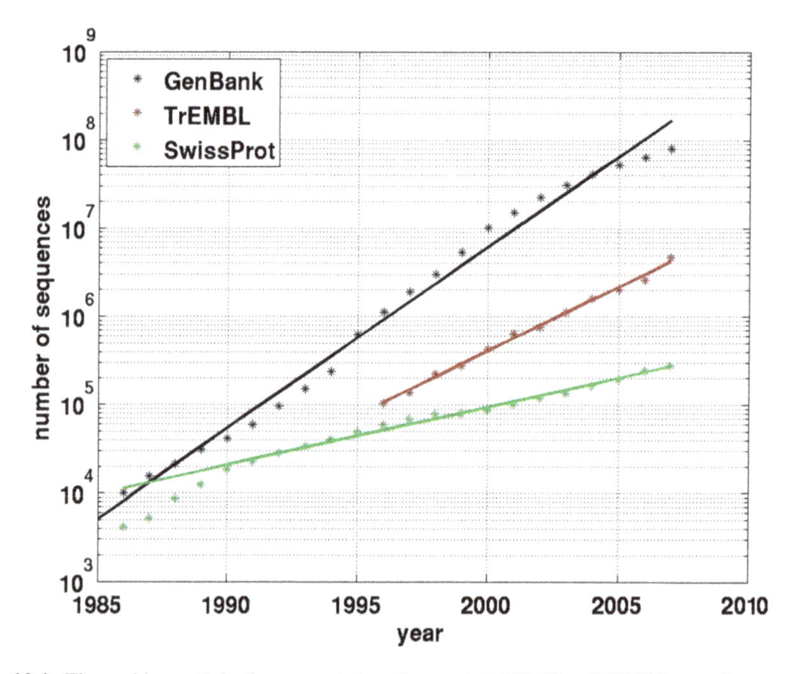


**Fig. 10.1** The rapid growth in the computationally annotated UniProt/TrEMBL protein sequence entries nearly tracks the exponential growth in GenBank sequences. The size, redundancy, and distribution of protein sequences can impede similarity searches. The UniProt Reference Clusters (UniRef) combines related protein sequences at three levels of sequence space resolution to increase the information obtained from homology searches. The smaller UniProt/Swiss-Prot consists of manually verified entries. It is difficult for manual curation to keep pace with computer-generated entries

elements include helices, sheets, and hyper-variable loops (Lesk 2001). Secondary elements come together to form the three-dimensional structure, also known as the *tertiary structure*.

Sanger and Thompson reported the first complete amino acid sequence for a protein in 1953 marking the beginning of protein sequence acquisition. Twelve years after Sanger and Thompson sequenced the insulin glycyl chain, the first edition of the Atlas of Protein Sequence and Structure was available on magnetic tape with 65 proteins (Dayhoff et al. 1965). In 1984, the Protein Information Resource (PIR) of the National Biomedical Research Foundation launched the first computerized protein sequence database, the PIR-International Protein Sequence Database (Barker et al. 1998). The UniProt Knowledgebase (UniProtKB) is now a central hub for the collection of functional information on protein sequences in the public domain (UniProt 2008a). The UniProt consortium includes the European Bioinformatics Institute, the Swiss Institute of Bioinformatics, and the Protein Information Resource (Leinonen et al. 2004). The primary database for manually curated protein sequences within UniProt is SwissProt (Boeckmann et al. 2005). Currently, the UniProt Knowledge Base contains approximately 1.8 billion amino acids; these are available in a compressed file of sequences in FASTA format (Pearson and Lipman 1988) and through specialized interfaces.

Digital collections of protein sequence data are rendered useful through their annotations. Efforts to classify protein sequences are ongoing (Camon et al. 2004; The Gene Ontology Consortium 2000; Kunin et al. 2003; Ouzounis et al. 2003; Raes et al. 2007; Rost and Valencia 1996). The challenges include computational tractability (Apweiler 2001; Moeller et al. 1999; Wieser et al. 2004), assessing and correcting the quality of function predictors (Friedberg 2006), and facilitating discovery within the protein sequence collections (Suzek et al. 2007). Contemporary computer-assisted analysis of protein function relies on sequence similarity (Fitch 1970; Pearson 1995; Altschul et al. 2005; Rost and Valencia 1996; Whisstock and Lesk 2003). The assessment of proteins through their sequence take into account metabolic pathways, protein secondary and quaternary structure, protein domains, and genome orientation.

At present, the majority of protein sequences in the public domain arise from computational predictions. The vast majority of these sequences are assessed for function through automated analysis with little to no human intervention. Protein sequences are routinely compared against large databases in these endeavors. The growth in sequence production over the past two decades has increased the demand for automated procedures to assess protein sequence similarity. In contrast to automated annotation, manual annotation of protein sequence data by human experts usually implies a critical review of experimental and computer-predicted assertions along with the literature and other lines of evidence. Current automated methods directed entirely by computer programs are less accurate in assessing protein function from sequence than expert manual curation. Nevertheless, the robust performance of automation addresses the wave of protein sequence data produced in genome and environmental studies.

## 10.2 Protein Classification Systems

The general aim of protein classification is to determine the role of proteins in biological processes. Protein classification based on structure, sequence, and function provides traction for organizing proteins in various ways. Ouzounis and colleagues argued that current classification methods based on similar sequence and structure (see following section) could benefit from unification into natural classification schemes that quantify physical relevance (Ouzounis et al. 2003).

The Gene Ontology Consortium provides the most widely used protein classification system, the ontologies collectively referred to as GO (The Gene Ontology Consortium 2000). GO represents the classification of proteins in three ontologies, each with a controlled vocabulary with well-defined interrelationships between the terms. The three ontologies distinguish molecular function, cellular component, and biological process. The association of a protein to one or more GO categories reflects the potential for multiple functions, locations, and processes. Each of the three ontologies takes the form of a directed acyclic graphic, where each node corresponds to a term that is more specialized than one or more parent nodes. UniProt and InterPro adapt and contribute to the GO terminology (Biswas et al. 2002), and both manual and automated methods utilize GO (Camon et al. 2004).

By the late 1950s, in a period when the number of known enzymes had increased to hundreds, the need for a systematic classification system was addressed. Toward this end, in 1956, The International Commission on Enzymes was established in consultation with the International Union of Pure and Applied Chemistry (IUPAC). The committee's charter was "*to consider the classification and nomenclature of enzymes and coenzymes, their units of activity, and standard methods of assay, together with the symbols used in the description of enzyme kinetics*" (NC-IUBMB 2008). The Enzyme Commission (EC) classification is oriented on the reactions catalyzed, not the enzymes. Whereas one goal of protein sequence database efforts is to map sequence and structure onto function, the EC classification system maps function onto proteins. EC numbers contain four fields in a four-level hierarchy. The first number specifies the enzyme class: oxidoreductases, transferases, hydrolases, lyases, isomerases, and ligases. The second and third numbers are context sensitive to the first number. To illustrate, for oxidoreductases, the second number describes the substrate and the third number describes the acceptor. The fourth number gives the specific enzymatic reaction catalyzed.

## 10.3 Analysis of Protein Sequences

The primary sequence-based analysis of proteins involves signatures, domains, and sequence similarity (Altschul et al. 1994; Durbin et al. 1998; Gribskov et al. 1987; Mulder 2007). In nearly all methods, sequence alignment is fundamental to the classification process. Many annotation systems begin by assigning proteins to families, groups of putative homologs (Gerlt and Babbitt 2001; Pontig 2001; Whisstock and Lesk 2003). Sequence and structural similarity with known proteins permit transfer of predicted function from characterized to unknown proteins. Current practice postulates that two proteins are homologous when sequence similarity is above a certain statistical threshold. Homology provides an evolutionary connection between proteins (Fitch 1970) which can be used as a basis to infer structural and functional similarities; such practice is with attendant caveats.

### *10.3.1 Alignment of Sequences*

The use of local sequence alignment is a standard approach to find similarity between protein sequences. Alignment algorithms model insertion, deletion, and substitution events using a scoring system based on biologically relevant probability distributions (Altschul 1991). Statistical interpretation of alignments can answer the question whether similarities between sequences signify an evolutionary or functional relationships, or simply arise by chance. A common practice is to orient an unknown protein using another protein of known function. Statistically significant sequence similarity of unknown protein A with annotated protein B can provide useful functional hypotheses for A based on B. The strengths of such hypotheses depend in part on the quality of the annotations of B. The BLAST program with rigorous statistical models is widely used for assessing protein-protein, protein-DNA, and DNA-DNA sequence similarity (Altschul et al. 1990; Altschul et al. 2005; Pearson 1995). Statistically significant alignments obtained from the BLAST search-and-align heuristic (Altschul et al. 1990) and other alignment methods (Pearson and Lipman 1988) can suggest an unknown protein's function through annotated homologs.

Manual classification by experts kept pace with protein discovery until the 1990s. Then, significant growth in key technological areas tipped the balance and manual curation fell significantly behind the mounting sequence collections (Fig. 10.1). Technological and intellectual advances in bioinformatics continue contemporaneously with the data surge. The quantitative assessment of sequence similarity is central among these advances. Early work in rigorous statistical models (Lipman and Pearson 1985; Karlin and Altschul 1990) together with the speed provided by the sequence-alignment program BLAST (Altschul et al. 1990) has had an extraordinary influence on creating, managing, and using protein-sequence data collections. Another pivotal development came in models to predict genes in genome sequence data (Burge and Karlin 1997; Koonin and Galperin 2002).

Although it is widely known that sequence does not necessarily determine function, it is a common practice to transfer function from one homolog to another. In an ideal world, one might assert that sequence implies structure and structure implies function, and therefore conclude that sequence implies function. However, this transitive relationship stands on shaky ground. The realization that divergent evolution can produce enzymes that catalyze different reactions (Gerlt and Babbitt 2001) is one of many reasons for the frequent qualification that sequence similarity does not necessarily imply function (Pearson and Wood 2001). Evidence for orthologs improves the case for inferring protein function from sequence (Fitch 1970; Mushegan 2007); however, the case is weaker for paralogs, as paralogs produced by a duplication event often diverge in function (Henikoff et al. 1997; Ganfornina and Sánchez 1999). The limitations attendant with sequence-based function inference (Whisstock and Lesk 2003) provide guidelines for its well-grounded use. Sangar and colleagues reported the encouraging result that, for proteins with greater than 50% sequence identity, the transfer of function between them will lead to an erroneous attribution in less than 6% of the cases (Sangar et al. 2007).

### *10.3.2 Domains and Motifs*

*Domains* are compact units of structure found in various combinations in proteins. Crucial to the definition of these conserved units is that they are capable of independently maintaining a characteristic structure apart from the bulk protein (Reddy and Bourne 2003). *Motifs* are short simple substructures identified within contiguous sequence. In contrast to the stand-alone character of domains, motifs usually depend on the overall protein context. The simple structure of motifs is straightforward to identify computationally (Durbin et al. 1998).Domains may appear as significant similarity in a local sequence alignment; this is true to a lesser extent for motifs. Pfam uses domains to group proteins (Bateman et al. 2002; Finn et al. 2006). Similarly, PROSITES categorizes proteins by motif (Sigrist et al. 2002; de Castro et al. 2006). Since biochemical functionality tends to be sensitive to overall protein integrity, function does not necessarily follow from the presence of a domain; however, domains are an important aspect of protein function prediction. In contrast to biochemical-process specificity where overall protein structure and binding capabilities are often essential, a single conserved domain can confer catalytic activity in a variety of contexts. It follows that, in divergent homologs, catalytic activity tends be better conserved than cellular and biochemical process functionality. Wilson and colleagues concluded that function was conserved relative to the EC classification system in pairs of single-domain proteins with $\geq$ 40% identical sequence (Wilson et al. 2000).

### *10.3.3 Profiles, Families, and Multiple Sequence Alignments*

Multiple sequence alignments are useful for identifying domains, motifs, and other features shared among a group of protein sequences. A multiple sequence alignment provides a basis for a profile spanning the length of a protein (Gribskov et al. 1987). Profile-based models are extremely useful for managing protein sequence data. Rapid search algorithms utilize profiles. An investigator armed with a query sequence can create a profile from protein sequence database with tools such as PSI-BLAST (Altschul et al. 1997). Curated profiles such as *hidden Markov model* representations of protein families (Eddy 1996; Durbin et al. 1998) provide a global strategy for organizing protein sequence data (Bateman et al. 2002). The PROSITE database of protein sequence patterns uses profiles in addition to motifs (Sigrist et al. 2002), the former similar to those reported by Gribskov and co-workers. Clustering methods unconcerned with specific protein features can be effective at meaningful groupings of protein sequences (Burke et al. 1999).

Methods to determine protein families vary. Pfam families use multiple sequence alignment, clustering, and statistical analysis of the sequence composition in protein domains and domain arrangement (Bateman et al. 2002). The structurally oriented SCOP database defines families consisting of proteins with 30% similarity, and proteins with lower sequence identity but with similar function and structure (Reddy and Bourne 2003). SCOP further delineates proteins into *superfamilies*; superfamilies form groups of distant homologs with potentially lower sequence identity than families, but sharing the same secondary structure in the same arrangement (Reddy and Bourne 2003). Pfam uses a concept similar to superfamilies to group families into *clans*. The clan relationship takes into account related structure and function and significant sequence similarity based on profile hidden Markov models (Finn et al. 2006). SwissProt semi-automatically curates families with expert interpretation of diverse evidence including the literature (Bairoch et al. 2004).

Gerlt and Babbit proposed a hierarchical classification of function for enzymes with the aim to incorporate structurally contextual definitions with chemistry-based relationships (Gerlt and Babbitt 2001). They observed that SCOP often groups enzymes by substrate specificity but may place diverged enzymes with similar active-site functional groups into separate superfamilies. In their classification, a *family* consists of homologous enzymes that catalyze the same reaction; a *superfamily* consists of homologous enzymes that catalyze similar reactions through different specificity or different overall reactions with common mechanistic attributes through conserved active-site residues; a *suprafamily* groups homologous enzymes that catalyze different reactions with different mechanistic attributes. One may envision evolutionary classes extended to include other physical and chemical traits.

## 10.4 Protein Sequence Databases

Protein sequence databases are numerous and growing. There has been considerable effort to integrate information associated with protein sequence data. Noteworthy developments include formation of the UniProt Consortium (UniProt Consortium 2008) together with integrated annotation methods by InterPro (Mulder et al. 2007). In this section, we introduce some key resources.

### *10.4.1 The UniProt Consortium*

The UniProt Consortium organizes principal protein-sequence repositories in the public domain (The UniProt Consortium 2007). Its databases consist of four components:

1. The UniProt Archive (*UniParc*) is oriented to provide a complete body of publically available protein sequence data without redundancy.
2. The central UniProt database of protein sequences is the UniProt Knowledgebase (*UniProtKB*) with highly curated and annotated sequences.
3. The UniProt Reference Clusters (*UniRef*) databases provide non-redundant views of sequences at varying levels of resolution.
4. The UniProt Metagenomic and Environmental Sequences (*UniMES*) database is a repository developed for high-throughput genome analysis for unclassified organisms.

The UniProtKB collection consists of the manually curated UniProt/Swiss-Prot and machine curated UniProt/TrEMBL. These sequence collections are available in compressed FASTA files.

**UniParc** is the main sequence warehouse for UniProt, a comprehensive repository aimed at maintaining all protein sequences at a single site. UniParc removes redundant sequences that are 100% identical and of the same length by merging them into single entries regardless of organism. The UniParc model intends sequences to be without annotation, from the viewpoint that annotation depends properly on biological context.

**SwissProt** is a manually curated, minimally redundant protein sequence database (Apweiler 2001; Boeckmann et al. 2005). This high-quality protein sequence database includes experimental findings, in addition to computed annotations. Many consider SwissProt to be one of the gold standards in the field.

**TrEMBL** represents the Translated EMBL Nucleotide Sequence Data Library. Introduced in 1996, it is the machine-annotated counterpart to SwissProt (Boeckmann et al. 2003). The purpose of TrEMBL is to accommodate the growth in protein sequences output from genome sequencing projects awaiting manual curation (Schneider et al. 2005). The number of sequences in TrEMBL has followed increases in DDBJ/EMBL/GenBank sequences over the past eight years (Fig. 10.1). InterPro uses an automated rule-based system with inferences deduced from SWISS-PROT to annotate UniProt TrEMBL (Biswas et al. 2002).

**UniRef** facilitates effective use of the large collection of translated CDS in TrEMBL (The UniProt Consortium 2007). It combines redundant sequences from UniProtKB and UniParc databases into three reduced sets. Currently these sets are UniRef100, 90, and 50. A single UniRef100 entry collects identical sequences and subfragments. UniRef100 is similar in scope and content to the NCBI RefSeq protein sequence database. Both UniRef and NCBI RefSeq non-redundant databases derive their protein sequences from similar repositories including DDBJ, EMBL, GenBank CDS translations, UniProtKB, and PDB. As of October 2006, UniRef100 contained 185,326 unique sequences from approximately 3.9 million proteins and protein fragments (Suzek et al. 2007). UniRef90 and UniRef50 loosen cluster criteria with 90 and 50% identity levels, resulting in fewer clusters and reduced data. The data size reduction from the core protein sequences in UniRef100, UniRef90, and UniRef50 are approximately 10, 40, and 65%, respectively (The UniProt Consortium 2007). Clusters are based on a similarity measure used by the Cd-hit program (Li and Godzik 2006).

**UniMES** was formed to accommodate sequences from organisms of unknown taxonomic origins; this is in contrast to UniProtKB, where entries are characterized with a known taxonomic source, making them unsuitable for the large-scale genomic analysis of uncharacterized microbes recovered from the recent the Global Ocean Sampling (GOS) expedition (Rusch et al. 2007) and other metagenomic sequencing projects.

### *10.4.2 NCBI and the Reference Sequence (RefSeq) Project*

The National Center for Biotechnology Information (NCBI) GenBank (Benson et al. 2000) is an international collection that includes data from the European

Molecular Biology Laboratories (EMBL) and the DNA Databank of Japan (DDBJ). NCBI protein resources also include data from UniProt. Data include translations of genome coding sections and cDNA sequences in GenBank (Benson et al. 2006), sequences from the Protein Research Foundation (PRF), UniProt, and the Protein Data Bank (PDB). The NCBI DNA-centric view links protein and nucleotide sequences to various resources including the NCBI genome browser (Pruitt et al. 2007) and the Entrez data retrieval system (Geer and Sayers 2003).

The NCBI Reference Sequence (RefSeq) database organizes protein sequence data in a genomic and taxonomical context with data from many hundreds of prokaryotes, eukaryotes, and viruses (Pruitt et al. 2007). The structure of RefSeq provides comprehensive annotated sequence data for species; a sequence is redundant in RefSeq if it matches some other sequence across 100% of its residues and is the same length. RefSeq is based on GenBank, but it is a distinct collection. NCBI uses both manual and automatic curation methods in RefSeq. At the time of writing, RefSeq contained approximately 5,120,043 protein sequences consisting of 181,848 *reviewed*, 87,245 *validated*, 2,776,342 *provisional*, 1,002,087 *predicted*, 534,070 *modeled*, and 554 *inferred*. Curated sequences have the status levels *validated* and *reviewed* with *reviewed* sequences at the highest level; other status levels are not curated (RefSeq 2008).

NCBI also maintains species-specific sequence collections for the sequence alignment interface (Benson et al. 2007). These collections are sub-categorized as non-redundant (nr), RefSeq, SwissProt, patented, Protein Data Bank, and environmental samples. NCBI provides BLAST formatted instances of the nr collection where sequences with 100% duplicate sequences represented once. RefSeq protein sequences and other NCBI sequence collections are available for FTP download.

### *10.4.3 The Protein Data Bank*

The Protein Data Bank (PDB) is a principal data repository for protein and DNA three-dimensional structures. Crystallographers established the PBD in the 1970s as a grassroot effort for a central repository of X-ray crystallography data (Berman et al. 2000, 2007). The historical underpinnings of the PDB continue to include the Research Collaboratory for Structural Bioinformatics (RCSB) PDB through to the Worldwide Protein Data Bank (wwPDB 2008; Berman et al. 2007; Henrick et al. 2008). Although the primary role of the PDB is to provide structural data, a component of the structure description is the protein sequence. In the PDB format, records for atoms are contained within a hierarchical structure that supplies the atom names and coordinates grouped by amino acid (or nucleic acids for DNA). In addition to the Cartesian coordinates, the PDB facilitates information about the entry's chemistry, origin, and other details (Berman et al. 2007). The number of structures (sequences with atomic coordinates) in the PDB number reached 50,000 on April 8, 2008 (RCSB PDB 2008). The majority of these structures result from X-ray crystallography; an increasing number result from NMR spectroscopy, with a small number of structures solved using electron microscopy (RCSB PDB 2008). Many of the structures are derived from biological contexts subject to post-translational modifications, whereas others may use recombinant expression techniques. The consumer of these data can determine the conditions of the source data from the annotations and the related literature.

### 10.4.4 Other Protein Sequence Databases

PROSITE archives a large collection of biologically meaningful signatures (motifs) expressed through regular-expression like patterns for short motif detection. Probabilistic profiles extend regular expressions in order to detect large domains (Finn et al. 2006); profiles are more adaptive to variation in sequence than the regular expressed based patterns. Signatures/patterns are determined manually from sequence alignment.

Pfam represents protein families using HMMs (Durbin et al. 1998). It maintains a hierarchical representation of protein sequence space based on families and clans (Finn et al. 2006). Pfam 22.0 contains 9,318 families, each represented by multiple-sequence alignments and hidden Markov models (HMMs). Pfam HMM profiles are available for downloading, and interactive web services provide access to Pfam alignments, domain analyses, and sequence similarity search engines. Simple client codes provided by Pfam allow for programmatic access.

The Protein Research Foundation (PRF) maintains the online Peptide/Protein Sequence Database (PRF/SEQDB) which includes amino acid sequences of peptides and proteins reported in the literature. The PRF website indicates that PRF/SEQDB contains some protein sequences not included in EMBL, GenBank, and SwissProt due to the literature based component (PRF 2008). PIR has developed the Super Family Classification System (PIRSF). The primary level for curation in the PIRSF model is based on the homeomorphic family, that is, proteins which are homologous and homeomorphic. These characteristics relate proteins evolved from a common ancestor and those sharing full-length sequence and domain similarity. The National Biomedical Research Foundation PIR established PIR in 1984 as a resource for the identification and interpretation of protein sequence data. Through its history, PIR has provided protein sequence databases and analysis tools to the scientific community. In 2002, it merged with the European Bioinformatics Institute and the Swiss Institute of Bioinformatics to form UniProt.

Data collections organized among specific organism-related groups (Benson et al. 2007; Gribskov et al. 2001; Pruitt et al. 2007) can facilitate interpretation of analyses. Organism clades variably group protein families (Finn et al. 2006; Schneider et al. 2005) using functional information (PlantsP 2008; The UniProt Consortium 2007; Ware et al. 2002). Organism clades may be biological kingdoms (Archae, Animalia, Eubacteria, Fungi, Protista, and Plantae), cellular structures (prokaryote and eukaryote), viral types, species, and others. Biochemical classifications include globular, membrane, and fibrous proteins (The UniProt Consortium 2007). The genomic resource for grasses, Gramene, contains a protein section searchable by PROSITE patterns and Pfam HMM profiles (Liang et al. 2008).

## 10.5 Current Trends

It can be computationally impractical to perform queries and similarity searches across unfiltered, comprehensive protein-sequence collections of the size of UniProt TrEMBL (Fig. 10.1). One problem centers on redundant sequences in the data collection. Since the expectation for an alignment score is proportional to the size of the database (Pearson 1995), removal of redundant or over-represented sequences can improve the interpretation of results. The orientation of some databases pivots on non-redundancy. UniProt/SwissProt (Boeckmann et al. 2005) is an important non-redundant protein database, whereas NCBI RefSeq (Pruitt et al. 2007) provides complementary DNA-centric views.

Protein families play a fundamental role in organizing collections. The general aim is to group homologs with attention toward distinguishing orthologs and paralogs (Mushegan 2007). Similarity methods used to classify sequences are numerous. These involve sequence alignment (Altschul et al. 1990), pattern finding (de Castro et al. 2006), clustering (Burke et al. 1999; Myers 1999), and machine learning (Yosef et al. 2008). Some clustering methods provide levels of views of cluster resolution with varying biological and computational motivation (Finn et al. 2006; Wu et al. 2006). The non-redundant UniProt/UniRef clusters sequences at several resolutions (Suzek et al. 2007). On average, the reach of protein similarity searches extends with a relaxation of clustering criteria. Computational requirements depend on the resolution of clustering and other representations.

The integration of classification methods among various protein sequence collections can improve annotation quality. For example, structure-based databases such as CATH (Orengo et al. 1999) and SCOP (Andreeva et al. 2004) assist primarily sequence-oriented annotation processes (Biswas et al. 2002; Mulder et al., 2007). InterPro integrates numerous approaches to strengthen protein-sequence classification. These include PROSITE patterns, Pfam hidden Markov model representations of protein domains, and structural classifications based on SCOP and CATH superfamilies (Biswas et al. 2002). The Protein Information Resource (PIR) has developed iProClass for functional characterization with integrated associative processing, which takes into account protein families, domains, and structure (Wu et al. 2004). It is becoming increasingly important to detect inconsistencies that occur through integrated methods (Biswas et al. 2002; Koonin and Galperin 2002; Wieser et al. 2004; Natale et al. 2005).

## References

Altschul SF (1991) Amino acid substitution matrices from an information theoretic prospective. J Mol Biol 219:555–565

Altschul SF, Gish W et al (1990) Basic local alignment search tool. J Mol Biol 215(3):403–410

Altschul SA, Boguski MS, Gish W, Wootton JC (1994) Issues in searching molecular sequence databases. Nat Genet 6:119–129
Altschul SF, Madden TL, Schaffer AA et al (1997) Gapped BLAST and PSI-BLAST: a new generation of protein database search programs. Nucleic Acids Res 25(17):3389–3402
Altschul SF, Wootton JC, Getz M et al (2005) Protein database searches using compositionally adjusted substitution matrices. FEBS J 272(20):5101–5109
Andreeva A, Howorth D, Brenner SE et al (2004) SCOP database in 2004: refinements integrate structure and sequence family data. Nucleic Acids Res 32:D226–D229
Apweiler R (2001) Functional information in Swiss-Prot: the basis for large-scale characterisation of protein sequences. Brief Bioinform 2:9–18
Bairoch A, Boeckmann B, Ferro S, Gasteiger E (2004) Swiss-Port: Juggling between evolution and stability. Briefings in Bioinformatics 5(1):39–55
Balaji S, Sujatha SN et al (2001) PALI-a database of alignments and phylogeny of homologous protein structures. Nucleic Acids Res 29:61–65
Barker WC, Garavelli JS, Haft DH et al (1998) The PIR-International protein sequence database. Nucleic Acids Res 26:27–32
Bateman A, Birney E, Cerruti L et al (2002) The Pfam protein families database. Nucleic Acids Res 30:276–280
Benson DA, Karsch-Mizarchi I, Lipman DJ, et al (2000) GenBank. Nucleic Acids Res 28(1):15–18
Benson DA, Karsch-Mizarchi I, Lipman DJ, et al (2007) GenBank. Nucleic Acids Res 36:D25–D30
Benson DA, Karsch-Mizarchi I, Karsch-Mizrachi I et al (2006) GenBank. Nucleic Acids Res 35:D21–D25
Berman HM, Westbrook J, Feng Z et al (2000) The protein data bank. Nucleic Acids Res 28:235–242
Berman HM, Henrick K, Nakamura H et al (2007) The Worldwide Protein Data Bank (wwPDB): Ensuring a single, uniform archive of PDB data. Nucleic Acids Res 35:D301–D303
Biswas M, O'Rourke JF, Camon E et al (2002) Applications of InterPro in protein annotation and genome analysis. Brief Bioinform 3(3):285–295
Boeckmann B, Bairoch A, Apweiler R et al (2003) The Swiss-Prot protein knowledgebase and its supplement TrEMBL. Nucleic Acids Res 31:365–370
Boeckmann B, Blatter MC, Farniglietti L et al (2005) Protein variety and functional diversity: Swiss-Prot annotation in its biological context. CR Biol 328:882–899
Burge C, Karlin S (1997) Prediction of complete gene structures in human genomic DNA. J Mol Biol 268:78–94
Burke J, Davison D, Hide W (1999) d2_cluster: A validated method for clustering EST and full-length cDNA sequences. Genome Res 9:1135–1142
Camon E, Magrane M, Barrell D et al (2004) The Gene Ontology Annotation (GOA database: sharing knowledge in UniProt with gene ontology. Nucleic Acids Res 32:D262–D266
Cantor CR, Schimmel PR (1980) Biophysical chemistry, Part I: The conformation of biological macromolecules. WH Freeman, San Francisco and Oxford
Dayhoff MO, Eck RV Chang M et al (1965) Atlas of protein sequence and structure, Vol 1. National Biomedical Research Foundation, Silver Spring, MD
de Castro E, Sigrist CJA, Gattiker A et al (2006) ScanProsite: detection of PROSITE signature matches and ProRule-associated functional and structural residues in proteins. Nucleic Acids Res 34:W362–W365
Durbin R, Eddy S, Krogh A, Mitchison G (1998) Biological sequence analysis. Cambridge University Press, Cambridge UK
Eddy SR (1996) Hidden Markov models. Curr Opin in Struct Biol 6:361–365
Finn RD, Mistry J, Schuster-Bockler B et al (2006) Pfam: clans, web tools and services. Nucleic Acids Res 34:D247–D251

Fitch WM (1970) Distinguishing homologous from analogous proteins. Syst Zool 19:99–113
Friedberg I (2006) Automated protein function prediction–the genomic challenge. Brief Bioinform 7(3):225–242
Ganfornina MD, Sánchez D (1999) Generation of evolutionary novelty by functional shift. BioEssays 21:432–439
Geer RC, Sayers EW (2003) Entrez: Making use of its power. Briefings in Bioinformatics 4(2):179–184
Gerlt JA, Babbitt PC (2001) Divergent evolution of enzymatic function: Mechanistically and functionally distinct suprafamilies. Annu Rev Biochem 70:209–246
Gribskov M, McLachlan AD, Eisenberg D (1987) Profile analysis: Detection of distantly related proteins. Proc Natl Acad Sci USA 84:4355–4358
Gribskov M, Fana F, Harper J et al (2001) PlantsP: a functional genomics database for plant phosphorylation. Nucleic Acids Res 29:111–113
Henikoff S, Greene SA, Piertrokovski S et al (1997) Gene families: The taxonomy of protein paralogs and chimeras. Science 278(5338):609–614
Henrick K, Feng Z, Bluhm WF (2008) Remediation of the protein data bank archive. Nucleic Acids Res 36:D426–D433
Karlin S, Altschul SF (1990) Methods for assessing the statistical significance of molecular sequence features by using general scoring schemes. Proc Natl Acad Sci USA 87:2264–2268
Koonin EV and Galperin MY (2002) Principles and methods of sequence analysis. In: Sequence–Evolution – Function, 1st edition. Kluwer, Waltham, MA
Kunin V, Cases I, Anton J et al (2003) Myriads of protein families, and still counting. Genome Biol 4:401
Leinonen R, Diez FG, Binns D et al (2004) UniProt Archive. Bioinformatics 20:3236–3237
Lesk AM (2001) Introduction to protein architecture. Oxford University Press, Oxford
Li W, Godzik A (2006) Cd-hit: a fast program for clustering and comparing large sets of protein or nucleotide sequences. Bioinformatics 22(13):1658–1659
Lipman DJ, Pearson WR (1985) Rapid and sensitive protein similarity searches. Science 227:1435–1441
Moeller S, Leser U, Fleischmann W, Apweiler R (1999) EDITtoTrEMBL: a distributed approach to high-quality automated protein sequence annotation. Bioinformatics 15:219–227
Moore GE (1965) Cramming more components onto integrated circuits. Electron Mag 38:8
Mulder NJ (2007) Protein family databases. Encyclopedia of life sciences Wiley, New York.
Mulder NJ, Apweiler R, Attwood TK et al (2003) The InterPro Database brings increased coverage and new features. Nucleic Acids Res 31(1):315–318
Mulder NJ, Apweiler R, Attwood TK et al (2007) New developments in the InterPro database. Nucleic Acids Res 35:D224–228
Mushegan AR (2007) Foundations of comparative genomics. Academic, Burlington, MA
Myers G (1999) A fast bit-vector algorithm for approximate string matching based on dynamic programming. J ACM 46:395–415
Natale DA, Vinakaya CR, Wu CH (2005) Large-scale, classification-driven, rule-based functional annotation of proteins. Encyclopedia Genet, Genomics, Proteomics Bioinform: . doi:10.1002/047001153X.g403314
NC-IUBMB (2008) Enzyme Nomenclature. http://www.chem.qmul.ac.uk/iubmb/enzyme/. Accessed 30 Apr 2008
Orengo CA, Peral FMG, Bray JE et al (1999) Assigigning genomic sequences to CATH. Nucleic Acids Res 28(1):277–282
Ouzounis CA, Coulson RMR, Enright AH et al (2003) Classification schemes for protein structure and function. Nat Rev Genet 4:508–519
Pearson WR (1995) Comparison of methods for searching protein sequence databases. Prot Sci 4:1145–1160

Pearson WR, Lipman DJ (1988) Improved tools for biological sequence analysis. Proc Natl Acad Sci USA 85:2444–2448
Pearson WR, Wood TC (2001) Statistical significance of biological sequence comparison. In: Bourne BE, Weissig H (eds) Handbook of statistical genetics. Wiley, West Sussex, England
PlantsP (2008) Functional genomics of plant phosphorylation. http://plantsp.genomics.purdue.edu/. Accessed 1 March 2008
Pontig CP (2001) Issues in predicting protein function from sequence. Brief Bioinform 2(1):19–29
PRF (2008) Protein Research Foundation. http://www.prf.or.jp/en/dbi.shtml/. Accessed 26 Oct 2008
Pruitt KD, Tatusova T, Maglott DR et al (2007) NCBI Reference Sequence (RefSeq): a curated non-redundant sequence database of genomes, transcripts and proteins. Nucleic Acids Res 35:D61–D65
Raes J, Harrington ED, Singh AH et al (2007) Protein function space: viewing the limits or limited by our view. Curr Opin Struct Biol 17:362–369
Reddy BVB, Bourne PE (2003) Protein structure evolution and the SCOP database. In: Bourne BE, Weissig H (eds) Structural bioinformatics, 1st edn. Wiley-Liss, Hoboken, NJ
RefSeq (2008) The National Center for Biotechnology Information: Reference Sequence database. http://www.ncbi.nlm.nih.gov/RefSeq/key.html#status/. Accessed 26 Feb 2008
Rost B, Valencia A (1996) Pitfalls of protein sequence analysis. Curr Opin Biotechnol 7:457–461
Rusch DB, Halpern AL, Sutton G et al (2007) The Sorcerer II Global Ocean Sampling expedition: Northwest Atlantic through Eastern tropical Pacific. PLoS Biol 5:398–431
Sangar V, Blankenberg DJ, Altman N et al (2007) Quantitative sequence-function relationship in proteins based on gene ontology. BMC Bioinform 8:294
Schneider M, Bairoch A, Wu CH et al (2005) Plant protein annotation in the UniProt Knowledgebase. Plant Physiol 138:59–66
Sigrist CJ, Cerutti L, Hulo N et al (2002) PROSITE: A documented database using patterns and profiles as motif descriptors. Brief Bioinform 3:265–274
Suzek BE, Huang H, McGarvey P et al (2007) UniRef: comprehensive and non-redundant UniProt reference clusters. Bioinformatics 23:1282–1288
The Gene Ontology Consortium (2000) Gene Ontology: tool for the unification of biology. Nat Genet 25:25–29
The UniProt Consortium (2007) The Universal Protein Resource (UniProt). Nucleic Acids Res 35:D193–D197
The UniProt Consortium (2008a) The Universal Protein Resource (UniProt). Nucleic Acids Res 35:D190–D195
The UniProt Consortium (2008b) The Universal Protein Resource (UniProt). Nucleic Acids Res 36:D190–D195
UniProt (2008) http://www.uniprot.org/. Accessed 30 Apr 2008
Ware D, Jaiswal P, Ni J et al (2002) Gramene: a resource for comparative grass genomics. Nucleic Acids Res 30:103–105
Whisstock JC, Lesk AM (2003) Prediction of protein function from protein sequence and structure. Q Rev of Biophys 36:307–340
Wieser D, Kretschmann E, Apweiler R (2004) Filtering erroneous protein annotation. Bioinformatics 20(1):i342–i347
Wilson CA, Kreychman J, Gerstein M (2000) Assessing annotation transfer for genomics: Quantifying the relations between protein sequence, structure and function through traditional and probabilistic scores. J Mol Biol 297:233–249
Wu CH, Nikolskaya A, Huang H et al (2004) PIRSF: family classification system at the Protein Information Resource. Nucleic Acids Res 32:D112–D114

Wu CH, Apweiler R, Bairoch A et al. (2006) The Universal Protein Resource (UniProt): an expanding universe of protein information. Nucleic Acids Res 34:D187–D191
wwPDB (2008) Worldwide Protein Data. http://www.wwpdb.org/. Accessed 8 Sept 2008
Yosef N, Sharan R, Noble WS (2008) Improved network-based identification of protein orthologs. Bioinformatics 24(16):i200–i206

# Chapter 11
# Protein Structure Prediction

**Sitao Wu and Yang Zhang**

## 11.1 Introduction

Owing to significant efforts in genome sequencing over nearly three decades (McPherson et al. 2001; Venter et al. 2001), gene sequences from many organisms have been deduced. Over 100 million nucleotide sequences from over 300 thousand different organisms have been deposited in the major DNA databases, DDBJ/EMBL/GenBank (Benson et al. 2003; Miyazaki et al. 2003; Kulikova et al. 2004), totaling almost 200 billion nucleotide bases (about the number of stars in the Milky Way). Over 5 million of these nucleotide sequences have been translated into amino acid sequences and deposited in the UniProtKB database (Release 12.8) (Bairoch et al. 2005). The protein sequences in UniParc triple this number. However, the protein sequences themselves are usually insufficient for determining protein function as the biological function of proteins is intrinsically linked to three dimensional protein structure (Skolnick et al. 2000).

The most accurate structural characterization of proteins is provided by X-ray crystallography and NMR spectroscopy. Owing to the technical difficulties and labor intensiveness of these methods, the number of protein structures solved by experimental methods lags far behind the accumulation of protein sequences. By the end of 2007, there were 44,272 protein structures deposited in the Protein Data Bank (PDB) (www.rcsb.org) (Berman et al. 2000) – accounting for just one percent of sequences in the UniProtKB database (http://www.ebi.ac.uk/swissprot). Moreover, the gap between the number of protein sequences and the number of structures has been increasing as indicated in Fig. 11.1.

One of the major efforts in protein structure determination in recent years is the structural genomics (SG) project initiated at the end of last century (Sali 1998; Terwilliger et al. 1998; Burley et al. 1999; Smaglik 2000; Stevens et al. 2001). The SG project aims to obtain 3D models of all proteins by an optimized combination of experimental

Y. Zhang (✉)
Center for Bioinformatics and Department of Molecular Bioscience, University of Kansas, Lawrence, KS, 66047
e-mail: yzhang@ku.edu

D. Edwards et al. (eds.), *Bioinformatics: Tools and Applications*,
DOI 10.1007/978-0-387-92738-1_11, © Springer Science+Business Media, LLC 2009

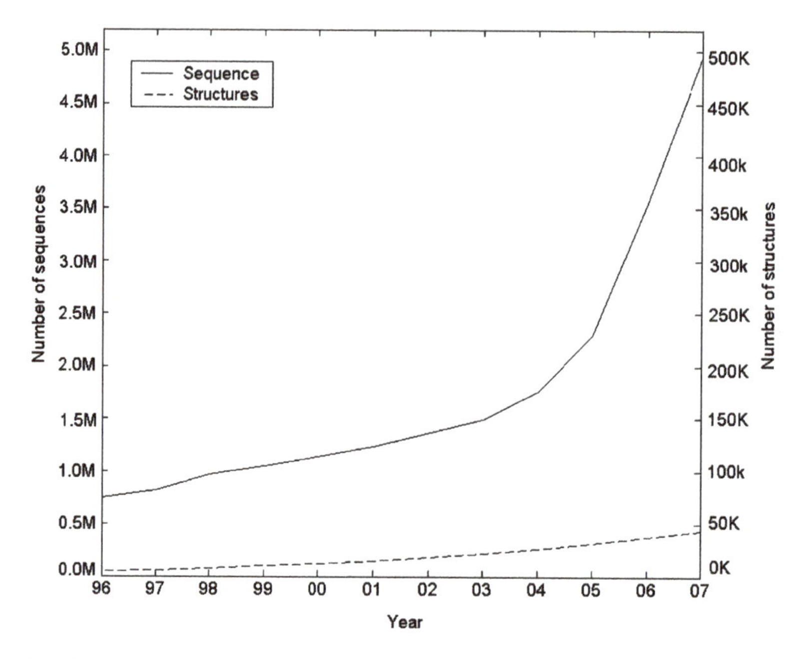


**Fig. 11.1** Determination of amino acid sequences (left-hand scale) is outpacing that of 3D structures (right-hand scale) by a factor of 100. Data are taken from PDB (Berman et al. 2000) and UniProtKB (Bairoch et al. 2005)

structure determination and comparative model (CM) building (Pieper et al. 2006). One of the key aspects of the SG project is the selection of key target proteins for structure determination, so that the majority of sequences can be within a CM distance to solved structures. Using a sequence identity of 30% with 80% alignment coverage as the CM distance cutoff, Vitkup et al. (2001) estimated that at least 16,000 new structures need to be determined by experiments to ensure that the CM represents 90% of protein domain families. Without optimal coordination of target selection, as many as 50,000 structure determinations may be required.

Currently, 36% of Pfam families (Bateman et al. 2004) contain at least one member with the solved structure, allowing comparative modeling of other family members. According to Chandonia and Brenner (Chandonia and Brenner 2006), the SG project solved 1,887 protein structures between 2000 and 2005, 294 of which are the first solved structures in their respective Pfam families. During 2004, around half of the PDB structures with new Pfam family annotations were because of the efforts of the SG centers (Chandonia and Brenner 2006). Determination of these new Pfam structures has dramatically extended the range of computer-based predictions using Comparative Model (CM) techniques (Sali 1998; Pieper et al. 2006). For example, based on 53 newly solved proteins from SG projects, Sali and coworkers (Pieper et al. 2004) built reliable models for domains in 24,113 sequences from the UniProtKB database with their CM tool MODELLER (Sali and Blundell 1993). These models have been deposited in a comprehensive CM model database, MODBase (http://salilab.org/modbase). In February 2008, MODBase contained around 4.3 million models or fold assignments for domains from 1.34 million sequences. In this study, the structure assignments were based on an all-against-all search of the amino acid sequences in UniProtKB using the solved protein structures in PDB (Berman et al. 2000). Structural genomics can also benefit from improvements in high-resolution structure prediction algorithms. Vitkup et al. (2001) estimated that "a 10% decrease in the threshold needed for accurate modeling, from 30 to 20% sequence identity, would reduce the number of experimental structures required by more than a factor of two".

There are two critical problems in the field of protein structure prediction. The first problem is related to the template-based modeling: How to identify the most suitable templates from known protein structures in the PDB library? Furthermore, following template structure identification, how can the template structures be refined to better approximate the native structure? The second major problem is related to free modeling for the target sequences without appropriate templates: How can a correct topology for the target proteins be constructed from scratch? Progress made in these areas has been assessed in recent CASP7 experiments (Moult et al. 2007) under the categories of template based modeling (TBM) and free modeling (FM), respectively.

In the following sections, current protein structure prediction methods will be reviewed for both template-based modeling and free modeling. The basic ideas and advances of these directions will be discussed in detail.

## 11.2 Template-Based Predictions

For a given target sequence, template-based prediction methods build 3D structures based on a set of solved 3D protein structures, termed the template library. The canonical procedure of template-based modeling consists of four steps: (1) finding known structures (templates) related to the sequence to be modeled (target); (2) aligning the target sequence on the template structures; (3) building the structural framework by copying the aligned regions, or by satisfying spatial restraints from templates; (4) constructing the unaligned loop regions and adding side-chain atoms. The first two steps are usually performed as a single procedure because the correct selection of templates relies on their accurate alignment with the target. Similarly, the last two steps are also performed simultaneously since the atoms of the core and loop regions interact closely.

Historically, template-based methods can be categorized into two types: (1) comparative modeling (CM) and (2) threading. CM builds models based on evolutionary information between target and template sequences, while threading is designed to match target sequences directly onto 3D structures of templates with the goal to detect target-template pairs even without evolutionary relationships. The schematic overview of CM and threading is depicted in the upper part of Fig. 11.2. In recent years, as a general trend in the field, the borders between CM and threading are becoming increasingly blurred since both comparative modeling and threading methods rely on evolutionary relationships, e.g. both use sequence profile-based

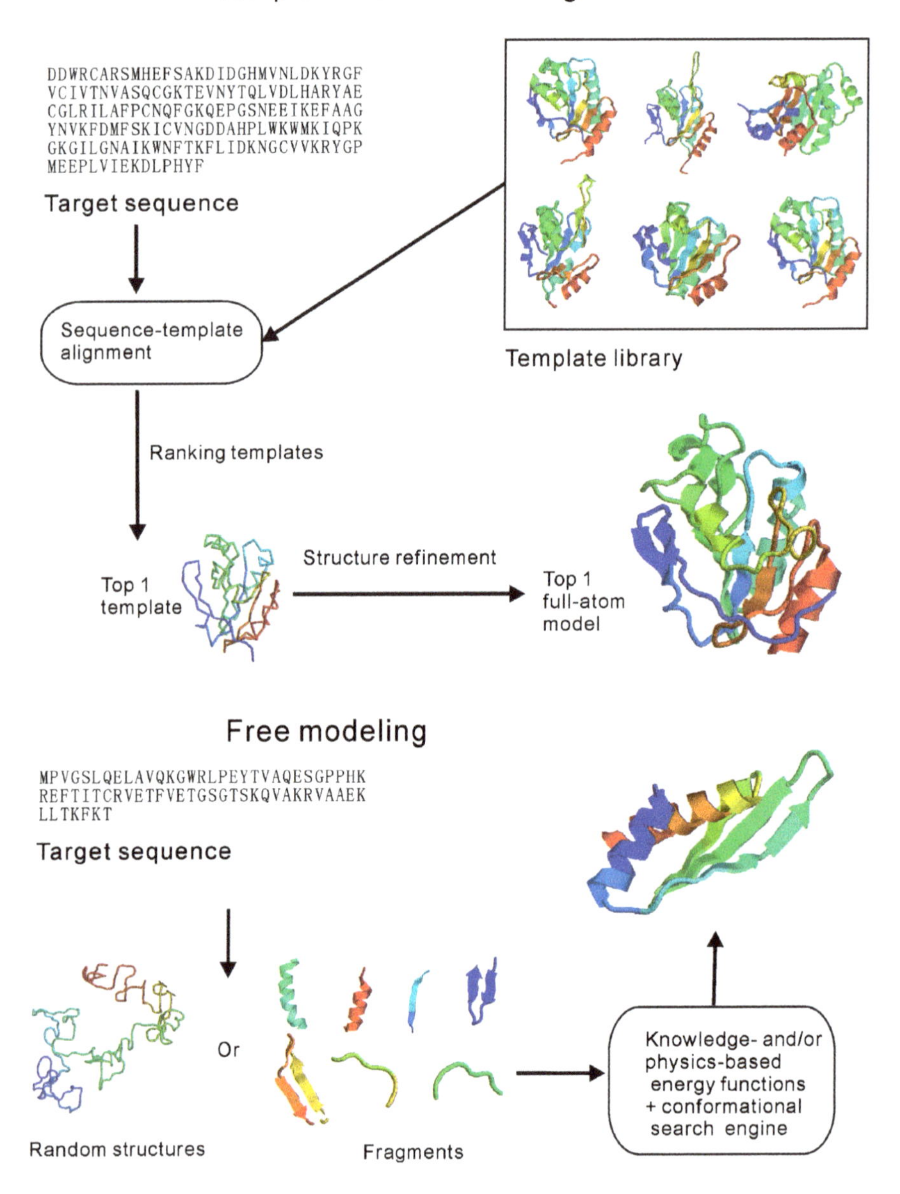


**Fig. 11.2** Schematic overview of the methodologies employed in template-based and free modeling

alignments (Marti-Renom et al. 2000; Skolnick et al. 2004; Zhou and Zhou 2005; Wu and Zhang 2008). In this chapter, we put them in the same category of template-based modeling without explicitly distinguishing them unless necessary.

### 11.2.1 *Completeness of the PDB Template Library*

The existence of similar structures to the target in the PDB is a precondition for successful template-based modeling. An important concern is thus the completeness of the current PDB structure library. Figure 11.3 shows a distribution of the best templates found by the structural alignment (Zhang and Skolnick 2005b) for 1,413 representative single-domain proteins between 80 and 200 residues.

Remarkably, even excluding the homologous templates of sequence identity, >20%, all the proteins have at least one structural analog in the PDB with a $C_\alpha$ root-mean-squared deviation (RMSD) to the target <6 Å covering >70% of regions. The average RMSD and coverage are 2.96 Å and 86% respectively. Zhang and Skolnick (2005a,b) recently showed that high quality full-length models can be built for all the single-domain proteins with an average RMSD of 2.25 Å when using the best possible templates in the PDB. These data demonstrate that the structural universe of the current PDB library is likely to be complete for solving the protein structure for at least single-domain proteins. However, most of the target-template pairs have only around 15% sequence identity, which are difficult to

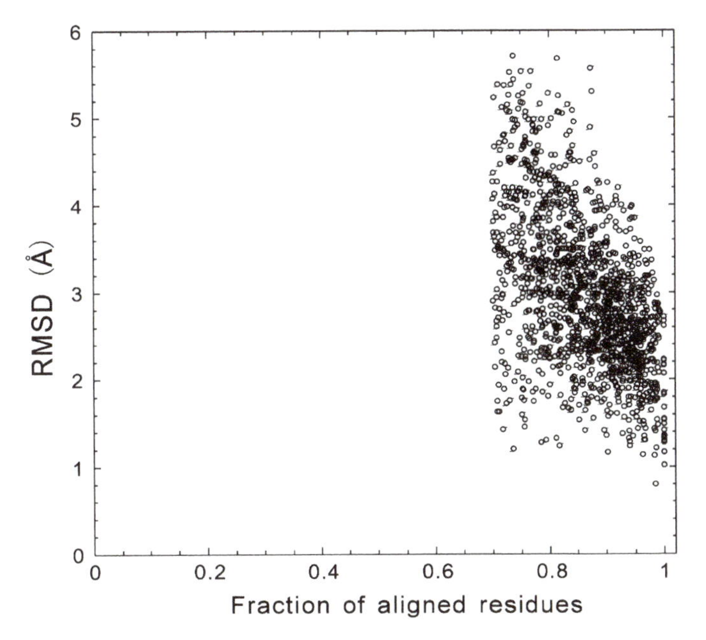


**Fig. 11.3** Structural superposition results of 1,413 representative single-domain proteins on their structural analogs in the PDB library. The structural analogs are found using a sequence-independent structural-alignment tool, TM-align (Zhang and Skolnick 2005b), and ranked by a TM-score (a structural similarity measure balancing RMSD and coverage) (Zhang and Skolnick 2004b). All structural analogs with a sequence identity >20% to the target are excluded. If the analog with the highest TM-score has a coverage below 70%, the first structural analog with the coverage >70% is presented. As a result, all the structural analogs have a root-mean-squared deviation (RMSD) <6 Å; 80% have a RMSD <4 Å with >75% of regions covered

recover by current threading approaches. In fact, after excluding templates with a sequence identity >30%, current threading techniques could only assign templates with correct topologies (average RMSD ~4 Å) to 2/3 of the proteins (Skolnick et al. 2004). Here, the role of structural genomics is to bridge the target-template gap for the remaining 1/3 proteins as well as improve the alignment accuracy of the 2/3 proteins by providing evolutionarily closer template proteins.

### 11.2.2 Template Structure Identification Using Threading Programs

Since its first application in the early 1990s (Bowie et al. 1991; Jones et al. 1992), threading has become one of the most active areas in proteins structure prediction. Numerous algorithms have been developed during the previous 15 years for the purpose of identifying structure templates from the PDB. Threading techniques include sequence profile–profile alignments (Ginalski et al. 2003; Skolnick et al. 2004; Jaroszewski et al. 2005; Zhou and Zhou 2005), structural profile alignments (Shi et al. 2001), hidden Markov models (HMM) (Karplus et al. 1998; Soding 2005), and machine learning (Jones 1999; Cheng and Baldi 2006) among others.

The sequence profile–profile alignment (PPA) is probably the most often-used and robust threading approach. Instead of matching the single sequences of target and template, PPA aligns a target multiple sequence alignment (MSA) with a template MSA. The alignment score in the PPA is usually calculated as a product of the amino-acid frequency at each position of the target MSA with the log-odds of the matching amino-acid in the template MSA, though there are also alternative methods for calculating the profile–profile alignment scores (Sadreyev and Grishin 2003). Profile–profile alignment based methods demonstrated advantages in several recent blind tests (Fischer et al. 2003; Rychlewski and Fischer 2005; Battey et al. 2007). In LiveBench-8 (Rychlewski and Fischer 2005), for example, the top four servers (BASD/MASP/MBAS, SFST/STMP, FFAS03, and ORF2/ORFS) are all based on sequence profile–profile alignment. In CAFASP (Fischer et al. 2003) and the recent CASP Server Section (Battey et al. 2007), several sequence profile based methods were ranked at the top of single threading servers. Wu and Zhang (2008) recently showed that the accuracy of the sequence profile–profile alignments can be further improved by about 5–6% by incorporating a variety of additional structural information.

In CASP7, HHsearch (Soding 2005), a HMM–HMM alignment method, was distinguished as the best single threading server. The principles of the HMM–HMM alignments and the profile–profile alignments are similar in that both attempt pair-wise alignments of the target MSA with the template MSA. Instead of representing the MSAs by sequence profiles, HHsearch uses profile HMMs which can generate the sequences with certain probabilities determined by the product of the amino acid emission and insertion/deletion probabilities. HHsearch aligns the target and template HMMs by maximizing the probability that two models co-emit the same amino acid sequence. In this way, amino acid frequencies and insertions/deletions of both HMMs are matched in an optimum way (Soding 2005).

### *11.2.3 Consensus of Various Threading Programs: Meta-Servers*

Although average performance differs among threading algorithms, there is no single threading program which outperforms all others on every target. This naturally leads to the popularity of the meta-server (Fischer 2003; Wu and Zhang 2007), which collects and combines results from a set of existing threading programs. There are two ways to generate predictions in meta-servers. One is to build a hybrid model by cutting and pasting the selected structure fragments from the templates identified by threading programs (Fischer 2003). The combined model has on average larger coverage and better topology than any single template. One defect is that the hybrid models often have non-physical local clashes. The second approach is to select the best models based on a variety of scoring functions or machine-learning techniques. This approach has emerged as a new research area called Model Quality Assessment Programs (MQAP) (Fischer 2006). Despite considerable efforts in developing various MQAP scores, the most robust score turns out to be the one based on the structure consensus, i.e. the best models are those simultaneously hit by different threading algorithms. The idea behind the consensus approach is simple: there are more ways for a threading program to select a wrong template than a right one. Therefore, the chance for multiple threading programs working collectively to make a commonly wrong selection is lower than the chance to make a commonly correct selection.

The meta-server predictors have dominated the server predictions in previous experiments (e.g. CAFASP4, Livebench8, and CASP6). However, in the recent CASP7 experiment (Battey et al. 2007) Zhang-Server (an automated server based on profile–profile threading and I-TASSER structure refinement (Wu et al. 2007; Zhang 2007)) clearly outperforms others (including the meta-servers which included it as an input (Wallner and Elofsson 2007)). A list of the top ten automated servers in the CASP7 experiment is shown in Table 11.1. This data, highlights the challenge of the MQAP methods in correctly ranking and selecting the best models; while the success of the composite threading plus refinement servers (as Zhang-Server, ROBETTA, and MetaTasser) demonstrates the advantage of the structure refinement in the TBM prediction, which is discussed in the next section.

### *11.2.4 Template Structure Assembly/Refinement*

The goal of protein structure assembly/refinement is to draw the templates closer to the native structure. This has proven to be a non-trivial task. Until only a few years ago, most of the TBM procedures either kept the templates unchanged or drove the templates away from the native structures.

Early efforts on template structure refinement have relied on molecular dynamics (MD) based atomic-level simulations; these attempt to refine low-resolution models using classic MD programs such as AMBER and CHARMM. However, with the

**Table 11.1** Top 10 servers in CASP7 as ranked by the accumulative GDT-TS score. Multiple servers from the same lab are represented by the highest rank one

| Servers | # of targets | GDT-TS | Server type and URL address |
|---|---|---|---|
| Zhang-Server | 124 | 76.04 | Threading, refinement and free modeling http://zhang.bioinformatics.ku.edu/I-TASSER |
| HHpred2 | 124 | 71.94 | HMM–HMM alignment (single threading server) http://toolkit.tuebingen.mpg.de/hhpred |
| Pmodeller6 | 124 | 71.69 | Meta threading server http://pcons.net |
| CIRCLE | 124 | 71.09 | Meta threading server http://www.pharm.kitasato-u.ac.jp/fams/fams.html |
| ROBETTA | 123 | 70.87 | Threading, refinement and free modeling http://robetta.org/submit.jsp |
| MetaTasser | 124 | 70.77 | Threading, refinement and free modeling http://cssb.biology.gatech.edu/skolnick/webservice/MetaTASSER |
| RAPTOR-ACE | 124 | 69.70 | Meta threading server http://ttic.uchicago.edu/~jinbo/RAPTOR_form.htm |
| SP3 | 124 | 69.38 | Profile–profile alignment (single threading server) http://sparks.informatics.iupui.edu/hzhou/anonymous-fold-sp3.html |
| beautshot | 124 | 69.26 | Meta threading server http://inub.cse.buffalo.edu/form.html |
| UNI-EID-expm | 121 | 69.13 | Profile–profile alignment (single threading server) (not available) |

exception of some isolated instances, this approach has not achieved systematic improvements.

Encouraging template refinements have been achieved by combining the knowledge- and physics-based potentials with spatial restraints from templates (Zhang and Skolnick 2005a; Misura et al. 2006; Chen and Brooks 2007). Misura et al. (2006) first built low-resolution models with ROSETTA (Simons et al. 1997) using a fragment library enriched by the query-template alignment. The $C_\beta$-contact restraints are used to guide the assembly procedure, and the low-resolution models are then refined by a physics-based atomic potential. As a result, in 22 out of 39 test cases, the ten lowest-energy models were found closer to the native structure than the template.

A more comprehensive test of the template refinement procedure, based on TASSER simulation combined with consensus spatial restraints from multiple templates, was reported by Zhang and Skolnick (2004a,b, 2005a,b). For 1,489 test cases, TASSER reduced the RMSD of the templates in the majority of cases, with an average RMSD reduction from 6.7 Å to 4.4 Å over the threading-aligned regions. Even starting from the best templates identified by the structural alignment, TASSER refines the models from 2.5 Å to 1.88 Å in the aligned regions. Here, TASSER built the structures based on a reduced model (specified by $C_\alpha$ and side-chain center of mass) with a purely knowledge-based force field. One of the major contributions to these refinements is the use of multiple threading templates, where the consensus restraint is more accurate than that from the individual template.

In addition, the composite knowledge-based energy terms have been extensively optimized using large-scale structure decoys (Zhang et al. 2003) which helps coordinate the complicated correlations of different interaction terms.

The recent CASP7 experiment assessed the progress of threading template refinements. The assessment team compared the predicted models with the best possible structural template (or "virtual predictor group") and commented that "The best group in this respect (24, Zhang) managed to achieve a higher GDT-TS score than the virtual group in more than half the assessment units and a higher GDT-HA score in approximately one-third of cases." (Kopp et al. 2007) This comparison may not entirely reflect the template refinement ability of the algorithms because the predictors actually start from threading templates rather than the best structural alignments; the latter requests the information of the native structures, which were not available when the predictions were made. On the other hand, a global GDT score comparison may favor the full-length model because the template alignment has a shorter length than the model. In a direct comparison of the RMSD over the same aligned regions, we found that the first I-TASSER model is closer to the native than the best initial template in 86 out of 105 TBM cases, while the other 13 (6) cases are worse than (or equal to) the template. The average RMSD is 4.9 Å and 3.8 Å for the templates and models, respectively, over the same aligned regions (Zhang 2007).

## 11.3 Free Modeling

When structural analogs do not exist in the PDB library or could not be detected by threading (which is more often the case as demonstrated by the data shown in Fig. 11.3), the structure prediction has to be generated from scratch. This type of prediction has been termed ab initio or de novo modeling, a term that may be easily understood as modeling "from first principle". Since CASP7, it is termed *free modeling*, which more appropriately reflects the status of the field, since the most efficient methods in this category still consider hybrid approaches including both knowledge-based and physics-based potentials. Evolutionary information is often used in generating sparse spatial restraints or identifying local structural building blocks.

### *11.3.1 Physics-Based Free Modeling*

Compared to template-based approaches, the purely physics-based ab initio methods – all-atom potential functions, like AMBER (Weiner et al. 1984), CHARMM (Brooks et al. 1983) and OPLS (Jorgensen and Tirado-Rives 1988), combined with molecular dynamics (MD) conformational sampling – have been less successful in protein structure prediction. Significant efforts have been made on the purely physics-based protein folding. The first widely recognized milestone of successful ab initio protein folding is the 1997 work of Duan and Kollman, who folded the *villin headpiece* (a 36-mer). This work used MD simulations in explicit solvent for 2 months on parallel supercomputers with models up to 4.5 Å (Duan and Kollman 1998). With the help of the worldwide-distributed computers, this small protein was recently folded by Pande and coworkers (Zagrovic et al. 2002) to 1.7 Å with a total simulation time of 300 μs or approximately 1,000 CPU years. Despite this remarkable effort, physics-based folding is far from routine for general protein structure prediction of normal size proteins, mainly because of the prohibitive computing demand.

Another niche for physics-based simulation is protein-structure refinement. This approach starts from low-resolution structures with the goal to draw the initial models closer to the native structure. Because the starting models are usually not far away from the native, the conformational change is relatively small and the simulation time is much less than in ab initio folding. One of the earliest MD-based protein structure refinements was for the GCN4 leucine zipper (a 33 residue dimer) (Nilges and Brunger 1991; Vieth et al. 1994). In that work, a low resolution coiled-coil dimer structure (2 ~ 3 Å) was first assembled using Monte Carlo simulation. With the help of the helical dihedral-angle restraints, Skolnick and coworkers (Vieth et al. 1994) refined the GCN4 structure with a backbone RMSD below 1 Å using CHARMM (Brooks et al. 1983) with the TIP3P water model (Jorgensen et al. 1983). Using AMBER 5.0 (Case et al. 1997) and the same explicit water model (Jorgensen et al. 1983), Lee et al. (2001) attempted to refine 360 low-resolution models generated using ROSETTA (Simons et al. 1997) for 12 small proteins (<75 residues), but concluded that there was no systematic structure improvement (Lee et al. 2001). Later, Fan and Mark (2004) tried to refine 60 ROSETTA models for 11 small proteins (<85 residues) using GROMACS 3.0 (Lindahl et al. 2001) with explicit water (Berendsen et al. 1981) and reported that 11/60 models had 10% RMSD reduction and 18/60 had increased RMSD after refinement. Recently, Chen and Brooks (2007) used CHARMM22 (MacKerell et al. 1998) to refine five CASP6 CM targets with lengths in the 70–144 residue range. In four cases, considerable refinements with up to 1 Å RMSD reduction were achieved. One of the major differences of this work is that an implicit solvent force field based on the generalized Born (GB) approximation (Im et al. 2003) was exploited, which significantly speeds up the MD simulations, while the spatial restraints extracted from the initial models are used to guide the refinement procedure (Chen and Brooks 2007). A particularly noteworthy observation was recently made by Summa and Levitt (Summa and Levitt 2007) who exploited different molecular mechanics (MM) potentials (AMBER99 (Wang et al. 2000; Sorin and Pande 2005), OPLS-AA (Kaminski et al. 2001), GROMOS96 (van Gunsteren et al. 1996), and ENCAD (Levitt et al. 1995)) on the refinement of 75 proteins by in vacuo energy minimization. The authors found that a knowledge-based atomic contact potential outperforms all the traditional MM potentials in moving almost all the test proteins closer to the native state, while all the MM potentials, except for AMBER99, essentially drive the decoys away from the native. The vacuum simulation without solvation may be part of the reason for the failure of the MM potentials. But this observation demonstrates the potential of combining knowledge-based potentials with physics-based force field in protein structure refinement.

Another use of the physics-based potential is in the discrimination of the native/near-native structures from structure decoys. For example, Lazaridis and Karplus (1999) exploited CHARMM19 (Neria et al. 1996) and EEF1 (Lazaridis and Karplus 1999) solvation potential to discriminate the native structure from the decoys generated by threading the native sequences on other protein structures. They found the energy of the native states is lower than that of the decoys in most cases. Later, Dominy and Brooks (2002), and Feig and Brooks (2002) used CHARMM plus GB, Felts et al. (2002) used OPLS plus GB, Lee and Duan (2004) used AMBER plus GB, and Hsieh and Luo (2004) used AMBER plus Poisson–Boltzmann solvation potential on the Park–Levitt decoy set (Park and Levitt 1996), Baker decoy set (Tsai et al. 2003), Skolnick decoy set (Kihara et al. 2001; Zhang et al. 2003), and CASP decoys set (Moult et al. 2001). Similar results were obtained by all the authors, i.e. the native structure can be distinguished from non-native decoys by the physics-based potentials. Recently, however, Wroblewska and Skolnick (2007) showed that the AMBER plus GB potential can only discriminate the native structure from roughly minimized TASSER decoys (Zhang and Skolnick 2004a). After a 2-ns MD simulation, none of the native structures have lower energy than decoys, and the energy-RMSD correlation was close to zero. This result partially explains the widely-reported discrepancy between the decoy-discrimination ability of the physics-based potentials and less-successful folding/refinement results (Wroblewska and Skolnick 2007).

In contrast, fast Monte Carlo simulations on the physics-based potentials have enjoyed considerable success in both protein structure prediction and refinement. For example, Scheraga and coworkers (Liwo et al. 1999) successfully built models of 4.2 Å for a fragment of 61 residues based on the MC optimization of a physics-based united-residue force field (Liwo et al. 1993) combined with the atomic ECEPP potential (Nemethy et al. 1992). Using ASTRO-FOLD (Klepeis and Floudas 2003) on the ECEPP optimization, Floudas and coworkers (Klepeis et al. 2005) constructed a model of 5.2 Å for a four-helical bundle protein of 102 residues. In the recent development of ROSSETA (Bradley et al. 2005; Das et al. 2007), the authors also cooperated the physics-based atomic potential in the final stage of Monte Carlo structure refinement, which is discussed in the next section.

### 11.3.2 Knowledge-Based Free Modeling

Probably the most well-known approach for efficient free-modeling was pioneered by Bowie and Eisenberg, who assembled new tertiary structures using small fragments (mainly 9-mers) cut from other PDB proteins (Bowie and Eisenberg 1994). Based on this idea, Baker and coworkers later developed ROSETTA (Simons et al. 1997), which works extremely well for free modeling in the CASP experiments, and popularized the fragment assembly approach in the field. In new developments with ROSETTA (Das et al. 2007), the authors first assemble structures in a reduced knowledge-based model with conformations specified by the heavy backbone atoms and $C_\beta$. In the second stage, Monte Carlo simulations with an all-atom physics-based potential are performed to refine the details of the low-resolution models. An exciting achievement was demonstrated in CASP6 by generating a model for a small hard target T0281 (70 residues) that is 1.6 Å away from the crystal structure. In CASP7, the atomic ROSETTA built a model for T0283 (112 residues) with RMSD = 1.8 Å over 92 residues (see Fig. 11.4). Despite significant success, the computer cost of the procedure (~150 CPU days for a small protein <100 residues) is still too expensive for routine use.

Another successful free modeling approach, called TASSER by Zhang and Skolnick (2004a,b), constructs 3D models based on a purely knowledge-based approach. Continuous fragments with various sizes are excised from threading alignments and used to reassemble protein structures in an on-and-off lattice system. A newer version of I-TASSER was recently developed by Wu et al. (2007), which refines the TASSER cluster centroids by iterative Monte Carlo simulations. Although the procedure uses structural fragments and spatial restraints from threading

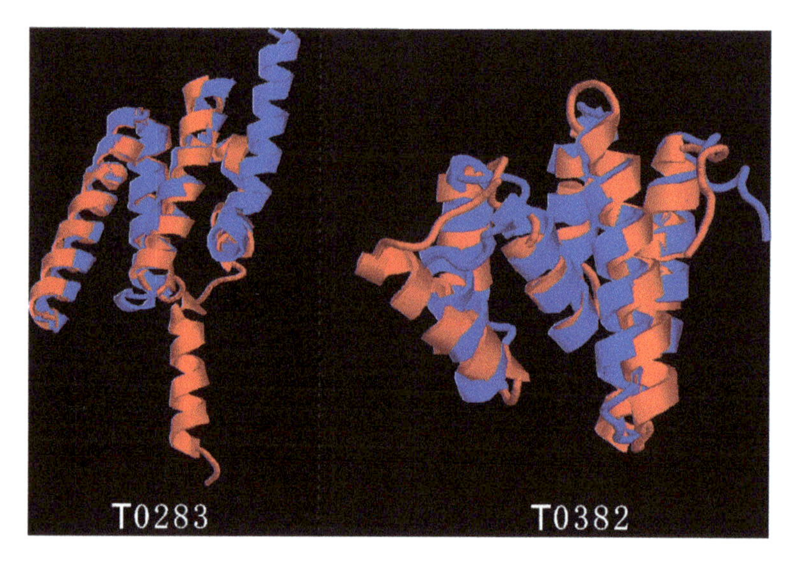


**Fig. 11.4** Representative examples of free modeling in CASP7 generated by two different approaches. T0283 (left panel) is a TBM target (from *Bacillus halodurans*) of 112 residues; but the model is generated by all-atom ROSETTA (a hybrid knowledge- and physics-based approach) (Das et al. 2007) based on free modeling, which gives a TM-score 0.74 and a RMSD 1.8 Å over the first 92 residues (the overall RMSD is 13.8 Å mainly because of the misorientation of C-terminal). T0382 (right panel) is a FM/TBM target (from *Rhodopseudomonas palustris* CGA009) of 123 residues; the model is generated by I-TASSER (a purely knowledge-based approach) (Zhang 2007) with a TM-score 0.66 and a RMSD 3.6 Å. Blue and red represent the model and the crystal structure representatively

templates, it often constructs models of correct topology even when the topologies of individual templates are incorrect. In CASP7 (Zhang 2007), among 19 FM and FM/TBM targets, I-TASSER builds correct topology (~3–5 Å) for 7 cases with sequences up to 155 residues long. In the right panel of Fig. 11.4, we show an example of T0382 (123 residues), where all initial templates have incorrect topologies (>9 Å); but the final model by I-TASSER is 3.6 Å away from the X-ray structure. Recently, Helles made a comparative study of 18 different ab initio prediction algorithms in the literature and concluded that I-TASSER is currently the best method in the balance of modeling accuracy and CPU cost (Helles 2008). However, as indicated by the fact that no high-resolution model has been predicted in the CASP7 New Fold category (Jauch et al. 2007), I-TASSER modeling has a resolution limit by the inherent reduced potential. One of the on-going efforts is to extend the reduced I-TASSER modeling to the atomic representation with the goal to improve the modeling accuracy in the atomic-level (Zhang 2007).

## 11.4 Conclusion

Since a detailed physicochemical description of protein folding principles does not yet exist, the most accurate structure predictions are generated based on evolutionary relationships between the target and solved structures in the PDB library. For the proteins with close templates, full-length models can be constructed by copying the template framework. Recent studies show that using the best possible template structures in PDB, the state-of-the-art modeling algorithms could build high-quality full-length models for almost all single-domain proteins with an average RMSD ~2.3 Å. This suggests that the current PDB structure universe is essentially complete for solving protein structure prediction problem (Zhang and Skolnick 2005a). However, most of the target-template pairs are evolutionarily too distant to be detected with current threading approaches.

The development of efficient threading algorithms to detect weakly/distant structure templates has been a central theme in the field and may persist as a principal direction; the gap between threading and the best structural alignment is obvious and tempting. However, progress in reducing this gap progresses slowly. As mentioned above, there is no single threading method that outperforms all others on every target. Consequently, meta-servers and MQAP have been used to generate predictions by collecting and selecting models from a set of different threading programs. In contrast, the template structure refinement has enjoyed promising progress. In the recent CASP7 experiment (Battey et al. 2007), automated threading plus structure refinement servers outperform the threading and MQAP based meta-servers by a noticeable margin. Nevertheless, template refinement mainly occurs at the topology level. The demand for atomic-level models, which can generate models of real use for new drug screening and biochemical function inference, is keener than ever as more template structures become available through the structure genomics and traditional structural biology.

Free modeling is the 'Holy Grail' of protein structure prediction because its success would mark the eventual solution to a problem manifested at genome scales. Although a purely physics-based ab initio simulation has the advantage in revealing the pathway of protein folding, the best current free-modeling results come from those which combine both knowledge-based and physics-based approaches. While there are consistent successes in building correct topologies (3 ~ 6 Å) for small proteins, the more exciting high-resolution free modeling (<2 Å) is much rarer and computationally more expensive. There is evidence that the current atomic potentials have the lowest energy near the native state, and the bottleneck of high-resolution folding seems to be insufficient conformational sampling (Bradley et al. 2005). However, a golf-hole-like energy landscape without middle range funnel is far from the one taken in nature and this can be a deeper reason for failures in conformational searches. Thus, the bottleneck for free modeling comes from the lack of both funnel-like force fields and efficient space searching methods, especially for proteins of larger sizes.

**Acknowledgment** We are grateful to Dr. Terry Clark for reading the manuscript and giving us helpful comments. The project is partly supported by KU Start-up Fund 06194. Y.Z. is supported by the Alfred P. Sloan Foundation.

## References

Bairoch A, Apweiler R, Wu CH, Barker WC, Boeckmann B, Ferro S et al (2005) The Universal Protein Resource (UniProt). Nucleic Acids Res 33(Database issue):D154–D159

Bateman A, Coin L, Durbin R, Finn RD, Hollich V, Griffiths-Jones S et al (2004) The Pfam protein families database. Nucleic Acids Res 32(Database issue):D138–D141

Battey JN, Kopp J, Bordoli L, Read RJ, Clarke ND, Schwede T (2007) Automated server predictions in CASP7. Proteins 69(S8):68–82

Benson DA, Karsch-Mizrachi I, Lipman DJ, Ostell J, Wheeler DL (2003) GenBank. Nucleic Acids Res 31(1):23–27

Berendsen HJC, Postma JPM, van Gunsteren WF, Hermans J (1981) Interaction models for water in relation to protein hydration. Intermolecular forces, Reidel, Dordrecht, The Netherlands

Berman HM, Westbrook J, Feng Z, Gilliland G, Bhat TN, Weissig H et al (2000) The Protein Data Bank. Nucleic Acids Res 28(1):235–242

Bowie JU, Eisenberg D (1994) An evolutionary approach to folding small alpha-helical proteins that uses sequence information and an empirical guiding fitness function. Proc Natl Acad Sci U S A 91(10):4436–4440

Bowie JU, Luthy R, Eisenberg D (1991) A method to identify protein sequences that fold into a known three-dimensional structure. Science 253:164–170

Bradley P, Misura KM, Baker D (2005) Toward high-resolution de novo structure prediction for small proteins. Science 309(5742):1868–1871

Brooks BR, Bruccoleri RE, Olafson BD, States DJ, Swaminathan S, Karplus M (1983) CHARMM: a program for macromolecular energy, minimization, and dynamics calculations. J Comput Chem 4(2):187–217

Burley SK, Almo SC, Bonanno JB, Capel M, Chance MR, Gaasterland T et al (1999) Structural genomics: beyond the human genome project. Nat Genet 23(2):151–157

Case DA, Pearlman DA, Caldwell JA, Cheatham TE, Ross WS (1997) AMBER 5.0. University of California, San Francisco, CA

Chandonia JM, Brenner SE (2006) The impact of structural genomics: expectations and outcomes. Science 311(5759):347–351

Chen J, Brooks CL III (2007) Can molecular dynamics simulations provide high-resolution refinement of protein structure? Proteins 67(4):922–930

Cheng J, Baldi P (2006) A machine learning information retrieval approach to protein fold recognition. Bioinformatics 22(12):1456–1463

Das R, Qian B, Raman S, Vernon R, Thompson J, Bradley P et al (200) Structure prediction for CASP7 targets using extensive all-atom refinement with Rosetta@home. Proteins 69(S8):118–128

Dominy BN, Brooks CL (2002) Identifying native-like protein structures using physics-based potentials. J Comput Chem 23(1):147–160

Duan Y, Kollman PA (1998) Pathways to a protein folding intermediate observed in a 1-microsecond simulation in aqueous solution. Science 282(5389):740–744

Fan H, Mark AE (2004) Refinement of homology-based protein structures by molecular dynamics simulation techniques. Protein Sci 13(1):211–220

Feig M, Brooks CL, 3rd (2002) Evaluating CASP4 predictions with physical energy functions. Proteins 49(2):232–245

Felts AK, Gallicchio E, Wallqvist A, Levy RM (2002) Distinguishing native conformations of proteins from decoys with an effective free energy estimator based on the OPLS all-atom force field and the Surface Generalized Born solvent model. Proteins 48(2):404–422

Fischer D (2003) 3D-SHOTGUN: a novel, cooperative, fold-recognition meta-predictor. Proteins 51(3):434–441

Fischer D (2006) Servers for protein structure prediction. Curr Opin Struct Biol 16(2):178–182

Fischer D, Rychlewski L, Dunbrack RL Jr, Ortiz AR, Elofsson A (2003) CAFASP3: the third critical assessment of fully automated structure prediction methods. Proteins 53(Suppl 6):503–516

Ginalski K, Pas J, Wyrwicz LS, von Grotthuss M, Bujnicki JM, Rychlewski L (2003) ORFeus: Detection of distant homology using sequence profiles and predicted secondary structure. Nucleic Acids Res 31(13):3804–3807

Helles G (2008) A comparative study of the reported performance of ab initio protein structure prediction algorithms. J R Soc Interface 5(21):387–396

Hsieh MJ, Luo R (2004) Physical scoring function based on AMBER force field and Poisson-Boltzmann implicit solvent for protein structure prediction. Proteins 56(3):475–486

Im W, Lee MS, Brooks CL III (2003) Generalized born model with a simple smoothing function. J Comput Chem 24(14):1691–1702

Jaroszewski L, Rychlewski L, Li Z, Li W, Godzik A (2005) FFAS03: a server for profile–profile sequence alignments. Nucleic Acids Res 33(Web Server issue):W284–W288

Jauch R, Yeo HC, Kolatkar PR, Clarke ND (2007) Assessment of CASP7 structure predictions for template free targets. Proteins 69(Suppl 8):57–67

Jones DT (1999) GenTHREADER: an efficient and reliable protein fold recognition method for genomic sequences. J Mol Biol 287(4):797–815

Jones DT, Taylor WR, Thornton JM (1992) A new approach to protein fold recognition. Nature 358(6381):86–89

Jorgensen WL, Chandrasekhar J, Madura JD, Impey RW, Klein ML (1983) Comparison of simple potential functions for simulating liquid water. J Chem Phys 79:926–935

Jorgensen WL, Tirado-Rives J (1988) The OPLS potential functions for proteins. Energy minimizations for crystals of cyclic peptides and crambin. J Am Chem Soc 110:1657–1666

Kaminski GA, Friesner RA, Tirado-Rives J, Jorgensen WL (2001) Evaluation and reparametrization of the OPLS-AA force field for proteins via comparison with accurate quantum chemical calculations on peptides. J Phys Chem B 105:6474–6487

Karplus K, Barrett C, Hughey R (1998) Hidden Markov models for detecting remote protein homologies. Bioinformatics 14:846–856

Kihara D, Lu H, Kolinski A, Skolnick J (2001) TOUCHSTONE: An ab initio protein structure prediction method that uses threading-based tertiary restraints. Proc Natl Acad Sci U S A 98:10125–10130

Klepeis JL, Floudas CA (2003) ASTRO-FOLD: a combinatorial and global optimization framework for Ab initio prediction of three-dimensional structures of proteins from the amino acid sequence. Biophys J 85(4):2119–2146

Klepeis JL, Wei Y, Hecht MH, Floudas CA (2005) Ab initio prediction of the three-dimensional structure of a de novo designed protein: a double-blind case study. Proteins 58(3):560–570

Kopp J, Bordoli L, Battey JN, Kiefer F, Schwede T (2007) Assessment of CASP7 predictions for template-based modeling targets. Proteins 6(S8):38–56

Kulikova T, Aldebert P, Althorpe N, Baker W, Bates K, Browne P et al (2004) The EMBL nucleotide sequence database. Nucleic Acids Res 32(Database issue):D27–D30

Lazaridis T, Karplus M (1999) Effective energy function for proteins in solution. Proteins 35(2):133–152

Lee MR, Tsai J, Baker D, Kollman PA (2001) Molecular dynamics in the endgame of protein structure prediction. J Mol Biol 313(2):417–430

Lee MC, Duan Y (2004) Distinguish protein decoys by using a scoring function based on a new AMBER force field, short molecular dynamics simulations, and the generalized born solvent model. Proteins 55(3):620–634

Levitt M, Hirshberg M, Sharon R, Daggett V (1995) Potential-energy function and parameters for simulations of the molecular-dynamics of proteins and nucleic-acids in solution. Comput Phys Commun 91(1–3):215–231

Lindahl E, Hess B, van der Spoel D (2001) GROMACS 3.0: A package for molecular simulation and trajectory analysis. J Mol Modeling 7:306–317

Liwo A, Lee J, Ripoll DR, Pillardy J, Scheraga HA (1999) Protein structure prediction by global optimization of a potential energy function. Proc Natl Acad Sci U S A 96(10):5482–5485

Liwo A, Pincus MR, Wawak RJ, Rackovsky S, Scheraga HA (1993) Calculation of protein backbone geometry from alpha-carbon coordinates based on peptide-group dipole alignment. Protein Sci 2(10):1697–1714

MacKerell AD Jr, Bashford D, Bellott M, Dunbrack RL, Evanseck JD, Field MJ et al (1998) All-atom empirical potential for molecular Modeling and dynamics studies of proteins. J Phys Chem B 102(18):3586–3616

Marti-Renom MA, Stuart AC, Fiser A, Sanchez R, Melo F, Sali A (2000) Comparative protein structure modeling of genes and genomes. Annu Rev Biophys Biomol Struct 29:291–325

McPherson JD, Marra M, Hillier L, Waterston RH, Chinwalla A, Wallis J et al (2001) A physical map of the human genome. Nature 409(6822):934–941

Misura KM, Chivian D, Rohl CA, Kim DE, Baker D (2006) Physically realistic homology models built with ROSETTA can be more accurate than their templates. Proc Natl Acad Sci U S A 103(14):5361–5366

Miyazaki S, Sugawara H, Gojobori T, Tateno Y (2003) DNA Data Bank of Japan (DDBJ) in XML. Nucleic Acids Res 31(1):13–16

Moult J, Fidelis K, Kryshtafovych A, Rost B, Hubbard T, Tramontano A (2007) Critical assessment of methods of protein structure prediction-Round VII. Proteins 69(Suppl 8):3–9

Moult J, Fidelis K, Zemla A, Hubbard T (2001) Critical assessment of methods of protein structure prediction (CASP): round IV. Proteins Suppl 5:2–7

Nemethy G, Gibson KD, Palmer KA, Yoon CN, Paterlini G, Zagari A et al (1992) Energy Parameters in Polypeptides. 10. Improved geometric parameters and nonbonded interactions for use in the ECEPP/3 algorithm, with application to proline-containing peptides. J Phys Chem B 96:6472–6484

Neria E, Fischer S, Karplus M (1996) Simulation of activation free energies in molecular systems. J Chem Phys 105(5):1902–1921

Nilges M, Brunger AT (1991) Automated modeling of coiled coils: application to the GCN4 dimerization region. Protein Eng 4(6):649–659

Park B, Levitt M (1996) Energy functions that discriminate X-ray and near native folds from well-constructed decoys. J Mol Biol 258(2):367–392

Pieper U, Eswar N, Braberg H, Madhusudhan MS, Davis FP, Stuart AC et al (2004) MODBASE, a database of annotated comparative protein structure models, and associated resources. Nucleic Acids Res 32(Database issue):D217–D222

Pieper U, Eswar N, Davis FP, Braberg H, Madhusudhan MS, Rossi A et al (2006) MODBASE: a database of annotated comparative protein structure models and associated resources. Nucleic Acids Res 34(Database issue):D291–D295

Rychlewski L, Fischer D (2005) LiveBench-8: the large-scale, continuous assessment of automated protein structure prediction. Protein Sci 14(1):240–245

Sadreyev R, Grishin N (2003) COMPASS: a tool for comparison of multiple protein alignments with assessment of statistical significance. J Mol Biol 326(1):317–336

Sali A (1998) 100, 000 protein structures for the biologist. Nat Struct Biol 5(12):1029–1032

Sali A, Blundell TL (1993) Comparative protein modelling by satisfaction of spatial restraints. J Mol Biol 234(3):779–815

Shi J, Blundell TL, Mizuguchi K (2001) FUGUE: sequence-structure homology recognition using environment-specific substitution tables and structure-dependent gap penalties. J Mol Biol 310(1):243–257

Simons KT, Kooperberg C, Huang E, Baker D (1997) Assembly of protein tertiary structures from fragments with similar local sequences using simulated annealing and Bayesian scoring functions. J Mol Biol 268(1):209–225

Skolnick J, Fetrow JS, Kolinski A (2000) Structural genomics and its importance for gene function analysis. Nat Biotechnol 18(3):283–287

Skolnick J, Kihara D, Zhang Y (2004) Development and large scale benchmark testing of the PROSPECTOR 3.0 threading algorithm. Protein 56:502–518

Smaglik P (2000) Protein structure groups seek to draft common ground rules. Nature 403(6771):691

Soding J (2005) Protein homology detection by HMM-HMM comparison. Bioinformatics 21(7):951–960

Sorin EJ, Pande VS (2005) Exploring the helix-coil transition via all-atom equilibrium ensemble simulations. Biophys J 88(4):2472–2493

Stevens RC, Yokoyama S, Wilson IA (2001) Global efforts in structural genomics. Science 294(5540):89–92

Summa CM, Levitt M (2007) Near-native structure refinement using in vacuo energy minimization. Proc Natl Acad Sci U S A 104(9):3177–3182

Terwilliger TC, Waldo G, Peat TS, Newman JM, Chu K, Berendzen J (1998) Class-directed structure determination: foundation for a protein structure initiative. Protein Sci 7(9):1851–1856

Tsai J, Bonneau R, Morozov AV, Kuhlman B, Rohl CA, Baker D (2003) An improved protein decoy set for testing energy functions for protein structure prediction. Proteins 53(1):76–87

van Gunsteren WF, Billeter SR, Eising AA, Hunenberger PH, Kruger P, Mark AE et al (1996) Biomolecular Simulation: The GROMOS96 Manual and User Guide. Vdf Hochschulverlag AG an der ETH Zürich, Zürich

Venter JC, Adams MD, Myers EW, Li PW, Mural RJ, Sutton GG et al (2001) The sequence of the human genome. Science 291(5507):1304–1351

Vieth M, Kolinski A, Brooks CL III, Skolnick J (1994) Prediction of the folding pathways and structure of the GCN4 leucine zipper. J Mol Biol 237(4):361–367

Vitkup D, Melamud E, Moult J, Sander C (2001) Completeness in structural genomics. Nat Struct Biol 8(6):559–566

Wallner B, Elofsson A (2007) Prediction of global and local model quality in CASP7 using Pcons and ProQ. Proteins 69(S8):184–193

Wang JM, Cieplak P, Kollman PA (2000) How well does a restrained electrostatic potential (RESP) model perform in calculating conformational energies of organic and biological molecules? J Comput Chem 21(12):1049–1074

Weiner SJ, Kollman PA, Case DA, Singh UC, Ghio C, Alagona G et al (1984) A new force field for molecular mechanical simulation of nucleic acids and proteins. J Am Chem Soc 106:765–784

Wroblewska L, Skolnick J (2007) Can a physics-based, all-atom potential find a protein's native structure among misfolded structures? I. Large scale AMBER benchmarking. J Comput Chem 28(12):2059–2066

Wu S, Skolnick J, Zhang Y (2007) Ab initio modeling of small proteins by iterative TASSER simulations. BMC Biol 5:17

Wu S, Zhang Y (2007) LOMETS: a local meta-threading-server for protein structure prediction. Nucleic Acids Res 35(10):3375–3382
Wu S, Zhang Y (2008) MUSTER: Improving protein sequence profile-profile alignments by using multiple sources of structure information. Proteins 72(2):547–556
Zagrovic B, Snow CD, Shirts MR, Pande VS (2002) Simulation of folding of a small alpha-helical protein in atomistic detail using worldwide-distributed computing. J Mol Biol 323(5):927–937
Zhang Y (2007) Template-based modeling and free modeling by I-TASSER in CASP7. Proteins 69(Suppl 8):108–117
Zhang Y, Kolinski A, Skolnick J (2003) TOUCHSTONE II: A new approach to ab initio protein structure prediction. Biophys J 85:1145–1164
Zhang Y, Skolnick J (2004a) Automated structure prediction of weakly homologous proteins on a genomic scale. Proc Natl Acad Sci U S A 101:7594–7599
Zhang Y, Skolnick J (2004b) Scoring function for automated assessment of protein structure template quality. Proteins 57(4):702–710
Zhang Y, Skolnick J (2005a) The protein structure prediction problem could be solved using the current PDB library. Proc Natl Acad Sci U S A 102:1029–1034
Zhang Y, Skolnick J (2005b) TM-align: a protein structure alignment algorithm based on the TM-score. Nucleic Acids Res 33(7):2302–2309
Zhou H, Zhou Y (2005) Fold recognition by combining sequence profiles derived from evolution and from depth-dependent structural alignment of fragments. Proteins 58(2):321–328

# Chapter 12
# Classification of Information About Proteins

**Amandeep S. Sidhu, Matthew I. Bellgard, and Tharam S. Dillon**

## 12.1 Introduction

The use of advanced high throughput technology applied to proteomics results in the production of large volumes of information rich data. This data requires considerable knowledge management to allow biologists and bioinformaticians to access and understand the information in the context of their experiments. As the volume of data increases, the results from these high throughput experiments will provide the foundations for advancing proteome biology.

In this chapter, we consider the challenges of information integration in proteomics from the perspective of researchers using information technology as an integral part of their discovery process. We firstly describe the information about proteins that is collected from high throughput experimentation and how this is managed. We then describe how protein ontologies can be used to classify this information. Finally we discuss some of the uses of protein classification systems and the biological challenges in proteomics which they help to resolve.

## 12.2 Why Are Proteins Important?

Proteins play a variety of roles in cellular processes including structural functions (viral coat proteins, molecules of the cytoskeleton, epidermal keratin); catalytic reactions (the enzymes); transport and storage (hemoglobin, myoglobin, ferritin); regulation (e.g., Hormones and transcription factors); as well as complex recognition roles such as the immune system or cell–cell recognition and signaling. Proteins can organize themselves in three dimensions and the system that produces them can create heritable structural variations, conferring the ability to evolve. The amino acid sequences of proteins predominantly dictate their three-dimensional structures.

A.S. Sidhu (✉)
Centre for Comparative Genomics, Murdoch University, Perth, Australia
e-mail: asidhu@ccg.murdoch.edu.au

D. Edwards et al. (eds.), *Bioinformatics: Tools and Applications*,
DOI 10.1007/978-0-387-92738-1_12, © Springer Science+Business Media, LLC 2009

Although known structural data is not as complete as sequence data, detailed atomic structures are now available for over 55,000 proteins and these structures reveal the great variety of spatial patterns and functional domains. An oxy T state hemoglobin protein structure (Paoli et al. 1996) is shown in the Fig. 12.1 below:

Research into proteins provides a number of scientific challenges:

- *Interpretation of mechanisms of function of individual proteins:* The catalytic activity of an enzyme can be explained in terms of physical-organic chemistry on the basis of interactions of residues of the protein.
- *The Protein Folding Problem:* Under physiological conditions of solvent and temperature, most proteins fold spontaneously to an active native state. The mechanism and kinetics of folding remain poorly understood and is the focus of intensive research.
- *Prediction of Protein Structures:* A majority of proteins, or their amino acid sequence dictates their three-dimensional structure. However, predicting the structure of a protein from the amino acid sequence remains a challenge. This problem is addressed by Wu and Zhang in this volume.
- *Patterns of Molecular Evolution:* There are several families of protein structures for which we know dozens or even hundreds of amino acid sequences, and at least 20 structures; for example, the globins, the cytochromes c, and serine protease.

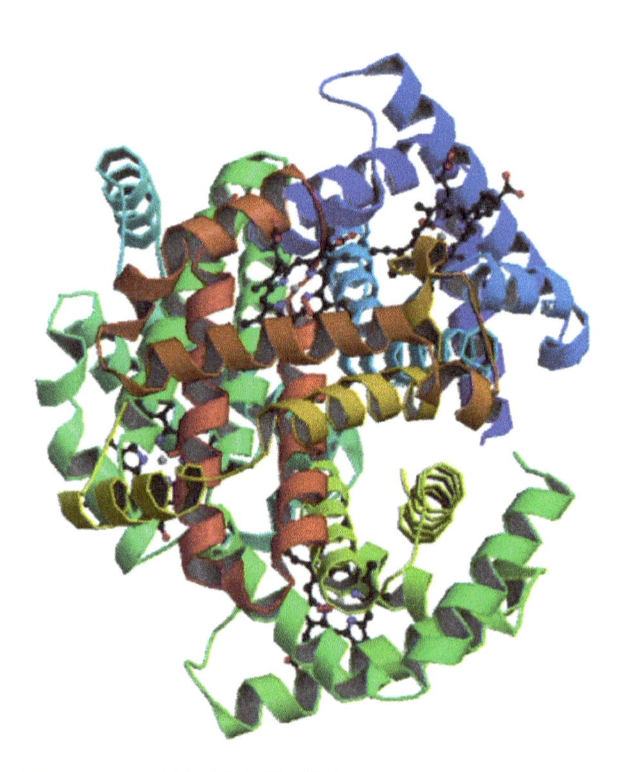

**Fig. 12.1** Oxy T state haemoglobin (PDB ID: 1GZX)

For these proteins, it has been possible to analyze the mechanism of evolution, in that we can observe the structural and functional roles of the sets of residues that are strongly conserved and those that vary relatively freely, and we can describe the structural consequences of changes in the amino acid sequence.

- *Protein Engineering:* Using the techniques of genetic engineering, it is possible to design and test modifications of known proteins and to design novel proteins. Potential applications include: (1) modifications to probe the mechanisms of protein function such as the method of "alanine scanning," (2) attempts to enhance thermostability by optimizing the choice of amino acids, (3) clinical applications, such as the transfer of the active site from a rat antibody to human antibody framework, and (4) modifying antibodies to give them catalytic ability.
- *Drug Design:* There are many proteins specific to pathogens that could potentially be deactivated by drugs. With the known structure of HIV-1 Protease (Yamazaki et al. 1996), it may be possible to design molecules that will bind tightly and specifically to an essential site on these molecules, to interfere with their function (Fig. 12.2).

Now that the structures of many proteins have been determined, questions of structure, function and evolution can be addressed by examining and comparing the positions of individual atoms. The primary events in the generation of biological diversity are the mutation, insertion and deletion of nucleotides of genomic DNA. If a gene produces a functional protein product, a mutant gene may produce an alternative protein of equivalent function; a protein that carries out the same function but at an altered rate; a protein with an altered function; or a protein that does not function at all. Examination of homologous genes and proteins in different species has shown that evolutionary variation and divergence occur at the molecular level and that proteins from related species often have similar but not identical amino acid sequences and studies have established relationships between divergence of sequence and divergence of structure.

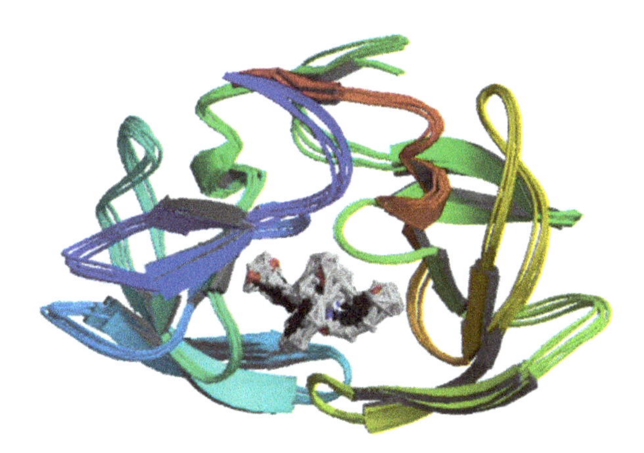

**Fig. 12.2** HIV-1 Protease-DMP323 complex in solution (PDB ID: 1BVE)

In the last half of the twentieth century, a highly focused, hypothesis-driven approach known as reductionist molecular biology gave scientists the tools to identify and characterize molecules and cells, the fundamental building blocks of living systems. However, there is an increasing awareness of the ability to undertake a systems biology approach to study complex biological problems. Systems analysis demands not just knowledge of parts – genes, proteins, and other macromolecular entities – but also knowledge of the connection of these molecular parts and how they work together. In other words, the pendulum of bioscience is now swinging away from the reductionist approach and towards the synthetic approach characteristic of systems biology and of an integrated biology capable of quantitative and/or detailed qualitative predictions. A synthetic or integrated view of biology will depend critically on information integration from a variety of data sources. Over the past two decades, research in evolutionary biology has come to depend on sequence comparisons at the gene and protein levels, and in the future it will depend more and more on tracking not just DNA sequences but on how entire genomes evolve over time (Pennisi 1998).

The connections and interactions among areas of genomics, gene expression profiles, proteomics, and systems biology depend on the integration of experimental procedures with databases and applications of computational algorithms and analysis tools. As the degree of complexity of the biological processes under study increases, our understanding at each level depends in a significant way on the levels beneath it. At every step, the computational analyses of data are an integral part of the discovery process. As we choose complex systems for study, experimentally generated data must be combined with data maintained within databases and computationally derived models or simulations for best interpretation. Modeling and simulation of protein–protein interactions, protein pathways, genetic regulatory networks, biochemical and cellular processes, and physiological states are in their infancy and need more experimental observations to fill in missing quantitative details. In these interactions, the boundaries between experimentally generated data and computationally generated data are blurred. Thus, accelerating progress now requires multidisciplinary teams to integrate their approaches. An information infrastructure, coupled with continued advances in experimental methods, will facilitate a greater understanding of biology.

## 12.3 Capturing Information about Proteins

The advent of high-throughput technologies has led to an ever-increasing rate of data acquisition and exponential growth of data volume. However, the most striking feature of data in life science is not its volume but its diversity and variability.

Mass spectrometry (MS) has increasingly become the method of choice for analysis of complex protein samples. MS-based proteomics is a discipline made possible by the availability of protein sequence databases and technical and conceptual advances in many areas, most notably the discovery and development of protein ionization methods, as recognized by the 2002 Nobel prize in chemistry to John B. Fenn, Koichi Tanaka and Kurt Wüthrich awarded"for the development of methods for identification and structure analyses of biological macromolecules." Electrospray ionization (ESI) and matrix-assisted laser desorption/ionization (MALDI) are the two techniques most commonly used to volatize and ionize proteins or peptides for mass spectrometric analysis (Fenn et al. 1989; Karas and Hillenkamp 1988; Pandey and Mann 2000). ESI ionizes the analytes out of a solution and is therefore readily coupled to liquid-based separation tools. MALDI sublimates and ionizes the samples out of a dry, crystalline matrix via laser pulses.

The results of these experiments are a list of peptides which are compared to known proteins. Because protein identification relies on matches with sequence databases, high-throughput proteomics is currently restricted largely to those species for which comprehensive sequence databases are available. MS-based proteomics has established itself as an indispensable technology to interpret the information encoded in genomes. So far, protein analysis (primary sequence, post-translational modifications (PTMs) or protein–protein interactions) by MS has been most successful when applied to small sets of proteins isolated in specific functional contexts. The systematic analysis of the much larger number of proteins expressed in a cell, an explicit goal of proteomics, is now rapidly advancing, due mainly to the development of new experimental approaches.

## 12.4 Ontologies for Proteins

There are a number of publicly available databases that host knowledge about proteins. The development of individual databases has generated a large variety of formats in their implementations. There is consensus that a common language for protein information should be valuable, but this goal has proved difficult to achieve. Attempts to unify data formats have included the application of a Backus–Naur based syntax (George et al. 1987), the development of an object-oriented database definition language (George et al. 1993) and the use of Abstract Syntax Notation 1 (Ohkawa et al. 1995; Ostell 1990). None of these approaches have achieved the hoped for degree of acceptance. Underlying questions of intercommunication between databases of different structure and format is the need for common semantic standards and controlled vocabulary in annotations (Pongor 1998; Rawlings 1998).(Please check the meaning of the sentence.) The technical problems of standardization may be addressed more easily in the context of a more general logical structure. As noted by Hafner (Hafner and Fridman 1996), general biological data resources are databases rather than knowledge bases: they describe miscellaneous objects according to the database schema, but no representation of general concepts or their relationships is given. Schulze-Kremer addressed this problem by developing ontologies for knowledge sharing in molecular biology (Schulze-Kremer 1998). The term ontology is originally a philosophical term referred to as "the object of existence." The computer science community borrowed the term ontology to refer to a “specification of conceptualization” for knowledge sharing in artificial intelligence (Gruber 1993). Ontologies provide a conceptual framework for a structured representation of the meaning, through a common vocabulary, in a given domain – in this case, biological or medical – that can be used by either humans or automated software agents in the domain. This shared vocabulary usually includes concepts, relationships between concepts, definitions of these concepts and also the possibility of defining ontology rules and axioms, in order to define a mechanism to control the objects that can be introduced in the ontology and to apply logical inference. Ontologies in biomedicine have emerged because of the need for a common language for effective communication across diverse sources of biological data and knowledge. In this section, we review a selection of ontologies used in the biomedical domain that relate to genes and the proteins that they encode.

### 12.4.1 The Gene Ontology

In 1998, efforts to develop the Gene Ontology began, leading ontological development in the genetic area (Ashburner et al. 2001; Lewis 2004). The Gene Ontology is a collaborative effort to create a controlled vocabulary describing genes and proteins, addressing the need for consistent descriptions of gene products in different databases. The GO collaborators are developing three structured, controlled vocabularies (ontologies) that describe gene products in terms of their associated biological processes, cellular components and molecular functions in a species-independent manner. The GO consortium was initially a collaboration among the Mouse Genome Database (Blake et al. 1998), FlyBase (Ashburner 1993), and the Saccharomyces Genome database (Schuler et al. 1996) efforts. GO is now a part of the Unified Medical Language System (UMLS), and the GO consortium is a member of the Open Biological Ontologies consortium discussed later in this section. One of the important uses of GO is the prediction of gene function based on patterns of annotation. For example, if annotations for two attributes tend to occur together in the database, then the gene holding one attribute is likely to have the other attribute as well (King et al. 2003). In this way, functional predictions can be made by applying prior knowledge to infer the function of the new entity (either a gene or a protein).

GO consists of three distinct ontologies, each of which serves as an organizing principle for describing gene products. The intention is that each gene product should be annotated by classifying it within each ontology (Fraser and Marcotte 2004). The three GO ontologies are:

1. Molecular Function: This ontology describes the biochemical activity of the gene product. For example, a gene product could be a transcription factor or DNA helicase.
2. Biological Process: This ontology describes the biological goal to which a gene product contributes. For example, mitosis or purine metabolism. An ordered assembly of molecular functions accomplishes such a process.

3. Cellular Component: This ontology describes the location in a cell in which the biological activity of the gene product is performed. Examples include the nucleus, telomere, or an origin recognition complex.

GO is the result of an effort to model concepts used to describe genes and gene products. The central unit of description in GO is a concept. Each concept consists of a unique identifier and one or more strings (referred to as terms) that provide a controlled vocabulary for unambiguous and consistent naming. Concepts exist in a hierarchy of IsA and part of?? relations in a directed acyclic graph (DAG) that locates all concepts in the knowledge model with respect to their relationships with other concepts.

GO is now clearly defined and is a model for numerous other biological ontology projects that aim similarly to achieve structured, standardized vocabularies for describing biological systems. There are many measures demonstrating the success of GO. The characteristics of GO that led to its success include: community involvement, clear goals, limited scope, simple, intuitive structure, continuous evolution, active curation, and early use. Within the genome community it has become the accepted standard for functional annotation.

### *12.4.2 The MGED Ontology*

The MGED Ontology (MO) was developed by the Microarray Gene Expression Data (MGED) Society. MO provides terms for annotating all aspects of a gene expression or a microarray experiment from the design of the experiment and array layout, to preparation of the biological sample and protocols used to hybridize the RNA and analyze the data (Whetzel et al. 2006). MO is a species-neutral ontology that focuses on commonalities among experiments rather than differences between them. MO is primarily an ontology used to annotate microarray experiments; however, it contains concepts that are universal to other types of functional genomics experiments. The major component of the ontology involves biological descriptors relating to samples or their processing. MO version 1.2 contains 229 classes, 110 properties and 658 instances.

### *12.4.3 The Protein Ontology*

We built the Protein Ontology (PO) (Sidhu et al. 2005a, b, 2007) to integrate protein data formats and provide a structured and unified vocabulary to represent protein synthesis concepts. PO provides an integration of heterogeneous protein and biological data sources, and converts the enormous amounts of data collected by geneticists and molecular biologists into information that biologists can use to more easily understand the mapping of relationships inside protein molecules, the interaction between two protein molecules, and interactions between proteins and other macromolecules at cellular level. The PO consists of concepts (or classes), which are data descriptors for proteomics data and the relationships among these concepts. PO has:

1. A hierarchical classification of concepts represented as classes, from general to specific
2. A list of properties related to each concept, for each class
3. A set of relationships between classes to link concepts in ontology in more complicated ways than implied by the hierarchy, to promote reuse of concepts in the ontology; and
4. A set of algebraic operators for querying protein ontology instances. In this section, we will briefly discuss various concepts and relationships that make up the Protein Ontology

### 12.4.4 Generic Concepts of Protein Ontology

There are seven concepts of PO, called generic concepts that are used to define complex PO concepts: {Residues, Chains, Atoms, Family, Atomic Bind, Bind, and Site Group}, and these generic concepts are reused in defining complex PO concepts. Details and properties of residues in a protein sequence are defined by instances of the residues concept. Instances of chains of residues are defined in the chains concept. All the three dimensional structure data of protein atoms are represented as instances of the atoms concept. Defining chains, residues and atoms as individual concepts has the advantage that any special properties or changes affecting a particular chain, residue or atom can be added easily. The family concept represents the protein super family and family details of proteins. Data about binding atoms in chemical bonds such as hydrogen bond, residue links, and salt bridges are entered into the ontology as an instance of the atomic bind concept. Similarly, data about binding residues in chemical bonds such as disulphide bonds and cis peptide bonds are entered into the ontology as an instance of the bind concept. When defining the generic concepts of atomic bind and bind in PO we again reuse the generic concepts of chain, residue, and atom. All data related to site groups of the active binding sites of proteins are defined as instances of the site group concept. In PO, notions classification, reasoning, and consistency are applied by defining new concepts from the defined generic concepts. The concepts derived from generic concepts are placed precisely into a class hierarchy of the PO to completely represent information defining a protein complex.

### 12.4.5 Derived Concepts of Protein Ontology

The PO provides a description of protein data that can be used to describe proteins in any organism using derived concepts formed from the generic concepts.

#### 12.4.5.1 Derived Concepts for Protein Entry Details

The PO describes the protein complex entry, and the molecules contained in a protein complex are described using the entry concept and its sub-concepts of description, molecule and reference. Molecule reuses the generic concepts of chain to represent the linkage of molecules in the protein complex to the chain of residue sequences.

#### 12.4.5.2 Derived Concepts for Protein Sequence and Structure Details

Protein sequence and structure data are described using structure concept in PO with the sub-concepts atom sequence and unit cell. Atom sequence represents protein sequence and structure and is made of the generic concepts of chain, residue and atom. Protein crystallography data is described using the unit cell concept.

#### 12.4.5.3 Derived Concepts for Structural Folds and Domains in Proteins

Protein structural folds and domains are defined in PO using the derived concept of structural domains. The family and super family of the organism in which protein is present are represented in structural domains by reference to the generic concept of family. Structural folds in proteins are represented by the sub-concepts of helices, sheets and other folds. Each definition of structural fold and domains also reuses the generic concepts of chain and residue for describing the secondary structure of proteins. Helix, which is a sub-concept of helices, identifies a helix?. Helix has a sub-concept helix structure that gives the detailed composition of the helix. In this way, PO distinguishes concepts for the identification and the structure of secondary structures in a protein. Other secondary structures of proteins such as sheets and turns (or loops) are represented in a similar way. Sheets have a sub-concept sheet that identifies a sheet?. Sheet has a sub-concept strands that describes the detailed structure of a sheet. Similarly, turns in protein structures are repeated in PO using the other folds concept. Turn is a sub-concept of other folds that identifies a turn; and turn structure describes its structure. Turns in protein structure are categorized as other folds in the protein ontology as they are less frequent than helices and sheets in protein structure.

#### 12.4.5.4 Derived Concepts for Functional Domains in Proteins

PO is the first functional domain classification model for proteins defined using the derived concept of functional domains. In a similar way to structural domains, the family and super family of the organism in which protein is present, are represented in functional domains by reference to the generic concept of family. Functional domains describe the cellular and organic?sm source of a protein using the source cell sub-concept, the biological functionality of protein using the biological function sub-concept, and describes active binding sites in proteins using the active binding sites sub-concept. Active binding sites are represented in the PO as a collection of various site groups, defined using site group generic concept.

#### 12.4.5.5 Derived Concepts for Chemical Bonds in Proteins

Various chemical bonds used to bind various substructures in a complex protein structure are defined using chemical bonds concept in PO. Chemical bonds are defined by their respective sub-concepts and are: disulphide bond, cis peptide, hydrogen bond, residue link, and salt bridge. They are defined using the generic concepts of bind and atomic bind. Chemical bonds that have binding residues (disulphide bond, cis peptide) reuse the generic concept of bind. Similarly the chemical bonds that have binding atoms (hydrogen bond, residue link, and salt bridge) reuse the generic concept of atomic bind.

#### 12.4.5.6 Derived Concepts for Constraints affecting the Protein Structural Conformation

Various constraints that affect the final protein structural conformation are defined using the constraints concept of PO. The constraints described in PO at the moment are: monogenetic and polygenetic defects present in genes that are present in molecules making proteins, and these are described using the genetic defects sub-concept, hydrophobic properties of proteins are described using the hydrophobicity sub-concept, and modification in residue sequences due to changes in chemical environment and mutations are described using the modified residue sub-concept.

### *12.4.6 Relationships Protein Ontology*

Semantics in protein data is normally not interpreted by annotating systems, since they are not aware of the specific structural, chemical and cellular interactions of protein complexes. A PO framework provides a specific set of rules to cover these application specific semantics. The rules only use the relationships whose semantics are predefined in PO to establish correspondence among terms. The set of relationships with predefined semantics is: {SubClassOf, PartOf, AttributeOf, InstanceOf, and ValueOf}. The PO conceptual modeling encourages the use of strictly typed relations with precisely defined semantics. Some of these relationships (such as SubClassOf, InstanceOf) are somewhat similar to those used in the resource description framework (RDF) schema (W3C-RDFSchema 2004), but the set of relationships that have defined semantics in our conceptual PO model is too small to maintain the simplicity of the model. The following is a brief description of the set of pre-defined semantic relationships in our common PO conceptual model: the SubClassOf relationship is used to indicate that one concept is a specialization of another concept; the AttributeOf relationship indicates that a concept is an attribute of another concept; the PartOf relationship indicates that a concept is a part of another concept; the InstanceOf relationship indicates that an object is an instance of the concept; and the ValueOf relationship is used to indicate the value of an attribute of an object. By themselves, the relationships described above do not impose order among the children of the node, rather, we defined a special relationship called sequence(s) in PO to describe and impose order in complex concepts defining structure, structural folds and domains and chemical bonds of proteins.

### *12.4.7 Protein Ontology as a Structured Hierarchy*

The PO consists of a hierarchical classification of concepts, discussed above, represented as classes, from general to specific. The concepts derived from generic concepts are placed precisely into the class hierarchy of Protein Ontology, as depicted in Fig. 12.3 below. Further details on PO are available on the website (http://www.proteinontology.org.au/).

## 12.5 Advantages of a Consistent Protein Ontology

### *12.5.1 Annotation and Retrieval*

Research into biological systems use different organisms chosen specifically because they are amenable to advancing these investigations. For instance, the rat is a good model for the study of human heart disease. For each of these model systems, there is a database employing curators who collect and store the body of biological knowledge of that organism. However, querying heterogeneous, independent databases in order to draw these inferences is difficult: The different database projects may use different terms to refer to the same concept and the same terms to refer to different concepts. Furthermore, typically,these terms are not formally linked with each other in any way. The PO provides a structured vocabulary that can be used to describe proteins, and can be shared between various protein data sources. This facilitates querying protein data that share biologically meaningful attributes, whether from separate databases or within the same database through??. Atoms of a protein structure described in protein data bank (PDB) format are converted to an instance of the atom concept stored in PO instance store (Fig. 12.4) represented using the web ontology language (OWL). As the OWL representation used in PO is an abbreviated extensible markup language (XML) notation, it can easily be transformed to the corresponding RDF and XML formats using the available converters. The PO instance store currently consists of various species of proteins from bacteria and plant to human proteins in OWL format. Such a generic representation using PO shows the strength of the PO format representation.

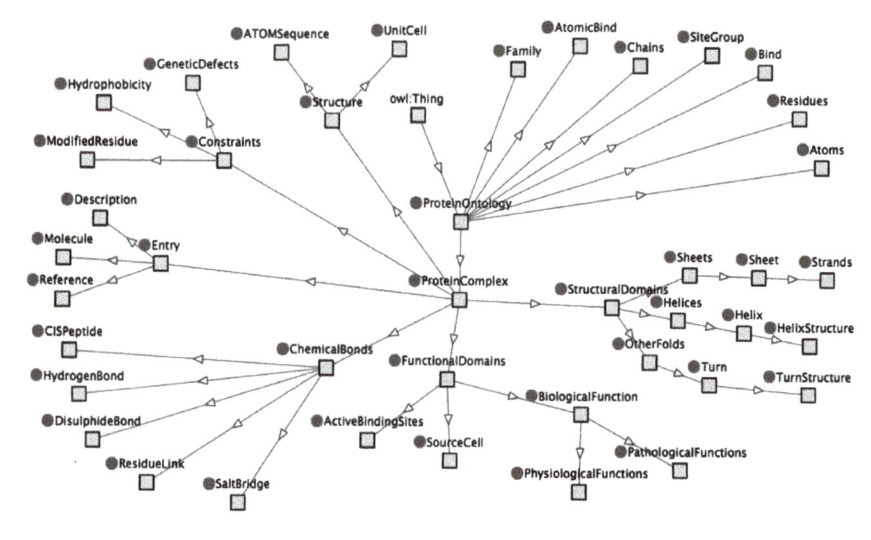


**Fig. 12.3** Class hierarchy of protein ontology

```
<Atoms rdf:about="http://www.owl-ontologies.com/ProteinOntology.owl#AtomInstance6131">
  <AtomID rdf:datatype="http://www.w3.org/2001/XMLSchema#int">2231</AtomID>
  <ATOMResSeqNum rdf:datatype="http://www.w3.org/2001/XMLSchema#int">99</ATOMResSeqNum>
  <Y rdf:datatype="http://www.w3.org/2001/XMLSchema#float">-6.229</Y>
  <TempratureFactor rdf:datatype="http://www.w3.org/2001/XMLSchema#float">20.48</TempratureFactor>
  <Element rdf:datatype="http://www.w3.org/2001/XMLSchema#string">O</Element>
  <Occupancy rdf:datatype="http://www.w3.org/2001/XMLSchema#float">1.0</Occupancy>
  <X rdf:datatype="http://www.w3.org/2001/XMLSchema#float">-5.864</X>
  <Atom rdf:datatype="http://www.w3.org/2001/XMLSchema#string"> O </Atom>
  <Z rdf:datatype="http://www.w3.org/2001/XMLSchema#float">-8.455</Z>
</Atoms>
```

**Fig. 12.4** Instance of the atom concept from the protein ontology instance store

## *12.5.2 Data Mining*

We compared efficiency of some of the standard hierarchical and tree mining algorithms on the human prion protein data (Tan et al. 2006; Hadzic et al. 2006) extracted from PO Instance Store in XML format (Sidhu et al. 2004a, b). We compared our MB3-Miner (MB3) algorithm with X3-Miner (X3), VTreeMiner (VTM) and PatternMatcher (PM) for mining embedded subtrees, and our IMB3-Miner (IMB3) with FREQT (FT) for mining induced subtrees. Figure 12.5 shows the time performance of different algorithms. Our original MB3 has the best time performance for this data.

Also, as can be seen in Fig. 12.6 with the prion dataset of PO, the number of frequent candidate subtrees generated, is identical for all the major data mining algorithms. This demonstrates that the conceptual framework of PO provides a powerful hierarchical classification of protein data, which provides consistency and accuracy in observations of various data analysis methodologies.

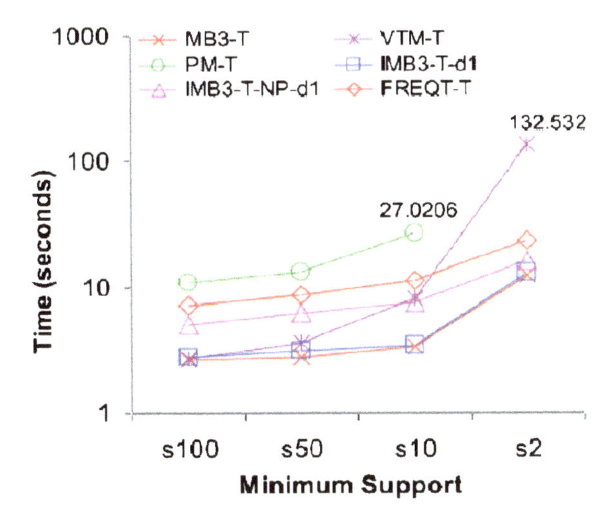


**Fig. 12.5** Time performance for the human prion proteins data in PO

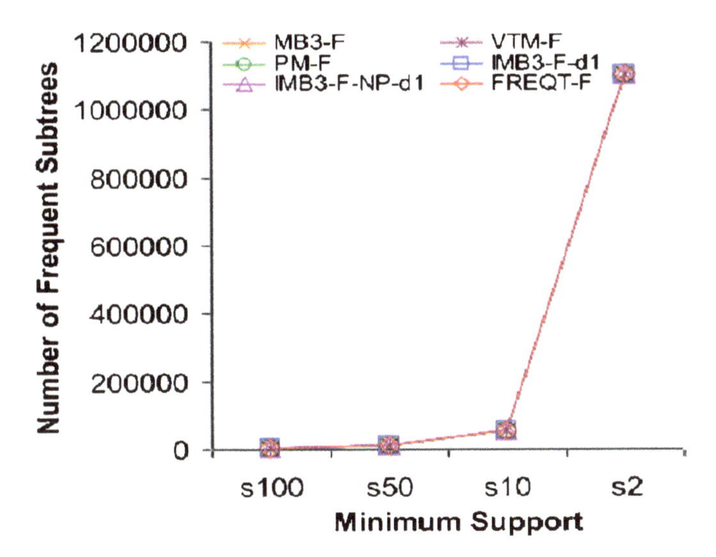


**Fig. 12.6** Frequent candidate subtrees generated by all algorithms

### *12.5.3 Advanced Reasxoning*

PO also provides a specific set of rules to cover application specific semantics over the PO framework (Sidhu et al. 2006). The rules only use the relationships whose semantics are predefined to establish a correspondence among terms in PO. These rules help in defining semantic query algebra for PO to efficiently reason and query

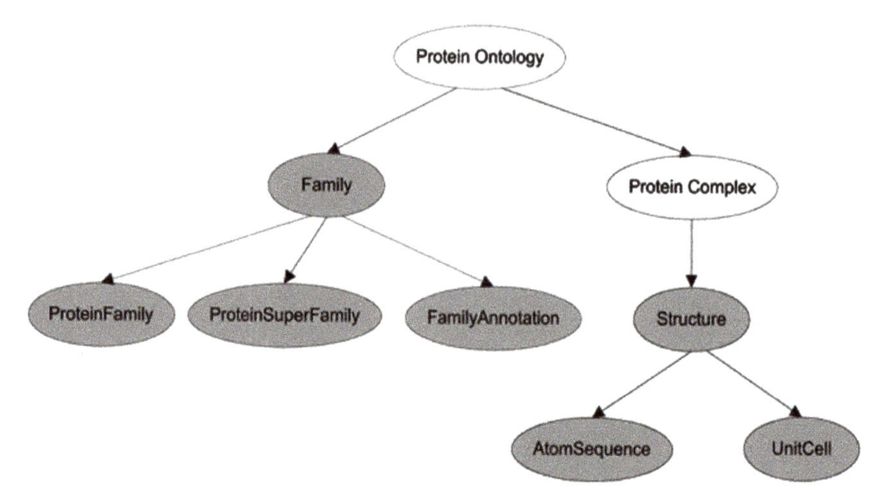


**Fig. 12.7** Depicting UNION operation

the underlying PO instance store. Let us consider that a user requires all the information available in the PO with regards to protein families and protein structure. In this case, all the information highlighted in Fig. 12.7 is displayed. The UNION operator is used for this purpose (Family ∪ Structure).

## 12.6 Conclusion

High throughput technologies can produce a large amount of information about all the proteins in a particular biological system and we will increasingly require computational approaches to mine this data. In this chapter, we addressed this issue by providing definitions of all the major biological concepts of protein synthesis and the relationships between them using a protein ontology. This ontology provides a unified structured vocabulary both for annotating data types and for annotating data and will enable informatics agents to use the data in more intelligent ways.

## References

Ashburner M (1993) FlyBase. Genome News 13:19–20

Ashburner M, Ball CA, Blake JA, Butler H, Cherry JC, Corradi J, Dolinski K (2001) Creating the gene ontology resource: design and implementation. Genome Res 11:1425–1433

Blake JA, Eppig JT, Richardson JE, Davisson MT (1998) The Mouse Genome Database (MGD): a community resource. Status and enhancements. The Mouse Genome Informatics Group. Nucleic Acids Res 26:130–137

Collins FS, Morgan M, Patrinos A (2003) The Human Genome Project: lessons from large-scale biology. Science 300:286–290

Fenn JB, Mann M, Meng CK, Wong SF, Whitehouse CM (1989) Electrospray ionization for the mass spectrometry of large biomolecules. Science 246:64–71
Fraser AG, Marcotte EM (2004) A probabilistic view of gene function. Nat Genet 36:559–564
Frazier ME, Johnson GM, Thomassen DG, Oliver CE, Patrinos A (2003a) Realizing the Potential of Genome Revolution: The Genomes to Life Program. Science 300:290–293
Frazier, M. E., Thomassen, D. G., Patrinos, A., Johnson, G. M., Oliver, C. E. & Uberbacher, E. (2003b) Setting Up the Pace of Discovery: the Genomes to Life Program. 2nd IEEE Computer Society Bioinformatics Conference (CSB 2003). Stanford, CA, USA, IEEE CS Press.
George DG, Mewes H-W, Kihara H (1987) A standardized format for sequence data exchange. Protein Seq Data Anal 1:27–29
George DG, Orcutt BC, Mewes H-W, Tsugita A (1993) An object-oriented sequence database definition language (sddl). Protein Seq Data Anal 5:357–399
Gruber TR (1993) A Translation Approach to Portable Ontology Specifications. Knowledge Acquisition 5:199–220
Hadzic, F., Dillon, T.S., Sidhu, A.S., Chang, E. and Tan, H. (2006) Mining Substructures in Protein Data. 2006 IEEE Workshop on Data Mining in Bioinformatics (DMB 2006) in conjunction with 6th IEEE ICDM 2006. IEEE Computer Society, Hong Kon
Hafner, C. D. & Fridman, N. (1996) Ontological foundations for biology knowledge models. 4th International Conference on Intelligent Systems for Molecular Biology. St. Louis, AAAI.
Jenssen TK, Laegreid A, Komorowski J, Hovig E (2001) A literature network of human genes for high-throughput analysis of gene expression. Nat Genet 28:21–28
Karas M, Hillenkamp F (1988) Laser desorption ionization of proteins with molecular masses exceeding 10, 000 daltons. Anal Chem 60:2299–2301
King OD, Foulger RE, Dwight S, White J, Roth FP (2003) Predicting gene function from patterns of annotation. Genome Res 13:896–904
Lewis SE (2004) Gene Ontology: looking backwards and forwards. Genome Biol 6:103.1–103.4
Li, Q., Shilane, P., Noy, N. F. & Musen, M. A. (2000) Ontology acquisition from on-line knowledge sources. AMIA 2000 Annual Symposium. Los Angeles, CA.
Mani, I., Hu, Z. & Hu, W. (2004) PRONTO: A Large-scale Machine-induced Protein Ontology. 2nd Standards and Ontologies for Functional Genomics Conference (SOFG 2004). UK.
Ohkawa, H., Ostell, J. & Bryant, S. (1995) MMDB: an ASN.1 specification for macromolecular structure. 3 rd International Conference on Intelligent Systems for Molecular Biology. Cambridge, United Kingdom, AAAI.
Ostell, J. (1990) GenInfo ASN.1 Syntax: Sequences. NCBI Technical Report Series. National Library of Medicine, NIH.
Pandey A, Mann M (2000) Proteomics to study genes and genomes. Nature 405:837–846
Paoli M, Liddington R, Tame J, Wilkinson A, Dodson G (1996) Crystal structure of T state haemoglobin with oxygen bound at all four haems. J Mol Biol 256(4):775–792
Pennisi E (1998) Genome data shake tree of life. Science 280:672–674
Pongor S (1998) Novel databases for molecular biology. Nature 332:24–24
Rawlings CJ (1998) Designing databases for molecular biology. Nature 334:447–447
Schuler GD, Boguski MS, Stewart EA, Stein LD, Gyapay G, Rice K et al (1996) A gene map of the human genome. Science 274:540–546
Schulze-Kremer, S. (1998) Ontologies for Molecular Biology. Pacific Symposium of Biocomputing. Hawaii, PSB 1998 Electronic Proceedings.
Sidhu AS, Dillon TS, Sidhu BS, Setiawan H (2004a) A unified representation of protein structure databases. In: Reddy MS, Khanna S (eds) Biotechnological approaches for sustainable development. Allied Publishers, India, pp 396–408
Sidhu AS, Dillon TS, Sidhu BS, Setiawan H (2004b) An XML based semantic protein map. In: Zanasi A, Ebecken NFF, Brebbia CA (eds) 5th International Conference on Data Mining, Text Mining and their Business Applications (Data Mining 2004). WIT Press, Malaga, Spain, pp 51–60
Sidhu, A. S., Dillon, T. S. & Chang, E. (2005a) An Ontology for Protein Data Models. 27th Annual International Conference of the IEEE Engineering in Medicine and Biology Society 2005 (IEEE EMBC 2005). Shanghai, China, IEEE Press

Sidhu, A. S., Dillon, T. S., Chang, E. & Sidhu, B. S. (2005b) Protein ontology: vocabulary for protein data. IN He, X., Hintz, T., Piccardi, M., Wu, Q., Huang, M. & Tien, D. (Eds.) 3rd International IEEE Conference on Information Technology and Applications, 2005 (IEEE ICITA 2005). Sydney, IEEE CS Press

Sidhu, A.S., Dillon, T.S. and Chang, E. (2006) Towards Semantic Interoperability of Protein Data Sources. 2nd IFIP WG 2.12 & WG 12.4 International Workshop on Web Semantics (SWWS 2006) in conjunction with OTM 2006. Springer, France, 1835-1843

Sidhu AS, Dillon TS, Chang E (2007) Protein ontology. In: Chen J, Sidhu AS (eds) Biological database modeling. Artech House, New York

Tan, H., Dillon, T.S., Hadzic, F., Chang, E. and Feng, L. (2006) IMB3-Miner: Mining Induced/Embedded Subtrees by Constraining the Level of Embedding. 10th Pacific-Asia Knowledge Discovery and Data Mining Conference (PAKDD 2006). Springer, Singapore, 450-461.

W3C-RDFSCHEMA (2004) RDF Vocabulary Description Language 1.0: RDF Schema. In Brickley, D., Guha, R. V. & Mcbride, B. (Eds.) W3C Recommendation 10 February 2004. World Wide Web Consortium.

Westbrook J, Ito N, Nakamura H, Henrick K, Berman HM (2005) PDBML: the representation of archival macromolecular structure data in XML. Bioinformatics 21:988–992

Whetzel PL, Parkinson H, Causton HC, Fan L, Fostel J, Fragoso G, Game L, Heiskanen M, Morrison N, Rocca-Serra P, Sansone S, Taylor C, White J, Stoeckert CJ (2006) The MGED Ontology: a resource for semantics-based description of microarray experiments. Bioinformatics 22:866–873

Yamazaki T, Hinck AP, Wang YX, Nicholson LK, Torchia DA, Wingfield P, Stahl SJ, Kaufman JD, Chang CH, Domaille PJ, Lam PY (1996) Three-dimensional solution structure of the HIV-1 protease complexed with DMP323, a novel cyclic urea-type inhibitor, determined by nuclear magnetic resonance spectroscopy. Protein Sci 5(3):495–506

# Chapter 13
# High-Throughput Plant Phenotyping – Data Acquisition, Transformation, and Analysis

**Matthias Eberius and José Lima-Guerra**

## 13.1 Introduction

The aim of applied plant biology has always been to understand how and why plants grow the way they do. In most cases, the target of research is to find correlations and dependencies between distinct factors of the biological system. Biologists often seek better measuring technologies to facilitate their research. To compensate for the invariably limited possibilities, they often developed tedious but nevertheless very successful methods of gaining and increasing knowledge about plants. Thus over long periods of time they produce exemplary results. These steps gradually create a more comprehensive model of how plants actually work and still form a broad basis of the majority of research. Nevertheless, it remains possible to gain deeper insights into "biological variation" and see how the newly discovered principles could be applied to a broad set of plants under differing conditions (light, soil, water, nutrients, plant genes, epigenetic plant history etc.).

Due to the restricted elementary knowledge on "how plants work," breeders devoted to plant improvement perform their tasks on a very empirical basis – in some cases more closely related to artisan craftsmanship than to research. Looking back on some thousand years of successful breeding, this approach has been immensely successful.

M. Eberius (✉)
Lemna Tec GmbH, Schumanstr. 18, 52146 Wuerselen, Germany
e-mail: matthias.eberius@lemnatec.com

J. Lima-Guerra (✉)
Keygen N.V., Agrobusiness Park 90, Wageningen, 6708 PW, The Netherlands
e-mail: jose.guerra@keygen.com

D. Edwards et al. (eds.), *Bioinformatics: Tools and Applications*,
DOI 10.1007/978-0-387-92738-1_13, © Springer Science+Business Media, LLC 2009

## 13.2 The Technological Approach to Plant High-Throughput Phenotyping

### *13.2.1 Biological Demands for Modern Plant Phenotyping*

A technology is now available to assist in the generation of high-throughput plant phenotype data. To be successfully applied, this technology requires the ability to gather large amounts of information for each plant (high-content) from large numbers of plants (high-throughput) over long periods of time with reasonable time resolution (nondestructively), and the ability to transform these raw data into qualitative and quantitative results.

While looking for important basic aspects of development, represented by large amounts of detailed measurements, all this information needs to be seen in the comprehensive context of complete plant development in an integrated approach. This is important, as the aim of breeding and long-term research is in the end directed at complex plant traits such as yield, drought tolerance, or disease resistance.

### *13.2.2 Technological Basis of High-Throughput Phenotyping*

A set of key technologies is required to fulfil the above goals.

1. High-resolution images, digitally stored to document the growth process and reactions to time-dependent stress factors. In this context "images" includes any kind of spatially resolved measurement in the full range of wavelengths technically available, including visible light (VIS), near-infrared (NIR), infrared (IR), X-rays or other imaging methods such as magnetic resonance imaging (MRI) or terahertz scanning.
2. Automation technology to allow the transport of plants under high-throughput conditions through multisensor detection systems, permitting the screening of thousands of plants per day.
3. Image processing algorithms to enable the extraction of hundreds of parameters from a set of images.
4. Molecular biology, genomics, proteomics, and metabolomics technologies.

These points describe the context in which automatized and image-based plant phenotyping is applied. The following sections detail this approach, with applications describing principles.

It is particularly important for data analysis experts to get a clear idea about phenotyping data production, as the character of the data largely determines the best methods of analysis.

### *13.2.3 The Role of Bioinformatics in High-Throughput Screening Experiments*

The task of designing and physically performing high-throughput screening experiments is hugely important in reaching the final aim of biologically significant data interpretation. This requires good co-operation between people trained in biology, agriculture, applied mathematics, and information technology. Efficient teamwork requires two general features:

1. Everyone involved needs a clear insight in what the others do, to develop a certain understanding about how others organize their part in the project.
2. Each part of the workflow needs to be assigned to one or the other group or be developed in close co-operation at the borders of the different areas of expertise.

Following the general scheme of research based on division of labour, the biological branch provides everything from seeds to soil and plants, and defines the growth conditions according to the specific aims of the tests. If things work out well, the statisticians are asked at this point to set values for control and replicate numbers and make the schemes, for example, for appropriate randomized block designs. The plant experts define which parameters are measured to produce the raw data. After measurement they transfer these data to the mathematically skilled people from the statistics or information technology department. The decision about how to transform raw data before use should be taken after careful consideration. The following sections highlight some important aspects of data production, handling, and transformation before statisticians can start their analysis.

## 13.3 Technical Data Acquisition Under High-Throughput Phenotyping Conditions

### *13.3.1 Some Practical Implications of Larger Plant Numbers*

High-throughput phenotyping may mean analysing up to several thousand units per day. In this context one unit could be equivalent to a single plant in a pot, a tray with up to 20 smaller plants such as *Arabidopsis*, or a multiwell plate with up to 96 wells per plate. While such numbers do not look impressive at first glance, measuring 1,000 pots of maize, each weighing 3 kg, would mean moving 3 tons of material each day. In many cases, each pot has to be imaged at least three times in each imaging unit (one top image and two side images by turning the pot through 90°), and up to four imaging units representing different wavelength ranges (VIS, NIR, fluorescence imaging, IR) may be aligned. This adds up to around 12,000 images each day or roughly one image every 2 seconds, requiring a daily disk storage capacity of around 10–20 GB. Should only 100 parameters be extracted from each image, 1,200,000 data points would need to be processed each day.

This brief example clearly demonstrates how closely high-throughput imaging and high-content analysis resemble industrial conveyor processes rather than conventional biological experiments. It becomes evident that many concepts and approaches used for small-scale tests will need a critical revision before the upscaling of experiments.

Here are some of the consequences and challenges resulting from this analysis:

1. A comprehensive documentation of results will only ever be possible by controling as many parameters as possible. In cases where full control is not possible (e. g., with growth conditions), these factors should be included in data transformation protocols as is done by growth rate calculation and subsequent normalization based on control plants.
2. The dataflow needs to be highly structured and designed to avoid any wrong assignment of data and samples.
3. The number of data points is too large to be handled manually by general spreadsheet programs such as Excel. Advanced data analysis methods are important right from the first analysis of raw data to validating results.

### 13.3.2 How High-Throughput Influences the Value and Character of Each Data Point

Traditionally, researchers measure a small number of biological objects to characterize them, and it makes sense to invest effort to measure parameters that are well defined. Generally, a small number of parameters are measured for each plant, and each value should be as significant as possible. For automated high-throughput measurements, biological relevance of the measurement parameters is still important, but the larger number of replicates and values for each plant and the repeated measurement of each plant over time make a difference to the analysis. Each data point in a series of time-related measurements is statistically backed by its neighbouring values. Variation between plants within a treatment group is limited if the same plants are measured in the course of time, thus reducing standard deviations, particularly if the dynamics of parameter change are properly considered (e. g., measuring growth rate instead of final biomass).

But even for very sensitive measurements, such as water content in leaves measured by near-infrared imaging, or leaf temperature measured by infrared cameras, values gain greater statistical relevance if large numbers of plants are measured. In this way, even difficult parameters such as leaf temperature may yield significant results. Again appropriate biological data transformation, such as deviation of single plant values from the moving average, enables appropriate measurements where traditional studies often fail. In addition, another important feature of image analysis is that a large set of different data extraction methods can be applied to each image.

### *13.3.3 The Character of Complex Plant Imaging Parameters – What Shall Be Measured and How Do Parameters Correlate?*

When upscaling experiments from low-to high-throughput and high-content analysis, changes in the measurement protocol often become necessary. This can be illustrated by "biomass" measurements. Biomass is a very abstract parameter intuitively correlated to weight. Nevertheless, it is important to comprehend that there is no absolute definition of biomass, and this parameter is generally defined in a very pragmatic manner. As long as plant numbers are not large, and plants are small enough, dry weight can be used, while large numbers and big plants often allow only for fresh weight measurement. In cases where nondestructive measurement is required, plant height is used as a substitute for biomass. Depending on the organism's cell number, optical density, light scattering or the number of leaves may also be used to describe increasing biomass. All these parameters increase as long as the plant grows, but the simple quantitative correlation between them is limited, especially if various stressors are applied or different plant lines are used.

From this perspective, image-based biomass measurement is just an additional measurement protocol to actually determine growth or biomass. A correlation between the manual biomass or growth parameter and image-based measurements is often searched for. There are several reasons why such simple correlations fail to work and sometimes even diminish the informational output:

1. There is no scientific rationale for any two parameters based on totally different measurement principles to lead to the same result. Weight measurements integrate leaf thickness as well as leaf area, but plant imaging cannot measure leaf thickness.
2. Even a correlation determined in a validation experiment with control plants can be hugely misleading, as the calibration factor may depend on the specific variety, treatment or environment. For example, plants may change their water content significantly without changing their outer shape. Thus a calibration factor between image-based parameters and fresh weight made with well-watered plants will fail with plants which have a different watering regime. Even more important in this respect are comparisons between different plant lines. If one cultivar has thicker leaves than another, the calibration factor between projected leaf area and fresh weight will not correspond between cultivars.

This may sound like a drawback of image-based parameters, but having this highly differentiated dataset and using multiparameter approaches for statistical analysis can provide far more information than highly aggregated single value biomass results. While scientists need much more statistical competence to extract data from multiparameter approaches, describing the morphology can answer functional questions of plant growth much better and produce more significant relations to functional genetic data.

Another important point in evaluating measurement parameters is the precise definition of what is measured. For example, any image of a plant shows its leaves. This parameter is generally referred to as leaf area, but what is measured is the two-dimensional projection of leaf area in a certain direction. Combining projections from different perspectives (e. g., top and two side images with a 90° turn) can lead to a reproducible and comparable image-based bio volume representing biomass, but will never describe the "true" leaf area. The orientation of the leaves is structurally ignored and hidden leaves are not quantified at all. This again is a drawback of image-based parameters, however, true leaf area is only measurable through destructive methods, as all leaves need to be flattened and scanned. Even then, wrinkles, folding, and distortion make true leaf area measurement almost impossible, even if the leaves are not looked upon as fractals. A true leaf value represents only one value, and ignores information that image processing provides for shape parameters, including leaf orientation, shadowing, and compactness of a plant. With this in mind, the parameter "absolute leaf area" measured destructively can lose its importance when compared to a large set of image-based areas and calculated bio volumes combined with morphological parameters.

The complex process described above shows the importance of advanced data analysis and the relevance of well-developed and adapted bioinformatics approaches for the transformation of raw data to biologically significant parameters, with the subsequent interpretation of the results. The following section will focus on these different kinds of data transformation.

## 13.4 Biologically Relevant Data Transformation

The raw data of any image-based high-throughput phenotyping system are the images. Any image analysis, extracting colors, area length values or shapes out of the images is a major step of data transformation. This process reduces the information from several megabytes of image data to some hundred data points amounting to a few kilobytes. Image processing of biological objects and its technical details are not discussed here, because this is a separate, broad field considered elsewhere (Sonka et al. 2008, Gonzalez and Woods 2008, Davies 2005, Russ 1999). The main focus of this section is on the results of image processing, and their subsequent processing.

### *13.4.1 Calculation of Dynamic, Time-Dependent Parameters from Multiple Datasets of One Plant – Growth Rate Data Transformation*

It is important to keep in mind that any use of data implies specific models and assumptions, even if these implications or hypotheses are not always obvious and data are taken "as measured." The following example regarding growth rate transformations is one case where data transformation is of high importance.

In many classical approaches, growth measurement is performed by measuring a maximum size-value for the plant at the end of the test (size may mean fresh weight, dry weight, height etc.). This method is taken from a field trial perspective, where all plants grow till they reach their final size and are then harvested and evaluated at the end of the season. This kind of field approach is a "yield" approach. By contrast, many screening designs do not let the plant grow to its final size. Thus time becomes a parameter, and absolute "*size*" must be substituted by "growth speed." Additionally, it can often be observed in greenhouse experiments that standard deviations of measured biomass increase during the test duration (Fig. 13.1).

The graphs show a model derived from "Arabidopsis" growth patterns. The longer the plants grow, the bigger the differences between lines become. An additional phenomenon is that the reproduction of a test at a later time shows a relatively similar ranking of growth intensities, but large deviations of absolute values between the two experiments. In numerous cases, particularly with fast growing plants, differences between lines are large at times of high average growth of all plants, but lower, for example, in wintertime when the artificial light cannot replace decreasing sunlight completely or annual rhythms of plants play a role. This can have a major influence on comparisons within large screenings where different lines need to be cultivated over long periods with different seasonal conditions. In cases where the greenhouse experts provide data to the statisticians, the statisticians have only a limited chance of understanding the reasons for these deviations.

Plants finally need to be compared either by significance testing or by a percentage value relative to a control. The comparison based on %-values relative to a control will be considered here in detail, especially with respect to how inappropriate (non) transformation can lead to mathematical artefacts.

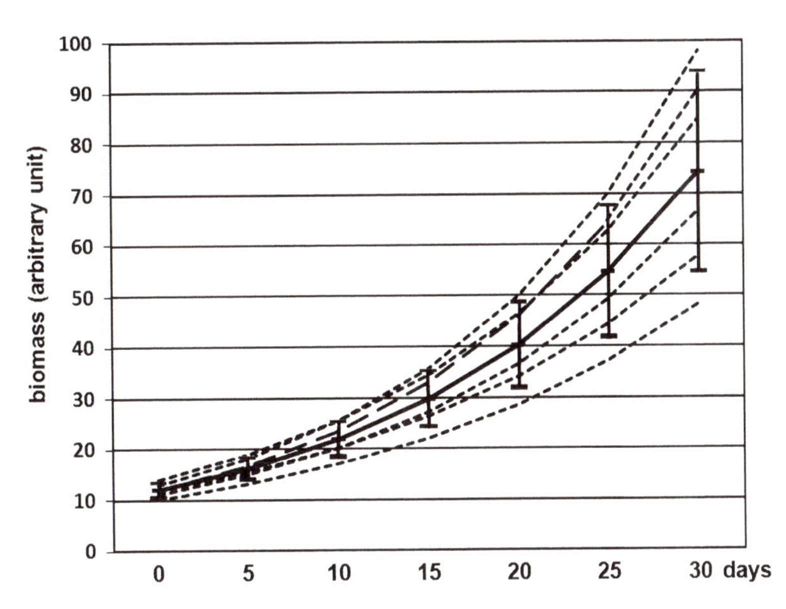


**Fig. 13.1** Model growth curves (*dotted lines*) of either single plants or mean values of different breeding lines or cultivars. The *full line* represents the mean value including the increasing standard deviation

Comparison of biomass values on a percentage scale seems very neutral and unbiased, but includes one important hypothesis: As the comparison is performed on a linear scale, the parameter to be compared with must also act as on a linear scale to avoid misinterpretation. As a result it is important to always determine if growth in the respective example is linear or not. Just by measuring the growth parameters nondestructively, for example by image processing of a leaf area over the whole growth period, it becomes evident that over long periods plants show a pattern that is closer to exponential than to linear (Nyholm 1985, Nyholm 1990). Illustrating the data of Fig. 13.1, a logarithmic scale clearly shows that growth is near to exponential (Fig. 13.2), resulting in linear growth curves after log transformation.

Following data transformation, the results are much easier to interpret. In addition, standard deviations caused by different biomass at the beginning of the measurements remain nearly constant over time. Deviations from this scheme occur in practice if maximum size is approached, stressors or nutrients are limiting the growth, or the rate of development changes, but in all cases the basis for comparison or control is generally the non-limited plant showing near to exponential growth. Thus applying logarithmic growth rates is always appropriate, as long as the control group shows such a logarithmic growth pattern. Additionally, normalizing all experiments to the growth rate of the control plant in the respective experiment allows compensation for mathematical artefacts that may inhibit comparisons between screens under seasonally differing environmental conditions (Eberius et al. 2002).

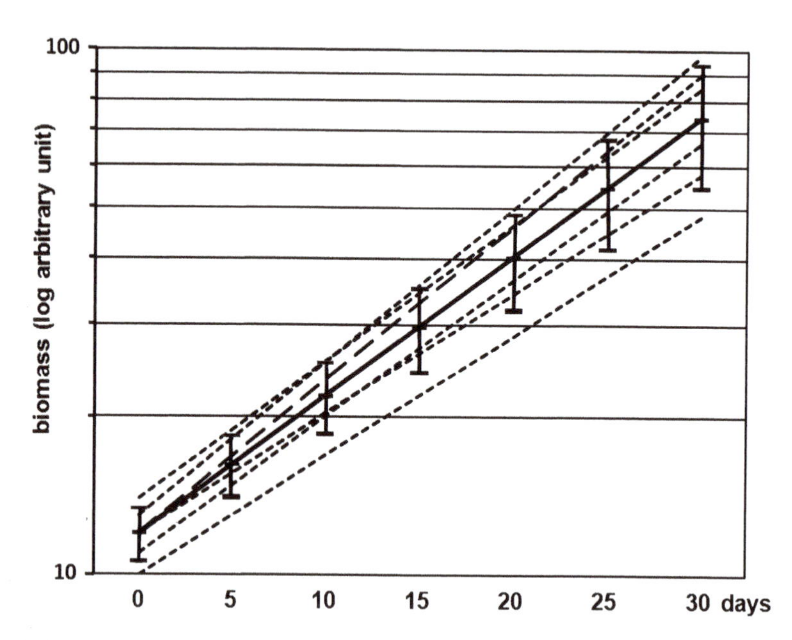


**Fig. 13.2** Model growth curves (*dotted lines*) of either single plants or mean values of different breeding lines or cultivars. Logarithmic presentation of the biomass values allows the slope of the growth curves to display the exponential growth rate and shows only small differences between the curves. The *full line* represents the mean value including the increasing standard deviation

Nevertheless, there may be some growth parameters whose behavior differs from others, for example, leaf area may grow exponentially while leaf length or plant height may not. To provide appropriate measurement parameters, relevant biological knowledge is essential. By using appropriate data models and data transformation, the interpretational quality of results can be greatly enhanced. In the case of growth rates, phenomena such as the one described above, with high standard deviations between replicates, can be massively reduced.

### *13.4.2 Transforming Raw Phenotype Data Resulting from Image Analysis into Biologically Relevant Parameters*

There are generally two approaches to the use of raw data resulting from basic image analysis.

The first approach simply employs the raw data to perform a statistical analysis. While this approach may appear rather unbiased and seems to minimize any preset hypothesis, it could be difficult to provide a biological interpretation of results. Apart from statistical correlation, the biologist or breeder works on a qualitative level, where a certain measurement parameter must represent facts of biological relevance in order to define breeding or screening targets. The second approach first transforms the raw data into biologically relevant values that allow for a better interpretation of results and a better understanding of the correlations observed. This approach involves significantly more work at the image processing level and with the transformation of raw data. More importantly, it requires a deeper understanding of the biological or breeding aims of the study. Two examples of data transformation are explained below.

#### 13.4.2.1 Static Shape Phenotyping – Leaf Angle

To extract biologically relevant information, it is important to understand how biologists or plant breeders look at plants and why they assess specific parameters. When starting a corn screening experiment, one of the biologist's requirements may be to measure "leaf angle." But from an image analysis perspective, it is usually difficult to find the algorithmic rules of how to measure leaf angle, considering greenhouse corn plants and their generally bent leaves (Fig. 13.3).

The term leaf angle is poorly defined, even in specialist literature. It is important to know why experienced breeders are interested in this measurement. When it is made clear that smaller angles between leaves and stem allow higher plant densities by reducing leaf interference and shading between neighboring plants, the agricultural traits on which this parameter is based become obvious. Breeders traditionally chose this trait for practical reasons as they rarely view plants from the top. Thus, image-based screening of width to height ratios, as well as leaf area ratios (side and top) could show a comparable result fulfilling the screening aims.

Another option may be the integration of leaf orientations based on skeleton measurement (Fig. 13.4).

This example shows how distinct imaging parameters that back each other up could form a stable representation of an important agronomic trait. The discussion about why plants in greenhouses grow differently from those in the field could be an incentive to think about how greenhouse experiments could be made more relevant. This may also result in further automation (LemnaTec 2008).

#### 13.4.2.2 Dynamic Shape Phenotyping – Leaf Rolling

In drought stress experiments of corn plants, it is important to be able to understand and quantify the relationship between water deficiency and plant response. The physiological responses include closing the stomata and leaf rolling to minimize the surface area. If leaves close their stomata they stop evaporating water and heat up, and leaf temperature can be imaged in the infrared spectrum. Leaf rolling is a visible parameter assessable by image analysis. Despite the easy visibility of this trait in the field, the lower magnification and resolution of images from three perspectives (top, broad side, small side) makes leaf rolling difficult to see. Rolling may be more or less severe and a quantitative measurement rather than the semi-quantitative measurement in the field would be preferable. In the case of corn plants, detailed analysis of data has shown that no single value from one image provides stable correlations with visual assessment. Looking for changes between images, for example, projected leaf area from the top shows that rolled leaves have

**Fig. 13.3** Broad side projected leaf area images of a greenhouse grown corn plant (*left*) and a field grown corn plant (*right*). The field grown plant has long, almost straight leaves showing a distinct angle between stem and leaf. The greenhouse plant has bent leaves due to the lower growing density

**Fig. 13.4** Broad side projected leaf area image of a greenhouse grown corn plant. Image processing can provide skeleton lines of the leaves, which can be used to calculate integrated leaf orientation values or distributions up to characteristic points of the leaf

a smaller area value, but this kind of assessment requires high frequency measurements (several times a day), which are impractical to carry out if thousands of plants need to be imaged on a daily basis. An assessment of data revealed that the ratio between projected top and broad side area of corn plants shows the best representation of leaf rolling. Due to the structure of corn leaves which fold slightly along the centre-line and simultaneously start rolling from the edges, the projected area on the broad side image remains almost constant (Fig. 13.5). This is a good normalization measurement to allow comparison of different plants and growth stages for leaf rolling, while only requiring two images from one point in time.

This example shows that attributing meaning to a certain ratio of image analysis parameters can help with the understanding of data and minimize the need for parameter validation. The meaning of image parameters is plant specific due to different morphology. For each plant species, different parameters and ratios need to be identified.

### *13.4.3 How High-Content Documentation of Images and Data Can Change the Way Biological Experiments Are Carried Out*

In previous sections, the focus was on performing one experiment and interpreting it immediately afterwards. However, comprehensive digitization of large screenings using image-based phenotyping can produce a perspective for further use of the raw data beyond the immediate evaluation of results. The data could be used to model real experiments in silico. Multisensor imaging and image storage allows for conservation not only of the plants, but also of the raw data. As a result, it may be

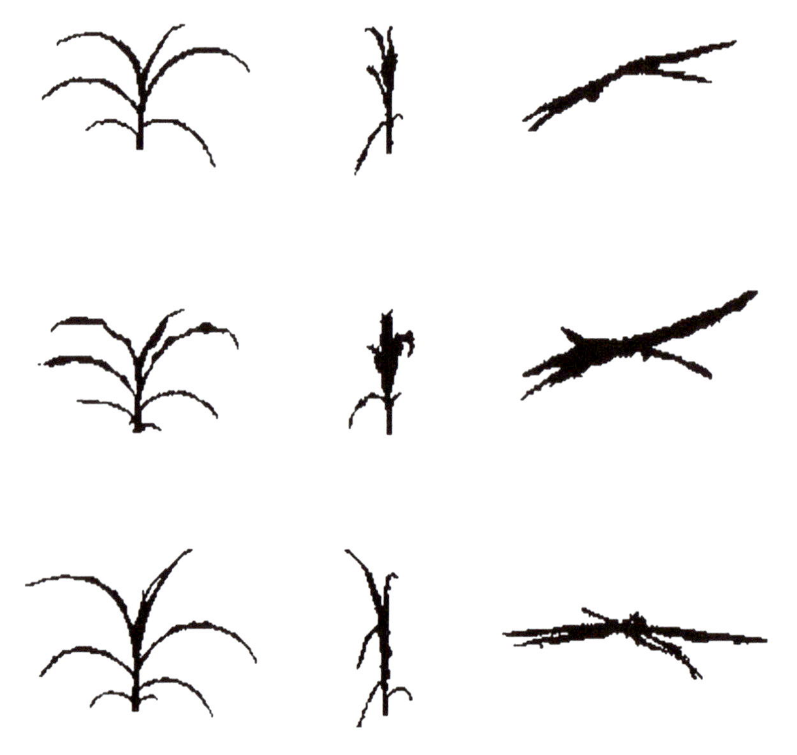

**Fig. 13.5** Top and side images of one corn plant before leaf rolling (*top row*), with rolled leaves (*intermediate row*) and after recovery (*lower row*) showing that the top area changes while the side area stays nearly constant

possible to begin an experiment with a data mining process of existing data. Once a comprehensive data archive has been developed, biological research can start to go in silico.

### 13.4.4 Statistical Knowledge and Tools

High-content data analysis requires a solid biological background to fully understand the experiment. In addition, good skills in statistical analysis adapted to the specific needs of biological screening procedures are required. Biologists are expected to gain basic knowledge of the tools needed to prepare the data, at least to a degree where they can hand them over to more specialized statisticians and bioinformaticians and communicate with them on an appropriate level. The following sections deal with some basic data handling approaches which have proved to be successful for the interpretation of high-content data from screening experiments. This includes the change of perspective from a biological to a mathematical focus without losing sight of the common aim, which is important when dealing with complex questions.

## 13.5 The Data Mining Perspective

While the first part of this chapter mainly dealt with the biological and breeding perspective, the second part will highlight the issues from the viewpoint of data mining.

### *13.5.1 Data Mining: Fundamental Approaches*

Data mining is a broad term. It covers a range of techniques being used in a variety of industries. New plant phenomics technologies and developments in the fields of automation and image processing are generating large amounts of data. This in turn is creating a requirement for novel data mining applications.

### *13.5.2 Successful Data Mining with High-Content, High-Throughput Data*

This section will not deal with the statistical theories underlying data modelling techniques since this is a topic already well covered by a number of other books. Instead it focuses on the fundamental techniques to handle multiple variables that can be applied to biology, with specific applications for image processing. The success of any data mining or modelling project requires a sound understanding of the methodologies, knowledge of the data, its applications, and overall project objectives. The goal of this section is to provide readers with a basic understanding of the skills and processes needed for the successful use of data models. The examples chosen will use the statistical software SAS® and R®.

#### 13.5.2.1 Objective 1: Define Goals

The goals must be clearly defined. This initial step is the most fundamental because it will influence all other steps. Once the goal is defined, the data that increases the likelihood of achieving the goal needs to be selected, and all other steps will follow, cascading down from the initial goal. Analysts need to understand these goals, and communication and comprehension skills are the key to success.

#### 13.5.2.2 Objective 2: Select the Data

The objective here is to select from all available data or generate new data in order to meet the goals defined in step 1. With the increased growth of data and the possibility to re-analyze it there come the need to efficiently store the wealth of information. The primary term for this is *data warehousing*, but it is also referred to as a data mart, central repository, meta data, etc. A well-designed data warehouse provides efficient access to multiple sources of data. Selecting the best data for model development requires a thorough understanding of the goals set up in step 1. Although the tools are important, the quality of the data used is the decisive basis of all interpretation of information. Quality may refer to reproducibility and accuracy as well as data transformation. For example, if leaf area values are used directly, the data quality will be worse than after transformation to growth rates, as demonstrated above. Data remain the limiting factor of statistics, and high-throughput phenotyping is helping to overcome this.

#### 13.5.2.3 Objective 3: Prepare the Data

The saying "garbage in, garbage out" highlights the importance of this step. Users should understand how data is classified. A fixed format can be easier to read and understand. The disadvantage of the fixed format is that it uses space for blank fields. Therefore, if many fields have values missing, data storage can be uneconomical (Scripts 13.1 and 13.2).

It is important to understand the source and format of the data. Once the data has been loaded into the appropriate statistical environment, in this case SAS®, there are several routines that enable users to quantify the data properties. The frequency procedures of SAS® can visualize missing data patterns, and the univariate procedures can perform outlier detection. The proper handling of missing data and outliers will require further interaction between statisticians and experimenters. Here, the option to look directly at the images to find reasons for outliers is often far more efficient than a numerical procedure that handles outliers merely on a statistical basis.

```
Data one
infile 'C:\sampledata.txt' missover recl=22;
input
id 1-9 /*unique prospect identifier from digit 1 to 9*/
sex $ 10 /*the sex start on digit 10*/
loc $ 11-15 /*five digit location code*/
ext $ 16-19 /*four digit far extension*/
/*0 --5-- 10 -- 15 -- 20 /*--
000001101M800143437
011000002F19380 CFD
000000113M008083522
000000014F945912441
100000005M696441001
```

**Script 13.1** SAS® example of fixed format

```
Data one;
infile 'C:\sampledata.txt' missover recl=22;
input
id 1-9 /*unique prospect identifier from digit 1 to 9*/
sex $ 10 /*the sex start on digit 10*/
loc $ 11-15 /*five digit location code*/
ext $ 16-19 /*four digit far extension*/
/*0 --5-- 10 -- 15 -- 20 --*/
001001001,M,80014,3437
000100002,F,19380, ,C,F,D
000000013,M,008083522
000000054,F,94591,2441
000300005,M,69644,1001
```

**Script 13.2** SAS® example of variable format (SAS®. 2004)

All potentially suitable data is then integrated into the final dataset ready to be imported into the statistical software. There are several methods to compute basic statistical calculations that can be incorporated into the hardware generating the data. For example, Keygene N. V. has developed a tool with R® (R®-Project 2008; Bioconductor 2008) to automate parts of their phenotype analysis. *Phenopipe* is a standardized approach to statistically analyze phenotypic data. Datasets are restricted to certain constraints and consist of observation matrices from different individuals under different aspects, such as location. The tool is not fully automatic, but will ask for input through a simple user interface. *Phenopipe* is currently constrained to five statistical models, considered the most frequently used by Keygene for the analysis of agronomical trials, and it has a built-in decision making function to connect data with the appropriate statistical model. Table 13.1 shows a decision tree for the models, always assuming multiple lines (genotypes).

All models are based on the plot average; therefore it is assumed that only one line is present in each plot. Once again, the extent of the importance of detailed knowledge of the data character to decide if averaging is appropriate becomes evident. Averages make only sense for homogeneously distributed data which is not the case for biomass data from plants having an exponential growth pattern, and such data need to be logarithmically transformed in advance. The functionality of *Phenopipe* enables its users to investigate data structure and test statistical hypotheses. Its statistical analysis is based on plot averages with a decision making function to guide the data to the correspondent model. The statistical analysis includes an analysis of variance, outlier detection, model goodness of fit, corrected means, and least significant differences. All functionality is implemented in R® and pipelined to the user via a web-based interface. R® is becoming the most frequently used tool for the dissemination of new methods in statistical computing. *Phenopipe* automatically performs a check for data structure, outliers, correlation analysis and analysis of variance, and any statistical procedure available in the CRAN and Bioconductor can be incorporated into *Phenopipe*.

**Table 13.1** Decision tree for models assuming multiple genotypic lines

| In case of multiple environments | (more sites) | |
|---|---|---|
| Multiple blocks | – model 1.1 | (more blocks) |
| Single block | | (one block) |
| Replication of lines (plots) | – model 1.2 | (more plots per line) |
| Single line observation | – model 1.3 | (one plot line) |
| Single environment | (one site) | |
| Multiple blocks | – model 1.4 | (more blocks) |
| Single block | | (one block) |
| Replication of lines (plots) | – model 1.5 | (more plots per line) |
| Single line observation | | (one plot per line: no analysis) |

#### 13.5.2.4 Objective 4: Select and Transform the Variables

In various real situations, for example with the LemnaTec image analysis software system, hundreds of variables are computed. One of the basic aims is to find image-based variables (phenotypes) that discriminate between two genotypes. The main problem of this relatively simple approach is that if it is performed in multivariate space, the number of potential combinations can increase rapidly as the number of variables increases. Once the data is considered correct and missing values have been handled, the next step is to look for opportunities to derive new variables. In this situation knowledge of the data is critical. Combining variables by summation or division can improve predictive power. Ratios between variables are also useful for certain types of prediction as demonstrated above for leaf rolling, and this is a general approach transferred from engineering, related to the Pi-theorem (because a "circle" always has a very specific ratio between circumference and diameter; Hart (Hart 1995; Kline 1986).

A basic approach is suggested for researchers dealing with large numbers of variables intended for modelling. Performing an in-depth analysis on each variable is not a time-efficient approach as many of the variables are correlated with each other. If some that have predictive power are eliminated, others will usually be able to fill the predictive gap. For cases such as these there is a handy trick: Use a dependent variable y and 3,000 independent *x*-variables, then do a few simple truncations with "if and then statements" in SAS® to create some binary variables. For example, if y is greater than the average value of all *y*-variables, then Yind = 1 else Yind = 0 or if greater than one standard deviation. One advantage of this strategy is that it can accommodate nonlinear effects and also enable researchers to truncate the truly relevant distribution sections of y. A robust procedure to use under these circumstances is logistic regression, which is used as a fitness function of some machine learning processes. This is implemented in SAS® in a procedure called PROC LOGISTIC, and can be used for diverse statistical scenarios in many areas, for example animal sciences and epidemiology. The procedure can be easily implemented with the code shown in Script 13.3.

The Ods-html will request that all output be created in *html*-format, so that it can easily be opened in Excel®© for further processing. The step wise selection option

```
Ods html;
proc logistic data=one descending;
model y = x1 - x3000
/selection= stepwise maxstep=1 details;
run;
end;
```

**Script 13.3** Logistic regression (SAS 1995, Stokes et al. 2000, Hosmer and Lemeshow 1989)

will cause each variable to be considered one at a time. With the $\chi^2$ probability value for their significance, these probabilities can be further processed in SAS® to do family-wise multiple testing corrections as well as false discovery testing. However, for an inexperienced user, filtering the variables with a very stringent *p*-value for the $\chi^2$ test ($p<0.0001$) could be a more conservative and safer option.

For more detail, an example is given here where the number of variables has been reduced to 100 independent variables with one dependent variable, such as manual dry weight measurement. In this case, it is time-consuming to measure the dependent variable, and requires the destruction of the sample. The 100 independent variables are digital phenotypes based on image analysis. An analytical model should be created so that pictures can be taken, and the time-consuming dry weight measurement predicted instead of measured. Assuming that the data has been gathered, the task now is to reduce this set to the most predictive set of variables.

For this step, the cluster procedure of SAS® called PROC VARCLUS (PROC VARCLUS 2008) is used. This procedure segregates the independent variables into disjoint or hierarchical clusters. Associated with each cluster is a linear combination of the variables in the cluster. This is an iterative procedure that reassigns the independent variables to different clusters. Each iteration computes a cluster and each variable is assigned to the cluster with which it has the highest squared correlation. The second phase is a search algorithm where each variable is tested to see if assigning it to a different cluster increases the coefficient of determination (Rsq) of the cluster model. If a variable is reassigned during the search phase, the Rsq of the new cluster model is recomputed before restarting the whole procedure. This iterative process is repeated until a convergence level is achieved. A helpful guide through this procedure can be found in the help section of SAS®. SAS® is stable and reliable and therefore an excellent choice, especially for less experienced users. It is useful to insert the data as a correlation matrix so that the correlation procedures of SAS® (PROC CORR) can be applied. The final result is a group of clusters where the most correlated variables cluster together and clusters are made in a way that variables in different clusters are as uncorrelated as possible. From each cluster, the variables with the highest Rsq with their own cluster are used. Therefore the number of variables that are equal to the number of clusters can be reduced. The pitfall of this approach is that the process requires interpretation of the clusters by researchers, but it can still be highly efficient in a multidisciplinary environment. Another option is to truncate the variables and use the logistic regression again (Script 13.4).

```
proc logistic data=one descending;
weight smp_wgt;
model y = x1-x100;
/selection = stepwise maxstep = 2 details;
run;
```

**Script 13.4** Model building one

```
dataone;
set one;
x_sq = x_est3**2; /*squared*/
x_cu = x_est3**3; /*cubed*/
x_sqrt = sqrt(x_est3); /*square root*/
x_curt = x_est3**.3333; /*cube root*/
x_log = log(max(.0001,x_est3)); /*log*/
x_exp = exp(max(.0001,x_est3)); /*exponent*/
x_tan = tan(x_est3); /*tangent*/
x_sin = sin(x_est3); /*sine*/
x_cos = cos(x_est3); /*cosine*/
x_inv = 1/max(.0001,x_est3); /*inverse*/
x_sqi = 1/max(.0001,x_est3**2); /*squared inverse*/
x_cui = 1/max(.0001,x_est3**3); /*cubed inverse*/
x_sqri = 1/max(.0001,sqrt(x_est3)); /*square root inv*/
x_curi = 1/max(.0001,x_est3**.3333); /*cube root inverse*/
x_logi = 1/max(.0001,log(max(.0001,x_est3))); /*log inverse*/
x_expi = 1/max(.0001,exp(max(.0001,x_est3))); /*exponent inv*/
x_tani = 1/max(.0001,tan(x_est3)); /*tangent inverse*/
x_sini = 1/max(.0001,sin(x_est3)); /*sine inverse*/
x_cosi = 1/max(.0001,cos(x_est3)); /*cosine inverse*/
run;
```

**Script 13.5** Functional modelling

With maxstep = 2, one will look for the two best forms of the independent variables $x$ that influence y. Different strategies can be used by changing the maxstep option.

The final step can be taken once a small group of variables is available. Nonlinear forms of these variables need to be looked for and their influence on the $y$-variable assessed. Script 13.5 produces 22 forms of the variable $x$. Logistic regression can be used again after truncation to look for the ideal mathematical form of the $x$-variables. A similar step can be created to compute multiplicative variables determining potential conditional associations, for example, interactions. This fine-tuning of the model can be very powerful, but should be done with care to avoid over-parameterization.

#### 13.5.2.5 Objective 5: Implementation

This final step is designed to avoid model over-fitting. It is recommended that the original data should be cross-validated in two steps: (1) recreate several thousands of copies of the original data with replacement sampling. This should be continued until 1/3 of the data is recreated. The predictive power of the model and its importance are measured in the newly formed cross-validated datasets. (2) Rebuild the model with a random sample of half the data and use it to predict the other half. These two strategies in combination and with different variants should provide a final evaluation of the model.

## 13.6 Conclusion

This example of a data mining application to image-based biological data and data from other sources (such as genetic data or chemical measurements) shows how intense the co-operation between high statistical expertize and biological knowledge needs to be, at each step of the development of a full pipeline from plant seed to final result. Biological systems are extremely complex and not all correlations can be either verified or falsified, due to a limited knowledge of other interactions. As a result, the use of appropriate biological and mathematical models is the best way to extract and analyze data efficiently. Nevertheless, an open mind for the unexpected should be a constant aim of any screening program that seeks to deliver new results and information. Only the detailed knowledge of many different researchers can help to understand the various aspects of plant development required to breed plants for our future needs.

## References[1]

Agresti A (2002) Categorical data analysis. Wiley, New York
Bernardo JM, Smith AFM (2002) Bayesian theory. Wiley, New York
Bioconductor (2008) http://www.bioconductor.org/. Accessed 13 July 2008
Breslow NR, Clayton DG (1993) Approximate inference in generalized linear mixed models. J Am Stat Assoc 88:9
Davies ER (2005) Machine vision. Morgan Kaufmann/Elsevier, San Francisco, CA
Devore JL (2000) Probability and statistics for engineering and the sciences. ITP, CA
Eberius M, Mennicken G, Reuter I, Vandenhirtz J (2002) Sensitivity of different growth inhibition tests – just a question of mathematical calculation? Ecotoxicology 11:293–297
Falconer DS (1989) Introduction to quantitative genetics. Longman, London
Gonzalez RC, Woods RE (2008) Digital image processing. Pearson Prentice Hall, Upper Saddle River, NJ
Hart GW (1995) Multidimensional analysis: algebras and systems for science and engineering. Springer-Verlag, New York
Hosmer DW, Lemeshow S (1989) Applied logistic regression. Wiley, New York

[1] The references listed below are mainly intended for further reading and should help to understand the various aspects of image-based high-throughput screening from acquisition to final phenome analysis.

Jeffreys H (1961) Theory of probability. Clarendom, Oxford

Kline SJ (1986) Similitude and approximation theory. Springer, New York

Koch CG, Carr GJ, Strokes ME, Uryniak TJ (1990) Categorical data analysis. In: Berry DA (ed) Statistical methodology in pharmaceutical sciences. Marcel Dekker, New York

LemnaTec (2008) http://www.lemnatec.de/Dokumente/2007/The moving field concept.pdf. Accessed 21 May 2008

Littell RC, Milliken GA, Shroup WW, Wolfinger RD (1996) SAS® systems for mixed models. SAS Institute, Cary NC

McCllagh P, Nelder JA (1989) Generalized linear models. Chapman and Hall, London

Nyholm N (1985) Response variable in algal growth inhibition test – biomass or growth rate? Water Res 19:273–279

Nyholm N (1990) Expression of results from growth inhibition toxicity tests with algae. Arch Environ Contam Toxicol 19:518–522

PROC VARCLUS (2008) http://www2.sas.com/proceedings/sugi26/p261-26.pdf. Accessed 19 June 2008

R®-Project (2008) http://www.r-project.org. Accessed 19 June 2008

Royall R (2000) Statistical evidence: a likelihood paradigm. Chapman & Hall/CRC, London

Russ JC (1999) The image processing handbook. CRC, Boca Raton, FL

SAS (1995) Logistic regression examples using the SAS system. SAS Institute, Cary, NC

SAS® (2004) SAS/STAT® help and documentation (release 9.1.3). SAS Institute, Cary, NC

Sonka M, Hlavac V, Boyle R (2008) Image processing, analysis and machine vision. Thomson, Toronto, Canada

Stokes M, Davis CS, Koch GG (2000) Categorical data analysis using the SAS system, 2nd edn. SAS, Cary, NC

Wedderburn RWM (1974) Quasilikelihood methods, generalized linear models, and the Gauss-Newton method. Biometrika 61:339

Weir BS (1996) Genetic data analysis II. Sinauer, Sunderland, MA

# Chapter 14
# Phenome Analysis of Microorganisms

**Christopher M. Gowen and Stephen S. Fong**

## 14.1 Introduction

Many terms used in systems biology and bioinformatics are loosely defined and may be interpreted differently depending upon the individual. The introductory section will detail our working definition of a phenome and phenomics and describe some ways in which microbial phenomics may differ from phenomic studies in other organisms.

### *14.1.1 Phenomics*

A promise of the genomic age is the ability to connect phenotype and genotype. Accurate and quantitative descriptions of an organism's phenotype are crucial to delivering on this promise and are the realm of phenomics. Strictly speaking, "phenomics" refers to the study of all the phenotypes of an organism that are heritable and a result of the genetic code. The "phenome" itself is the fully observed heritable traits and phenotypes of an organism. This broad definition includes gene expression, protein translation, and even epigenetic regulation. Transcriptomics, proteomics, metabolomics, fluxomics, and many other "omic" data types refer to quantitative measurement of biochemical and cellular processes, "phenomics" refers to organismal phenotypes and traits: macro-scale phenotyes such as growth rate, temperature or pH tolerance, substrate consumption, production of valuable byproducts, and pathogenicity to name a few. In this chapter we discuss the study of phenomics of microorganisms in this same context, covering benefits and challenges unique to working with microorganisms, as well as experimental and analytical techniques that a researcher is most likely to encounter as part of this field.

C.M. Gowen (✉)
Department of Chemical and Life Science Engineering, Virginia Commonwealth University, P.O. Box 843028, Richmond, VA, 23284, USA
e-mail: gowencm@vcu.edu

D. Edwards et al. (eds.), *Bioinformatics: Tools and Applications*,
DOI 10.1007/978-0-387-92738-1_14, © Springer Science+Business Media, LLC 2009

## *14.1.2 Microbial Phenomics*

Microorganisms are used as model organisms for studying a wide range of basic biological questions. Many microorganisms are considered as some of the simplest self-sustaining organisms, and for many, an entire microorganism is a single cell. The unicellular nature of many microorganisms facilitates studying the connection between genotype and phenotype. In the microbial context, it is theoretically possible to directly correlate the genotype of an organism to its phenotype. The fact that a single cell is a single organism is one of the largest benefits of studying phenomics using microorganisms. The following sections give details about additional benefits to studying phenomics in microbes.

### 14.1.2.1 Facility of Growth

Microorganisms have long been favored for the study of fundamental biological concepts due to their ease of culture and resilience, and many of these advantages carry over to the study of phenomics. The ability of microorganisms to grow (and grow well) in many environments makes them easier and cheaper to work with by permitting the use of inexpensive media formulations. Many microorganisms also survive freezing and thawing, thus making experimentation more flexible and reproducible. Additionally, many species can be cultivated in liquid suspension at a relatively high density. The benefits of these growth attributes are twofold and very significant. First, cell culture can be reduced to liquid handling, greatly facilitating automation and increasing throughput. Second, high density growth means that a relatively large quantity of biomass is present in a given volume, greatly improving experimental measurability. Their versatility also permits researchers to precisely and consistently vary culture conditions such as carbon source, temperature, and oxygen supply to determine an organism's ability to respond to changes in its environment. This ease of cultivation makes it relatively straightforward to grow microorganisms in standardized high-throughput platforms such as 96- or 384- well plates. Colorimetric or fluorometric assays are then especially easy to perform with several replicates for each condition being tested. In addition to being easy to please, microorganisms typically proliferate very quickly when compared to animal cell lines (not to mention live animals or plants). Doubling times of roughly 6 h are common and some organisms such as viruses and bacteria rapidly grow with a doubling time well under an hour. Compare this to the weeks or months taken to propagate plants or breed mice or years taken to follow health statistics for humans. Clearly microorganisms provide a very good platform for studying the fundamental connections between genetics and phenotype. This connection is even more direct in prokaryotes because epigenetic regulation is less complex in these organisms. Table 14.1 summarizes the benefits offered by microorganisms for the study of phenomics (Elena and Lenski 2003).

**Table 14.1** Benefits of using microorganisms for the study of phenomics

| Characteristic of microorganism | Benefits for studying phenomics |
|---|---|
| Ease of culture / resilience | Easy to control culture conditions with precise, consistent variations |
| | Simpler, cheaper media formulations |
| | Conducive to high throughput experiments for more conditions tested and more repetitions |
| | Usually culturable in suspension, allowing simpler assays for certain characteristics and higher cell densities |
| | High growth densities enhances measurability and therefore improves reliability of quantitative data |
| | Survive freezing and thawing well |
| Short life-cycle | Very quick to grow, reducing time required for experiments and stock culture maintenance |
| Less complex transcriptional regulation | More direct connection between genetic and phenomics characteristics |
| Relatively small genomes | Abundant sequence information available for many species |
| | Easier and less expensive to sequence |

#### 14.1.2.2 Quantitative Phenotypes

The rapid and reproducible growth characteristics of microorganisms allow researchers to very reliably quantify phenotypes to a very fine degree, and it is of course crucial that this data is managed properly. Experimental design should include sufficient biological replicates (independently established experiments) to generate reliable and significant results. Growth rate, biomass yield, or productivity of a given metabolite are most valuable to researchers studying microorganisms, especially in industry where a typical goal is to improve the performance of an organism in a bioprocess. By repeat sampling across multiple times and across different genotypes, it is possible to find correlations in genetic changes at the nucleic acid level (DNA mutation or change in mRNA transcript level) to a corresponding change in performance/phenotype. Due to the overwhelming complexity of biological systems, this type of information will be crucial in piecing together and refining models which could eventually predict phenotype from the genetic level. In contrast to these quantitative, continuous variables, a few phenomic characteristics such as morphology and pathogenicity are discrete in nature (i.e. true/false, bacillus/coccus, etc.) and it can be useful to collect data on the proportions of a population that have a certain defined characteristic. This type of study can be experimentally difficult, although advances in technologies such as flow cytometry and antibody or other labeling methods have made this more feasible. Still, when possible, it is often advisable to design experiments to minimize or eliminate variations within a sample so that the entire sample can be counted discretely.

## *14.1.3 Microbial physiology*

Phenomics is concerned with carefully establishing the latter half of the genotype–phenotype relationship. To complete the genotype–phenotype picture, two additional areas of information are required: genomics (the starting point) and intermediate components (the mechanisms).

### 14.1.3.1 Microbial Metabolism

In the context of studying the phenome of an organism, it is often important to consider the mechanisms that an organism utilizes to produce the observed phenotype. With the genome as the starting point, the three levels of mechanisms used to bring about a phenotype that we are most concerned with are mRNA transcripts, proteins, and pathways (reactions), shown in Fig. 14.1. The biochemical conversions that occur as part of metabolism dictate much of an organism's phenotype and thus represent a connecting mechanistic link between genotype and phenotype.

Metabolic networks typically are composed of thousands of biochemical reactions. Each of these reactions can be connected to many other reactions due to shared metabolites (substrates) that are facilitated by catalysts (enzymes). Enzymes are expressed proteins and are encoded as genes in an organism's genome. The knowledge that links metabolic reactions to enzymes is inferred from mapping genetic mutations in strains which have lost the ability to complete a particular metabolic step – early work that led to the 1958 Nobel Prize in Physiology or Medicine. Without the connection of enzymes to metabolic reactions it is much more difficult to directly link changes in genotype and phenotype.

It is painstaking and infeasible to determine the metabolic network of all organisms, but work done in a key set of model organisms, notably microbes like the

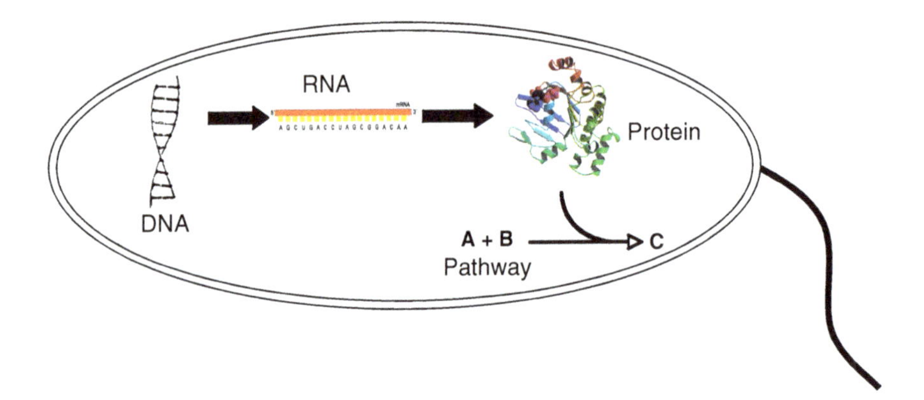


**Fig. 14.1** General schematic depicting the central dogma of molecular biology. Genetic information from DNA (genotype) is propagated through intracellular components of RNA and proteins to allow biochemical conversions. Collectively, these molecular events give rise to integrated cellular function.

bacterium *Escherichia coli* and baker's yeast *Saccharomyces cerevisiae*, has enabled detailed determination of genes encoding proteins involved in most essential cellular metabolic reactions. Much of what is known has been compiled into centralized online databases such as the Kyoto Encyclopedia of Genes and Genomes (KEGG) (Kanehisa and Goto 2000; Kanehisa et al. 2006; Kanehisa et al. 2008), Biocyc (Karp et al. 2005), or the National Microbial Pathogen Database Resource (McNeil et al. 2007).

A second challenge associated with studying metabolic networks as a connection between genotype and phenotype is the complexity of these networks. With thousands of metabolites participating in thousands of reactions, it is difficult to intuitively determine the overall effects of discrete parameters (chemical concentrations, genetic changes, temperature changes, etc.). In this respect, the problem is one of developing or implementing appropriate analytical tools that can extract the relative signal from the noisy background. To date, some of the most effective approaches have used computational modeling to solve this problem.

#### 14.1.3.2 Microbial Transcriptional Regulation

Biological networks of biochemical reactions are large, complex, and ultimately dictate organism function. In a sense, the collection of reactions outline all of the capabilities of an organism, however, these reactions occur at different, often precisely orchestrated intervals and conditions. Many organisms function with internal control mechanisms that inhibit or foster production of specific mRNA transcripts that can influence protein production and reaction activity. The collection of control mechanisms are referred to as the transcriptional regulatory network. Regulatory mechanisms play a critical role in determining an organism's specific phenotype by effectively shutting down or activating specific pathways. While metabolic networks are generally understood reasonably well, transcriptional regulatory networks are still being elucidated in many organisms. Eventually, information about transcriptional regulation will likely be as well-characterized as metabolic networks and become aggregated and available through online databases such as RegulonDB (Huerta et al. 1998; Salgado et al. 2001).

## 14.2 Experimental Methods for Microbial Phenomics

Phenomics is not only concerned with studying the array of phenotypes expressed by an organism, but attempts to do so in a quantitative manner. Historically, establishing precise and accurate quantitative measurements for various physiological functions has been challenging. Fortunately, advances in sensor technology and automation have continued to improve measurements enabling phenomics. We outline here several common assays, typical analytical approaches, and currently available technologies.

### 14.2.1 Microbial Growth

Growth rate is, for perhaps obvious reasons, one of the most often measured characteristics of microorganisms. Often growth rate itself is the variable of interest, especially when studying adaptation to specific environments and the regulation of anabolism. At other times, it is necessary to measure growth rate in the process of monitoring other characteristics such as biomass yield, and specific productivity. There are a number of ways to measure microbial growth and a thorough investigation of them is beyond the scope of this chapter. The most common situation is to study microorganism growth rate in liquid suspension using optical density (OD). Because OD varies linearly with cell concentration at low densities, it is straight forward to calibrate these measurements with either diluted plate counts or dry weight of biomass. Optical density measurements are commonly used because these measurements are very amenable to automation. There are a number of high throughput plate readers (e.g. VersaMax™ by Molecular Devices, Sunnyvale, CA and Bioscreen C™ by Growth Curves, Ltd., Piscataway, NJ) with a spectrophotometric or turbidometric function. These systems essentially are multiplexed, miniature incubator-shaker-plate readers, thus measurements often can be correlated directly to measurements that are conducted by hand. The throughput of these types of growth rate measurements is only limited by the number of wells on a multiwell plate (Fig. 14.2).

In addition to growth rate, it is often of interest to determine simply whether an organism can or cannot grow on a given substrate. This discrete information can be collected for a wide range of medium variations and can be a valuable tool for identifying organisms or determining the specific effects of individual gene deletions. To assist this, Biolog, Inc. (Hayward, CA) manufactures premade multiwell plates that test for metabolic activity on a very large range of carbon sources. An organism's characteristic "metabolic fingerprint" (essentially its phenome as evaluated by this system) can be used to quickly identify or characterize it. The same company also produces Phenotype Microarrays™, in the same 96-well format, that are designed to fingerprint phenotypic responses to carbon, nitrogen, and other metabolic sources, pH, or chemical sensitivities. These types of plates can be used to compare a cell line with or without gene deletions under a broad range of environmental chemical stimuli to determine ge ne function (Bochner et al. 2001).

### 14.2.2 Oxygen Consumption

Oxygen consumption rate (OCR) is often of interest because, when compared with growth rate, it will affect the redox reactions of a cell and can give an indication of the efficiency with which a cell is operating. Oxygen consumption is inherently difficult to measure directly. First of all, working with any gas is more difficult than liquids or solids. It is particularly difficult to measure oxygen consumption as the quantities of oxygen dissolved in liquid media are low and thus any measurement requires a high level of sensitivity to detect changes due to consumption by an organism. Respirometers which can measure oxygen uptake rate as well as production rates of off-gases such as $CO_2$ are available at many different scales and using many different measurement techniques. However, most respirometers are used in large-scale industrial and waste treatment applications, so they are not especially well suited for small-scale laboratory work where many different strains would need to be tested. Designing devices for a small-scale laboratory application is difficult again because the sensors must be extremely precise. Instech Laboratories (Plymouth Meeting, PA) offers one solution which uses fiber optic sensors with fluorescent dyes to sense OCR in samples as small as 170μL. These devices are some of the most reliable sensors available, but they are also expensive and have relatively low throughput. BD Biosciences (San Jose, CA) produces an Oxygen Biosensor System™ (OBS) in 96- or 384- well format in which a fluorescent dye responds in an inversely proportional manner to dissolved oxygen concentration. A sensor embedded in the bottom of each well increases in fluorescence as available oxygen decreases. By controlling oxygen partial pressure at the surface and formulating the medium so that oxygen is not the limiting substrate, OCR can be estimated for a number of cell types or over a range of environments. High-throughput measurements of OCR using fluorescent markers can be read in a flourimetric plate reader such as the SPECTRAMax™ by Molecular Devices, Sunnyvale, CA or the VICTOR³™ by PerkinElmer, Waltham, MA (formerly Wallac). Due to limitations in our ability to control oxygen partial pressure at the surface of each well, this approach cannot achieve the same precision as some respirometers, but the huge improvement in throughput can in some cases outweigh these drawbacks.

### *14.2.3 Substrate consumption*

In many cases it is necessary to directly determine the rate at which an organism can utilize a particular substrate other than oxygen. It should be noted that this data is distinguishable from the "metabolic fingerprint" information discussed above. In the "metabolic fingerprint" case, the actual substrate consumption was not detected; instead, general metabolic activity was used as a signal that a particular substrate was able to support growth. However, when measuring substrate consumption, quantitative information is being measured about how much substrate is being consumed to produce a given amount of byproduct or to support a given number of cells. Many of these questions are commonly studied in conjunction with metabolomics, often setting the boundary conditions for understanding the intermediate metabolic reactions. In some cases, however, this information in itself can be of interest, especially to researchers interested in bioremediation of toxic substances or the production of chemical compounds. With huge improvements over the past few decades in high performance liquid chromatography (HPLC) – and auto-injectors for these systems – experimentation for these variables has become much less labor intensive. Hundreds of samples can be collected over a range of time points, and relative quantities of metabolites of interest can be readily detected, providing valuable information about the consumption of individual substrates. As technologies improve, more researchers are moving towards detecting chemical compounds using different mass spectrometry (MS) systems (gas chromatography-MS, liquid chromatography-MS, gel electrophoresis-MS). While there is improved sensitivity with MS detection and quantitation, other factors such as ease of sample preparation and overall cost still favor HPLC.

### *14.2.4 Motility/Adhesion*

Motility and adhesion are other phenotypes that hold interest for researchers with relevance to process designs in industry or infectious capacity and pathogenicity in medicine. Both of these characteristics depend on spatial measurements, and so automation for high-throughput assays can be difficult. A standard method for measuring cell motility is to inoculate a set amount of cells into semi-solid agar and to allow the cells to migrate over a period of time. Typically these assays employ a semi-solid agar called Chemical-Gradient Motility Agar (CGMA) that eliminates convective currents but still allows microbes to move through them. The degree of motility can be measured quantitatively by measuring the spread of cells (Fig. 14.2).

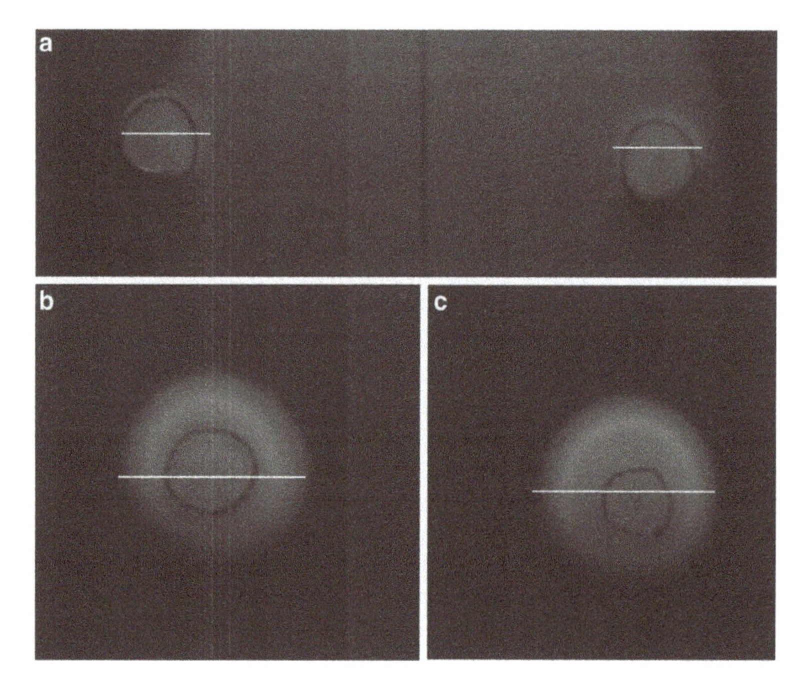


**Fig. 14.2** Sample results from a motility assay using semi-solid agar. Cells are initially inoculated into the semi-solid agar and the initial boundary of the inoculum is marked (black circle). Cells are incubated for a set amount of time (e.g. 12 h) and the motility of the cells is evaluated by measuring the spread of the population (diameter measured by the white line). In these results, cells with low motility (**a**) are shown in comparison to cells with moderate motility (**b**) or high motility (**c**)

A chemical gradient can also be applied across the agar in order to test for chemotaxis, which is an organism's ability to move in response to chemical signals. Chemotactic movement is mediated by cell-surface receptors, and it can be in response to positive (e.g. glucose) or negative (e.g. antibiotics) signals, and these assays can clearly be very useful in determining the genetic causes and regulations for this type of phenotype.

Current advances in phenomics are broadly connected to developments in microscopy and microfluidics. From the microscopy perspective, studying cell motility is challenging as the problem requires the cells to be (1) kept alive on a microscope stage, (2) analyzed and tracked over time, and (3) maintained in the field of view and in focus. Several different microscopy solutions such as the Cellomics system (Thermo Scientific, Pittsburgh, PA) and a variety of others are being developed to address these issues. Coupled with microscopy advances is the use of microfluidic devices that can be used to create controlled chemical environments (and gradients) for the microenvironment surrounding cells. A well-designed microfluidic device not only generates a defined chemical environment, but it can also help alleviate some of the microscopy-related problems.

### *14.2.5 Fluorescence*

While not typically considered to be a macroscopic phenotype analogous to those mentioned in the previous sections, whole-cell fluorescence measurements are broadly used for a number of different purposes. Most applications use a fluorescent protein, such as green fluorescent protein (GFP), as a means of labeling a protein of interest. Thus, these studies typically involve the generation of GFP-fusion proteins that can then be analyzed by either fluorescence microscopy or by a flow cytometer. For phenomic studies, flow cytometry is a very useful experimental tool as it can generate large amounts of quantitative data on individual cells. Typical parameters measured by a flow cytometry experiment would be forward scatter (related to cell volume/size), side scatter (related to the composition of cells), and total fluorescence of the labeled protein.

## 14.3 Analytical Tools for Microbial Phenomics

The current age of genome biology has produced a vast array of biological data (starting with whole genome sequences). This abundance of available data allows researchers to study biological problems from a different perspective than was previously possible, however, it has come with its own challenges. The sheer volume of data now almost precludes any careful item-by-item analysis. If a specific scientific question is not carefully formulated or appropriate analytical tools are not available, the analysis and interpretation of any of these large data sets (an organism's phenome included) would be a daunting, almost impossible task. The following section describes some of the analytical tools available to the study of phenomics.

### *14.3.1 Statistics*

Microbial phenomics has the capability to produce huge volumes of quantitative and qualitative data. Therefore, statistical analysis must play a prominent role in the planning of experiments, as well as the evaluation of their results. Experiments must be planned with sufficient replicates to provide reliable conclusions, typically with three biological replicates per sample usually being a good guideline. Sample size also should depend on the expected precision of the measurements being taken. In cases where instrumentation is adequate and very high precision can be expected, smaller sample sizes (i.e., fewer replicates) are possible while still providing statistically significant results. In contrast, when high precision is not available or when variations between sample populations are very small, more replicates should be planned in order to obtain statistically significant results.

The analysis of experimental data using statistics typically involves the same methodology as other scientific disciplines, although there are some additional measures as a result of the volume of data that is typically being managed. When growth rates are being measured, for example, linear regression can be applied to the logarithmic transform of the exponential growth phase, and the resulting slope is then recorded for each individual sample. This growth rate, having the units (1 / time) can then be analyzed as a separate variable, either with further linear regression if being studied as a response to another variable (e.g. substrate or inhibitor concentration) or with analysis of variance if discrete sample groups are being tested. Figure 14.3 shows a schematic of such an analysis in which ten different cell lines are allowed to evolve on the same substrate for a period of up to 50 days. The growth rates of the evolved strains are compared to those of the original strains. Every sample is repeated in duplicate with several blanks to detect variation within the 96-well plate. The growth curves recorded from this are clearly different (Fig. 14.3b), but it is more valuable to quantify that difference. To accomplish this, the growth curves are transformed onto a logarithmic scale. A straight line on a log scale plot represents ideal exponential growth, and the slope is equivalent to the exponential growth rate. A simple linear regression of the linear portion of each curve can therefore provide the exponential growth rate for each sample group. Each growth rate can then be treated as a continuous statistic and analyzed using classical statistical methods. In this case, a box plot clearly indicates that each sample group is different and that the growth rate has increased as the organism has evolved, and, in fact, analysis of variance confirms this observation. This analysis is not unlike traditional analysis of scientific data; however, it is important to point out that the format of the experiment (in 96-well plates) lends itself nicely to automation, allowing researchers to answer more questions more reliably in less time.

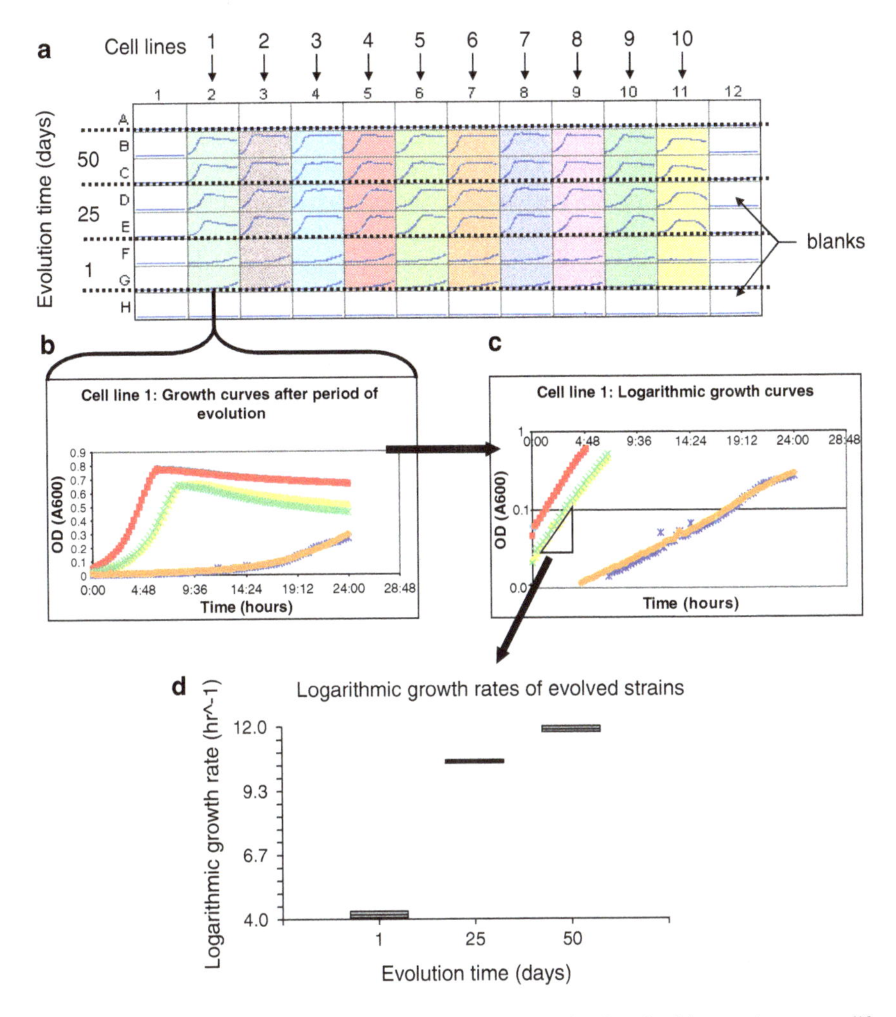


**Fig. 14.3** Schematic of typical statistical analysis of phenomics data. In this experiment, ten different cell lines were allowed to evolve on the same substrate for 25 or 50 days. Two replicates of each sample group were grown in a 96-well plate (**a**) and optical density (OD) was measured over the course of 24 h to capture the growth curves of each sample. The resulting graphs of OD over time for cell line 1 (**b**) was transposed onto a logarithmic scale (**c**), and the linear portion of each curve was fitted using linear regression. The slope of each line was recorded as the exponential growth rate for that sample. The growth rates are seen to vary with evolution time in a box plot (**d**). Analysis of variance (not shown) confirms that each sample group is statistically, significantly different from every other group from the same cell line

## *14.3.2 Modeling*

Phenomics is most useful when viewed in the context of the entire cellular system, that is, in conjunction with genomics, gene expression, protein transcription, and metabolism, as well as with the regulation of all of these processes. It can be said that a primary goal of bioinformatics and systems biology is the feat of linking all of these internal mechanisms with observed cellular phenotypes (the complete picture of linking genotype to phenotype). With this perspective, a number of computational modeling techniques have been developed in an attempt to understand the complex interactions between genetic, proteomic, metabolic, and phenomic data. At the metabolic level, very robust analytical techniques such as biochemical systems theory (Savageau et al., 1987a, b), metabolic control analysis (Cascante et al. 2002), inverse flux analysis (Delgado and Liao 1997), flux balance analysis (Varma and Palsson 1994; Edwards et al. 2002), and extreme pathway analysis (Schilling et al. 2000) have been developed. Many of these analytical techniques require detailed information about reaction kinetics and enzyme activities, and so the implementation of these theories to genome-scale systems is difficult, and to this point has not been done for any method relying on detailed enzyme kinetic information. Flux balance analysis (FBA) and extreme pathway analysis (ExPA), in contrast, require information about which reactions are available to an organism, independent of kinetics, and therefore have been successfully applied to genome-scale metabolic networks (Edwards et al. 2001; Schilling and Palsson 2000). As an example of the use of modeling to predict metabolic traits, FBA will be briefly discussed below, but the reader is encouraged to examine the methodologies employed by the other analytical techniques.

Metabolic flux balance analysis is based on the construction of a comprehensive tabulation of the all biochemical reactions that are available to a cell by means of its unique collection of enzymes, as dictated by the organism's genome. The stoichiometries of these reactions constrain the phenotype of a cell, based on mass balances around the entire network. Such an analysis creates a solution space within which a cell could conceivably operate, but it is limited in the ability to supply deterministic information about how the cell will *actually* behave. Linear optimization, therefore, should be applied to detect the boundaries of the resulting solution space. To this end, an objective function that will yield some valuable information can be selected to be maximized or minimized. For example, one objective that is commonly used is a biomass function (associated with growth rate), which is a combination of the fluxes (including energy production) required for the anabolism of new biomass. When the metabolic network is optimized for maximum biomass, the result is a numerical prediction of the best possible growth rate that an organism could achieve with its given set of genes in a specified growth environment. Studies have shown that organisms will actually experimentally adapt/evolve towards this maximum value over several generations. A subtle consequence of this phenomenon is that FBA can avoid the need to account for regulatory effects, since adaptive evolution will effectively modify these controls as needed to reach the maximum possible growth rate.

Another extension of this methodology is that since the presence of an available metabolic reaction does not depend on the activity of an enzyme or the degree to which it is expressed, a direct link can be made between a gene that encodes an enzyme and the reaction which it catalyzes. This assumes, of course, that the organism in question is able to express the gene, and the resulting protein can be translated and assembled, and can function normally in the given microenvironment. This is not a trivial assumption as any genetic engineer can explain; however, it allows us to make an important step in the efforts to design engineered organisms. Deletions or additions of specific genes can be simulated *in silico*, allowing researchers to make better decisions about which genes to target for a given engineering goal. Sophisticated computational algorithms (Burgard et al. 2003; Pharkya et al. 2004) can be used to automate this process by searching possible deletions for the combinations at which the highest production of a desired metabolic byproduct coincides with the maximum growth rate. In this way, improved production is coupled to improved growth rate, thereby utilizing the mechanisms already present for adaptive evolution (Fong et al. 2005; Hua et al. 2006). This type of simulation represents one facet of connecting genotype to phenotype through computational analysis as modifications are suggested at the genetic level and consequences are predicted at the phenotypic level.

## References

Bochner BR, Gadzinski P, Panomitros E (2001) Phenotype microarrays for high-throughput phenotypic testing and assay of gene function. Genome Res 11(7):1246–1255

Burgard AP, Pharkya P, Maranas CD (2003) Optknock: a bilevel programming framework for identifying gene knockout strategies for microbial strain optimization. Biotechnol Bioeng 84(6):647–657

Cascante M, Boros LG, Comin-Anduix B, de Atauri P, Centelles JJ, Lee PWN (2002) Metabolic control analysis in drug discovery and disease. Nat Biotechnol 20(3):243–249

Delgado J, Liao JC (1997) Inverse flux analysis for reduction of acetate excretion in Escherichia coli. Biotechnol Prog 13(4):361–367

Edwards JS, Ibarra RU, Palsson BO (2001) In silico predictions of *Escherichia coli* metabolic capabilities are consistent with experimental data. Nat Biotechnol 19:125–130

Edwards JS, Covert M, Palsson B (2002) Metabolic modelling of microbes: the flux-balance approach. Environ Microbiol 4(3):133–140

Elena SF, Lenski RE (2003) Evolution experiments with microorganisms: the dynamics and genetic bases of adaptation. Nat Rev Genet 4(6):457–469

Fong SS, Burgard AP, Herring CD, Knight EM, Blattner FR, Maranas CD et al (2005) In silico design and adaptive evolution of *Escherichia coli* for production of lactic acid. Biotechnol Bioeng 91(5):643–648

Hua Q, Joyce AR, Fong SS, Palsson B (2006) Metabolic analysis of adaptive evolution for in silico designed lactate-producing strains. Biotechnol Bioeng 95(5):992–1002

Huerta AM, Salgado H, Thieffry D, Collado-Vides J (1998) RegulonDB: a database on transcriptional regulation in *Escherichia coli*. Nucleic Acids Res 26(1):55–59

Kanehisa M, Goto S (2000) KEGG: kyoto encyclopedia of genes and genomes. Nucleic Acids Res 28(1):27–30

Kanehisa M, Goto S, Hattori M, Aoki-Kinoshita KF, Itoh M, Kawashima S et al (2006) From genomics to chemical genomics: new developments in KEGG. Nucleic Acids Res 34(Database issue):D354–D357

Kanehisa M, Araki M, Goto S, Hattori M, Hirakawa M, Itoh M et al (2008) KEGG for linking genomes to life and the environment. Nucleic Acids Res 36(Database issue):D480–D484

Karp PD, Ouzounis CA, Moore-Kochlacs C, Goldovsky L, Kaipa P, Ahren D et al (2005) Expansion of the BioCyc collection of pathway/genome databases to 160 genomes. Nucleic Acids Res 33(19):6083–6089

McNeil LK, Reich C, Aziz RK, Bartels D, Cohoon M, Disz T et al (2007) The National Microbial Pathogen Database Resource (NMPDR): a genomics platform based on subsystem annotation. Nucleic Acids Res 35(Database issue):D347–D353

Pharkya P, Burgard AP, Maranas CD (2004) OptStrain: a computational framework for redesign of microbial production systems. Genome Res 14(11):2367–2376

Salgado H, Santos-Zavaleta A, Gama-Castro S, Millen-Zarate D, Diaz-Peredo E, Sanchez-Solano F et al (2001) RegulonDB (version 3.2): transcriptional regulation and operon organization in *Escherichia coli* K-12. Nucleic Acids Res 29(1):72–74

Savageau MA, Voit EO, Irvine DH (1987a) Biochemical systems theory and metabolic control theory: I. Fundamental similarities and differences. Math Biosci 86:127–145

Savageau MA, Voit EO, Irvine DH (1987b) Biochemical systems theory and metabolic control theory: II. The role of summation and connectivity relationships. Biosciences 86:147–169

Schilling CH, Palsson BO (2000) Assessment of the metabolic capabilities of *Haemophilus influenzae* Rd through a genome-scale pathway analysis. J Theor Biol 203(3):249–283

Schilling CH, Letscher D, Palsson BO (2000) Theory for the systemic definition of metabolic pathways and their use in interpreting metabolic function from a pathway-oriented perspective. J Theor Biol 203(3):229–248

Varma A, Palsson BO (1994) Metabolic flux balancing: basic concepts, scientific and practical use. Bio/Technology 12:994–998

# Chapter 15
# Standards for Functional Genomics

**Stephen A. Chervitz, Helen Parkinson, Jennifer M. Fostel, Helen C. Causton, Susanna-Assunta Sanson, Eric W. Deutsch, Dawn Field, Chris F. Taylor, Philippe Rocca-Serra, Joe White, and Christian J. Stoeckert**

## 15.1 Introduction

Fuelled by the fruits of the genome sequencing projects that are defining the complete sets of genes, transcripts, and proteins within an organism and the advent of highly multiplex technologies capable of measuring thousands to millions of biomolecules per sample in one assay, functional genomics studies are enabling new approaches for studying biological systems. A single experiment can generate very large amounts of raw data as well as summaries in the form of lists of sequences, genes, proteins, metabolites, SNPs, etc. which have been identified by various analytical tests. Managing, reporting, and integrating the results from these experiments present challenges to researchers and bioinformaticians in this relatively young field because the standards and conventions developed for single-gene or single-protein studies do not accommodate the needs of functional genomics studies (Boguski 1999).

Functional genomics technologies and their applications are evolving rapidly, and there is widespread awareness of the need for, and value of, standards in the life sciences community. Not only do the widely-adopted standards help scientists and data analysts utilize the ever-growing mountain of functional genomics data sets better, they also are essential for the application of functional genomics approaches in healthcare environments. This chapter provides an introduction to the major functional genomics standards initiatives in the domains of genomics, transcriptomics, proteomics, and metabolomics, thereby providing a summary of goals, example applications, and references for further information. It also covers the application of standards in healthcare settings, where functional genomics technologies are having an increasing impact. New standards and organizations may come along in the future that will augment or supersede the ones described here. Interested readers are invited to further explore the standards mentioned in this chapter (as well as others not mentioned) and keep up with the latest developments by visiting the website http://biostandards.info.

S.A. Chervitz (✉)
Affymetrix, Inc., Santa Clara, CA 95051, USA
e-mail: Steve_Chervitz@affymetrix.com

D. Edwards et al. (eds.), *Bioinformatics: Tools and Applications*,
DOI 10.1007/978-0-387-92738-1_15, © Springer Science+Business Media, LLC 2009

### 15.1.1 Goals and Motivations for Standards in the Life Sciences

The use of standards within a scientific domain have the potential to provide a uniformity and consistency to the data generated by different researchers, organizations, and technologies thereby facilitating more effective re-use, integration, and mining of that data by other researchers and third-party software applications, and enabling easier collaboration between different groups. Standards-compliant data sets have increased value for scientists who must interpret and build on earlier efforts and for software tools that have been designed to process data that conforms to those standards. Standard laboratory procedures and reference materials enable the creation of guidelines, systems benchmarks, and laboratory protocols for quality assessment and cross-platform comparisons of experimental results that are needed in order to deploy a technology within research, industrial, or clinical environments. The value of standards in the life sciences for improving the utility of data from high-throughput post-genomic experiments has been widely discussed for some time (Brazma 2001; Stoeckert et al. 2002; Brooksbank and Quackenbush 2006; Rogers and Cambrosio 2007; Warrington 2008).

To understand how conclusions from a study were obtained, not only do the underlying data need to be available but also the details of how the data were generated need to be adequately described (i.e., samples, procedural methods, and data analysis). Depositing data in public repositories is necessary but not sufficient for this purpose. Reporting on "minimum information" standards are needed to ensure that submitted data are sufficient for clear interpretation and querying by other scientists. Standard data formats greatly reduce the amount of effort required to share and make use of data produced by different investigators. Standards for the terminology used to describe the study and how the data were generated enable not only improved understanding of a given set of experimental results but also improved ability to compare studies produced by different scientists and organizations. Standard physical reference materials as well as standard methods for data collection and analysis can also facilitate such comparisons as well as aid the development of reusable data quality metrics.

A major goal of any effort to set standards is to take into account its usability. A standard that is not widely used is not really a standard and the successful adoption of a standard by end-user scientists requires a reasonable cost-benefit ratio. The cost of developing and learning how to use the standard or generating standards-conforming data has to be outweighed by gains in the ability to publish experimental results, the ability to use other published results to advance one's own work, and the higher visibility of standards-compliant publications (Piwowar et al. 2008). Thus, a major focus of standards initiatives is minimizing usability barriers, typically done by educational outreach via workshops and tutorials as well as fostering the development of software tools that help scientists utilize the standard in their investigations. There must also be a means for incorporating feedback from the target community both at the initiation of standard development and for the standard to maintain a good fit to user needs that can change over time. Brazma et al. (2006) discuss additional factors that contribute to the success of standards in systems biology and functional genomics.

### *15.1.2 History of Standards for Functional Genomics*

The motivation for standards for functional genomics initially came from the parallel needs of the scientific journals, which wanted standards for data publication, and the needs of researchers, who recognized the value of comparing the large and complex data sets characteristic of functional genomics experiments. Such data sets, often with thousands of data points, required new data formats and publication guidelines. Workers using DNA microarrays for genome-wide gene expression analysis were the first to respond to these needs. In 2001, the MGED Society published the MIAME standard (Brazma et al. 2001), a guideline for the minimum information required to describe a DNA microarray experiment, specifying the information required to describe an experiment so that another researcher in the same discipline could either reproduce it or analyze the data de novo.

Adoption of the MIAME guidelines was expedited when a number of journals and funding agencies required compliance with the standard as a pre-condition for publication. In parallel with MIAME, a data modeling and exchange standard called MAGE, and a controlled vocabulary called the MGED Ontology (Whetzel et al. 2006b), were created. These standards facilitated the creation and growth of a number of interoperable databases and public data repositories and also led to the establishment of open-source software projects for DNA microarray data analysis. Resources such as the EBI's ArrayExpress (Parkinson et al. 2009) and the NCBI's Gene Expression Omnibus (GEO) (Barrett et al. 2007) and others were advertised as "MIAME-supportive" with some capable of importing data submitted in the XML-based MAGE-ML format (Spellman et al. 2002).

Minimum information guidelines akin to MIAME then arose within other functional genomics communities, for example the MIAPE guidelines for proteomics studies (Taylor et al. 2007) and the MIGs guidelines for genomic and metagenomic studies (Field et al. 2008). More recent initiatives have been directed towards technology-independent standards for reporting, modeling, and exchange that better support work spanning multiple "omics" technologies or domains and harmonization of related standards. These projects have, of necessity, required extensive collaboration across disciplines. The resulting standards have gained in sophistication, benefiting from insights in using and implementing earlier standards, the use of formalisms imposed by the need to make the data computationally tractable and logically coherent, and the need to engage multiple academic communities.

Increasingly, the drive for standards in functional genomics is shifting from the academic communities to include the biomedical and healthcare communities as well. As application of functional genomics technologies and data expands into the clinical and diagnostic arena, organizations such as the U.S. Food and Drug

Administration and technology manufacturers are becoming more involved in a range of standards efforts, for example the MAQC consortium brings together representatives of many such organizations (Shi et al. 2006). Quality control/assurance projects and reference standards that support comparability of data across different manufacturer platforms are of particular interest as functional genomics technologies mature and start to play an expanded role in healthcare settings.

## 15.2 Classification of Standards for Functional Genomics

Functional genomics standards are typically scoped to a specific aspect of a functional genomics investigation. Generally speaking, a given standard will cover either the description of a completed experiment, or will target some aspect of performing the experiment or analyzing results. Standards are further stratified to handle more specific needs, such as reporting data for publication, providing data exchange formats, or defining standard terminologies. Such scoping reflects a pragmatic decoupling that permits different standards groups to develop complementary specifications concurrently and allows different initiatives to attract individuals with relevant expertise or interest in the target area (Brazma et al. 2006).

As a result of this arrangement, a standard or standardization effort within functional genomics can be generally characterized by its domain and scope. The *domain* reflects the type of experimental data (transcriptomics, proteomics, metabolomics, etc.) while the *scope* defines the area of applicability of the standard or the methodology being standardized (experiment reporting, data exchange, etc.). Tables 15.1 and 15.2 list the different domains and scopes, , that characterize existing functional

**Table 15.1** Domains of functional genomics standards. The domain indicates the type of experimental data that the standard is designed to handle

| Domain | Description |
|---|---|
| Genomics | Genome sequence assembly, genetic variations, metagenomics, DNA modifications |
| Transcriptomics | Gene expression (transcription), alternative splicing, promoter activity |
| Proteomics | Protein levels, protein–protein interactions, post-translational modifications |
| Metabolomics | Metabolite profiling, pathway flux, pathway perturbation analysis |
| Healthcare and Toxicogenomics[a] | Clinical, diagnostic, or toxicological applications |
| Harmonization and Multiomics[a] | Cross-domain compatibility, interoperability |

[a]Healthcare, toxicological, and harmonization standards may be applicable to one or more other domain areas. These domains impose additional requirements on top of the needs of the pure "omics" domains

**Table 15.2** Scope of functional genomics standards. Scope defines the area of applicability or methodology to which the standard pertains. *Scope-General:* Standards can be generally partitioned based on whether they are to be used for describing or executing an experiment. *Scope-Specific:* The scope can be further narrowed to cover more specific aspects of the general scope. *Abbreviations: CV* controlled vocabulary; *QA/QC* quality control/quality assurance

| Scope – General | Scope – Specific | Description |
|---|---|---|
| Experiment description | Reporting (minimum information) | Documentation for publication or data deposition |
| | Data exchange and modeling | Communication between organizations and tools |
| | Terminology | Ontologies and CV's to describe experiments or data |
| Experiment execution | Physical standards | Reference materials, spike-in controls |
| | Data analysis and quality metrics | Analyze, compare, QA/QC experimental results |

genomics standardizations efforts respectively. Fig. 15.1 illustrates the areas where standards of different scope are applied within the general life cycle of a typical functional genomics experiment.

The remainder of this section describes the different scopes of functional genomics standards, listing the major standards initiatives and organizations relevant to each scope. The next section then surveys the standards by domain, providing more in-depth description of the relevant standards, example applications, and references for further information.

### *15.2.1 Experiment Description Standards*

Experiment description standards, also referred to generally as "data standards," are concerned with the development of guidelines, conventions, and methodologies for representing and communicating the raw and processed data generated by experiments as well as the metadata for describing how an experiment was carried out, including a description of all reagents, specimens, samples, equipment, protocols, controls, data transformations, software algorithms, and any other factors needed to accurately communicate, interpret, reproduce, or analyze the experimental results.

Functional genomics studies and the data they generate are complex. The diversity of application areas, experimental designs, and technology platforms creates a challenging landscape of data for any descriptive standardization effort. Even within a given domain and technology type, it is not practical for a single specification to encompass all aspects of describing an experiment. Certain aspects of an experiment ?? are more effectively handled separately; for example, a description of the essential elements to be reported for an experiment is independent of the specific data format in which that information should be encoded for import or export by software applications.

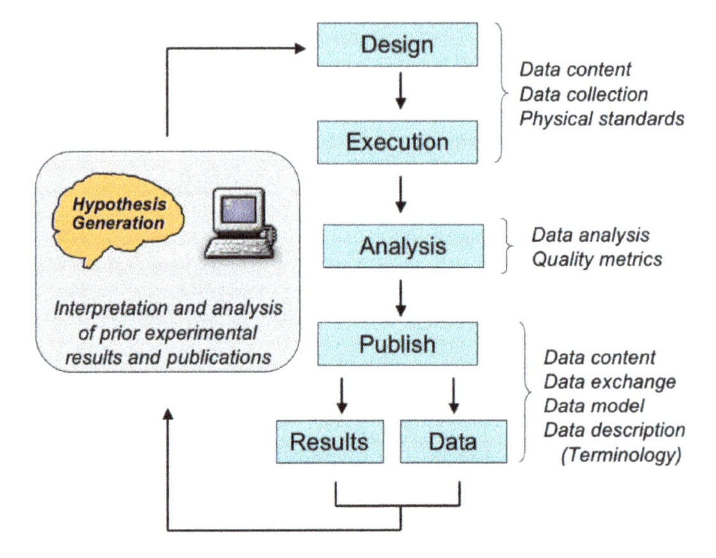


**Fig. 15.1** Application of standards during a functional genomics experiment lifecycle. Multiple standards, each with a different, complementary scope, are employed at different stages of a functional genomics experiment. Reporting standards and physical standards inform both the design and execution of the experiment. During analysis of the results, analytical and quality metrics standards provide guidance and standard terminologies can also assist in the analysis. When it is time to publish results, reporting standards recommend what should be included to ensure accurate interpretation and replication by third parties. Data exchange and modeling standards describe data structures and formats useful for sharing and computationally processing the data. Terminology standards provide common language for annotating the experimental data in a way that fosters more effective comparison and querying between experiments, investigators, and organizations. A standards-compliant published experiment facilitates interpretation, analysis, integration, and comparison of the results by other researchers and software tools, and this can lead to new hypotheses to be tested by further experiments

In recognition of this, experiment description standardization efforts within the functional genomics community are further scoped into more specialized areas that address distinct data handling requirements encountered during different aspects of or types of data encountered in a functional genomics study

- Reporting
- Data exchange and modeling
- Terminology

These different areas play complementary roles and together, provide a complete package for describing a functional genomics experiment within a given domain or technology platform. For example, a data exchange/modeling standard will typically have elements to satisfy the needs of a reporting standard with a set of allowable values for those elements to be provided by an associated standard terminology.

#### 15.2.1.1 Reporting Standards: Minimum Information Guidelines

A *reporting standard* is specifies the information required to unambiguously communicate experimental designs, treatments and analyses, to contextualize the data generated and underpin the conclusions drawn. Such standards are also known as *data content* or *minimum information* standards because they usually have an acronym beginning with "MI" standing for "minimum information" (e.g., MIAME). The motivation behind reporting standards is to enable an experiment to be independently reproduced and interpreted by other scientists. It provides guidance to investigators when preparing to report or publish their investigation or archive their data in a repository of experimental results. When an experiment is submitted to a journal for publication, compliance with a reporting standard can be valuable to reviewers, aiding them in their assessment of whether an experiment has been adequately described and thus worthy of approval for publication.

A reporting specification does not normally mandate a particular format in which to capture/transport information; but simply delineates the data and metadata that their originating community considers appropriate to sufficiently describe how a particular investigation was carried out. Although a reporting standard does not have a specific data formatting requirement, the expectation is that the data should be provided using a technology-appropriate standard format where feasible. Data repositories may impose such a requirement as a condition for data submission.

Functional genomics experiments, in addition to their novelty, can be quite complex in their execution, analysis, and reporting. These minimal information guidelines help in this regard by providing a consistent framework for scientists think about and report essential aspects of their experiments, with the ultimate aim of ensuring the usefulness of the results to future scientists who want to understand or reproduce the study. Such guidelines also help by easing compliance with a related data exchange standard, which is often designed to support the requirements of a reporting standard (discussed below). Depending on the nature of a particular investigation, information in addition to what is specified by a reporting standard may be provided as desired by the authors of the study or as deemed necessary by reviewers of the study.

Table 15.3 lists the major reporting standards for different functional genomics domains. The MIBBI project (discussed later in this chapter) catalogues these and many other reporting standards and provides a useful introduction (Taylor et al. 2008).

Publishers and reporting standards compliance

For some publishers, compliance with a reporting standard is increasingly becoming an important criterion for accepting or rejecting a submitted functional genomics manuscript (DeFrancesco 2002). The journals Nature, Cell, and The Lancet lead the way in the enforcement of compliance for DNA microarray experiments by requiring submitted manuscripts to demonstrate compliance with the MIAME guidelines as a condition of publication. Now, many journals that publish such experiments have generally adopted this policy.

**Table 15.3** Existing reporting standards for functional genomics. Acronyms and definitions of the major reporting standards efforts are shown, indicating their target domain and the maintaining organization, which are as follows: *MGED* MGED Society, http://mged.org. *GSC* Genomic Standards Consortium, http://gensc.org. *HUPO-PSI* Human Proteome Organization Proteomics Standards Initiative, http://www.psidev.info. *MSI* Metabolomics Standards Initiative, http://msi-workgroups.sourceforge.net

| Acronym | Full name | Domain | Organization |
|---|---|---|---|
| CIMR | Core Information for Metabolomics Reporting | Metabolomics | MSI |
| MIAME | Minimum Information about a Microarray Experiment | Transcriptomics | MGED |
| MIAPE | Minimum Information about a Proteomics Experiment | Proteomics | HUPO-PSI |
| MIGS-MIMS | Minimum Information about a Genome/Metagenomic Sequence/Sample | Genomics | GSC |
| MINSEQE | Minimum Information about a high-throughput Nucleotide Sequencing Experiment | Genomics, Transcriptomics | MGED |
| MIMIx | Minimum Information about a Molecular Interaction eXperiment | Proteomics | HUPO-PSI |
| MISFISHIE | Minimum Information Specification For In Situ Hybridization and Immunohistochemistry Experiments | Transcriptomics | MGED |
| MIBBI | Minimum Information for Biological and Biomedical Investigations | Multiomics | Multiple organizations |

#### 15.2.1.2 Data Exchange and Modeling Standards

A *data exchange standard* defines an encoding format for use in sharing data between researchers and organizations, and for exchanging data between software programs or information storage systems. A data exchange standard delineates what data types can be encoded and the particular way they should be encoded (e.g., tab-delimited columns, XML, binary, etc.) but does not specify what the document should contain in order to be considered complete. There is an expectation that the content will be constructed in accordance with a community-approved reporting standard and the data exchange standard itself is typically designed so that users can construct documents that are compliant with a particular reporting standard (e.g., MAGE-ML and MAGE-TAB contain placeholders that are designed to hold the data needed for the production of MIAME-compliant documents).

A data exchange standard is often designed to work in conjunction with a *data modeling standard*, which defines the attributes and behaviors of key entities and concepts (objects) that occur within a functional genomics data set. The model is intended to capture the exchange format-encoded data for the purpose of storage or downstream data mining by software applications. The data model itself is designed to be independent of any particular software implementation (database schema, XML file, etc.) or programming language (Java, C++, Perl, etc.). The implementation decisions are thus left to the application programmer, to be made using the most appropriate technology(s) for the target user base. This separation of the model (or "platform-independent model") and the implementation (or "platform-specific implementation") was first defined by the Object Management Group's Model Driven Architecture (http://www.omg.org/mda) and offers a design methodology that holds promise for building software systems that are more interoperable and adaptable to technological change. Such extensibility has been recognized as an essential feature of data models for functional genomics experiments (Table 15.4) (Jones and Paton 2005).

#### 15.2.1.3 Terminology Standards

A *terminology* standard defines of a standard vocabulary describing the entities and concepts along with their properties and relationships within an application area or

**Table 15.4** A sampling of data exchange and modeling standards for functional genomics. Acronyms and names of some of the major data exchange standards efforts are shown, indicating their target domain and the maintaining organization, which are as described in the legend to Table 15.3 with the following additions: *RSBI* Reporting Structure for Biological Investigations, http://www.mged.org/Workgroups/rsbi. *FuGE* Functional Genomics Experiment, http://fuge.sourceforge.net. *GEN2PHEN* Genotype to phenotype databases, http://www.gen2phen.org. *CDISC* Clinical Data Interchange Standards Consortium, http://www.cdisc.org. Additional proteomics exchange standards are described on the HUPO-PSI website, http://www.psidev.info

| Acronym | | | | |
|---|---|---|---|---|
| Data format | Object model | Full Name | Domain | Organization |
| FuGE-ML | FuGE-OM | Functional Genomics Experiment Markup Language/Object Model | Multiomics | FuGE |
| ISA-TAB | – | Investigation Study Assay – Tabular | Multiomics | RSBI |
| MAGE-ML | MAGE-OM | MicroArray and Gene Expression Markup Language | Transcriptomics | MGED |
| MAGE-TAB | – | MicroArray and Gene Expression Tabular Format | | |
| MIF (PSI-MI XML) | – | Molecular Interactions Format | Proteomics | HUPO-PSI |
| mzML | – | Mass Spectrometry Data Markup Language | Proteomics | HUPO-PSI |
| PML | PAGE-OM | Polymorphism Markup Language/Phenotype and Genotype Object Model | Genomics | GEN2PHEN |
| – | SDTM | Study Data Tabulation Model | Healthcare | CDISC |

type of investigation. A terminology standard provides terms (controlled vocabularies) for documenting (annotating) a particular investigation for publication, including procedures, materials, and results of an experiment. More extensive terminologies, such as the Gene Ontology (Ashburner and Lewis 2002), provide a means of capturing biological knowledge, accumulated over many experiments by different organizations. The primary goal of a terminology standard is to promote consistent use of standard terms within a community and thereby facilitate knowledge integration by enabling better querying and data mining within and across data repositories as well as across domain

Use of standard terminologies by scientists working in different functional genomics domains can enable interrelation of experimental results from diverse data sets. For example, annotating results with standard terminologies could help correlate the expression profile of a particular gene, assayed in a transcriptomic experiment, to its protein modification state, assayed in a separate proteomic experiment. Using a suitably annotated metabolomics experiment, the gene/protein results could then be linked to the activity of the pathway(s) in which they operate, or to a disease state documented in a patient's sample record.

Consistent use of standard vocabularies such as GO has enabled data integration by permitting relational database-type queries over diverse data sets that are annotated using terms from such well-adopted terminologies. Numerous tools that do this are listed at the Gene Ontology site http://www.geneontology.org/GO.tools.microarray.shtml.

There are many publicly available terminologies in the life sciences. Key to their success is adoption by scientists, bioinformaticians, and software developers for use in the annotation of functional genomics data. However, the proliferation of ontologies which are not interoperable can be a barrier to integration (Smith et al. 2007). The OBO Foundry targets this area and is delineating best practices underlying the construction of terminologies, maximizing their internal integrity, extensibility, and re-use. This is discussed in more detail in the "Standards Harmonization" section of this chapter.

Table 15.5 some of the ontologies or controlled vocabularies relevant to functional genomics. For a more comprehensive list of ontologies, see the OBO Foundry website (http://www.obofoundry.org/).

## 15.2.2 Experiment Execution Standards

### 15.2.2.1 Physical Standards

The scope of a *physical standard* pertains to the development of standard reagents for use as spike-in controls in assays. A physical standard serves as a stable reference point that can facilitate the quantitation of experimental results and the comparison of results between different runs, investigators, organizations, or technology platforms. Physical standards are essential for quality metrics purposes and are

**Table 15.5** Terminology standards. Acronyms and names of some of the major terminology standards in use with functional genomics data are shown, indicating their target domain and the maintaining organization, which are as described in the legends to Tables 15.3 and 15.4 with the following additions: *GOC* Gene Ontology Consortium, http://geneontology.org/GO.consortiumlist.shtml. *NCI* National Cancer Institute, www.cancer.gov. *NCBO* National Center for Biomedical Ontology, http://bioontology.org. *OBI* Ontology Biomedical Investigations, http://purl.obofoundry.org/obo/obi

| Acronym | Full name | Domain | Organization |
|---|---|---|---|
| EVS | Enterprise Vocabulary Services | Healthcare | NCI |
| GO | Gene Ontology | Multiomics | GOC |
| MO | MGED Ontology | Transcriptomics | MGED |
| OBI | Ontologies for Biomedical Investigators | Multiomics | OBI |
| OBO | Open Biomedical Ontologies | Multiomics | NCBO |
| PSI-MI | Proteomics Standards Initiative Molecular Interactions ontology | Proteomics | HUPO-PSI |
| sepCV | Sample processing and separations controlled vocabulary | Proteomics | HUPO-PSI |
| SO | Sequence Ontology | Multiomics | GOC |

especially important within applications of functional genomics technologies in regulated environments such as clinical or diagnostic settings.

Within the early days of DNA microarray-based gene expression experiments, results from different investigators, laboratories, or array technology were notoriously hard to compare despite the use of reporting and data exchange standards (Salit 2006). The advent of physical standards and the improved metrology promises to increase the degree of cross-platform and cross-investigator comparability of functional genomics experimental results. Such improvements are necessary for the adoption of functional genomics technologies in clinical and diagnostic applications within the regulated healthcare industry (Table 15.6).

#### 15.2.2.2 Data Analysis and Quality Metrics

The scope of a *data analysis* or *quality metrics standard* is the delineation of best practices for algorithmic and statistical approaches to processing experimental results as well as methods to assess and assure data quality. Methodologies for data analysis cover the following areas:

**Table 15.6** Organizations involved in the creation of physical standards relevant to functional genomics experiments

| Acronym | Full Name | Domain | Website |
|---|---|---|---|
| ERCC | External RNA Controls Consortium | Transcriptomics | http://www.cstl.nist.gov/biotech/Cell&TissueMeasurements/GeneExpression/ERCC.htm |
| LGC | Laboratory of the Government Chemist | Transcriptomics, Proteomics | http://www.lgc.co.uk |
| NIST | National Institute for Standards Technology | Transcriptomics | http://www.cstl.nist.gov/biotech/Cell&TissueMeasurements/Main_Page.htm |
| NMS | National Measurement System (NMS) Chemical and Biological Metrology | Multiomics | http://www.nmschembio.org.uk/ |
| ATCC | American Type Culture Collection Standards Development Organization | Healthcare | http://www.atcc.org/Standards/ATCCStandardsDevelopmentOrganizationSDO/tabid/233/Default.aspx |

**Table 15.7** Data analysis and quality metrics projects. BioConductor's arrayQualityMetrics: http://bioconductor.org/packages/2.3/bioc/html/arrayQualityMetrics.html (Gentleman et al. 2004). CAMDA is managed by a local organizing committee at different annual venues: http://camda.bioinfo.cipf.es. EMERALD's NTO: http://www.microarray-quality.org/ontology_work.html. MAQC is described in Sect. 15.3.2.6 of this chapter

| Acronym | Full name | Domain | Organization |
|---|---|---|---|
| arrayQualityMetrics | Quality assessment software package | Transcriptomics | BioConductor |
| CAMDA | Critical Assessment of Microarray Data Analysis | Transcriptomics | n/a |
| MAQC | Microarray Quality Control Project | Transcriptomics | FDA |
| NTO | Normalization and Transformation Ontology | Transcriptomics | EMERALD |

- Data transformation (normalization) protocols
- Background or noise correction
- Clustering
- Hypothesis testing
- Statistical data modeling

Analysis procedures have historically been developed in a tool-specific manner by commercial vendors, and users of these tools rely on the manufacturer for guidance. Yet, efforts to define more general guidelines and protocols for data analysis best practices are emerging. Driving some of these efforts is the need for consistent approaches to measure data quality, which is critical for determining one's confidence in the results from any given experiment and for judging the comparability of results obtained under different conditions (days, laboratories, equipment operators, manufacturing batches, etc.). Data quality metrics rely on data analysis standards as well as the application of physical standards. Collecting or assessing data quality using quality metrics is facilitated by having data conforming to widely adopted reporting standards and available in common data exchange formats. A number of data analysis and quality metrics efforts are listed in Table 15.7.

## 15.3 Survey of Functional Genomics Standards by Domain

Here we review some of the more prominent standards and initiatives within the main functional genomics domains: genomics, transcriptomics, proteomics, and metabolomics. Of these, transcriptomics is the most mature in terms of standards development and community adoption, though proteomics is a close second.

### *15.3.1 Genomic Standards*

This section describes the standards and organizations related to technologies that collect, assemble, or analyze genomic sequences. Genomics standards pertain to the following technologies or types of investigations:

- Genome and metagenomic sequencing
- Genotyping and polymorphisms

#### 15.3.1.1 MIGS/MIMS

Minimum Information About a Genome Sequence/ Minimum Information about a Metagenomic Sequence/Sample: MIGS/MIMS

Home: http://gensc.org

MIGS/MIMS is a reporting standard for describing metadata about a genomic or metagenomic sequence, such as the complete assembly of a bacterial or eukaryotic genome (Field et al. 2008). MIMS is an extension of MIGS to support the needs of metagenomics experiments in which environmental samples are being analyzed by DNA sequencing for the purpose of organism identification. The MIMS extension specifies additional parameters about the environment from which a sample is taken for sequencing. These standards go beyond the information that is traditionally captured in genome annotations and aim to facilitate such things as comparative genomics and data mining from sequence data repositories which contain information from many organisms, contributed by different laboratories using a variety of technologies.

The Genomic Contextual Data Markup Language (GCDML) (Kottmann et al. 2008) is an XML markup language that implements the MIGS/MIMS standard and also allows capture of richer sets of contextual data describing genomic and metagenomic studies.

#### 15.3.1.2 PaGE-OM and PML

Phenotype And Genotype Experiment Object Model (PaGE-OM) and the Polymorphism Markup Language (PML).

PaGE-OM Home: http://www.pageom.org
PML Home: http://www.openpml.org

PML was approved as a XML-based data format for exchange of genetic polymorphism data (e.g., SNPs) in June 2005. It was designed to facilitate data exchange among different data repositories and researchers that produce or consume this data. PAGE-OM is a platform independent model for representing genotypic, phenotypic data and the correlations between them. PAGE-OM is an updated, broader version of the PML standard and provides a richer object model and incorporates phenotypic information. It was approved as a standard by the OMG in March 2008. Further refinements of the PaGE-OM object model, harmonization with object models from other domains, and generation of exchange formats are underway at the time of writing.

## *15.3.2 Transcriptomic Standards*

This section describes the organizations and standards related to technologies that measure transcription, gene expression or its regulation on a genomic scale.

Transcriptomics standards pertain to the following technologies or types of investigations:

- Gene expression via DNA microarrays or ultra high-throughput sequencing
- Genome tiling arrays to measure gene expression or promoter binding (ChIP-chip, ChIP-seq)
- In-situ hybridization studies of gene expression

### 15.3.2.1 MIAME

Minimum Information About a Microarray Experiment (MIAME).

Home: http://www.mged.org/Workgroups/MIAME/miame.html

The goal of MIAME is to permit the unambiguous interpretation, reproduction, and verification of the results of a microarray experiment. MIAME was the original reporting standard which inspired similar "minimum information" guidelines in other functional genomics domains (Brazma et al. 2006).

MIAME defines the following six elements as essential for achieving these goals:

1. The raw data for each hybridization.
2. The final processed data for the set of hybridizations in the experiment.
3. The essential sample annotation, including experimental factors and their values.
4. The experiment design including sample data relationships.
5. Sufficient annotation of the array design.
6. Essential experimental and data processing protocols.

#### Example Application of MIAME

The MIAME standard has proven useful for microarray data repositories which have used it both as a guideline to data submitters and as a basis for judging the completeness of data submissions. The ArrayExpress database, for example, provides a service to publishers of microarray studies wherein ArrayExpress curators assess a dataset on the basis of how well it satisfies the MIAME requirements (Brazma et al. 2006). A publisher can then choose whether to accept or reject a manuscript on the basis of the assessment.

ArrayExpress judges the following aspects of a report to be the most critical toward MIAME compliance:

1. Sufficient information about the array design (e.g., reporter sequences for oligonucleotide arrays or database accession numbers for cDNA arrays).
2. Raw data as obtained from the image analysis software (e.g., CEL files for Affymetrix technology, or GPR files for GenPix).
3. Processed data for the set of hybridizations.
4. Essential sample annotation, including experimental factors (variables) and their values (e.g., the compound and dose in a dose response experiment).
5. Essential experimental and data-processing protocols.

#### 15.3.2.2 MINSEQE

Minimum Information about a high-throughput Nucleotide SEQuencing Experiment (MINSEQE)

Home: http://www.mged.org/minseqe/

MINSEQE provides a reporting guideline akin to MIAME that is applicable to high-throughput nucleotide sequencing experiments used to assay biological state. It does not pertain to traditional sequencing projects, where the aim is to assemble a chromosomal sequence or resequence a given genomic region, but rather to applications of sequencing in areas such as transcriptomics where high-throughput sequencing is being used to compare the populations of sequences between samples derived from different biological states, for example, sequencing cDNAs to assess differential gene expression. Here, sequencing provides a means to assay the sequence composition of different biological samples, analogous to the way that DNA microarrays have traditionally been used.

#### 15.3.2.3 MAGE

MicroArray Gene Expression (MAGE).

Home: http://www.mged.org/Workgroups/MAGE/mage.html

The MAGE project aims to provide a standard for the representation of microarray gene expression data to facilitate the creation of software tools for exchanging microarray information between different users and data repositories. The MAGE family of standards does not have direct support for capturing the results of higher-level analysis (e.g., clustering of expression data from a microarray experiment).

It include the following sub-projects:

- MAGE-OM – MAGE Object Model
- MAGE-ML – MAGE Markup Language
- MAGEstk – MAGE Software Toolkit
- MAGE-TAB – MAGE Tabular Format

MAGE-OM is a platform independent model for representing gene expression microarray data. MAGE-OM has been implemented by MAGE-ML (an XML-based format) as well as MAGE-TAB (tab-delimited values format). Both formats can be used for annotating and communicating data from microarray gene expression experiments in a MIAME-compliant fashion. MAGE-TAB evolved out of a need to create a simpler version of MAGE-ML that would be easier to use and thus be more accessible to a wider cross section of the microarray-based gene expression community, which has struggled with the often large, structured XML-based MAGE-ML documents. A limitation of MAGE-TAB is that only single values are permitted for certain data slots that may in practice be multi-valued. Data that cannot be adequately represented by MAGE-TAB can be described using MAGE-ML, which is quite flexible.

MAGEstk is a collection of Open Source packages that implement the MAGE Object Model in various programming languages (Spellman et al. 2002). The toolkit is meant for bioinformatics users that develop their own applications and need to integrate functionality for managing an instance of a MAGE-OM. The toolkit facilitates easy reading and writing of MAGE-ML to and from the MAGE-OM, and all MAGE-objects have methods to maintain and update the MAGE-OM at all levels. What MAGE-stk doesn't implement is the glue between a software application and the standard way of representing DNA microarray data in MAGE-OM as a MAGE-ML file.

#### 15.3.2.4 MAGE-TAB

MAGE-TAB is a simple tab delimited format that is used to represent gene expression and other high throughput data such as high throughput sequencing; it is the main submission format for ArrayExpress and is supported by the BioConductor package ArrayExpress. There are also converters available to MAGE-TAB from GEO soft format, from MAGE-ML to MAGE-TAB and an open source template generation system (Rayner et al. 2009). A complete list of applications using MAGE-TAB is maintained by the MGED community *www.mged.org/mage-tab*.

#### 15.3.2.5 MO

MGED Ontology (MO).

Home: http://mged.sourceforge.net/ontologies/index.php

The MGED Ontology (MO) provides standard terms for describing the different components of a microarray experiment (Whetzel et al. 2006b). MO is complementary to the other MGED standards, MIAME and MAGE, which specify what information should be provided and how that information should be structured respectively. The specification of the terminology used for labeling that information has been left to MO. MO is an ontology with defined classes, instances and relations. A primary motivation of MO was to provide terms where ever needed in the MAGE Object Model leading to MO being organized along the same lines as the MAGE-OM packages. A feature of MO is that it provides pointers to other resources as appropriate to describe sample or biomaterial characteristics and treatment compounds used in the experiment (e.g., NCBI Taxonomy, ChEBI) rather than import, map, or duplicate those terms.

A major revision of MO (currently at version 1.3.1.1 released in February 2007) was planned to address structural issues. Effort has been placed instead in incorporating MO into the Ontology for Biomedical Investigations (OBI).

Example Applications of MO

The primary usage of MO has been for the annotation of microarray experiments. MO terms can be found incorporated in a number of microarray databases (e.g., ArrayExpress (Parkinson et al. 2009), RAD (Manduchi et al. 2004), caArray (*caarray.nci.nih.gov/*). SMD (Gollub et al. 2003), maxD (*www.bioinf.manchester.ac.uk/microarray/maxd/*), MiMiR (Navarange et al. 2005)) enabling retrieval of studies consistently across these different sites. MO terms have also been used as part of column headers for MAGE-TAB (Rayner et al. 2006), a tab-delimited form of MAGE.

Example terms from MO v.1.3.1.1

- BioMaterialPackage (MO_182): Description of the source of the nucleic acid used to generate labeled material for the microarray experiment. (an abstract class taken from MAGE to organize MO).
- BioMaterialCharacteristics (MO_5): Properties of the biomaterial before treated in any manner for the purposes of the experiment. (a subclass of BioMaterialPackage).
- CellType (MO_135): CellType, the type of cell used in the experiment if non mixed, if mixed the TargetedCellType should be used, example of instances, epithelial, glial etc. (a subclass of BioMaterialCharacteristics, uses property has_database to point to CellTypeDatabase).
- CellTypeDatabase (MO_141): Database of cell type information. (a subclass of Database).
- eVOC (MO_684): Ontology of human terms that describe the sample source of human cDNA and SAGE libraries. (An instance of CellTypeDatabase).

#### 15.3.2.6 MAQC

MicroArray Quality Control (MAQC) Project.

Home: http://www.fda.gov/nctr/science/centers/toxicoinformatics/maqc/

The MAQC project aims to develop best practices for executing microarray experiments and analyzing results in a manner that maximizes consistency between different vendor platforms. The effort is spearheaded by the FDA and has participants spanning the microarray industry. The work of the MAQC project is providing guidance to develop quality measures and procedures that will facilitate the reliable use of microarray technology within clinical practice and regulatory decision-making, and thereby help realize the promises of personalized medicine (Allison 2008).

The project consists of two phases:

1. MAQC-I demonstrated the technical performance of microarray platforms in the identification of differentially expressed genes (Shi et al. 2006).
2. MAQC-II is aimed at reaching consensus on best practices for developing and validating predictive models based on microarray data. This phase of the project includes genotyping data as well as expression data, which was the focus of MAQC-I. MAQC-II is currently in progress with results expected in early 2009 (http://www.fda.gov/nctr/science/centers/toxicoinformatics/maqc).

#### 15.3.2.7 ERCC

External RNA Control Consortium (ERCC).

Home: http://www.cstl.nist.gov/biotech/Cell&TissueMeasurements/GeneExpression/ERCC.htm

The ERCC aims to create well-characterized and tested RNA spike-in controls for gene expression assays. They have worked with NIST to create certified reference materials useful for evaluating sample and system performance, to facilitate standardized data comparisons among commercial and custom microarray gene expression platforms as well as by an alternative expression profiling method such as qRT-PCR.

The ERCC originated in 2003 and has grown to include more than 90 organizations spanning a cross-section of industry and academic groups from around the world. The controls developed by this group were based on contributions from member organizations and have undergone rigorous evaluation to ensure efficacy across different expression platforms.

### *15.3.3 Proteomic Standards*

This section describes the standards and organizations related to technologies that measure protein-related phenomena on a genomic scale.

#### 15.3.3.1 HUPO-PSI

Human Proteome Organization (HUPO) Proteomics Standards Initiative (PSI).

Home: http://www.psidev.info/

The largest standards organization in this domain is the HUPO-PSI, which has an official process for drafting, reviewing, and accepting proteomics-related standards (Orchard and Hermjakob 2008). As with other standardization efforts, the PSI creates and promotes both minimum information standards, which define what metadata about a study should be provided as well as data exchange standards, which provide the standardized, computer-readable format for conveying the information.

The standards promoted by the PSI are organized by the working group, which define standards in the proteomics domains such as the following technologies or types of investigations:

- Gel electrophoresis
- Mass spectrometry
- Proteomics Informatics
- Molecular interactions
- Protein modifications
- Sample processing

Example Applications of Proteomic Standards

HUPO-PSI standards have been used to drive the design of proteomics databases, such as the Proteomics IDEntification (PRIDE) database (Jones et al. 2008) and Jones et al. 2008 software tools such as the Trans Proteomic Pipelines for processing mass spectrometry data (Keller et al. 2005). Consistent application of standard terminologies for the annotation of experiment metadata in the PRIDE database has recently enabled a re-analysis of high-throughput proteomics experiments, which would have otherwise not been possible without such annotations (Klie et al. 2008).

#### 15.3.3.2 MIAPE

Minimum Information About a Proteomics Experiment (MIAPE).

Home: http://www.psidev.info/index.php?q=node/91

MIAPE is a reporting standard analogous to MIAME for proteomics experiments. The primary MIAPE publication (Taylor et al. 2007) describes the precepts of the MIAPE specifications, and then each sub-domain (e.g., sample processing, column chromatography, mass spectrometry, etc.) has a separate MIAPE module, which specifies the information needed for each component of the study being described.

#### 15.3.3.3 Proteomics Experiment Data Exchange Formats

Several data formats to encode data related to proteomics experiments have emerged since 2003 (Jones et al. 2008). Some early XML-based formats originating from the Institute for Systems Biology such as mzXML (Pedrioli et al. 2004) and pepXML/protXML (Keller et al. 2005) were widely adopted and became de-facto standards. More recently, the PSI has built up on these formats to develop mzML (Deutsch 2008) for mass spectrometer output, GelML for gel electrophoresis, and AnalysisXML for the bioinformatic analysis results from such data, and others. See Deutsch et al. (2008) for a review of some of these formats. Accompanying these formats are controlled vocabularies, validators, example instance documents, and in some cases software libraries to enable adoption of these standards.

#### 15.3.3.4 Molecular Interactions

Molecular Interactions (PSI-MI) Working Group Standards

Home: http://www.psidev.info/index.php?q=node/277

The PSI's Molecular Interactions (MI) Working Group has defined several standards to enable better sharing of molecular interaction information. MIMIx (Orchard et al. 2007) is the minimum information standard that defines what information must annotate a list of molecular interactions. The PSI-MI XML (or MIF) standard is an XML-based data exchange format for sharing data from molecular interactions experiments. The format relies heavily on a controlled vocabulary (PSI-MI CV) that insures that terms to describe and annotate interactions are used consistently. In addition to the XML format, a simpler tab-delimited data exchange format MITAB2.5 is available that supports a subset of the PSI-MI XML functionality and can be read using widely available spreadsheet software (Kerrien et al. 2007).

### *15.3.4 Metabolomic Standards*

This section describes the standards and organizations related to the study of metabolomics, which studies low molecular weight metabolites profiles on a comprehensive, genomic scale within a biological sample. Metabolomic standards initiatives are not as mature as those in the transcriptomic and proteomic domains, though there is a growing community interest in this area. (Note that no distinction is made in this text between metabolomics vs. metabonomics, using "metabolomics" to refer to both types of investigations, in so far as a distinction exists. For further information, see http://en.wikipedia.org/wiki/Metabolomics).

Metabolomic standards pertain to the following technologies or types of investigations:

- Metabolic profiling of all compounds in a specific pathway
- Biochemical network modeling
- Biochemical network perturbation analysis (environmental, genetic)
- Network flux analysis

The metabolomics research community is engaged in the development of a variety of standards, coordinated by the Metabolomics Standards Initiative (Fiehn et al. 2007a; Sansone et al. 2007b). Under development are reporting "minimum information" standards (Fiehn et al. 2006, 2007b; Goodacre et al. 2007), data exchange formats (Hardy and Taylor 2007), data models (Jenkins et al. 2004, 2005; Spasić et al. 2006), and standard ontologies (Sansone et al. 2007a). A number of specific experiment description-related projects for metabolomics are described below.

#### 15.3.4.1 CIMR

Core Information for Metabolomics Reporting (CIMR).

Home: http://msi-workgroups.sourceforge.net/

CIMR is being developed as a minimal information guideline for reporting metabolomics experiments. It is expected to cover all metabolomics application areas and analysis technologies. The MSI is also involved in collaborative efforts to develop ontologies and data exchange formats for metabolomics experiments.

#### 15.3.4.2 MeMo

Metabolic Modelling (MeMo).

Home: http://dbkgroup.org/memo/

MeMo defines a data model and XML-based data exchange format for metabolomic studies in yeast (Spasić et al. 2006).

#### 15.3.4.3 ArMet

Architecture for Metabolomics (ArMet).

Home: http://www.armet.org

ArMet defines a data model for plant metabolomics experiments and also provides guidance for data collection (Jenkins et al. 2004, 2005).

### *15.3.5 Healthcare Standards*

The health care community has a long history of using standards to drive data exchange and submission to regulatory agencies. Within this setting, it is vital to ensure that data from assays pass quality assessments and can be transferred without loss of meaning and in a format that can be easily used by common tools. The drive to translate functional genomics approaches from a research to a clinical setting has provided strong motivation for the development of physical standards and guidelines for their use in this setting in particular. Functional genomics technologies hold much promise to improve our understanding of the molecular basis of diseases and develop improved diagnostics and therapeutics tailored to individual patients (Kumar 2007; Biomed Central Genome Medicine Journal announcement 2008; Warrington 2008).

Looking forward, the health care community is now engaged in numerous efforts to define important standards for clinical, diagnostic, and toxicological applications of data from high-throughput genomics technologies. The types and amount of data from a clinical trial or toxico-genomics study are quite extensive, incorporating data from multiple Omics domains. Standards development for electronic submission of this data is on-going with best practices yet to emerge. While it is likely that high-throughput data will be summarized prior to transmission, it is anticipated that the raw files should be available for analysis if requested by regulators and other scientists.

Standards-related activities pertaining to the use of functional genomics technologies within a health care setting can be roughly divided into three main focus areas: experiment description standards, reference materials, and laboratory procedures.

#### 15.3.5.1 Healthcare Experiment Description Standards

Orthogonal to the experiment description standards efforts in the basic research and technical communities, clinicians and biologists have identified the need to describe the characteristics of an organism or specimen under study in a way that is understandable to scientists and clinicians. Under development within these biomedical communities are reporting standards to codify the data that should be captured and the data exchange format that must be used to permit re-use of the data by others. As with the other minimum information standards, the goal is to create a common way to describe characteristics of the objects of a study, and identify and include the essential characteristics when publishing the study. Parallel work is underway in the arena of toxico-genomics (Fostel 2008; Taylor et al. 2008). Additionally, standard terminologies in the form of thesauri or controlled vocabularies and systematic annotation methods are also under development.

It is envisioned that clinically relevant standards will be used in conjunction with the experiment description standards being developed by the basic research communities that study the same biological objects and organisms. For example, ISA-TAB (described below) is intended to complement existing biomedical formats such as the Study Data Tabulation Model (SDTM), a U.S. Food and Drug Administration-endorsed data model created by CDISC to organize, structure, and format both clinical and nonclinical (toxicological) data submissions to regulatory authorities (http://www.cdisc.org/models/sds/v3.1/index.html). It is inevitable that some information will be duplicated between the two frameworks, but this is not generally seen as a major problem. Links between related components of ISA-TAB and SDTM could be created using properties of the subject source, for example (Table 15.8).

**Table 15.8** Summary of healthcare experiment description standards initiatives

| Acronym | Full name | Description | Scope | Website |
|---|---|---|---|---|
| BIRN | Biomedical Informatics Research Network | Collaborative informatics resources medical/clinical data | Data analysis; Terminology | www.nbirn.net |
| CDISC | Clinical Data Interchange Standards Consortium | Regulatory submissions of clinical data | Data exchange and modeling | www.cdisc.org |
| CONSORT | Consolidated Standards of Reporting Trials | Minimum requirements for reporting randomized clinical trials | Reporting | www.consort-statement.org |
| EVS | Enterprise Vocabulary Services | Controlled vocabulary by the NCI in support of cancer | Terminology | www.cancer.gov/cancertopics/terminologyresources |
| HL7 | Health Level 7 | Programmatic data exchange for healthcare applications | Data exchange | www.hl7.org |
| SEND | Standards for Exchange of Preclinical data | Regulatory submissions of preclinical data; based on CDISC | Data exchange and modeling | www.cdisc.org/standards |
| ToxML | Toxicology XML | Toxicology data exchange; based on controlled vocabulary | Data exchange; Terminology | www.leadscope.com/toxml.php |

#### 15.3.5.2 Reference Materials

Developing industry-respected standard reference materials, such as a reagent for use as a positive or negative control in an assay, is essential for any work in a clinical or diagnostic setting. Reference materials are physical standards (see above) that provide an objective way to evaluate the performance of laboratory equipment, protocols, and sample integrity, and the lack of suitable reference materials and guidelines for their use has been a major factor slowing the adoption of functional genomics technologies such as DNA microarrays within clinical and diagnostic settings (Warrington 2008).

The ERCC (described above) and the LGC (http://www.lgc.co.uk) are the key organizations working on development of standard reference materials, currently targeting transcriptomic experiments.

#### 15.3.5.3 Laboratory Procedures

Standard protocols providing guidance in the application of reference materials, experiment design, and data analysis best practices are essential for performing high-throughput functional genomics procedures in clinical or diagnostic applications.

The Clinical Laboratory Standards Institute (CLSI, http://www.clsi.org/) is an organization that provides an infrastructure for ratifying and publishing guidelines for clinical laboratories. Working with organizations such as the ERCC (described above), they have produced a number of documents applicable to the use of multiplex, whole-genome technologies such as gene expression and genotyping within a clinical or diagnostic setting (Table 15.9).

## 15.4 Standards Harmonization

The field of functional genomics is not suffering from lack of interest in standards development, as the number of different standards reviewed in this chapter attests. Such a complex landscape can have adverse effects on data sharing, integration, and systems interoperability – the very things that the standards are intended to help

**Table 15.9** CLSI documents most relevant to functional genomic technologies

| Document | Description | Status |
|---|---|---|
| MM12-A | Diagnostic Nucleic Acid Microarrays | Approved guideline |
| MM14-A | Proficiency Testing (External Quality Assessment) for Molecular Methods | Approved guideline |
| MM15-A | Use of External RNA Controls in Gene Expression Assays | Approved guideline |
| MM17-A | Verification and Validation of Multiplex Nucleic Acid Assays | Approved guideline |

(Quackenbush 2006). To address this, there are a number of projects in the research and biomedical communities engaged in so called "harmonization" activities, which focus on integrating standards with related or complementary scope and aim to enhance interoperability in the reporting and analysis of data generated by different technologies or within different functional genomics domains (Nature Cell Biology Editorial 2008).

Some standards facilitate harmonization by having a sufficiently general-purpose design, so they can accommodate data from experiments in different domains. Such "multiomics" standards typically have a mechanism that allows them to be extended as needed in order to incorporate aspects specific to a particular application area. The use of these domain- and technology-neutral frameworks is anticipated to improve the interoperability of data analysis tools that need to handle data from different types of functional genomics experiments as well as to reduce wheel reinvention by different standards groups with similar needs. Harmonization and multiomics projects are collaborative efforts, involving participants from different domain-specific standards developing organizations with shared interests. Indeed, the success of these efforts depends on continued broad-based community involvement (Table 15.10).

### 15.4.1 FuGE

Functional Genomics Experiment (FuGE).

Home: http://fuge.sourceforge.net/

**Table 15.10** Existing functional genomics standards harmonization projects and initiatives. P³G covers harmonization between genomic biobanks and longitudinal population genomic studies including technical, social, and ethical issues: http://www.p3gconsortium.org. The other projects noted in this table are described further in this chapter

| Acronym | Full name | Scope | Organization |
|---|---|---|---|
| FuGE-ML<br>FuGE-OM | Functional Genomics Experiment Markup Language/Object Model | Data exchange and modeling | FuGE |
| ISA-TAB | Investigation Study Assay Tabular Format | Data exchange | RSBI, GSC, MSI, HUPO-PSI |
| HITSP | Healthcare Information Technology Standards Panel | (various) | ANSI |
| MIBBI | Minimum Information for Biological and Biomedical Investigations | Reporting | MIBBI |
| OBI | Ontologies for Biomedical Investigations | Terminology | OBI |
| OBO | Open Biomedical Ontologies | Terminology | NCBO |
| P³G | Public Population Project in Genomics | (various) | International Consortium |

The FuGE project aims to build generic components that capture common facets of different functional genomics domains (Jones et al. 2007). Its contributors come from different standards efforts, primarily MGED and HUPO-PSI, reflecting the desire to build components that provide properties and functionalities that are common across different functional genomics technologies and application areas.

The vision of this effort is that, using FuGE-based components, a software developer will be better able to create and modify tools for handling functional genomics data, without having to reinvent the wheel for common tasks in potentially incompatible ways. Further, tools based on such shared componentry are expected to be more interoperable.

FuGE has the following sub-projects which include the FuGE Object Model (FuGE-OM) and the FuGE Markup Language (FuGE-ML)—a data exchange format. Technology-specific aspects can be added by extending the generic FuGE components, and building on the common functionalities. For example, a microarray-specific framework equivalent to MAGE could be derived by extending FuGE, deriving microarray-specific objects from the FUGE object model. Guidelines have emerged for using and extending FUGE data model, and FUGE-supportive software tools have been developed (Jones et al. 2009).

### 15.4.2 HITSP

The Healthcare Information Technology Standards Panel (HITSP) is a public–private sector partnership of standards developers, healthcare providers, government representatives, consumers, and vendors in the healthcare industry. It is administered by the American National Standards Institute (ANSI, http://www.ansi.org) to harmonize healthcare-related standards and improve interoperability of healthcare software systems. It produces recommendations and reports contributing to the development of a Nationwide Health Information Network for the United States (NHIN, http://www.hhs.gov/healthit/healthnetwork/background).

The HITSP is driven by use cases issued by the American Health Information Community (AHIC, http://www.hhs.gov/healthit/community/background). A number of use cases have been defined on a range of topics, such as personalized healthcare, newborn screening, and consumer adverse event reporting (http://www.hhs.gov/healthit/usecases).

### 15.4.3 ISA-TAB

Investigation Study Assay Tabular format (ISA-TAB)

Home: http://isatab.sourceforge.net

The ISA-TAB format is a general purpose framework with which to communicate both data and metadata from experiments involving a combination of functional technologies (Sansone et al. 2008). ISA-TAB therefore has a broader applicability and more extended structure compared to a domain-specific data exchange format such as MAGE-TAB. An example where ISA-TAB might be applied would be an experiment looking at changes both in (1) the metabolite profile of urine, and (2) gene expression in the liver in subjects treated with a compound inducing liver damage, using both mass spectrometry and DNA microarray technologies, respectively.

The general motivation for this work stems from the needs of the BioInvestigation Index project at EBI (http://www.ebi.ac.uk/bioinvindex) to create a common structured representation of the metadata required to interpret an experiment for the purpose of combined submission to experimental data repositories such as ArrayExpress, PRIDE, and an upcoming metabolomics repository. Additional motivation comes from a group of collaborative systems, part of the MGED's RSBI group (Sansone et al. 2006), either committed to pipelining omics-based experimental data into EBI public repositories or willing to exchange data among themselves, or to enable their users to import data from public repositories into their local systems.

#### 15.4.3.1 Relating ISA-TAB to Other Formats and Requirements

ISA-TAB has a number of additional features that make it a more general framework that can comfortably accommodate multi-domain experimental designs. ISA-TAB builds on the MAGE-TAB paradigm, and shares its motivation for the use of tab-delimited text files i.e., they can easily be created, viewed and edited by researchers, using spreadsheet software such as Microsoft Excel. ISA-TAB also employs MAGE-TAB syntax as far as possible, to ensure backward compatibility with existing MAGE-TAB files. It was also important to align the concepts in ISA-TAB with some of the objects in the FuGE model. The ISA-TAB format could be seen as competing with XML-based formats such as the FuGE-ML. However, ISA-TAB addresses the immediate need for a framework to communicate multiomics experiments, whereas all existing FuGE-based modules are still under development. When these become available, ISA-TAB could continue serving those with minimal bioinformatics support, as well as finding utility as a user-friendly presentation layer for XML-based formats (via an XSL transformation); i.e., in the manner of the HTML rendering of MAGE-ML documents.

Initial work has been carried out to evaluate the feasibility of rendering FuGE-ML files (and FuGE-based extensions, such as GelML and Flow-ML) in the ISA-TAB format. Examples are available at the ISA-TAB website under the document section, along with a report detailing the issues faced during these transformations. When finalized, the XSL templates will also be released, along with Xpath expressions and a table mapping FuGE objects and ISA-TAB labels. Additional ISA-TAB-formatted examples are available, including a MIGS/MIMS-compliant dataset (see http://isatab.sourceforge.net/examples.html).

The decision on how to regulate the use of the ISA-TAB (marking certain fields mandatory, or enforcing the use of controlled terminology) is a matter for those who will implement the format in their system. Although certain fields would benefit from the use of controlled terminology, ISA-TAB files with all fields left empty are syntactically valid, as are those where all fields are filled with free text values rather than controlled vocabulary or ontology terms.

### *15.4.4 MIBBI*

Minimal Information for Biological and Biomedical Investigations (MIBBI)

Home: http://mibbi.org (developers: http://mibbi.sourceforge.net )

Experiments in different functional genomics domains typically share some reporting requirements (for example, specifying the source of a biological specimen). The MIBBI project aims to work collaboratively with different groups to harmonize and modularize their minimum information checklists (e.g., MIAME, MIGS/MIMS, etc.) refactoring the common requirements, to make it possible to use these checklists in combination (Taylor et al. 2008). Additionally, the MIBBI project provides a comprehensive web portal providing registration of and access to different minimum information checklists for different types of functional genomics (and other) experiments.

### *15.4.5 OBI*

Ontology for Biomedical Investigations (OBI).

Home: http://purl.obofoundry.org/obo/obi

From the OBI home page: The Ontology for Biomedical Investigations (OBI) project is developing an integrated ontology for the description of biological and medical experiments and investigations. This includes a set of "universal" terms that are applicable across various biological and technological domains, and domain-specific terms relevant only to a given domain. This ontology will support the consistent annotation of biomedical investigations, regardless of the particular field of study. The ontology will model the design of an investigation, the protocols and instrumentation used, the material used, the data generated and the type of analysis performed on it. This project was formerly called the Functional Genomics Investigation Ontology (FuGO) project (Whetzel et al. 2006a).

OBI is a collaborative effort of many communities representing particular research domains and technological platforms (http://obi-ontology.org/page/Consortium). OBI is meant to serve very practical needs rather than be an academic exercise. Thus it is very much driven by use cases and validation questions. An example OBI use case is provided in Fig. 15.2.

The OBI user community provides valuable feedback about the utility of OBI and acts as a source of terms and use cases. As a member of the OBO Foundry (described below), OBI has made a commitment to be interoperable with other biomedical ontologies. Each term in OBI has a set of annotation properties, some of which are mandatory (minimal metadata defined at http://obi-ontology.org/page/

**Entities:**
mouse
portion of blood
glucose meter, an instrument used to measure glucose concentration
collection tubes
syringe

**Processes:**
implementation of a study design
analyte-measuring assay
material separation

**Roles and Relations:**

| Entity | Role | Realized by implementation of |
|---|---|---|
| Mouse | Study subject role | study design (describing how and when mouse will be handled) |
| Urine | Subject specimen role | study design (describig how and when urine is to be collected) |
| Urine | Evaluant role | process specifying it as an input to an assay designed to measure some quality of the evaluant (in this case the assay of measuring glucose concentration in mouse urine) |
| Glucose | Analyte role | assay (in this case an assay designed to measure glucose in mouse urine) |
| Standard curve reagent | Reference role | |

entities with role "assay participant" include urine, reagents used to measure (if any), device used to measure, tubes used to collect, prepare and assay the urine, the standard curve reagents, specified outputs.

concentration datum is_an Information Content entity.
concentration datum has_units mg/dL
concentration datum is_about (mouse has_quality urine glucose)
concentration datum has_magnitude 100

**Fig. 15.2** An OBI use case of measuring the concentration of glucose in the blood from a particular mouse that is the subject in a study (additional details of this example are available at http://obi.svn.sourceforge.net/viewvc/obi/trunk/docs/developer/images/assay.pdf)

OBI_Minimal_metadata). These include the term preferred name, definition source, editor, and curation status.

### *15.4.6 OBO Consortium and the NCBO*

Open Biomedical Ontologies (OBO) Consortium and the National Center for Biomedical Ontology (NCBO).

- OBO Foundry: http://www.obofoundry.org
- NCBO: http://bioontology.org

The OBO Consortium, a voluntary, collaborative effort among different OBO developers, has developed the OBO Foundry as a way to avoid the proliferation of incompatible ontologies in the biomedical domain (Smith et al. 2007). The OBO Foundry provides validation and assessment of ontologies to ensure interoperability. It also defines principles and best practices for ontology construction such as the Basic Formal Ontology, which serves as a root-level ontology from which other domain-specific ontologies can be built, and the relations ontology, which defines a common set of relationship types (Smith et al. 2005). Incorporation of such elements within OBO is intended to facilitate interoperability between ontologies (i.e., for one OBO Foundry ontology to be able to import components of other ontologies without conflict) and the construction of "accurate representations of biological reality."

The NCBO supports the OBO Consortium by providing tools and resources to help manage the ontologies and to help the scientific community access, query, visualize, and use them to annotate experimental data (Rubin et al. 2006). The NCBO's BioPortal website provides searches across multiple ontologies and contains a large library of these ontologies spanning many species and many scales, from molecules to whole organism. The ontology content comes from the model organism communities, biology, chemistry, anatomy, radiology, and medicine.

Together, the OBO Consortium and the NCBO are helping to construct a consistent arsenal of ontologies to promote their application in annotating functional genomics and other biological experiments. This is the sort of community-based ontology building that holds much potential to help the life science community convert the complex and daunting functional genomics data sets into new discoveries that expand our knowledge and improve human health.

## 15.5 Conclusion

A key motivation behind functional genomics standards is to foster data sharing, re-use, and integration with the ultimate goal of producing new biological insights (within basic research environments) and better medical treatments (within healthcare environments). Widely adopted minimum information guidelines for publication and formats for data exchange are leading to more reporting of results and submission of experimental data into public repositories, and more effective data mining of large functional genomics data sets. Standards harmonization efforts are in progress to improve data integration and interoperability of software within both basic research setting as well as within healthcare environments. Standard reference materials and protocols for their use are also under active development and hold much promise for improving data quality, systems benchmarking, and facilitating the use of functional genomics technologies within clinical and diagnostic settings.

### 15.5.1 Challenges for Functional Genomics Standards in Basic Research

A major challenge facing functional genomics standards is proving their value to a significant fraction of the user base and facilitating widespread adoption. Given the relative youth of the field of functional genomics and standardization efforts, the main selling point to use a standard is that it will benefit future scientists and application/database developers, with limited added value for the users who are being asked to comply with the standard at publication time. Regardless of how well designed the standard is, if complying with it is perceived as being difficult or complicated, widespread adoption is unlikely to occur. Some degree of enforcement of compliance by publishers and data repositories will most likely be required to inculcate the standard and build a critical mass within the targeted scientific community that then sustains its adoption. Significant progress has been achieved here: for DNA microarray gene expression studies, for example, most journals now require MIAME compliance and there is a broad recognition of the value of this standard within the target community.

Here are some of the "pressure points" any standard will experience from its community of intended users:

- Domain experts who want to ensure comprehensiveness of the standard
- End-user scientists who want the standard to be easy to comply with
- Software developers who want tools for encoding and decoding standards-compliant data
- Standards architects who want to ensure formal correctness of the standard

Satisfying all of these interests is not an easy task. One complication is that the various interested groups may not be equally involved in the development of the standard. Balancing these different priorities and groups is the task of the group responsible for maintaining a standard. This is an ongoing process that must remain responsive to user feedback. The MAGE-TAB data exchange format in the DNA microarray community provides a case in point: it was created largely in response to users that found MAGE-ML difficult to work with.

### 15.5.2 Challenges for Functional Genomics Standards in Healthcare Settings

The handling of clinical data adds additional challenges on top of the intrinsic complexities of functional genomics data. Investigators must respect certain regulations imposed by regulatory authorities. For example, the Health Insurance Portability Accountability Act (HIPAA) mandates the de-identification of patient data to protect an individual's privacy. Standards and information systems used by the healthcare community therefore must be formulated to deal with such regulations (e.g., Bland et al. 2007).

While the use of open standards poses risks to the release of protected health information, the removal of detailed patient metadata about samples can present barriers to research (Ferris et al. 2002; Meslin 2006). Enabling effective research while maintaining patient privacy remains an on-going issue (Joe White, personal communication).

### *15.5.3 Current and Future Directions*

High-throughput functional genomics experiments, with their large and complex data sets, have posed many challenges to the creation and adoption of standards, but in recent years, the standards initiatives in this field have risen to the challenge and continue to engage their respective communities to improve the effectiveness of the standards to user and market needs.

Functional genomics communities have recognized that standards-compliant software tools can go a long way towards enhancing the adoption and usefulness of a standard by enabling ease-of-use. For data exchange standards, such tools can "hide the technical complexities of the standard and facilitate manipulation of the standard format in an easy way" (Brazma et al. 2006). Some tools can themselves become part of standard practice when they are widely used throughout a community. Efforts are underway within organizations such as MGED and HUPO PSI to enhance the usefulness of tools for end user scientists working with standard data formats, in order to ease the process of data submission, annotation, and analysis.

The widespread adoption of some of the more mature functional genomics standards by large numbers of life science researchers, data analysts, software developers and journals has had a number of benefits. It has promoted data sharing and reanalysis, facilitated publication, and spawned a number of data repositories to store data from functional genomics experiments. A higher citation rate and other benefits have been detected for researchers who share their data (Piwowar et al. 2008; Piwowar and Chapman 2008). Estimates of total volume of high-throughput data available in the public domain are complex to calculate, but a list of databases maintained by the NAR journal (http://www3.oup.co.uk/nar/database/a/) contain more than 1,000 databases in areas ranging from nucleic acid sequence data to experimental archives and specialist data integration resources (Galperin and Cochrane 2009). The volume will undoubtedly rise asmore public databases appear every year and as technologies change so that deep sequencing of genomes and transcriptomes becomes more cost effective.

Consistent annotation of this growing volume of functional genomics data using interoperable ontologies and controlled vocabularies will play an important role in enabling collaborations and re-use of the data by other third parties. More advanced forms of knowledge integration that rely on standard terminologies are beginning to be explored using semantic web approaches (Sagotsky et al. 2008; Stein 2008)

While adherence to standards by public data repositories is expected to facilitate data querying and re-use, even in the absence of strict standards-compliant requirements on data submission, useful data mining can be performed from large bodies of raw data originating from the same technology platform (Day et al. 2007). Approaches such as this may help researchers better utilize the limited levels of consistently annotated data in the public domain.

It was recently noted that only a fraction of data generated is deposited in public data repositories (Ochsner et al. 2008). Improvements in this area can be anticipated through the proliferation of better tools for bench scientists that make it easier for them to submit their data in a consistent, standards-compliant manner. The full value of functional genomics research will only be realized once scientists in the laboratory and the clinic are able to share and integrate large amounts of functional genomics data as easily as they can now do with primary biological sequence data.

**Acknowledgments** SAC acknowledges financial support received from Affymetrix, Inc. during the preparation of this manuscript. The following people provided useful feedback: Nigel Hardy, Henning Hermjakob, Janet Warrington, and the OBI-developers mailing list.

## References

Allison M (2008) Is personalized medicine finally arriving? Nat Biotechnol 26(5):509–517
Ashburner M, Lewis S (2002) On ontologies for biologists: the Gene Ontology – untangling the web. Novartis Found Symp 247:66–80 discussion 80-3, 84-90, 244-52
Barrett T, Troup DB, Wilhite SE, Ledoux P, Rudnev D, Evangelista C et al (2007) NCBI GEO: mining tens of millions of expression profiles – database and tools update. Nucleic Acids Res 35(Database issue):D760–D765
Biomed Central Genome Medicine Journal announcement (2008) Personalized medicine: Innovative online journal leads the way. From http://www.eurekalert.org/pub_releases/2008-11/bc-pmi111208.php
Bland PH, Laderach GE, Meyer CR (2007) A web-based interface for communication of data between the clinical and research environments without revealing identifying information. Acad Radiol 14(6):757–764
Boguski MS (1999) Biosequence exegesis. Science 286(5439):453–455
Brazma A (2001) On the importance of standardisation in life sciences. Bioinformatics 17(2):113–114
Brazma A, Hingamp P, Quackenbush J, Sherlock G, Spellman P, Stoeckert C et al (2001) Minimum information about a microarray experiment (MIAME)-toward standards for microarray data. Nat Genet 29(4):365–371
Brazma A, Krestyaninova M, Sarkans U (2006) Standards for systems biology. Nat Rev Genet 7(8):593–605
Brooksbank C, Quackenbush J (2006) Data standards: a call to action. OMICS 10(2):94–99
Day A, Carlson MR, Dong J, O'Connor BD, Nelson SF (2007) Celsius: a community resource for Affymetrix microarray data. Genome Biol 8(6):R112
DeFrancesco L (2002) Journal trio embraces MIAME. News from The Scientist. 3:20021010-05
Deutsch E (2008) mzML: A single, unifying data format for mass spectrometer output. Proteomics 8(14):2776–2777
Deutsch EW, Lam H, Aebersold R (2008) Data analysis and bioinformatics tools for tandem mass spectrometry in proteomics. Physiol Genomics 33(1):18–25
Ferris, T. A., G. M. Garrison and H. J. Lowe (2002). A proposed key escrow system for secure patient information disclosure in biomedical research databases. Proc AMIA Symp: 245-9.
Fiehn O, Kristal B, van Ommen B, Sumner LW, Sansone SA, Taylor C et al (2006) Establishing reporting standards for metabolomic and metabonomic studies: a call for participation. OMICS 10(2):158–163

Fiehn O, Robertson D, Griffin J, van der Werf M, Nikolau B, Morrison N et al (2007a) The metabolomics standards initiative (MSI). Metabolomics 3(3):175–178

Fiehn O, Sumner L, Rhee S, Ward J, Dickerson J, Lange B et al (2007b) Minimum reporting standards for plant biology context information in metabolomic studies. Metabolomics 3(3):195–201

Field D, Garrity G, Gray T, Morrison N, Selengut J, Sterk P, et al (2008) The minimum information about a genome sequence (MIGS) specification. Nat Biotechnol 26(5):541–547

Fostel JM (2008) Towards standards for data exchange and integration and their impact on a public database such as CEBS (Chemical Effects in Biological Systems). Toxicol Appl Pharmacol 233(1):54–62

Galperin MY, Cochrane GR (2009) Nucleic Acids Research annual Database Issue and the NAR online Molecular Biology Database Collection in 2009. Nucleic Acids Res 37:D1–D4

Gentleman RC, Carey VJ, Bates DM, Bolstad B, Dettling M, Dudoit S et al (2004) Bioconductor: open software development for computational biology and bioinformatics. Genome Biol 5(10):R80

Gollub J, Ball CA, Binkley G, Demeter J, Finkelstein DB, Hebert JM, Hernandez-Boussard T, Jin H, Kaloper M, Matese JC, Schroeder M, Brown PO, Botstein D, Sherlock G (2003) The Stanford Microarray Database: data access and quality assessment tools. Nucleic Acids Res 31(1):94–96

Goodacre R, Broadhurst D, Smilde A, Kristal B, Baker J, Beger R et al (2007) Proposed minimum reporting standards for data analysis in metabolomics. Metabolomics 3(3):231–241

Hardy N, Taylor C (2007) A roadmap for the establishment of standard data exchange structures for metabolomics. Metabolomics 3(3):243–248

Jenkins H, Hardy N, Beckmann M, Draper J, Smith AR, Taylor J et al (2004) A proposed framework for the description of plant metabolomics experiments and their results. Nat Biotechnol 22(12):1601–1606

Jenkins H, Johnson H, Kular B, Wang T, Hardy N (2005) Toward supportive data collection tools for plant metabolomics. Plant Physiol 138(1):67–77

Jones AR, Lister AL, Hermida L, Wilkinson P, Eisenacher M, Belhajjame K, Gibson F, Lord P, Pocock M, Rosenfelder H, Santoyo-Lopez J, Wipat A, Paton NW (2009) Modelling and managing experimental data using FUGE. Omics 13(3):239–251

Jones AR, Paton NW (2005) An analysis of extensible modelling for functional genomics data. BMC Bioinformatics 6:235

Jones AR, Miller M, Aebersold R, Apweiler R, Ball CA, Brazma A et al (2007) The Functional Genomics Experiment model (FuGE): an extensible framework for standards in functional genomics. Nat Biotechnol 25(10):1127–1133

Jones P, Côté RG, Cho SY, Kile S, Martens L, Quinn AF, Thorneycroft D, Hermjakob H (2008) PRIDE: New developments and new data sets. Nucleic Acids Res 36 (Database issue): D878–D883.

Keller A, Eng J, Zhang N, Li XJ, Aebersold R (2005) A uniform proteomics MS/MS analysis platform utilizing open XML file formats. Mol Syst Biol 1(2005):0017

Kerrien S, Orchard S, Montecchi-Palazzi L, Aranda B, Quinn AF, Vinod N et al (2007) Broadening the horizon – level 2.5 of the HUPO-PSI format for molecular interactions. BMC Biol 5:44

Kile S, Martens L, Vizcaíno JA, Côté R, Jones P, Apweiler R, Hinneburg A. Hermjakob H (2008) Analyzing large-scale proteomics projects with latent semantic indexing. J Proteome Res 7(1):182–191

Kottmann R, Gray T, Murphy S, Kagan L, Kravitz S, Lombardot T, Field D, Glöckner FO (2008) A standard MIGS/MIMS compliant XML schema: toward the development of the Genomic Contextual Data Markup Language (GCDML). Omics 12(2):115–121

Kumar D (2007) From evidence-based medicine to genomic medicine. Genomic Med 1(3–4):95–104

Manduchi E, Grant GR, He H, Liu J, Mailman MD, Pizarro AD et al (2004) RAD and the RAD Study-Annotator: an approach to collection, organization and exchange of all relevant information for high-throughput gene expression studies. Bioinformatics 20(4):452–459

Meslin EM (2006) Shifting paradigms in health services research ethics. Consent, privacy, and the challenges for IRBs. J Gen Intern Med 21(3):279–280
Nature Cell Biology Editorial (2008) Standardizing data. Nat Cell Biol 10(10):1123–1124
Navarange M, Game L, Fowler D, Wadekar V, Banks H, Cooley N et al (2005) MiMiR: a comprehensive solution for storage, annotation and exchange of microarray data. BMC Bioinformatics 6:268
Ochsner SA, Steffen DL, Stoeckert CJ Jr, McKenna NJ (2008) Much room for improvement in deposition rates of expression microarray datasets. Nat Methods 5(12):991
Orchard S, Hermjakob H (2008) The HUPO proteomics standards initiative–easing communication and minimizing data loss in a changing world. Brief Bioinform 9(2):166–173
Orchard S, Salwinski L, Kerrien S, Montecchi-Palazzi L, Oesterheld M, Stumpflen V et al (2007) The minimum information required for reporting a molecular interaction experiment (MIMIx). Nat Biotechnol 25(8):894–898
Parkinson H, Kapushesky M, Kolesnikov N, Rustici G, Shojatalab M, Abeygunawardena N et al (2009) ArrayExpress update – from an archive of functional genomics experiments to the atlas of gene expression. Nucleic Acids Res 37:D868–D872
Pedrioli PG, Eng JK, Hubley R, Vogetzang M, Deutsch EW, Raught B, et al (2004) A common open representation of mass spectrometry data and its application to proteomics research. Nat Biotechnol 22(11):1459–1466
Piwowar HA, Chapman W (2008) Identifying data sharing in biomedical literature. AMIA Annu Symp Proc 6:596–600
Piwowar HA, Becich MJ, Bilofsky H, Crowley RS (2008) Towards a Data Sharing Culture: Recommendations for Leadership from Academic Health Centers. PLoS Med 5(9):e183
Quackenbush J (2006) Standardizing the standards. Mol Syst Biol 2(2006):0010
Rayner T, Rocca-Serra P, Spellman PT, Causton HC, Farne A, Holloway E, Liu J, Maier DS, Miller M, Petersen K, Quackenbush J, Sherlock G, Stoeckert C Jr, White J, Whetzel P, Wymore F, Parkinson H, Sarkans U, Ball C, Brazma A (2006) A simple spreadsheet-based, MIAME-supportive format for microarray data. BMC Bioinformatics 7:489
Rayner TF, Rezwan FI, Lukk M, Bradley XZ, Farne A, Holloway E et al (2009) MAGETabulator, a suite of tools to support the microarray data format MAGE-TAB. Bioinformatics 25(2):279–280
Rogers S, Cambrosio A (2007) Making a new technology work: the standardization and regulation of microarrays. Yale J Biol Med 80(4):165–178
Rubin DL, Lewis SE, Mungall CJ, Misra S, Westerfield M, Ashburner M et al (2006) National Center for Biomedical Ontology: advancing biomedicine through structured organization of scientific knowledge. OMICS 10(2):185–198
Sagotsky JA, Zhang L, Wang Z, Martin S, Deisboeck TS (2008) Life Sciences and the web: a new era for collaboration. Mol Syst Biol 4:201
Salit M (2006) Standards in gene expression microarray experiments. Methods Enzymol 411:63–78
Sansone SA, Rocca-Serra P, Tong W, Fostel J, Morrison N, Jones AR (2006) A strategy capitalizing on synergies: the Reporting Structure for Biological Investigation (RSBI) working group. OMICS 10(2):164–171
Sansone S-A, Schober D, Atherton H, Fiehn O, Jenkins H, Rocca-Serra P et al (2007a) Metabolomics standards initiative: ontology working group work in progress. Metabolomics 3(3):249–256
Sansone SA, Fan T, Goodacre R, Griffin JL, Hardy NW, Kaddurah-Daouk R et al (2007b) The metabolomics standards initiative. Nat Biotechnol 25(8):846–848
Sansone SA, Rocca-Serra P, Brandizi M, Brazma A, Field D, Fostel J et al (2008) The first RSBI (ISA-TAB) workshop: "can a simple format work for complex studies?". OMICS 12(2):143–149
Shi L, Reid LH, Jones WD, Shippy R, Warrington JA, Baker SC et al (2006) The MicroArray Quality Control (MAQC) project shows inter- and intraplatform reproducibility of gene expression measurements. Nat Biotechnol 24(9):1151–1161
Smith B, Ceusters W, Klagges B, Köhler J, Kumar A, Lomax J, Mungall C, Neuhaus F, Rector AL, Rosse C (2005) Relations in biomedical ontologies. Genome Biol 6:R46
Smith B, Ashburner M, Rosse C, Bard J, Bug W, Ceusters W et al (2007) The OBO Foundry: coordinated evolution of ontologies to support biomedical data integration. Nat Biotechnol 25(11):1251–1255

Spasić I, Dunn WB, Velarde G, Tseng A, Jenkins H, Hardy N et al (2006) MeMo: a hybrid SQL/XML approach to metabolomic data management for functional genomics. BMC Bioinformatics 7:281

Spellman PT, Miller M, Stewart J, Troup C, Sarkans U, Chervitz S et al (2002) Design and implementation of microarray gene expression markup language (MAGE-ML). Genome Biol 3(9):RESEARCH0046

Stein LD (2008) Towards a cyberinfrastructure for the biological sciences: progress, visions and challenges. Nat Rev Genet 9(9):678–688

Stoeckert CJ Jr, Causton HC, Ball CA (2002) Microarray databases: standards and ontologies. Nat Genet 32(Suppl):469–473

Taylor CF (2006) Minimum reporting requirements for proteomics: a MIAPE primer. Proteomics 6(Suppl 2):39–44

Taylor CF, Paton NW, Lilley KS, Binz PA, Julian RK Jr, Jones AR et al (2007) The minimum information about a proteomics experiment (MIAPE). Nat Biotechnol 25(8):887–893

Taylor CF, Field D, Sansone SA, Aerts J, Apweiler R, Ashburner M et al (2008) Promoting coherent minimum reporting guidelines for biological and biomedical investigations: the MIBBI project. Nat Biotechnol 26(8):889–896

Warrington JA (2008) Standard controls and protocols for microarray based assays in clinical applications. Book of Genes and Medicine. H. Aburatan, Osaka, Medical Do Co

Whetzel PL, Brinkman RR, Causton HC, Fan L, Field D, Fostel J et al (2006a) Development of FuGO: an ontology for functional genomics investigations. OMICS 10(2):199–204

Whetzel PL, Parkinson H, Causton HC, Fan L, Fostel J, Fragoso G et al (2006b) The MGED Ontology: a resource for semantics-based description of microarray experiments. Bioinformatics 22(7):866–873

# Chapter 16
# Literature Databases

**J. Lynn Fink**

## 16.1 Introduction

Published research is the foundation for all future research. Therefore, access to published research is a requirement for effectively engaging in scientific research. Scientists and policy-makers came to this realization several centuries ago and created the first scientific journals in 1665: the *Journal de Sçavans*, initiated by the French parliament member, Denis de Sallo[1]; and the *Philosophical Transactions of the Royal Society of London*, published by the Royal Society of London (Eisenstein 1979). These publications provided a common forum to help scientists disseminate observations and present theories to a much broader audience than was previously possible and thus established our current paradigm of scientific communication.

Only about 20 articles were published each month in the early *Transactions* covering all realms of science, which made it relatively easy for seventeenth century scientists to keep up with their particular interests. For example, articles in the first issue of the *Philosophical Transactions*[2] relating to the life sciences included Robert Boyle's description of a "Very Odd Monstrous Calf" "…whose Tongue was, *Cerberus*-like, triple, to each side of his Mouth one, and one in the midst…" (Boyle 1665) and an anonymous report of a man with "Milk Found in Veins, Instead of Blood" (1665). Today, thousands of articles have since been published on each of the probable culprits causing the trifurcated tongue and milky blood, craniofacial midline anomalies (Vandenhaute et al. 2000) and hyperlipidemia (Yuan et al. 2007), making it impossible for a scientist to read and recall all reported data concerning either topic. Current estimates suggest that about 65,000 articles are

J. Lynn Fink (✉)
Skaggs School of Pharmacy and Pharmaceutical Sciences, University of California, San Diego, CA, USA
e-mail: jlfink@ucsd.edu

[1] http://es.rice.edu/ES/humsoc/Galileo/Catalog/Files/sallo.html

[2] From the literature database J-STOR (http://www.jstor.org), an excellent resource, but not described here since it is not specific to the sciences

D. Edwards et al. (eds.), *Bioinformatics: Tovols and Applications*,
DOI 10.1007/978-0-387-92738-1_16, © Springer Science+Business Media, LLC 2009

now published per month and, overall, literature in the life sciences has swelled to a corpus of nearly 20 million articles from tens of thousands of scientific journals.[3] The need to make more intelligent use of this body of work is crucial to the progress of research.

Literature databases reach well beyond the printed periodical by employing cyber infrastructure to index, store, and serve scientific literature to an international community of scientists, clinicians, and interested non-scientists. These databases archive the legacies and current knowledge of innumerable research topics and enable a tracing of a topic's origin, impact, and associated data. Most databases also enhance their content with additional information, such as extracted metadata (performed by computers and/or humans) and links to relevant resources in order to assist readers and researchers in their quest for knowledge and new hypotheses. Of particular interest in the face of this overwhelming mass of data is the use of automated text-mining methods, discussed in the following chapter, for inferring biological relationships and extracting specific details from natural language in order to collapse plain article text into highly specific, information-rich quanta.

## 16.2 Publishing Models and Access to Literature

Any discussion of scientific literature in the twenty-first century must include mention of the current publishing models. The legal concept of copyright did not exist for the *Philosophical Transactions* during the first 45 years of publication (Feather 1980), but today few scientific articles are published without judicious regard to the copyright owner and the applicable licensing terms. Traditionally, in what is considered to be the *closed access* model, publishers hold the copyright for articles published in their journals and charge a usage fee for print or electronic access. However, in the last decade, *open access* literature has been established as a viable, and increasingly more popular, publishing model.

Open access was formally defined in the mid-2000s in the Budapest,[4] Bethesda,[5] and Berlin[6] public statements, collectively known as the BBB definition of open access. Essentially, they state that open access content must be available on the internet and free with regards to access and cost. There is room for interpretation regarding commercial and non-commercial use and whether derivative works are permitted. The practical definition of open access has changed over the years such that not all open access literature is truly open (MacCallum 2007) so licenses must be carefully reviewed.

---

[3]The *Journal de Sçavans* suffered the same fate as some its readers during the French Revolution, but the *Philosophical Transactions* now enjoys the standing of the oldest extant scientific journal.

[4]http://www.soros.org/openaccess/

[5]http://www.earlham.edu/~peters/fos/bethesda.htm

[6]http://oa.mpg.de/openaccess-berlin/berlindeclaration.html

For the average consumer of scientific literature the subtleties are probably unimportant, but for scientists who aim to use literature in a manner beyond that of reading for information, these subtleties can be crucial. RoMEO (http://www.sherpa.ac.uk/romeo/) is a useful resource for exploring copyright and usage policies for individual publishers and journals.

The concept of open access has been a controversial one, engendering much debate among policy-makers, publishers, and researchers. However, recent mandates from the US National Institutes of Health and a number of European funding agencies lend strong support to the open access publishing model, at least as it pertains to publicly-funded research. The NIH mandate, for example, requires that published NIH-funded research is to become open access within a year of publication. In response, many publishers are now offering the option to publish an article as open access. The Directory of Open Access Journals, http://www.doaj.org, provides a complete listing of journals that publish under open access licenses to assist authors in finding these publishers.

## 16.3 Literature Databases

Numerous databases archiving scientific literature exist and they differ based on the needs of the communities they serve. The primary reason most databases exist is to provide a searchable interface to article citations and abstracts in a given field. Bioinformaticians, however, have needs beyond those of the average literature database user. Access to large amounts of data via non-browser-based and batch download processes is desirable. Access to the full text of articles, especially in a machine-readable format, is also of interest. In addition, many bioinformaticians are interested in article metadata – very specific information that has been extracted from the article content. While the following is by no means a complete list, the databases that are likely of most use to bioinformaticians based on these criteria are described here.

### *16.3.1 PubMed: http://www.ncbi.nlm.nih.gov/pubmed/*

PubMed was developed by the National Center for Biotechnology Information (NCBI) at the National Library of Medicine (NLM) and is located at the U.S. National Institutes of Health (NIH) (Benson et al. 1990; McEntyre and Lipman 2001). It currently holds data for over 18 million articles from more than 34,000 journals dating back to the 1950s. It would be difficult to argue that PubMed is not the most popular source for literature data in the life sciences. This resource is freely accessible and indexes the majority of life sciences articles. Most PubMed records are enhanced with human-curated metadata, contain additional links to other resources, and are easily obtained via programmatic access and a web interface.

PubMed archives and indexes the abstracts of articles and provides links from each article to a number of related resources. These can include molecular biology data deposited in GenBank and links to other literature sources such as publishers and the companion database, PubMed Central (PMC) (described later in this chapter), which contains the full text of many of the articles in PubMed. Recent statistics suggest that an impressive number of PubMed users-one third-belong to the general public while the rest are scientific researchers or clinical personnel (McEntyre and Lipman 2001), thus underscoring the ubiquity and value of this resource because it reaches such a broad range of users.

One of the most valuable features offered by PubMed is the inclusion of metadata from MEDLINE, a freely accessible database of biomedical articles also created by the NLM. Articles indexed in MEDLINE are assigned terms from NLM's controlled vocabulary, the Medical Subject Headings (MeSH®), a hierarchical vocabulary of biomedical words and phrases. Table 16.1 shows a typical MEDLINE record from PubMed. Expert curators read each article and manually assign MeSH terms that best capture the content of that article. Assigned terms are then used for searching and indexing purposes. Because these terms are hierarchical, the ascendant-descendant relationship of terms can be used in various ways. PubMed, for example, takes advantage of this relationship during database searches by expanding a user-specified MeSH search term into all descendant terms. MeSH terms are annotated in the MEDLINE record of the article as English words or phrases, but these can be translated into tree numbers which define that term's position in the hierarchy; the entire MeSH vocabulary is available for download and all terms and tree numbers are defined and described therein.[7] Table 16.2 shows a typical MeSH term from the hierarchy with supporting data. There are also translations of MeSH terms into several other languages making these data particularly accessible to non-English-speaking bioinformaticians.

The most common types of PubMed searches appear to be a plain term (e.g., diabetes; hansen) or a term with a specific field identifier such as author (e.g., hansen d[au]) and PubMed ID (e.g., 18185982[pmid]) (see PubMed Help pages for a full description of field identifiers) (Herskovic et al. 2007). Indeed, it is rarely necessary in the course of research to build extremely sophisticated search queries, in part due to the specificity of scientific terminology and the extensive indexing of articles based on MeSH terms, author names, and words from the title and text of the paper. When a complex query is deemed necessary, PubMed allows this query to be saved for future re-use (obviating the need to recreate the logic and syntax anew) and saved queries can be automatically searched and results emailed according to a user-defined schedule. It can take a little practice to become fluent with the interface (PubMed helpfully supplies useful and extensive tutorials) but, once even a few skills are mastered, searches become powerful and direct. Alternately, the Related Articles section is helpful when approaching an unfamiliar topic for which the appropriate keywords are not necessarily obvious.

[7]http://www.nlm.nih.gov/mesh/filelist.html

**Table 16.1** The first table below details the main tags used to describe a journal article in MEDLINE

| Name | Description | Tag |
|---|---|---|
| PubMed Unique Identifier | Unique number assigned to each PubMed citation | PMID |
| Owner | Organization acronym that supplied citation data | OWN |
| Volume | Volume number of the journal | VI |
| Issue | The number of the issue, part, or supplement of the journal in which the article was published | IP |
| Publication Date | The date the article was published | DP |
| Title | The title of the article | TI |
| Pagination | The full pagination of the article | PG |
| Abstract | English language abstract taken directly from the published article | AB |
| Affiliation | Institutional affiliation and address of the first author | AD |
| Full Author Name | Full Author Names | FAU |
| Author | Authors | AU |
| Secondary Source Identifier | Identifies secondary source databanks and accession numbers of molecular sequences discussed in articles, e.g., GenBank, GEO, PubChem, ClinicalTrials.gov, ISRCTN. The field is composed of the source followed by a slash followed by an accession number and can be searched with one or both components | SI |
| Full Journal Title | Full journal title from NLM's cataloging data | JT |
| NLM Unique ID | Unique journal ID in NLM's catalog of books, journals, and audiovisuals | JID |
| EC/RN Number | Number assigned by the Enzyme Commission to designate a particular enzyme or by the Chemical Abstracts Service for Registry Numbers | RN |
| Subset | Journal or citation subset values representing specialized topics | SB |
| MeSH Terms | NLM's Medical Subject Headings (MeSH) controlled vocabulary; major MeSH terms are one of the main topics discussed in the article and these are denoted by an asterisk on the MeSH term or MeSH/Subheading combination | MH |
| PubMed Central unique identifier | Unique number assigned to this article in PubMed Central | PMC |
| Entrez Date | The date the citation was added to PubMed | EDAT |
| MeSH Date | The date MeSH terms were added to the citation. The MeSH date is the same as the Entrez date until MeSH are added | MHDA |
| Article Identifier | Article ID values supplied by the publisher may include the pii (controlled publisher identifier) or doi (digital object identifier) | AID |

(continued)

**Table 16.1** (continued)

| Tag | Content |
|---|---|
| PMID | 17997600 |
| VI | 3 |
| IP | 11 |
| DP | 2007 Nov |
| TI | A point mutation in a herpesvirus polymerase determines neuropathogenicity. |
| AB | Infection with equid herpesvirus type 1... |
| AD | Department of Microbiology and Immunology, Cornell University, Ithaca, New York, United States of America. |
| FAU | Goodman, Laura B |
| AU | Goodman LB |
| FAU | Loregian, Arianna |
| SI | GENBANK/AY665713 |
| SI | PDB/2GV9 |
| SI | SWISSPROT/P28858 |
| SI | SWISSPROT/Q6S6P1 |
| GR | 2T32AI007618-06A1/AI/United States NIAID |
| PT | Journal Article |
| PT | Research Support, N.I.H., Extramural |
| PT | Research Support, Non-U.S. Gov't |
| PL | United States |
| TA | PLoS Pathog |
| JT | PLoS pathogens |
| RN | 0 (Antiviral Agents) |
| RN | 38966-21-1 (Aphidicolin) |
| RN | EC 2.7.7.7 (DNA-Directed DNA Polymerase) |
| SB | IM |
| MH | Amino Acid Sequence |
| MH | Animals |
| MH | Antiviral Agents/pharmacology |
| MH | Aphidicolin/pharmacology |
| MH | Blotting, Western |
| MH | CD4-Positive T-Lymphocytes/virology |
| MH | Chromosomes, Artificial, Bacterial |
| MH | DNA-Directed DNA Polymerase/chemistry/drug effects/*genetics |
| MH | Female |
| MH | Genotype |
| MH | Herpesviridae Infections/pathology/*veterinary |
| MH | Herpesvirus 1, Equid/*enzymology/*genetics/*pathogenicity |
| MH | Horse Diseases/enzymology/*genetics |
| MH | Horses |
| MH | Mice |
| MH | Mice, Inbred BALB C |
| MH | Molecular Sequence Data |
| MH | Point Mutation |
| MH | Reverse Transcriptase Polymerase Chain Reaction |
| MH | Structure-Activity Relationship |

(continued)

Table 16.1 (continued)

| Tag | Content |
|---|---|
| PMC | PMC2065875 |
| EDAT | 11/14/07 9:00 |
| MHDA | 2/1/08 9:00 |
| PHST | 2007/06/11 [received] |
| PHST | 2007/09/17 [accepted] |
| AID | 07-PLPA-RA-0352 [pii] |
| AID | 10.1371/journal.ppat.0030160 [doi] |
| SO | PLoS Pathog. 2007 Nov;3(11):e160 |

Details of all tags can be found at: http://www.ncbi.nlm.nih.gov/books/bv.fcgi?rid=helppubmed.section.pubmedhelp.Search_Field_Descrip. The second table shows part of the MEDLINE record from PubMed for the article "A point mutation in a herpesvirus polymerase determines neuropathogenicity", PLoS Pathog. 2007 Nov;3(11):e160. Many bioinformaticians will be particularly interested in the SI, RN, and MH fields because these contain metadata describing the science in the paper. It is worth noting that the Secondary Source Identifiers are not included for all articles; many IDs may be mentioned in the article text but are not listed in the MEDLINE record.

PubMed offers much functionality beyond the scope of its digital archive. Links to offsite resources are plentiful (if occasionally difficult to find). When applicable, links to the full text version of the article are clearly presented as icons in the AbstractPlus view (default view) of a citation; these may include links to freely available full text in PMC and a link supplied by the publisher which resolves to the full text if the user or user's institution has a license for that content. The LinkOut feature, currently buried within the "Links" link to the right of the full text icons or as an option in the drop-down "Display" menu, can provide a path to additional information such as databases associated with the article content. To date, there are over 250 external resource links that are not related to publisher websites so it is worthwhile exploring the LinkOut list on articles of particular interest.

PubMed can be accessed via a web interface with sophisticated searching tools and data persistence features and via programmatic access in order to retrieve large sets of articles or to automate article retrieval via a separate web resource. PubMed offers E-utilities which provide programmatic access through either SOAP or HTTP protocols (see Fig. 16.1 for some examples). The data formats that are likely to be of particular interest in this context are the XML (eXtensible Markup Language) and MEDLINE records for each article. NCBI provides excellent help for the E-utilities on their help web pages.

### *16.3.2 PubMed Central: http://www.pubmedcentral.gov/*

PMC was founded in 2000 as an outgrowth of the open access movement. It is a digital archive which houses full text, open access literature in the life sciences. Developed and managed by NCBI, PMC benefits from strong integration with PubMed. Although the majority of PMC articles are also indexed in PubMed, it is

**Table 16.2** Typical MeSH term

| MeSH Tag Name | MeSH Tag | Content |
|---|---|---|
| RECORD TYPE | RECTYPE | D |
| MESH HEADING | MH | Mammals |
| ALLOWABLE TOPICAL QUALIFIERS | AQ | AB AH BL CF CL EM GD GE IM IN ME MI PH PS PX SU UR VI |
| ENTRY TERM, PRINT | ENTRY | Mammal |
| MESH TREE NUMBER | MN | B01.150.900.649 |
| MESH HEADING THESAURUS ID | MH_TH | NLM (1966) |
| SEMANTIC TYPE | ST | T015 |
| ANNOTATION | AN | avoid: too general: prefer specifics |
| PRE-EXPLOSION | PX | Mammals |
| GRATEFUL MED NOTE | GM | very general; consider specific mammals or + for all |
| MESH SCOPE NOTE | MS | Warm-blooded vertebrate animals belonging to the class Mammalia, including all that possess hair and suckle their young. It includes three major groups: placentals and marsupials, which are viviparous, and monotremes, which are oviparous. (Dorland, 28th ed.) |
| ONLINE NOTE | OL | pre-explosion |
| BACKFILE POSTINGS | MED | *419 |
| | MED | 2089 |
| | M90 | *557 |
| | M90 | 1745 |
| | M85 | *566 |
| | M85 | 1194 |
| | M80 | *420 |
| | M80 | 933 |
| | M75 | *381 |
| | M75 | 982 |
| | M66 | *578 |
| | M66 | 1812 |
| | M94 | *580 |
| | M94 | 2965 |
| MAJOR REVISION DATE | MR | 20040728 |
| DATE OF ENTRY | DA | 19990101 |
| DESCRIPTOR CLASS | DC | 1 |
| UNIQUE IDENTIFIER | UI | D008322 |

Medical Subject Headings listed in the MEDLINE record (MH tag) describe the content of the article. These terms are part of a hierarchical controlled vocabulary. When MeSH terms are used as search term on the PubMed website, they are automatically expanded to include all child terms. However, if MEDLINE records are being used independently of PubMed, a user must translate MeSH terms into MeSH tree numbers (MN tag) to be able to navigate through the hierarchy.

| Parameter Description | Database | Record Identifier | Retrieval Mode | Retrieval Type |
|---|---|---|---|---|
| Parameter | db | id | retmode | rettype |
| Options | pubmed | (PubMed ID) | xml | uilist<br>abstract |
| | | | html | uilist<br>abstract<br>citation<br>medline |
| | | | text | uilist<br>abstract<br>citation<br>medline |
| | | | asn.1 | |
| | pmc | (PubMed Central ID) | xml | uilist<br>abstract<br>citation<br>medline<br>full |
| | | | html | uilist |
| | | | text | uilist |

**Fig. 16.1** Using EFetch to retrieve information from PubMed and PMC. The NCBI E-utilities have several tools for accessing content in any of their supported databases. For retrieving literature data, the tool EFetch support the majority of queries. EFetch accepts PubMed IDs and PMC IDs and retrieves the corresponding records, which can then be parsed or archived as necessary. The PubMed IDs and PMC IDs, if not known, can be searched for using ESearch (not detailed here). EFetch supports a number of parameters, detailed below, which can be appended to the Base URL: http://eutils.ncbi.nlm.nih.gov/entrez/eutils/efetch.fcgi?

not strictly a subset of PubMed; there are many records that are unique to PMC. All content in this archive is accessible in electronic format with no restrictions. As mentioned previously, not all open access licenses grant the same freedom of re-use of an article so it is recommended that users read the copyright statement for each article if re-use is intended. However, nearly all full text articles with scientific content in PMC can be used in excerpted form.

**Table 16.2** (continued) For example, a researcher may be interested in all articles related to mammals. The article example shown in Table 16.1a relates to mammals (the mouse, in particular), but a machine cannot make this translation without knowledge of the vocabulary hierarchy. The entire MeSH tree is downloadable from the National Library of Medicine as both XML and ASCII files and the relationships between terms can be inferred from this tree. The ASCII MeSH tree record for "Mammals" is shown below. The term "Mice" from the MEDLINE record of PLoS Pathog. 2007 Nov;3(11):e160 can be translated to the tree number B01.150.900.649.865.635.505.500. This number can then be related to the term "Mammals" (B01.150.900.649) via the MeSH tree numbering system. The MeSH tree can also be used to translate terms into descriptions and synonyms.

In a display of remarkable foresight, the NLM developed a document type definition (DTD) to which all PMC articles conform[8]; full text articles are available as downloadable, machine-readable XML files. It is interesting to note that no other major literature database provides full text XML content, even in other disciplines which are more computationally-oriented; bioinformaticians are the fortunate beneficiaries of this development. While the freedom allowed by open access is helpful to a reader and consumer of scientific literature, the far more impressive benefit is the availability of a large corpus of parsable literature data for use by text miners. Many text-mining efforts use PubMed abstracts but are limited by the inherent incompleteness of the information. Text-mining full text articles promises a greater yield of information although there are a number of other associated challenges. Not all publishers use the DTD in the same way so, in spite of the attempt to standardize article mark-up, there are occasionally inconsistencies and missing data.

Full text XML files are downloadable via FTP as individual files or as a large archive of all of PMC's holdings. PMC content can also be accessed via NCBI's E-utilities. If using the E-utilities, it is important to note that PMC unique identifiers are different from PubMed unique identifiers for any given article although they are both digits starting in the 10,000s. Specifying the correct database in the query string is necessary. For many PMC articles, MeSH terms are available via the MEDLINE record.

Recently, a UK version of PMC was implemented, UK PMC (UKPMC, http://ukpmc.ac.uk/), which is based on the US version. It is run in partnership by the British Library, the University of Manchester, and the European Bioinformatics Institute (EBI), the latter of which aims to manage integration of the literature content with EBI databases.[9]

#### 16.3.2.1 PubMed Central Open Archives Service

In addition the NCBI's E-utilities, PMC data can be accessed via PubMed Central Open Archives (PMC-OAI). This service provides access to the metadata of all items in the PMC archive and to the full text of a subset of these items via the base URL http://www.pubmedcentral.nih.gov/oai/oai.cgi; full text articles cannot be retrieved via E-utilities. Help on using this service is available on PMC's website. PMC-OAI is an implementation of the Open Archives Initiative Protocol for Metadata Harvesting, a standard for retrieving metadata from digital archives.[10]

---

[8] http://dtd.nlm.nih.gov/publishing/

[9] http://www.bl.uk/news/2007/pressrelease20070105.html

[10] http://www.openarchives.org/OAI/2.0/openarchivesprotocol.htm

### 16.3.3 HighWire Press: http://highwire.stanford.edu/

HighWire Press, a division of the Stanford University Libraries, hosts nearly 5 million full text articles dating back to 1812 covering many science and social science topics. Almost half are freely available making it the largest free, full text archive available. Many of these articles are already available via PMC, a publisher's website, or other content providers; HighWire Press provides a convenient, unified interface for searching for this content, although programmatic access is not available. RSS feeds and email alerts are available.

### 16.3.4 DRIVER Project: http://www.driver-community.eu/

The "Digital Repository Infrastructure Vision for European Research" (DRIVER) project was initiated by an international consortium with the specific purpose of building infrastructure to manage and serve scientific information to European countries (Weenink et al, 2008). Although the project is funded by the European Commission, the project is not geographically exclusive; any willing institution can participate and several countries, such as India and China, are now involved. Much like the directives that launched the creation of PMC and PMC UK, DRIVER aims to make scientific information freely available via an open, standardized, Internet-based interface. The archived information includes articles, reports, data, and any other type of digital media that can be served over the Internet. The overall aim is to provide a unified interface through which a user can search or browse content, while allowing the information to be remotely hosted by many institutions rather than archiving the information in a single institution.

The content available through the DRIVER interface is too abundant to describe in detail due to the number and variety of source institutions, but there is a great deal of content that cannot be found in other databases and would be relevant to a bioinformatician. As such, the DRIVER repositories are an excellent complement to PubMed and PMC. As the initiative is fairly new, but has strong support, the content can only be expected to grow. The interface is freely available for use and users can create a personal profile which with to store search types of interest. For managers or curators of digital repositories, it is worthwhile considering becoming a DRIVER participant in order to gain a broader audience.

### 16.3.5 Web of Science®: http://isiknowledge.com

The Web of Science indexes a large number of articles – nearly one million, from about 8,700 journals – and has a powerful search interface. The corpus extends to the nineteenth century, covers the sciences and the arts, and includes access to some full text articles. The Web of Science is the resource of choice for statistics about the lives of articles after they are published. Scientists' impact on their field is a frequently-used criterion when evaluating prospective employees, principal investigators, and collaborators. One measure for evaluating impact is to track how often an author's papers are cited in subsequent publications, the idea being that frequently-cited articles indicate a greater contribution. Indeed, even in the first issues of the *Journal de Sçavans* and the *Philosophical Transactions of the Royal Society of London* the journals cited each other, referring to articles of particular merit or interest. Tracking citations and generating a quantitative measure of an author's merit is nearly becoming a science in its own right and the Web of Science provides data for calculating this measure by showing the number of times an article has been cited by other articles. (Whether this measure is useful or accurately reflects the impact of an article or author is a separate debate.) The Web of Science can similarly be used to trace the influences that led to the publication of an article or chart the course of an author's career; a user can navigate through cited references of a list of articles. These data are available as XML via web services. The Web of Science is a subscription-based service but is available at many major institutions.

### *16.3.6 arXiv: http://www.arxiv.org*

Pioneered by Paul Ginsparg of Cornell University, arXiv (think of the X as the Greek letter chi) is a digital repository and pre-print server with over half a million open access articles in physics, mathematics, computer science, quantitative biology, and statistics. arXiv is maintained by Cornell University and offers unrestricted access to its content. This resource is popular with the more quantitative branches of science and is virtually unheard of in the life sciences. It is an excellent resource for open access literature; articles are available in full text in formats including PDF, Postscript, DVI, and TeX. Generally, article pre-prints are deposited in arXiv and notes are added with citation information if/when the article is officially published in a journal. arXiv is affirming its position as an essential literature database by keeping pace with current evolving data standards; submissions are accepted in the newly minted .docx/OOXML format.

### *16.3.7 CiteSeer: http://citeseer.ist.psu.edu*

CiteSeer archives articles primarily in computer and information sciences and is a crucial resource in these fields (again, it is a resource virtually unknown in the life sciences). Originally developed at the NEC Research Institute, it is now hosted at Pennsylvania State University's College of Information Sciences and Technology. CiteSeer archives full-text articles as Postscript and PDF files and is available without restriction.

In addition to providing the traditional search interface, CiteSeer makes extensive use of the citations listed in articles. In fact, CiteSeer was the first digital archive to provide automated citation indexing and linking using the method of autonomous citation indexing (ACI). CiteSeer finds articles in electronic format on the Internet, converts them to text, indexes the full text of these articles (including citations), automatically extracts citation data, and establishes relationships between citations and articles. Using these relationships and the access to full article text, CiteSeer can show the context of a citation in a selected article in order to show the information the authors referred to. CiteSeer also provides links to other resources including computer science-related digital libraries whenever possible. CiteSeer also automatically extracts metadata from all indexed articles and these data can be downloaded as XML from one large archive. The full source code of the database is available for download (for non-commercial use only). Like PMC, CiteSeer supports OAI-PMH version 2.0.

### *16.3.8 Other Resources*

Identifying, relating, and labeling specific key concepts and terms in an electronic article text is the first step towards integrating article content into the semantic web (Berners-Lee et al. 2001; Berners-Lee et al. 2006), the next revolution in scientific publishing. Some publishers and researchers are joining this revolution and are beginning to develop electronic article text with semantic enrichment. The Royal Society of Chemistry (RSC) has developed the award-winning Project Prospect[11] which delivers articles enhanced with semantic information. Using existing ontologies and mark-up standards, RSC editors use both automated and manual methods to identify and label chemical compounds and related information in articles from all RSC-published journals. Enhanced articles are available on the publisher's website and as XML via RSS feeds.

The BioLit project[12] similarly delivers semantically enhanced full text article content, including markup of bio-ontological terms and biological database identifiers for all research articles in PMC; all mark-up is automatically generated, not human-curated (Fink et al. 2008). GoPubMed also uses ontology terms (from the Gene Ontology (Ashburner et al. 2000)) extracted from PubMed abstracts in order to generate a list of literature search results with higher relevance than an equivalent search of PubMed (Doms and Schroeder 2005). Search results are categorized by ontology terms to allow quick navigation through the returned abstracts.

Recognizing the need for programmatic access to, and use of, closed literature, Nature Publishing Group has developed the Open Text Mining Initiative (OTMI),[13] allowing content providers to translate article text into a machine-readable, but

[11] http://www.rsc.org/Publishing/Journals/ProjectProspect/

[12] http://biolit.ucsd.edu

[13] http://opentextmining.org/

non-human-readable, format in order to facilitate indexing and text-mining without giving away the full material. Articles translated into OTMI files will not necessarily be available for enhanced views of the full text, like those offered by the RSC and BioLit projects, but any enhanced or extracted data can be used for other purposes.

Finally, another resource that is an excellent complement to the described literature databases and resources is SciVee[14] (Fink and Bourne 2007). SciVee is an online science video community tool aimed at improving scientific communication via new media. With the development of cyber infrastructure, scientists have the opportunity to take advantage of media beyond that of static images and text in order to share their results and experiences. SciVee allows authors to upload a video associated with a published paper in which they describe the content of the paper. If the paper is open access, authors can link their video directly to the text of the paper in order to highlight specific points or figures. Use of video in this way gives the author the opportunity to describe aspects of the research that do not generally get included in articles or to provide more extensive explanations of difficult points. Video can bring the paper to life in a way that is not possible with the traditional article format. SciVee accepts videos from any published paper, including from the databases described here. It is well-integrated with PubMed and PMC and is listed in the PubMed LinkOut feature for applicable papers.

## 16.4 Conclusion

Retrieving literature data, especially full text articles, is frequently only the first step in the process of extracting literature-derived information. More sophisticated tools are frequently necessary to identify relevant sections of the text and transform that information into knowledge. The following chapter describes some existing literature-mining tools that can be employed in this task.

## References

Ashburner M et al (2000) Gene ontology: tool for the unification of biology. The Gene Ontology Consortium. Nat Genet 25:25–29

Benson D et al (1990) The National Center for Biotechnology Information. Genomics 6:389–391

Berners-Lee T et al (2006) Computer science. Creating a science of the Web. Science 313:769–771

Berners-Lee T et al (2001) The Semantic Web. Scientific American Magazine

Boyle R (1665) An account of a very odd monstrous calf. Phil Trans Royal Soc Lond 1:10

[14] http://www.scivee.tv

Doms A, Schroeder M (2005) GoPubMed: exploring PubMed with the gene ontology. Nucleic Acids Res 33:W783–786

Eisenstein E (1979) The printing press as an agent of change: communications and cultural transformations in early-modern Europe. Cambridge University Press, Cambridge, England

Feather J (1980) The Book Trade in Politics: The Making of the Copyright Act of 1710. Publishing His 8:19–44

Fink L, Bourne P (2007) Reinventing scholarly communication for the electronic age. CTWatch Q 3:26–31

Fink J et al. (2008) BioLit: Integrating biological literature with databases. Nucleic Acids Res 36:W385–W389

Herskovic JR et al (2007) A day in the life of PubMed: analysis of a typical day's query log. J Am Med Inform Assoc 14:212–220

MacCallum CJ (2007) When is open access not open access? PLoS Biol 5:2095–2097

McEntyre J, Lipman D (2001) PubMed: bridging the information gap. CMAJ 164:1317–1319

Vandenhaute B et al (2000) Epignathus teratoma: report of three cases with a review of the literature. Cleft Palate Craniofac J 37:83–91

Weenink K, Waaijers L, van Godtsenhoven K (2008) A DRIVER's Guide to European Repositories. Amsterdam University Press, Amsterdam

Yuan G et al (2007) Hypertriglyceridemia: its etiology, effects and treatment. CMAJ 176:1113–1120

Some anatomical observations of milk found in veins, instead of blood; and of grass, found in the wind-pipes of some animals. Phil Trans Royal Soc Lond 1:100–101

# Chapter 17
# Advanced Literature-Mining Tools

**Pierre Zweigenbaum and Dina Demner-Fushman**

## 17.1 Introduction

The complexity and wide range of current biomedical research is reflected in the number and scope of biomedical publications. Due to this abundance scientists are often no longer capable of keeping up with publications in their specific areas of research, let alone finding, reading, and analyzing potentially related scientific publications. Real advances in research, however, can be achieved only if a researcher can obtain an overview of the state of a given research question in a timely manner. This chapter presents methods to help researchers access the content of the biomedical literature. *Information Retrieval* (IR) identifies, in a large document database, the documents that are most relevant to a search topic provided by a user. *Natural Language Processing* (NLP) affords finer-grained access to more precise information contained in texts, which opens up a range of data analysis and knowledge synthesis functionalities. Powerful tools have been designed to exploit these techniques for the benefit of biomedical researchers, extracting millions of facts from the published literature and assisting *Literature-Based Discovery*.

This chapter is organized as follows. It first describes the current capacities of IR from the Medline® bibliographic database. A short introduction to the main concepts of Natural Language Processing follows. Tasks which build on Natural Language Processing are then presented: Information Extraction and its derivatives and Literature-Based Discovery. A review of some existing applications closes the chapter. The references cited in the text are supplemented by a list of textbooks and Web resources.

P. Zweigenbaum (✉)
LIMSI-CNRS, BP 133, 91403, Orsay Cedex, France
e-mail: pz@limsi.fr

D. Demner-Fushman
N ational Library of Medicine, Bethesda, Maryland, USA

D. Edwards et al. (eds.), *Bioinformatics: Tools and Applications*,
DOI 10.1007/978-0-387-92738-1_17, © Springer Science+Business Media, LLC 2009

## 17.2 Information Retrieval from the Biomedical Literature

Automated IR systems were developed to reduce the information overload by retrieving documents related to a user's request and thus reducing the number of publications to read and analyze. Biomedical researchers were among the first scientists to benefit from the availability of such systems. Medline, a database of life sciences and biomedical bibliographic information maintained by the U.S. National Library of Medicine, became available online in the 1970s (Dee 2007). In 1997, searching over 11 million documents in Medline using the Internet and PubMed® services became free. Owing to a sustained expansion of the published scientific literature, over 17 million Medline records were accessible in 2009.

### *17.2.1 PubMed Search*

PubMed is a Boolean search engine that indexes titles of scientific publications, their abstracts and metadata separately (Miles 1992). These indices allow users to specify which fields of the Medline bibliographic citations should be searched. PubMed draws on the Medical Subject Headings (MeSH), the controlled vocabulary used in manual indexing of Medline. PubMed automatically recognizes and translates controlled vocabulary terms, and expands identified MeSH headings. The automatic term mapping process matches query terms not tagged by a user against the entries in the following tables/indexes:

- MeSH Translation Table (contains MeSH terms, entry terms for MeSH terms, MeSH Subheadings, Publication Types, Pharmacologic action terms, Terms derived from the UMLS (Unified Medical Language Systems (Lindberg et al. 1993)) that have equivalent synonyms or lexical variants in English, Supplementary concept (substance) names and their synonyms)
- Journals Translation Table (contains Full journal title, Medline abbreviation, ISSN)
- Full Author translation table, Author index, the Full Investigator (Collaborator) translation table and an Investigator (Collaborator) index

When a match is found for a term or phrase in a translation table, the mapping process is complete and does not continue on to the next translation table. If a match is found in the MeSH Translation Table, the term will be searched as MeSH (that includes the MeSH term and any specific terms indented under that term in the MeSH hierarchy), and as a Text Word.

Owing to PubMed origins and the intent to serve a wide variety of users, ranging from general public to highly-specialized biomedical researchers and clinicians, PubMed searches produce best results when built by experienced medical librarians with intimate knowledge of Medline structure and indexing, knowledge of the MeSH structure, and knowledge of PubMed tags and Boolean operators. For example, to answer the question: What genes are induced by LPS in diabetic mice? an expert PubMed user constructed the following query:

(lipopolysaccharides OR lps) AND diabetes mellitus[mh] AND mice[mh] AND (gene OR genes OR ge[sh]) AND (free full text[sb]).

The query was then automatically translated in PubMed to:

((“lipopolysaccharides”[MeSH Terms] OR lipopolysaccharides[Text Word]) OR lps[All Fields]) AND “diabetes mellitus”[MeSH Terms] AND “mice”[MeSH Terms] AND (((“genes”[TIAB] NOT Medline[SB]) OR “genes”[MeSH Terms] OR gene[Text Word]) OR (“genes”[MeSH Terms] OR genes[Text Word]) OR “genetics”[Subheading]) AND “loattrfree full text”[sb] (Demner-Fushman et al. 2007)

The expert’s query illustrates her knowledge of the domain terminology (expansion of LPS to lipopolysaccharides) as well as her knowledge of Medline and PubMed: the expert requests looking up lipopolysaccharides in all PubMed indices, equates diabetic with diabetes mellitus and requests only the MeSH index look-up for this term (to avoid spurious matches in publications which do not focus on the disease but mention it in the abstract). This command is encoded using the [mh] tag.

### 17.2.2 *Specialized Biomedical Literature Retrieval Systems*

Clearly, not all users posses the knowledge of the expert librarian, or have access to services provided by such experts. This understanding led to search for IR algorithms capable of taking over the burden of query formulation and gave rise to quite a few *specialized biomedical literature retrieval systems*. One of the first systems, SAPHIRE (Hersh and Greenes 1990), allowed natural language to be used for query input through finding medical concepts in its text and converting them to canonical form. The observation that query terms are often conceptually related to terms in a document, but do not occur in the document text, motivated development of Essie (Ide et al. 2007), a phrase-based search engine with term and concept query expansion and probabilistic relevancy ranking. An approach alternative to development of a specialized search engine is to modify existing open source software. For example, MedSearch (Hliaoutakis et al. 2006) supports semantic retrieval of Medline citations using an open source search engine Lucene[1] as a base. The approach of modifying general purpose search engines or using them “as is” was demonstrated to be effective in several large-scale evaluations of biomedical text retrieval in the Genomics track within the Text Retrieval Conference evaluations (Hersh et al. 2007). Yet another approach, implemented in the eTBLAST system (Lewis et al. 2006), is based on the notion that it is much easier for a user to provide a relevant sample document than to employ the best search terms.

[1] http://lucene.apache.org/

### *17.2.3 Clustering Search Results*

Despite the differences in the expected queries and the internal processing of the queries, the commonality of the above search engines is in presenting a ranked list of documents as a result of searching one data source. Searching over multiple databases offers the potential for greater query power. In this model, a user is presented with a single query interface, which connects to federated knowledge sources. An example of such integrated text-based search and retrieval system is *Entrez*[2]. Realizing that even the best ranked list does not provide an overview of available information many developers focus on the organization of search results. Clustering and categorization were proposed to assist the exploration of search results. Figure 17.1 presents clusters generated for the *LPS in diabetic mice* query using HubMed (Eaton 2006), an alternative search interface to the Medline based on the PubMed web services API[3] and Vivisimo[4]. The idea behind clustering is to organize publications into groups by their similarity, for example, the number of words in common.

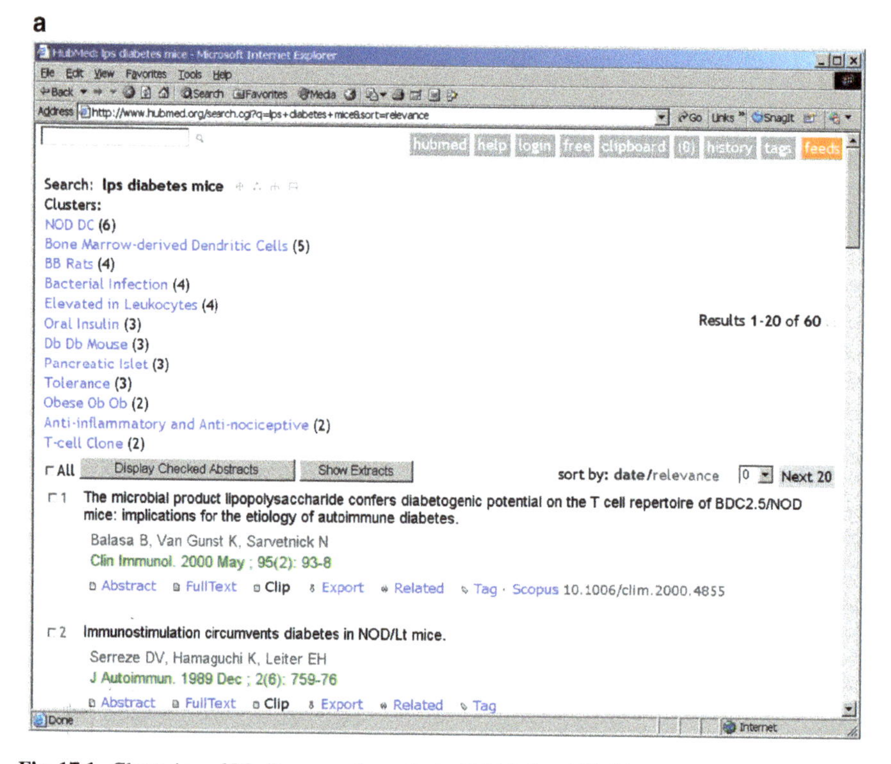


**Fig. 17.1** Clustering of Medline search results in HubMed and Vivisimo

[2] http://www.ncbi.nlm.nih.gov/sites/entrez?db=pubmed

[3] http://www.ncbi.nlm.nih.gov/entrez/query/static/eutils_help.html

[4] http://vivisimo.com/

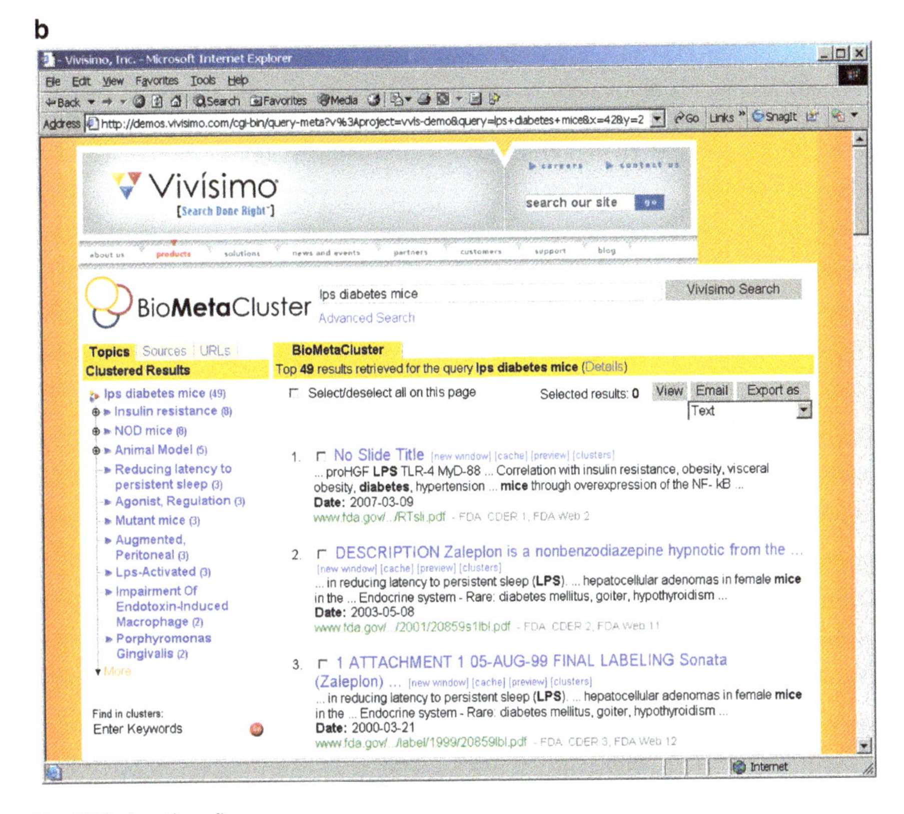


**Fig. 17.1** (continued)

The clusters will not be very useful in exploration of the information landscape without a concise representation of their contents. One of the widely used techniques is to describe cluster content using most frequently occurring key words. As these labels are not always meaningful (e.g., *Augmented, Peritoneal* or *Tolerance* in Fig. 17.1), dynamic categorization of search results is used to ensure the labels are valid biomedical domain concepts (Pratt et al. 1999).

The above techniques exhaust the possibilities of presenting available information without applying Natural Language Processing and using domain knowledge. Techniques and tools used to assist with information analysis and synthesis are presented in the next sections of the chapter.

## 17.3 A Brief Introduction to Natural Language Processing

For a computer program a text is simply a long string of characters. The goal of natural language processing is to design methods and programs that make the most of information conveyed by texts, ideally as well as humans do, seeing words and sentences and meaning in character strings. This section introduces the main concepts of natural language processing, with a focus on processing English biomedical texts. It is necessarily too dense and cursory; the interested reader is referred to handbooks on natural language processing for more detail (e.g., Mitkov 2003; Jurafsky and Martin 2000; Jackson and Moulinier 2002).

### 17.3.1 Lexical Level Processing

The *character* is the smallest unit in a (written) text. Early computer programs could only handle characters in the ascii *character set*, which is suitable for writing English but nearly no other language. More modern character sets include iso-latin-1, which contains letters with diacritics used in Western European languages (such as the cedilla in *Behçet*), and Unicode (e.g., the utf-8 encoding), which includes virtually all writing systems in use in the world today (The Unicode Standard 2007). A major advantage of Unicode is its ability to handle different writing systems in the same text, e.g., both *Behçet's syndrome* (with cedilla) and *TNF-α* or *NF-κB protein* (with Greek alpha and kappa). For instance, Medline uses the Unicode utf-8 encoding.

Sequences of characters delimited by spaces and various punctuation marks make up the elementary *tokens* of a text. The basic linguistic unit in an English text is the *word*. Whereas most tokens are words by themselves, some words are composed of several tokens (e.g., three tokens in *tug of war* or two in *TNF-α*). Words can vary in form through *inflection* (e.g., to mark plural as in *study/studies* or tense as in *investigate/investigated*). A list of known words can be compiled into a *lexicon*, together with their properties, e.g., *part-of-speech* (noun, verb, adjective, etc.) and *morphosyntactic features* (number, tense, etc.). A lexicon may additionally relate inflected forms (*studies*) to base forms (*study*), also called *lemmas*.

*Tokenization* is the process of breaking up a text into tokens. Tokenization generally simply relies on character properties (alphabetic, numeric, space, punctuation, etc.) to identify token boundaries. *Word segmentation* identifies word boundaries in a string or stream of tokens. It may need a lexicon to detect multi-token words. *Lemmatization* is the process of finding the base form (lemma) of an inflected form, generally to use it as a normalized form of the word. It may rely on a lexicon, but can sometimes resort to simple heuristics (e.g., replace final *-ies* with *-y*).

Some words are made up of smaller units called *morphemes*. For instance, *immunosuppressive* is composed of three morphemes: *immune, suppress* and the adjectival suffix *-ive*. The addition of a prefix or suffix to a base word builds derived words, whereas the combination of modern words (or Greek and Latin roots) builds compound words (or neoclassical compounds). Normalizing a derived word to its base may be convenient in many instances (e.g., from *intraperitoneal* to *peritoneum*). *Stemming* is an approximate, robust method to perform this task (Porter 1980). Stemmers generally only deal with suffixes, not with prefixes. A *morphological analyzer* may go further and decompose a derived or compound word into all its components (Namer and Baud 2007).

### 17.3.2 Terms and Concepts

Let us step aside for a second from this enumeration of linguistic units and consider the domain in which we want to represent information. This information may pertain to the *concepts* of the domain (diseases, genes, therapies, etc.) and their relations. *Terms* are the way concepts are expressed in a language. A term may consist of one or several words (*necrosis*, *tumor*, *macrophage*; *synovial fluid*, *inhibitory potential*, *myocardial infarction*, *aminoacyl-tRNA synthetase multienzyme complex*). A list of known terms may be compiled into a *terminology*, together with their concept identifiers, synonyms, broader and narrower terms. Given a terminology, *automatic term recognition* (ATR), or *controlled indexing*, is the process of identifying its terms in a text, possibly under variant forms (e.g., *myocardial infarction*, *cardiac infarction*, *infarction of heart*) (Jacquemin 2001; Aronson 2001). Relevant information may also pertain to individual *entities* in the domain, such as a given hospital, organization, person, or location. Reference to such entities in a text typically uses proper names (*Lhassa*, *Pasteur*, etc.) and more generally what is called *named entities* in the natural language processing literature (*Saint Jude Children's Research Hospital*, *Dr. Singh*, etc.). In a specialized domain, the distinction between terms and named entities may be somewhat blurred: does *ICAM-1* refer to a unique entity or to a concept? It looks like a named entity but has the function of a term (*intercellular adhesion molecule-1*). We return to named entities and their recognition in texts below in Sect. 17.4. The concepts and relations of a domain may be organized in a formal representation called an *ontology* (Bodenreider 2008) which allows programs to draw inferences, including computing whether a concept is more specific or more general than another concept.

### 17.3.3 Syntactic Level Processing

Words are assembled into phrases and sentences which are structured by the *syntax* of the language. Syntax rules the form of terms, most of which are *noun phrases* (or *NPs*); for instance, *malignant neoplasm of myocardium* follows the general rules for English NP formation (adjective *malignant* modifies noun *neoplasm*, which has a complement introduced by preposition *of*). Syntactic structure is often represented as a tree where a sentence is recursively broken into smaller constituents (clauses, verb phrases, noun phrases, prepositional phrases, etc.: *constituent structure*) or as a tree where words are linked to their dependents through syntactic relations (*dependency structure*). It is often difficult for a program to determine the correct syntactic structure of a sentence because of the presence of ambiguity: a given word form may be assigned multiple parts-of-speech (e.g., *controls* may be a verb or a noun) and a given sequence of words may often be assigned multiple syntactic structures. For instance, consider the reasoning necessary to decide that *and TNF-$\alpha$ release* is coordinated with *elevated levels of TNF-$\alpha$* rather than with *disease* in sentence (1):

*(1) Tumor necrosis factor alpha (TNF-α) is believed to play a significant role in the pathogenesis of dengue virus (DV) infection, with elevated levels of TNF-α in the sera of DV-infected patients paralleling the severity of disease and TNF-α release being coincident with the peak of DV production from infected monocyte-derived macrophages (MDM) in vitro.*

As a first step to compute the syntactic structure of a sentence, *part-of-speech tagging* is the task which aims to determine the correct part-of-speech of each word, in context, in a sentence. Most POS-taggers apply machine-learning methods (e.g., Hidden Markov Models) which are trained on human-annotated texts. A number of taggers are freely available, some of which have been specifically tuned for biomedical text (Smith et al. 2004).

Two classes of methods can be used to determine syntactic constituents and relations. Shallow, partial parsing methods perform a local analysis of sentence fragments (often called *chunks*), often based on *regular grammars*. Such grammars describe the (non-recursive) patterns which define these structures. They have limited expressiveness but are sufficient to identify many noun phrases and related constructs, and can be applied very efficiently to texts to locate these structures. They can be described with rules, regular expressions, or finite-state automata. Fuller parsing methods rely on more complete and expressive grammars, from context-free grammars to the more powerful unification grammars such as HPSG (head-driven phrase-structure grammar), which have a higher complexity but for which efficient algorithms have been designed (Miyao et al. 2006). Semantic grammars consist of syntactic rules which directly include semantic constraints, and are particularly suited to the analysis of sublanguages (Friedman et al. 2001).

### *17.3.4 Semantics and Discourse*

Syntactic structure includes the *grammatical relations* (subject, object, etc.) between verbs (more generally, predicates) and their arguments (noun phrases, verb phrases, etc.). *Thematic relations* (or *semantic roles*) express the meaning that an argument plays with respect to the action or state described by its predicate: *agent, patient, instrument, locative, temporal, manner, cause*, etc. They provide a more precise representation of the information content of a sentence (who does what to whom etc.), which is particularly suitable for relation extraction (see below Sect. 17.4). *Semantic role labeling* is the task which aims to recognize the semantic role of each argument (Carreras and Màrquez 2005).

*Discourse* is the next linguistic level; it is made of several sentences. Discourse structure represents the links between the events described in these sentences. Several expressions referring to the same given entity or event may be encountered in a discourse (coreferences). *Anaphora resolution* is the identification of the link between an expression (pronoun, noun phrase) and the antecedent with which it corefers in the same discourse (see e.g., Branco et al. 2005). An example of coreference is shown below in the section, Question-answering and summarization.

### 17.3.5 *Data Driven Analysis*

NLP methods, as artificial intelligence methods, typically break into two components: an algorithm, e.g., a parser, and a knowledge base, e.g., a grammar, which is crucial for the algorithm to perform its task. The knowledge base can be written by hand, relying on expert knowledge: this was the case for most NLP systems until the mid-nineties. It can also be learned from annotated data (even sometimes from un-annotated data), using machine-learning methods. When enough training data is available, machine-learning methods often meet or outperform expert-based methods. This is the case for part-of-speech tagging, which can be viewed as a *categorization* problem, where a supervised learning method is trained on a corpus where each word has been hand-tagged with its correct part-of-speech (Smith et al. 2004). A *language model* encodes the probabilities of encountering a sequence of tokens in a language. High-level grammatical formalisms can also be the support of machine-learning methods, as in the Enju parser (Miyao et al. 2006) where a probabilistic HPSG grammar is learned and then used to parse Medline abstracts. The proper assignment of predicate-argument relations (Surdeanu et al. 2003) or of semantic roles (Carreras and Màrquez 2005) based on a syntactic tree may be learned from a corpus where semantic roles have been manually annotated.

## 17.4 Information Extraction

Texts such as news articles or scientific papers are meant to convey information or knowledge to a reader. *Information Extraction* is the natural language processing task which consists in recovering predefined, interesting pieces of information from a collection of texts, typically to fill a database. This can be broken down into two main subtasks: recognizing entities of interest, *named entity recognition* (NER) and recognizing the relations described between these entities (*relation extraction:* who does what). A third, intermediate subtask (*entity normalization*) consists in resolving the found entities to unique identifiers (identifiers in a database or concepts in an ontology). These three subtasks can be compared to a typical curation pipeline for model organism databases (Morgan and Hirschman 2007).

To determine whether current text mining approaches are successful and practical in predicting interactions and accomplishing tasks needed for a successful prediction pipeline (entity recognition and normalization, etc.), the Second Critical Assessment for Information Extraction in Biology challenge (BioCreAtIvE) was held in 2006–2007 (Hirschman 2007). The best systems combined machine learning with manually derived rules to find articles for curation, find relevant genes and proteins, and identify specific interactions. The next challenge in accomplishing the above tasks is a system that will be easy and intuitive to use “out of the box” (Hirschman 2007).

### *17.4.1 Named Entity Recognition*

Entities and relations depend on the domain and the focus of the task at hand. Information Extraction was first introduced in the context of the MUC series of DARPA-sponsored challenges (Grishman and Sundheim 1996), where entities and relations of interest have ranged from those in terrorist events (who was the perpetrator, human target, etc.) to those in joint ventures (which companies created which new company to produce what) as reported in newswire or newspaper articles. Interesting entities in these domains occurred in the texts as proper names (names of persons, organizations, locations, etc.), numeric expressions (quantities, monetary amounts) and dates. Because of the importance of proper names in these domains, they have been called *named entities*.

Interesting entities in the biomedical domain include genes, proteins, diseases, drugs, body parts, etc. Although many of these are expressed by terms (*polyostotic fibrous dysplasia*) rather than proper names (*McCune-Albright*), the term *named entity* is also applied to them. NER is the task which consists in spotting all occurrences of such entities in a text (Ananiadou et al. 2004). A related task was introduced above: ATR aims at detecting all occurrences of the terms of a terminology in a text. In the remainder of the paper, we use the term NER indifferently for both these tasks. Below is an example passage where human annotators marked the boundaries of named entities within square brackets and additionally tagged them with a semantic type (shown within angle brackets after the entity).

> A MOST INTRIGUING EXAMPLE of [gain-of-function mutations]<*mutation event*>in [G protein α-subunits]<*protein*>causing [human]<*organism*>[disease]<*disease*>is the case of [McCune-Albright syndrome]<*disease*>([MAS]<*disease*>). [MAS]<*disease*>is a sporadic [disease]<*disease*>typified by [precocious puberty]<*disease*>, [monoostotic[5] [*fibrous dysplasia*]]<*disease*>or [polyostotic fibrous dysplasia]<*disease*>, [café au lait pigmentation]<*disease*>, and several [endocrinopathies]<*disease*>.

NER can be based on a dictionary of all known entities. The problem with this approach, however, is that dictionaries are seldom complete because of the existence of variants and new names. A given entity name may be written in different ways: *TNF-α, TNF-alpha, TNF alpha, Tumor necrosis factor alpha, Tumour necrosis factor alpha* illustrate an abbreviation (TNF), a Greek letter and its full spelling, and US/GB variant spellings. For instance, the Metamap system (Aronson 2001) recognizes occurrences of the biomedical terms of the UMLS Metathesaurus, directly and under variant forms: upper/lower case, plural, noun/adjective, word order, punctuation, etc. The Metathesaurus covers a very large part of medical entities (body parts, diseases, procedures, drugs) but its coverage of biological entities (cell components, genes, proteins, etc.) is less developed. As an example of state-of-the-art NER, the above-mentioned sentence, run through Metamap, is tagged as follows:

[5] The complete term here is *monoostotic fibrous dysplasia*, and was obtained through the analysis of adjective coordination in a noun phrase.

A MOST INTRIGUING EXAMPLE of gain-of-[function]<*Physiologic Function*> [mutations]<*Genetic Function*>in [G protein alpha-subunits]<*Amino Acid, Peptide, or Protein,Enzyme*>causing [human]<*human*>[disease]<*Disease or Syndrome*>is the case of [McCune-Albright syndrome]<*Disease or Syndrome*>([MAS][6]<=McCune-Albright syndrome:<*Disease or Syndrome*>>). [MAS]<=McCune-Albright syndrome:<*Disease or Syndrome*>>is a sporadic [disease]<*Disease or Syndrome*>typified by [precocious puberty]<*Disease or Syndrome*>, monoostotic or [polyostotic fibrous dysplasia]<*Disease or Syndrome*>, cafe [au]<*Gene or Genome*>lait [pigmentation]<*Physiologic Function*>, and several [endocrinopathies]<*Disease or Syndrome*>.

Moreover, new names are coined as new entities (e.g., genes, proteins, diseases, drugs) are discovered or created. Such names are understood in the text in which they occur, either because they are explicitly introduced (*a new gene called XYZ*) or marked by typical words (IGL *gene,* mitotic *inhibitor*), or because of the relations they have with other entities (*XYZ binds to UVW*). Methods must therefore be designed to cope with new terms or new variants dynamically. A study of entity occurrences may help unveil patterns such as those just cited. Alternatively, machine learning methods may be called upon to select the combinations of clues which best detect named entities. Hidden Markov Models (HMMs) and Conditional Random Fields (CRFs) are two methods which are suited to learn from sequential data: here, sequences of words and their features. The choice of features (e.g., capitalization, presence of numbers, presence of marker words, etc.) seems to be more important than the algorithm though (Yeh et al. 2005). Methods have also been proposed to constitute lists of entity names by observing the contexts in which they occur: words which occur in similar contexts tend to have similar meanings (Firth 1957; Habert and Zweigenbaum 2002). The context of occurrence of a word is represented by its neighboring words or by its syntactic relations with the rest of the sentence. Similarity between context profiles is then used by clustering methods to build sets of words with similar meanings: genes, proteins, etc. (Sandler et al. 2006). Table 17.1 shows two clusters built this way.

The BioCreAtIvE II Gene Mention (GM) task (Wilbur et al. 2007) evaluated the NER systems of 21 teams on the task of deciding which substrings of a set of

**Table 17.1** Clusters of words with similar meanings

| Cluster 1 | Cluster 2 |
|---|---|
| Mouse, rat, rabbit, hamster, dog, cat, monkey, lean, GSH-Px, rhesus, cynomolgus, virgin, F344, hind, calf, lamb, horse, pig, sheep, goat, cattle, swine, chicken, quail, pigeon, heron, goose, chick, frog, zebrafish, rerio, Xenopus, owl, trout, salmon, lamprey, muskox, ovine, bovine, equine, PEDF, Yankasa, Charolais, bull, spermatozoa, Lewis, Fischer, secretor, Duffy, TN | junB, UCP2, decorin, Myogenin, AR, GR, PFK-A, TH, uPA, rhis4, GnRH-R, ZK7, c-mos, SPT10, FUT1, LHR, HO-1, tyrosinase, psaL, IRF-1, HAC1, SOCS-3, T24-ras, schizonts, CYP1A1, matrilysin, VCAM1, FGF-2, ULK2, AZS, CLA-1, PAI-2, PRLR, proenkephalin, PGHS-2, HFHZ, 5-HT2C-R, c-fms, ICK2, ASN1, ASN2, mHIF-1 |

[6] MetaMap 2008 expands the abbreviation within the span of submitted text

20,000 sentences were mentions of genes. The highest F-measure obtained by a system was 0.872. Most of the top-scoring systems used POS-tagging, several used stemming, abbreviations or chunking; all top-scoring systems used machine-learning methods, including Conditional Random Fields and Support Vector Machines.

### *17.4.2 Entity Normalization*

*Entity normalization* (sometimes called *term identification*) consists of relating the mention of an entity in text to a unique identifier in a reference database. For instance, the BioCreAtIvE II Gene Normalization (GN) task (Morgan and Hirschman 2007) required identifying the EntrezGene identifier for each human gene or protein mentioned in a set of 262 Medline abstracts (a total of 785 gene identifiers). The highest F-measure obtained by a system in this task was 0.810, the highest recall was 0.875, and the highest precision 0.841. This task can indeed build on the results of an NER system, augmented by a step where candidate gene mentions are matched against a lexicon containing gene symbols, names and synonyms, or aliases.

Entity normalization should also normalize protein names across multiple species. This was done for instance by (Wang 2007), who tagged the species of each protein using a combination of machine learning and rule-based disambiguation, improving their system's entity normalization F-measure by 10 points compared to a baseline system.

### *17.4.3 Relation Extraction*

Two entities found in the same sentence often take part in a joint event. For instance, in the following sentence, there is a (negative) exclusion relation between *CRP measurement* and *deep vein thrombosis*:

*A normal serum CRP measurement does not exclude deep vein thrombosis.*

Identifying such relations can rely on an analysis of the sentence. Full syntactic analysis (Miyao et al. 2006) or syntactico-semantic analysis (Lussier et al. 2006) may be called upon. As the verb *exclude* has the subject *CRP measurement* and the object *deep vein thrombosis*, the predicate-argument structure *exclude(CRP, thrombosis)* can be found in this sentence (Miyao et al. 2006). Local parsing may often be sufficient: patterns are built to check whether the expression between the two entities (here, *does not exclude*) expresses a relation. Such patterns may be designed by hand (Miyao et al. 2006), checking for typical relation markers (e.g., *influence, effect, affect, role, response; mediate, regulate, regulation; induce, activate, activation*, etc. See Table 17.2) and more precise regular expressions. They can also be learned from an annotated corpus where entity boundaries have been previously identified and each such expression has been tagged by the relation that links the two entities (Haddow and Matthews 2007). Machine learning methods can then be

**Table 17.2** Sentences containing relation markers

| Sentence | Example relations |
|---|---|
| Nicotinic acetylcholine receptors (*AChRs*) mediate *rapid excitatory synaptic transmission* throughout the peripheral and central nervous systems | Mediates (AChRs, rapid excitatory synaptic transmission) |
| *Quercetin* but not luteolin suppresses the induction of *lethal shock* upon infection | Suppresses (quercetin, lethal shock) |
| High levels of the intracellular signalling molecule cyclic diguanylate (*c-di-GMP*) suppress *motility* and activate exopolysaccharide (*EPS*) *production* | Suppresses (c-di-GMP, motility) activate(c-di-GMP, EPS production) |
| Negative regulators of PGC-1alpha such as *RIP140* and *160MBP* suppress *mitochondrial biogenesis* | Suppresses (RIP140, mitochondrial biogenesis)<br>Suppresses (160MBP, mitochondrial biogenesis) |

trained on such a tagged corpus to learn which combinations of cues are the most effective in detecting relations.

### *17.4.4 Annotated Corpora*

Semantic role labeling can help further improve relation extraction. Based on an extract of the GENIA corpus where the predicate-argument structures of 30 frequently used biomedical verbs predicates were annotated, Tsai et al. (2006) trained a role labeling system. Their system was much more effective at extracting arguments in the biomedical text than a general-purpose (newswire-oriented) semantic role labeling system, and obtained a global F-measure of 87%.

Annotated corpora are therefore a key element to develop information extraction systems. This is essential for machine-learning methods, which need annotated corpora for training. This is also important in general to provide a common basis for the comparative evaluation of information extraction systems, whatever their methods. The BioCreAtIvE II challenge protein–protein interaction task (PPI) organizers Krallinger et al. (2007) prepared a corpus where each full-text article was associated with manually-derived protein–protein interaction pairs. The systems had to produce, for each article, a ranked list of protein–protein interaction pairs, where each protein must be uniquely identified by a UniProt accession number or ID. The corpus was divided into training and test sub-corpora of 740 and 358 articles. The highest F-measure (averaged over the articles) obtained by a system was 0.2885, the highest averaged precision and recall were 0.3893 and 0.3073. The top-performing systems used not only machine learning methods but also general-language and domain-specific resources. It must be noted that different corpora generally have different properties, e.g., contain a varying number of occurrences of protein–protein interactions. Airola et al. (2008) underscore that in general, F-measure cannot be used to draw comparisons meaningfully across different corpora.

### 17.4.5 Relation Extraction Challenges

Achieving perfect accuracy is a tall order for an NLP tool. The assertion *let-7 is a master temporal regulator* is relatively straightforward to analyze: the entities (*let-7* and *regulator*) and their potential roles (encoded as semantic types in an ontology such as UMLS) could be recognized, for example, by a lexicon-based NER tool such as MetaMap; and the IS_A relation between the entities is explicitly stated and could be extracted by a number of statistical or knowledge-based relations extraction systems. More sophisticated processing is required to automatically "understand," extract facts, and make inferences given more complicated text. Table 3 lists some manually derived relations potentially of interest to a biomedical researcher, which might be asserted and inferred given a passage of complex text.

The first challenge encountered by a NLP system in the text in Table 17.3 is to characterize the semantic relation that holds between *let-7* and *miRNA* in the noun–noun compound (Nakov and Hearst 2006). Similarly, *Ras guanosine triphosphatases* and *tumor cell lines* need to be identified as multi-word terms, the former needs to be correctly associated with the *Ras oncogene family*. The *location_of* and *part_of* relations need to be identified for the latter term. Extraction of interactions between biological entities (*let-7 NEGATIVELY_REGULATES Ras GTPases*) is a complex task that builds upon recognition of named entities, determination of the entities participating in an interaction, and determination of a relation type between the interacting entities.

The above-described techniques are necessary to obtain precise information from each sentence in a corpus. The sentence-level information may then be aggregated and filtered depending on the purpose of the information extraction process. Information may also be sought at the level of a whole corpus, e.g., to know whether there exists a relationship between two entities, for example, if somewhere in the corpus it is asserted that bone injuries are one of the neurofibromatosis type 1 (NF1) lesions and it is asserted elsewhere that statins accelerate bone healing, a *treat* relation might be inferred between statins and Nf1-related bone fractures. In that case, a global analysis may be sufficient, and the mere, repeated co-occurrence of the two entities in the same sentence or even in the same abstract may be enough to assert this relationship.

Methods for extracting drug–gene relations based on co-occurrence of drug and gene names in a sentence have been developed in pharmacogenomics

**Table 17.3** A clause and relations of potential interest to biomedical researcher

| | |
|---|---|
| The let-7 miRNA negatively regulates the oncogenic family of Ras guanosine triphosphatases in both Caenorhabditis elegans and human tumor cell lines | IS_A (let-7, miRNA)<br>NEGATIVELY_REGULATES(let-7,Ras GTPases)<br>PART_OF (Ras GTPases, Ras oncogene family)<br>LOCATION_OF (tumor cell, Ras GTPases)<br>PART_OF (tumor cell, Caenorhabditis elegans)<br>PART_OF (tumor cell, human)<br>CAUSES (Ras gene, tumors) |

(Rindflesch et al. 2000; Chang and Altman 2004). Wren and Garner (2004) establish a network of gene names associated with Gene Ontology (GO) categories by their co-occurrence within Medline records using fuzzy set theory. A graph-bigram traversal algorithm was proposed to identify and extracts point mutation associations from biomedical literature (Lee et al. 2007).

The global co-occurrence analysis method will be developed further in Section Literature-based discovery.

## 17.5 Building on Information Extraction

Natural Language Processing described above provides the foundation for more sophisticated text mining: deriving information from text through finding assertions and associations among entities, and inferencing, which ultimately serves as foundation for summarization, question answering, prediction, and discovery as well as automatic merging of knowledge derived from textual information with information extracted from knowledge bases.

### *17.5.1 Assisted Database Curation*

One of the goals of finding assertions such as *let-7 is a master temporal regulator* is to then extract these facts for insertion into databases increasingly used by researchers. For example, the above fact is used as a part of Gene Reference Into Function (GeneRIF) in the Entrez Gene database. GeneRIFs are added to the database to facilitate access to publications describing experiments that clarify gene functions. Such entries are mostly added to databases manually by database curators or suggested by volunteers. Manual database curation involves the following steps: (1) finding articles of interest; (2) finding and extracting facts (relations, events, associations, etc.) relevant to the database focus; and (3) converting extracted information into predefined standardized form. Text mining tools, although not currently ready to carry out the curation task on their own, are useful in assisting curators with information extraction (Rebholz-Schuhmann et al. 2005). A recent evaluation of such assistance in extracting protein–protein interactions found that a maximum speed-up of 1/3 in curation time can be expected if the tools are perfectly accurate (Alex et al. 2008).

As finding and predicting entity interactions is important in understanding the biological processes, many tools are dedicated to providing assistance with this task. Because of the complexity of the task and a need for high quality, these tools are specific to entity types, relation types, and organisms, and often developed for a specific database. For example, rather than modifying publicly available tools such as Textpresso (described in Section Applications) that focuses on *Caenorhabditis elegans* literature (Müller et al. 2004), or MedMiner developed for exploration of literature on gene–drug and gene–gene relationships (Tanabe et al. 1999), special tools are developed to assist FlyBase curation (Karamanis et al. 2007), or establish knowledge about the relationships among drugs, diseases and genes (Chang and Altman 2004; Ahlers et al. 2007).

### 17.5.2 Question-Answering and Summarization

Finding and extracting sentences containing interactions is an important and hard problem, but not the end-goal of text mining. Ideally, depending on the nature of researchers' questions and their background knowledge, a system would generate answers to the questions or summarize all extracted facts in personalized digests. First steps towards answering questions in the genomics domain were undertaken by a fairly large number of researchers within the Genomics track of the Text REtrieval Conference evaluations. Questions were collected in interviews with biomedical researchers. The evaluation participants had to extract passages of text from full text scientific publications to answer these questions. The track evaluators then manually extracted exact answers and their supporting statements from the passages submitted by participating systems (Hersh et al. 2007). As most systems relied on a typical two-stage question-answering process (finding relevant passages using an IR system and then looking for answers in these passages), the evaluation did not demonstrate any benefits of processing beyond passage retrieval. The collection of questions and answers generated in the process of Genomics track evaluations provides an opportunity to further research and develop biological question-answering systems, which ultimately will generate answers to a variety of biological questions.

In addition to recognition and extraction techniques described above, answer generation will require tools that reliably perform the following tasks:

- *Duplicate or similarity detection*. For example, given the following two statements: *miR-214 induces cell survival and cisplatin resistance by targeting PTEN* and *PTEN was shown to be a target of miR-214*, a tool capable of identifying the redundancy of the second statement will present the first one as an answer.
- *Controversy detection*. A slight modification of the above second statement to *PTEN was shown to be NOT a target of miR-214* would create a controversy that needs to be detected. Subsequently, both statements need to be presented to a researcher.
- *Negation detection*. Recognizing the above controversy requires negation detection along with other factors. General purpose negation detection tools (Mutalik et al. 2001), as well as tools developed specifically for the biomedical domain (Chapman et al. 2001) are extensively used in clinical text processing. Feasibility and complexity of detecting uncertainty and negation of protein–protein interaction were recently analyzed in a small scale study (Sanchez-Graillet and Poesio 2007). Further studies in this area will be facilitated by the BioScope corpus,

which consists of medical and biological texts annotated for negation, speculation, and their linguistic scope (Szarvas et al. 2008).

- *Anaphora resolution* (resolving what a pronoun or a noun phrase refers to). For example, it is impossible to know which proteins are discussed in the following sentence: *We show that human microRNA miR369-3 directs association of these proteins with the AREs to activate translation.* without relating *these proteins* to the entities recognized as proteins in the preceding sentence in the publication: *Upon cell cycle arrest, the ARE in tumor necrosis factor-alpha (TNFalpha) mRNA is transformed into a translation activation signal, recruiting Argonaute (AGO) and fragile X mental retardation-related protein 1 (FXR1), factors associated with micro-ribonucleoproteins (microRNPs).* A tool capable of resolving this reference will assist answering a question about *miR369-3* role in translation activation given the document from which the above sentences were extracted.

The above techniques are needed not only to answer questions, but also to synthesize summaries. Summarization is a long-standing and active area of research in NLP. Free and commercial tools that generate single or multi-document summaries are available to generate indicative (suggesting the contents of a document) or informative (describing the essence of the document) summaries of retrieval results. A current more ambitious goal in this research area is to move from generic, all-purpose summaries to generation of topic-oriented summaries for specific users with complex information needs.

The ultimate goal of text mining tools is to go beyond question answering and summarization and help researchers predict gene and protein functions, localization, and interactions, as well as discover interactions between proteins and drugs. This is the topic of the next section.

## 17.6 Literature-Based Discovery

Information extraction collects basic facts[7] from texts: relation between entities of interest, such as *mediates (ERbeta, ODD protection)*. Analyzing large repositories of scientific texts such as Medline can therefore help assemble large collections of potentially relevant facts. *Literature-based discovery* aims to build on such collections of basic facts to discover hidden, indirect connections which may constitute new knowledge.

Foundational research in this area was performed by Swanson (1986) who, by exploring the published literature, suggested that fish oil may be beneficial to patients with Raynaud's disease. The linking clues were an effect of fish oil on blood viscosity, blood platelet aggregation, and certain vasoreactive characteristics, three conditions which are affected in patients with this disease. Since then, a series

[7] Although these may actually be hypotheses, observations, results, etc., we refer to them uniformly as "facts."

of methods and systems have been proposed by Swanson and others to automate a large part of this process (Srinivasan and Libbus 2004; Swanson et al. 2006; Smalheiser et al. 2006; Hristovski et al. 2006; Jelier et al. 2008).

The general schema of Swanson's method considers a disease *C* (e.g., Raynaud's disease). It is known on the one hand, according to the literature on disease *C*, that this disease has a characteristic *B* (e.g., increased blood viscosity); and the literature on a substance *A* (e.g., fish oil), on the other hand, records that *A* affects characteristic *B*. In such a situation, substance *A* may be a candidate to treat disease *C*. This is generally implemented through one of the following two search strategies. In *closed discovery*, the researcher tests the hypothesis that *C* and *A* are connected, and *Bs* must be discovered which link *A* and *C*. In *open discovery*, only disease *C* is given, and *B* characteristics, then *A* substances, must be generated. See Fig. 17.6 in the section on Applications for an illustration.

Literature-based discovery must address two issues: how to obtain basic facts from the considered literatures and how to connect these facts. The easiest medium to access the biomedical literature is the Medline bibliographic database. Most Medline records contain MeSH keywords and free-text title and abstract; full-text articles are a promising avenue for further research, but not all of them are freely available. Each *MeSH keyword* which indexes an article in Medline encodes an important topic covered in the article. The literature for an input term *C* can be obtained through a PubMed search for *C*. The MeSH keywords found in the retrieved records are therefore associated with the *C* literature and make up the set of candidate *B* terms: a basic fact here is an association between the initial *C* search term and a *B* MeSH keyword which indexes some of the retrieved articles (Srinivasan and Libbus 2004). As there may exist numerous *B* terms, they must be filtered to select the strongest associations. The number of occurrences of each association (and derived formulas, such as the term frequency–inverse document frequency *(tf.idf)* used in IR) and the semantic type of the *B* MeSH keywords (e.g., *enzyme*, *gene*, etc., as found in the UMLS semantic network) are typical filtering criteria (Srinivasan and Libbus 2004). In the case of a closed-discovery strategy, the same process is applied to *A* terms, leading to a second set of candidate *B* terms, and the intersection of the two sets of *B* terms is computed. In the case of an open-discovery strategy, candidate *B* terms constitute new search terms from which associated literatures and attached indexing MeSH terms *A* are obtained. In the latter situation, additional filtering needs to be performed on the *A* terms, again taking into account their semantic types and combined frequency of occurrence in the *B* literatures.

Instead of using MeSH terms, Swanson et al. (2006) and Smalheiser et al. (2006) use words and phrases found in *article titles* to obtain *B* terms. Words which occur frequently in article titles must be removed to prevent irrelevant links from being collected; a list of such words (*stoplist*) of a few hundred to a few thousand words was used in their experiments.

More precise basic facts may be obtained by calling on NLP techniques as presented in the section on Information Extraction above. NER helps detecting biomedical entities in the input titles and abstracts (Jelier et al. 2007; Pospisil et al. 2006). Relation extraction based on full parsing identifies labeled relations between these entities (Rzhetsky et al. 2006; Hristovski et al. 2006). Relation extraction based on co-occurrence based methods is a simpler, less computationally demanding approach, which can nevertheless yield interesting results (Jelier et al. 2007). In the above-mentioned A–B–C scheme, these more precise relations help to label the A–B and B–C links and provide further filtering.

Another way to uncover implicit knowledge is to build larger networks of elementary relations extracted from the literature. Each elementary link in such a network is obtained by relation extraction as seen above. Paths of more than one link represent indirect relations between biomedical entities (Palakal et al. 2007). The chronological ordering of links may also be important: Rzhetsky et al. (2006) study the patterns of chronologically ordered chains of statements about published molecular interaction. Their model enables them to make several observations, e.g., that "*scientists are often strongly affected by prior publications in interpreting their own experimental data, while weighting their own private results [...] at least 10-fold as high as a single result published by somebody else.*"

More information can be found in the review article (Weeber et al. 2005) and in the recent book (Bruza and Weeber 2008).

## 17.7 Applications

The research activities and specialized tools described above allowed for implementation of quite a number of applications actively used by biologists and bioinformatics researchers. This section presents several of the publicly available text mining tools. As some of the enhanced search engines, NER and relation extraction tools were presented previously, only a few online applications that combine these capabilities are discussed here.

### *17.7.1 Sentence and Entity Extraction Tools*

A step beyond presenting a list of articles ranked by relevancy to the query is to present a ranked list of relevant sentences. Textpresso, an information extracting and processing package for biological literature associated with WormBase, provides access to over 80,000 publications covering various model organisms. Figure 17.2 presents the structured query and the results of a search for C. elegans genes involved in axon guidance. The top sentence extracted by the system lists genes (linked to WormBase entries) reported as acting in axonogenesis.

Textpresso's collection of the full text of scientific articles is split into individual sentences, on which Textpresso performs semantic searches enabled by an ontology populated with terms categorized into 33 classes describing biological processes, concepts (e.g., gene), and relations (e.g., regulation) (Müller et al. 2004).

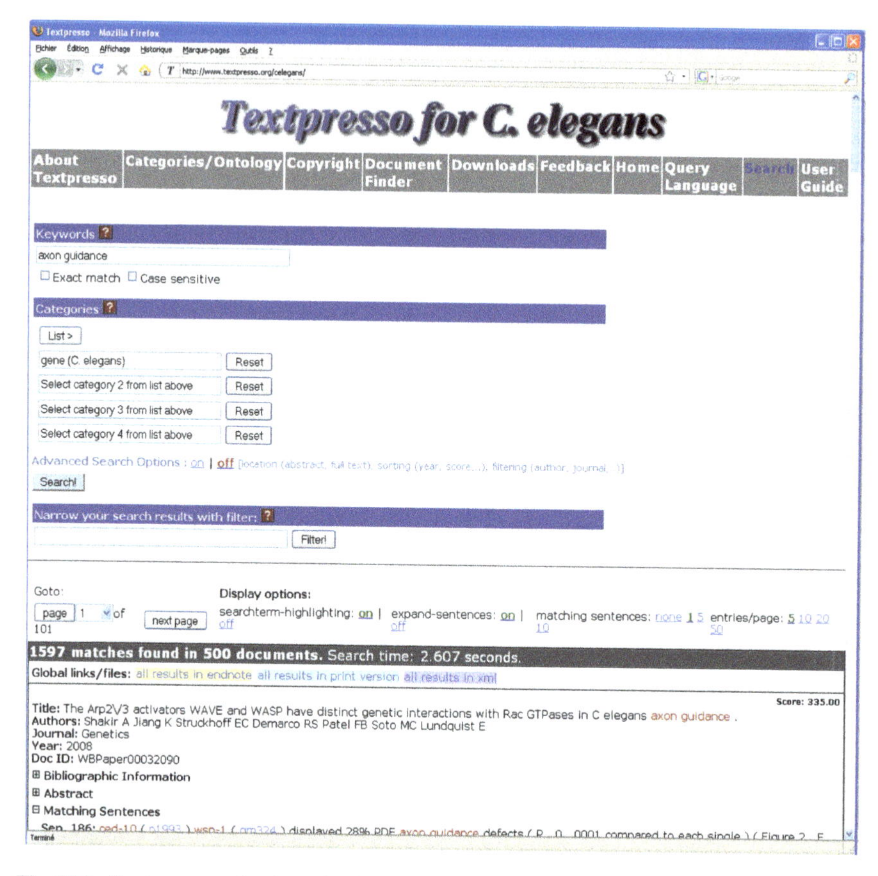


**Fig. 17.2** Textpresso retrieval results for C. elegans genes involved in axon guidance

EbiMed,[8] an EMBL-EBI service, allows searching a local copy of Medline. It then searches retrieval results for sentences containing biomedical terminology and generates a table that displays proteins, GO annotations, drugs, and species extracted from sentences in retrieved abstracts (see Fig. 17.3). The extracted sentences and terms are linked to corresponding entries in biomedical databases (Rebholz-Schuhmann et al. 2007).

### 17.7.2 Relation Extraction and Entity Linking Tools

A step beyond sentence and terminology extraction and linking to databases is to identify and extract associations between entities (such as protein–protein interactions). The iHOP system,[9] representative of the interactions detection systems,

[8]EbiMed – http://www.ebi.ac.uk/Rebholz-srv/ebimed/index.jsp

[9]iHOP – http://www.ihop-net.org/

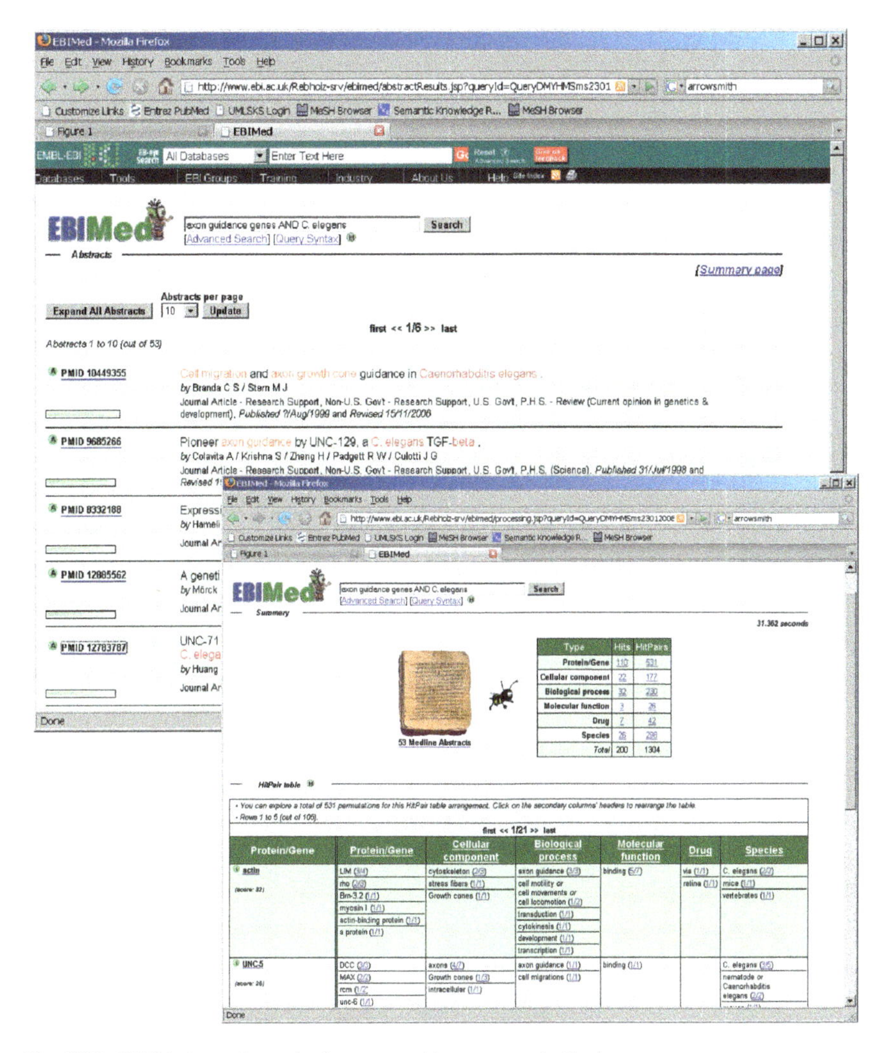

Fig. 17.3 EBIMed search results for axon guidance genes in C. elegans

is gene/protein-centric: it organizes relevant sentences extracted from literature search results using gene/protein names (Hoffmann and Valencia 2004). Starting with a search for a gene or protein of interest, for example, *unc-129*, a user will see search results in the form of sentences containing gene names, one of which is the gene that served as a query. In addition to information about the gene, search results provide an opportunity to navigate to information about all other genes and biomedical terms identified in the page and linked to external resources, and build an interaction network of genes/proteins that co-occur in sentences (see Fig. 17.4).

**Fig. 17.4** iHop search results

Chilibot (chip literature robot)[10] specializes in identification of relationships between genes, proteins, or arbitrary keywords in PubMed search results (Chen and Sharp 2004). Upon submission of two terms or two lists of terms of interest, Chilibot generates a network map that encodes the relations between the submitted terms. Figure 17.5 presents search results for *unc-129, unc-130 and axon guidance, C. elegans*.

Only abstract co-occurrence relations were identified between these terms, however, the system is capable of identifying four other relation categories (stimulatory,

[10]Chilibot – http://www.chilibot.net/

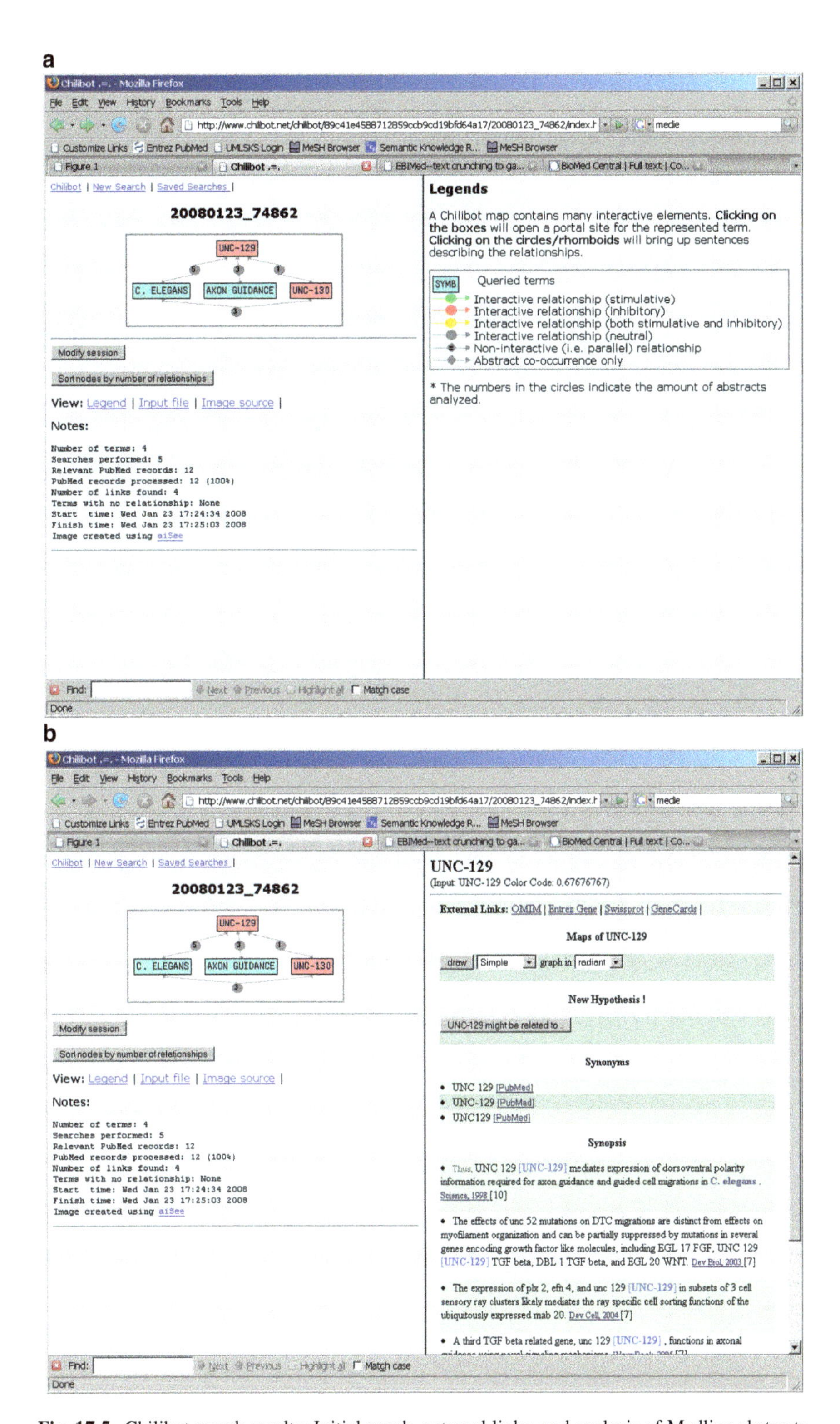


**Fig. 17.5** Chilibot search results: Initial graph, external links, and analysis of Medline abstracts

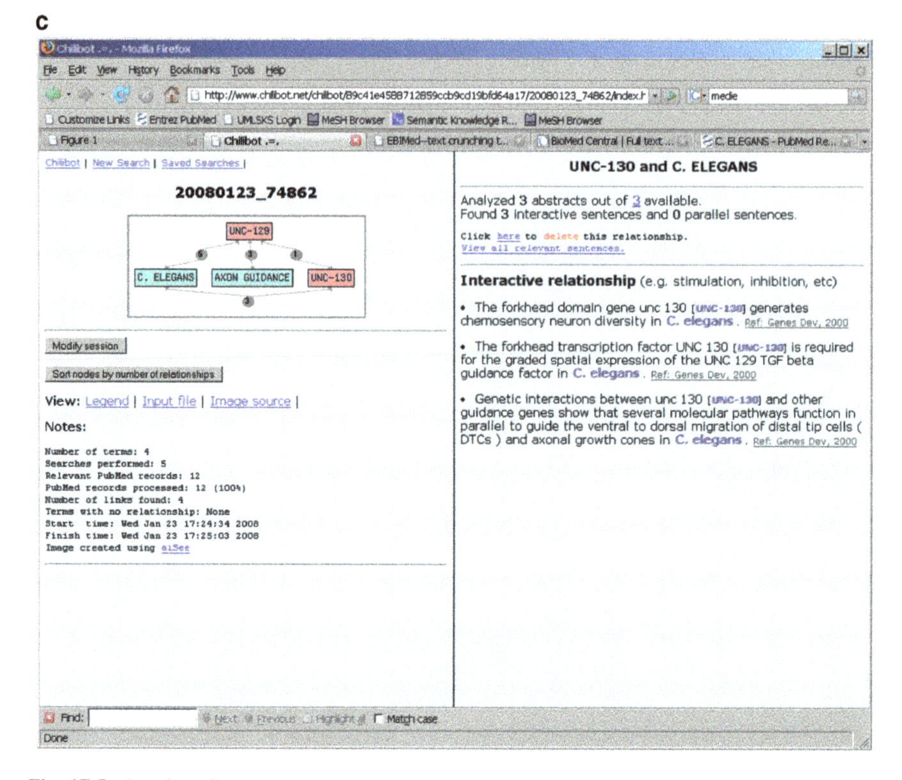


**Fig. 17.5** (continued)

inhibitory, neutral, and parallel) based on manually derived rules for each relationship type. Clicking on the symbol in the interactive network map brings up information about the term, a synopsis of search results for the term, its synonyms, and suggestions to generate hypotheses about the term's relationships with other search terms. Clicking on the circles in the map brings up sentences describing the identified relationships.

### *17.7.3 Literature Based Discovery Assistance*

Hypotheses generation is the ultimate goal of text mining tools. As described in the section on Literature-Based Discovery, the first literature based discovery was assisted by the ARROWSMITH[11] application (Smalheiser et al. 2006).

[11] ARROWSMITH – http://arrowsmith.psych.uic.edu/cgi-bin/arrowsmith_uic/start.cgi

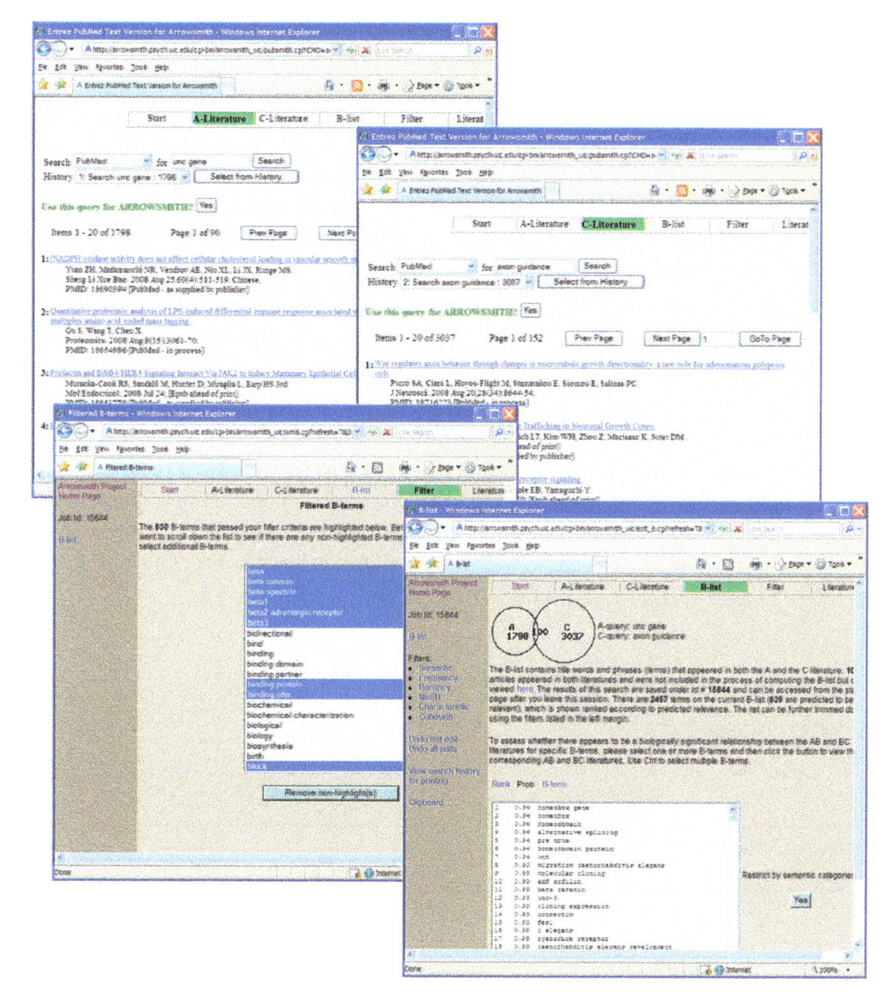


**Fig. 17.6** ARROWSMITH search results

Figure 17.6 demonstrates the steps in the ARROWSMITH process: selection of two sets of publications ("literature A" and "literature C"); an intersection of the literatures and extraction of the potentially interesting terms performed by the tool which results in the B-list displayed ranked by relevance; and restriction of terms to certain semantic categories per user's selection.

Another interactive literature-based biomedical discovery support system, BITOLA,[12] assists with both open and closed discovery types (Hristovski et al. 2005).

[12]BITOLA – http://www.mf.uni-lj.si/bitola/

### *17.7.4 Visualization Tools*

The form of presentation of text mining results is very important in assisting a scientist's analysis of information. Visualization as a means to lessen the cognitive load and provide an overview of available information is increasingly used in specialized text mining tools. These tools have to address the following issues: visualization type (e.g., graph, treemap, star tree, hyperbolic tree, etc.); layout (aesthetically pleasing and meaningful arrangements of the elements); visual cues (color coding, shapes, etc.); navigation; and interactivity.

In addition to visualization capabilities in the above tools, Arizona Network Visualizer (Leroy & Chen 2005) presents a framework for pathway-related knowledge integration and visualization. This demo application allows searching within the results of five broad PubMed searches, then selection of the extracted relations to present visually, and further interaction with the table and graph (Fig. 17.7).

Semantic MEDLINE[13] is another visual exploratory demonstration of the results of relations and entity extraction from 35 PubMed searches. Figure 17.8 presents the results of visualization of pharmacogenomics relations identified by this application in the literature retrieved in a PubMed search for metabolic syndrome.

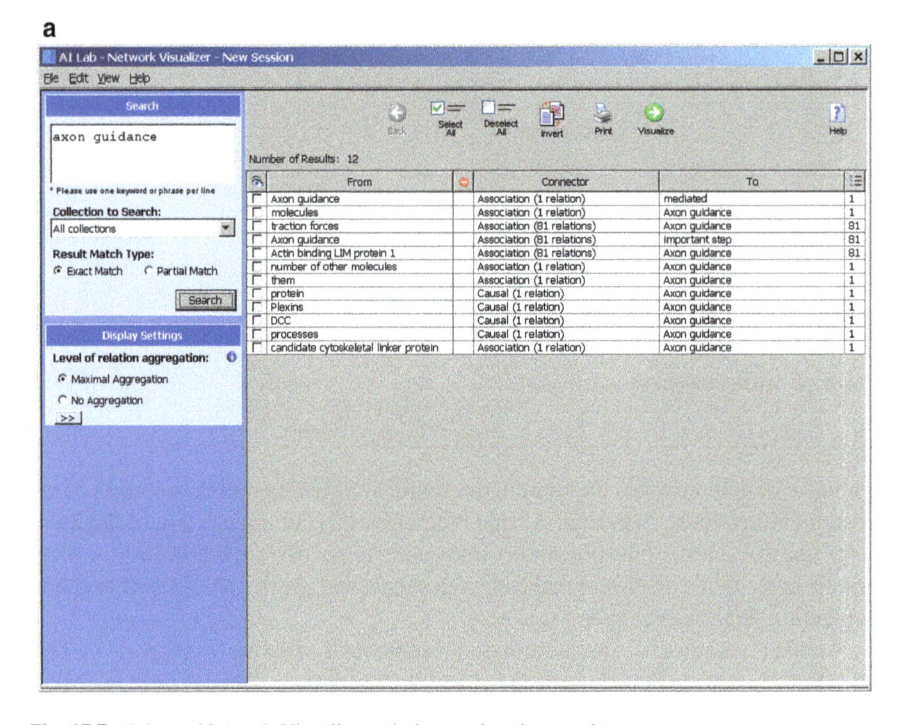


**Fig. 17.7** Arizona Network Visualizer relation exploration results

[13]Semantic MEDLINE – http://skr3.nlm.nih.gov/SemMedDemo/

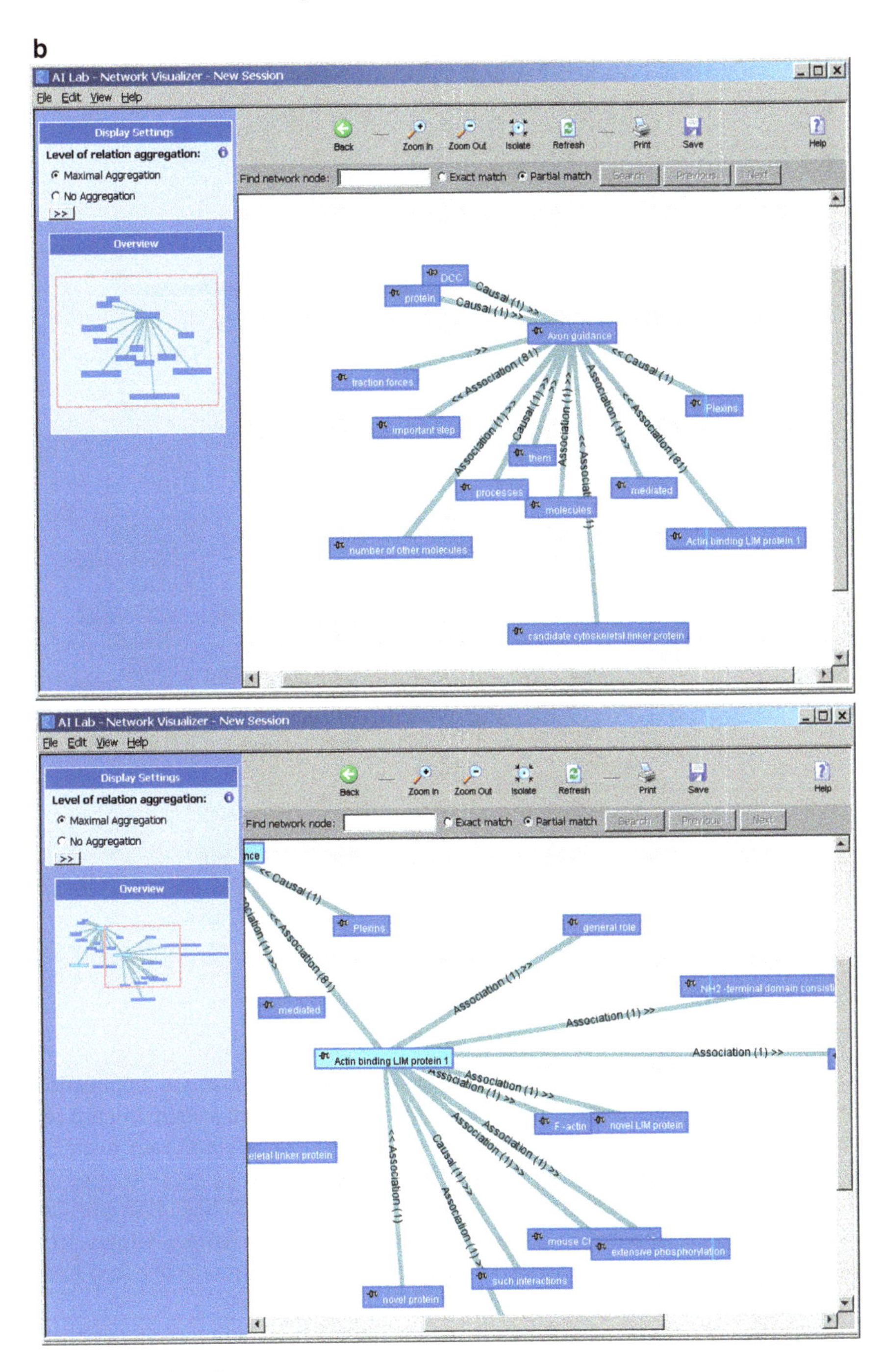


Fig. 17.7 (continued)

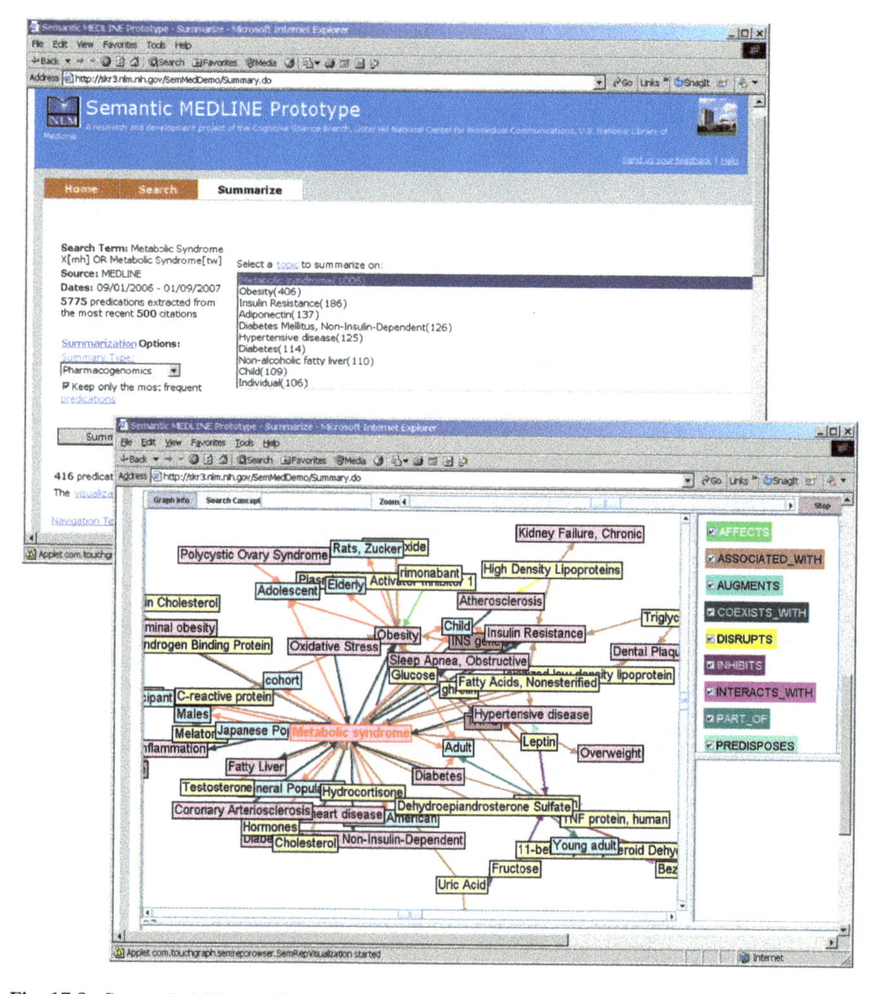


**Fig. 17.8** Semantic MEDLINE summarization and visualization results

Due to complexity of the underlying competation, some of the applications described in this section seem fairly slow compared to online search engines and interactive web-portals from which a user expects almost instantaneous response. Some of the biomedical text mining tools speed-up processing by restricting the searches to local collections in which entities are identified and linked off-line, and relations are pre-computed to achieve real-time online interactions. Others seem better optimized, yet others provide information about the current active step in the pipeline to let a user know the system is processing the request.

In general, the maturity of the tools engaged in the text mining pipelines paved the way for implementation of the sophisticated applications presented in this section and provided for many future improvements in terms of the scope of literature and knowledge bases coverage, data visualization, linking, and multi-media navigation possibilities, as well as complex question answering and focused personalized summarization.

This brief introduction to BioNLP aims to provide an overview of natural language processing methods currently applied to biomedical sub-language. The textbooks and other lists at the end of the chapter provide links to in-depth information about NLP methods, available resources and the growing dynamic set of tools and applications.

## 17.8 Appendix

## 17.9 Lists of Tools and Services

State of the art NLP tools:
http://aclweb.org/aclwiki/index.php?title=State_of_the_art
Resource list compiled by Kevin Bretonell Cohen
http://compbio.uchsc.edu/corpora/bcresources.html
Resource list compiled by Robert Futrelle:
http://www.bionlp.org/
BioCreAtIvE bio-NLP tools:
http://biocreative.sourceforge.net/bionlp_tools_links.html
NLP and Text Mining Research list at NaCTeM:
http://www.nactem.ac.uk/research.php?view=4
Arrowsmith:
http://arrowsmith.psych.uic.edu/arrowsmith_uic/tools.html
The Open Directory Project:
http://www.dmoz.org/Science/Biology/Bioinformatics/Software/
The National Centers for Biomedical Computing (NCBC) funded under the NIH Roadmap for Bioinformatics and Computational Biology:
http://www.ncbcs.org/

## 17.10 Gene and Protein Name Resources

Entrez Gene:
http://www.ncbi.nlm.nih.gov/sites/entrez?db=gene
FlyBase:
http://flybase.org/
HUGO Gene:
http://www.genenames.org/index.html
Model organisms:
http://www.nih.gov/science/models/
Mouse Genome Informatics:

http://www.informatics.jax.org/
Saccharomyces genome database:
http://www.yeastgenome.org/gene_list.shtml
The Worldwide Protein Data Bank:
http://www.wwpdb.org/
UniProt:
http://www.ebi.ac.uk/uniprot/

## 17.11 Biomedical Terminologies

The national center for biomedical ontology:
http://bioontology.org/
The open biomedical ontologies:
http://www.obofoundry.org/
Resources for Biomedical Terminology and Ontology:
http://www.ldc.upenn.edu/mamandel/itre/term.html#Dictionaries
Unified Medical Language System
http://www.nlm.nih.gov/research/umls/

## References

Ahlers CB, Fiszman M, Demner-Fushman D, Lang FM, Rindflesch TC (2007) Extracting semantic predications from MEDLINE citations for pharmacogenomics. In: Pac Symp Biocomput 12. Maui, Hawaii, pp 209–220
Airola A, Pyysalo S, Bjorne J, Pahikkala T, Ginter F, Salakoski T (2008) A graph kernel for protein–protein interaction extraction. In: Proceedings of the workshop on current trends in biomedical natural language processing (BioNLP'08). Association for Computational Linguistics, pp 1–9
Alex B, Grover C, Haddow B, Kabadjov M, Klein E, Matthews M et al (2008) Assisted curation: Does text mining really help? In: Pac Symp Biocomput 13. Big Island, Hawaii, pp 556–567
Ananiadou S, Friedman C, Tsujii JI (eds) (2004) Named entity recognition in biomedicine. J Biomed Inform 37(6):393–528
Aronson AR (2001) Effective mapping of biomedical text to the UMLS Metathesaurus: The MetaMap program. In: Proc AMIA Symp, pp 17–21
Bodenreider O (2008) Biomedical ontologies in action: Role in knowledge management, Data integration and decision support. IMIA Yearbook of Medical Informatics, pp 67–79
Branco A, McEnery T, Mitkov R (eds) (2005) Anaphora processing: Linguistic, cognitive and computational modelling. Current Issues in Linguistic Theory, vol 263. John Benjamins, Amsterdam and Philadelphia
Bruza P, Weeber M (eds) (2008) Literature-based discovery. Information Science and Knowledge Management, vol 15, Springer, Berlin Heidelberg New York
Carreras X, Màrquez L (2005) Introduction to the CoNLL-2005 Shared Task: Semantic Role Labeling. In: Proc 9th CoNLL. ACL, pp 152–164
Chang JT, Altman RB (2004) Extracting and characterizing gene-drug relationships from literature. Pharmacogenetics 14(9):577–586
Chapman WW, Bridewell W, Hanbury P, Cooper GF, Buchanan BG (2001) A simple algorithm for identifying negated findings and diseases in discharge summaries. J Biomed Inform 34(5):301–310

Chen H, Sharp BM (2004) Content-rich biological network constructed by mining PubMed abstracts. BMC Bioinform 8;5:147

Dee CR (2007) The development of the Medical Literature Analysis and Retrieval System (MEDLARS). J Med Libr Assoc 95(4):416–425. http://www.pubmedcentral.nih.gov/articlerender.fcgi?artid=2000779

Demner-Fushman D, Humphrey SM, Ide NC, Loane RF, Mork JG, Ruch P et al (2007) Combining resources to find answers to biomedical questions. In: The sixteenth text retrieval conference TREC-2007, Gaithersburg, MD, pp 205–215

Eaton AD (2006) HubMed: a web-based biomedical literature search interface. Nucleic Acids Res 1:34(Web Server issue):W745-7

Firth JR (1957) Papers in linguistics, 1934–1951. Oxford University Press, London

Friedman C, Kra P, Yu H, Krauthammer M, Rzhetsky A (2001) GENIES: A natural-language processing system for the extraction of molecular pathways from journal articles. Bioinformatics 17(Suppl 1):S74–S82

Grishman R, Sundheim B (1996) Message understanding conference – 6: A brief history. In: Proc 16th COLING. ACL, pp 466–471

Habert B, Zweigenbaum P (2002) Contextual acquisition of information categories: what has been done and what can be done automatically? In: Nevin BE, Johnson SM (eds) The Legacy of Zellig Harris: Language and information into the 21st Century, Mathematics and computability of language, vol 2. John Benjamins, Amsterdam, pp 203–231

Haddow B, Matthews M (2007) The Extraction of Enriched Protein-Protein Interactions from Biomedical Text. In: Proceedings of the workshop on Biological, translational and clinical language processing (BioNLP 2007). Association for Computational Linguistics. Prague, Chech Republic, pp 145–152

Hersh W, Cohen AM, Ruslen L, Roberts P (2007) TREC 2007 genomics track overview. In: The Sixteenth Text Retrieval Conference – TREC 2007. NIST

Hersh WR, Greenes RA (1990) SAPHIRE: An information retrieval system featuring concept matching, automatic indexing, probabilistic retrieval, and hierarchical relationships. Comput Biomed Res 23(5):410–425

Hirschman L (2007) The second biocreative evaluation: Lessons learned and future directions. In: Fifth Fraunhofer-symposium on text mining. Bonn, Germany, http://www.scai.fraunhofer.de/fileadmin/download/vortraege/tms_07/Lynette_Hirschmann.pdf

Hliaoutakis A, Varelas G, Petrakis EGM, Milios EE (2006) MedSearch: A retrieval system for medical information based on semantic similarity. In: Proc ECDL, Lecture Notes in Computer Science 4172. Springer, Berlin Heidelberg New York

Hoffmann R, Valencia A (2004) A gene network for navigating the literature. Nat Genet 36:664

Hristovski D, Friedman C, Rindflesch TC, Peterlin B (2006) Exploiting semantic relations for literature-based discovery. In: AMIA Annu Symp Proc, pp 349–353

Hristovski D, Peterlin B, Mitchell JA, Humphrey SM (2005) Using literature-based discovery to identify disease candidate genes. Int J Med Inform 74(2–4):289–298

Ide NC, Loane RF, Demner-Fushman D (2007) Essie: A concept based search engine for structured biomedical text. J Am Med Inform Assoc 14(3):253–263

Jacquemin C (2001) Spotting and discovering terms through NLP. MIT Press, Cambridge, MA

Jelier R, Jenster G, Dorssers LCJ, Wouters BJ, Hendriksen PJM, Mons B et al (2007) Text-derived concept profiles support assessment of DNA microarray data for acute myeloid leukemia and for androgen receptor stimulation. BMC Bioinform 8:14

Jelier R, Schuemie MJ, Veldhoven A, Dorssers LCJ, Jenster G, Kors JA (2008) Anni 2.0: A multipurpose text-mining tool for the life sciences. Genome Biol 9(6):R96

Karamanis N, Lewin I, Seal R, Drysdale R, Briscoe E (2007) Integrating natural language processing with flybase curation. In: Pac Symp Biocomput 12. Maui, Hawaii, pp 245–256

Krallinger M, Leitner F, Valencia A (2007) Assessment of the second BioCreative PPI task: Automatic extraction of protein–protein interactions. In: Proceedings of the BioCreAtIvE II Workshop, Madrid, pp 41–54

Lee LC, Horn F, Cohen FE (2007) Automatic extraction of protein point mutations using a graph bigram association. PLoS Comput Biol Feb 2;3(2):e16

Leroy G, Chen H (2005) Genescene: An ontology-enhanced integration of linguistic and co-occurrence based relations in biomedical texts. J Am Soc Inf Sci Technol (JASIST) 56(5):457–468

Lewis J, Ossowski S, Hicks J, Errami M, Garner HR (2006) Text similarity: an alternative way to search MEDLINE. Bioinformatics September 15;22(18):2298–2304

Lindberg DA, Humphreys BL, McCray AT (1993) The unified medical language system. Methods Inf Med 32(4):281–291

Lussier Y, Borlawsky T, Rappaport D, Liu Y, Friedman C (2006) PhenoGO: Assigning phenotypic context to Gene Ontology annotations with natural language processing. In: Pac Symp Biocomput 11. Maui, Hawaii, pp 64–75

Miles WD (1992) A history of the national library of medicine: The Nation's Treasury of Medical Knowledge. Bernan Assoc. http://www.nlm.nih.gov/hmd/manuscripts/miles/miles.pdf. Accessed 4 August 2008

Miyao Y, Ohta T, Masuda K, Tsuruoka Y, Yoshida K, Ninomiya T, et al (2006) Semantic retrieval for the accurate identification of relational concepts in massive textbases. In: Proceedings of the 21st international conference on computational linguistics and 44th annual meeting of the association for computational linguistics (COLING/ACL 2006). Sydney, Australia, pp 1017–1024

Morgan AA, Hirschman L (2007) Overview of BioCreative II gene normalisation. In: Proceedings of the BioCreAtIvE II Workshop, Madrid, pp 17–27

Müller HM, Kenny EE, Sternberg PW (2004) Textpresso: an ontology-based information retrieval and extraction system for biological literature. PLoS Biol 2(11):e309

Mutalik PG, Deshpande A, Nadkarni PM (2001) Use of general-purpose negation detection to augment concept indexing of medical documents: a quantitative study using the UMLS. J Am Med Inform Assoc 8(6):598–609

Nakov P, Hearst M (2006) Using verbs to characterize noun–noun relations. In: Proceedings of the twelfth international conference on artificial intelligence: Methodology, systems, applications (AIMSA), Bulgaria

Namer F, Baud R (2007) Defining and relating biomedical terms: Towards a cross-language morphosemantics-based system. Int J Med Inform 76(2–3):226–233

Palakal M, Bright J, Sebastian T, Hartanto S (2007) A comparative study of cells in inflammation, EAE and MS using biomedical literature data mining. J Biomed Sci 14(1):67–85

Porter MF (1980) An algorithm for suffix stripping. Program 14(3):130–137

Pospisil P, Iyer LK, Adelstein SJ et al. (2006) A combined approach to data mining of textual and structured data to identify cancer-related targets. BMC Bioinform 7(354)

Pratt W, Hearst M, Fagan L. A knowledge-based approach to organizing retrieved documents. In: Proceedings of 16th annual conference on artificial intelligence (AAAI-99). July 1999 Orlando, FL, pp 80–85

Rebholz-Schuhmann D, Kirsch H, Arregui M, Gaudan S, Riethoven M, Stoehr P (2007) EBIMed–text crunching to gather facts for proteins from Medline. Bioinformatics 23(2):e237–e244

Rebholz-Schuhmann D, Kirsch H, Couto F (2005) Facts from text – is text mining ready to deliver? PLoS Biol 3(2):e65

Rindflesch TC, Tanabe L, Weinstein JN, Hunter L (2000) EDGAR: Extraction of drugs, genes and relations from the biomedical literature. Pac Symp Biocomput 5:517–528

Rzhetsky A, Iossifov I, Loh JM et al. (2006) Microparadigms: Chains of collective reasoning in publications about molecular interactions. PNAS 103:4940–4945

Sanchez-Graillet O, Poesio M (2007) Negation of protein–protein interactions: Analysis and extraction. Bioinformatics 23(13):i424–i432

Sandler T, Schein AI, Ungar LH (2006) Automatic term list generation for entity tagging. Bioinformatics 22(6):651–657

Smalheiser NR, Torvik VI, Bischoff-Grethe A, Burhans LB, Gabriel M, Homayouni R, et al (2006) Collaborative development of the Arrowsmith two node search interface designed for

laboratory investigators. J Biomed Discov Collab 1(8). http://www.j-biomed-discovery.com/content/1/1/8
Smith L, Rindflesch T, Wilbur WJ (2004) MedPost: A part-of-speech tagger for bioMedical text. Bioinformatics 20(14):2320–2321
Srinivasan P, Libbus B (2004) Mining MEDLINE for implicit links between dietary substances and diseases. Bioinformatics 20(Suppl 1):i290–i296
Surdeanu M, Harabagiu S, Williams J, Aarseth P (2003) Using predicate-argument structures for information extraction. In: Proc 41st ACL. Sapporo, Japan, pp 8–15
Swanson DR (1986) Fish oil, Raynaud's syndrome, and undiscovered public knowledge. Perspectives in biology and medicine 30:7–18
Swanson DR, Smalheiser NR, Torvik VI (2006) Ranking indirect connections in literature-based discovery: The role of medical subject headings. J Am Soc Inf Sci Technol 57(11):1427–1439
Szarvas G, Vincze V, Farkas R, Csirik J (2008) The BioScope corpus: annotation for negation, uncertainty and their scope in biomedical texts. Proceedings of the 2008 Workshop on Biomedical Natural Language Processing (BioNL'08), Columbus, Ohio. June 2008. pp 38–45
Tanabe L, Scherf U, Smith LH, Lee JK, Hunter L, Weinstein JN (1999) MedMiner: An internet text-mining tool for biomedical information, with application to gene expression profiling. BioTechniques 27:1210–1217
The Unicode Standard (2007), Version 5.0 Addison-Wesley, Boston, MA
Tsai RTH, Chou WC, Lin YC, Sung CL, Ku W, Su YS, et al (2006) BIOSMILE: Adapting semantic role labeling for biomedical verbs: An exponential model coupled with automatically generated template features. In: HLT-NAACL BioNLP, pp 57–64
Wang X (2007) Rule-based protein term identification with help from automatic species tagging. In: Proceedings of the Conference on Intelligent Text Processing and Computational Linguistics (CICLING 2007), Lecture Notes in Computer Science 4394, Springer, Berlin Heidelberg New York, pp 288–298
Wilbur J, Smith L, Tanabe L (2007) BioCreative 2. Gene Mention Task. In: Proceedings of the BioCreAtIvE II Workshop 2007, Madrid, Spain, pp 7–16
Wren JD, Garner HR (2004) Shared relationship analysis: Ranking set cohesion and commonalities within a literature-derived relationship network. Bioinformatics 20(2):191–198
Yeh A, Morgan A, Colosimo M, Hirschman L (2005) BioCreAtIvE task 1A: gene mention finding evaluation. BMC Bioinform 6:(Suppl)1

## Textbooks and Introductions to NLP and BioNLP

Allen JF (1995) Natural language understanding, 2nd edn. Benjamin/Cummings, Menlo Park, CA
Ananiadou S, Kell DB, Tsujii JI (2006) Text mining and its potential applications in systems biology. Trends Biotechnol 24(12):571–579
Ananiadou S, McNaught J (2006) Text mining for biology and biomedicine. Artech House Publishers, Norwood, Massachusetts, USA
de Bruijn B, Martin J (2002) Getting to the (c) ore of knowledge: Mining biomedical literature. Int J Med Inform 67:7–18
Cohen AM, Hersh W (2005) A survey of current work in biomedical text mining. Brief Bioinform 6(1):57–71
Cohen KB, Hunter L (2004) Natural language processing and systems biology. In: Dubitzky W, Azuaje F (eds) Artificial intelligence methods and tools for systems biology. Springer, Norwell, MA, pp 147–174
Cohen KB, Hunter L (2008) Getting started in text mining. PloS Comput Biol 4(1):e20
Hearst MA (2003) What is text mining? Available online at http://www.ischool.berkeley.edu/~hearst/text-mining.html

Hunter L, Cohen KB (2006) Biomedical language processing: what's beyond PubMed? Mol Cell 21:589–594

Jackson P, Moulinier I (2002) Natural language processing for online applications: text retrieval, extraction, and categorization. John Benjamins Publishing Company

Jensen LJ, Saric J, Bork P (2006) Literature mining for the biologist: from information retrieval to biological discovery. Nat Rev Genet 7:119–129

Jurafsky D, Martin JH (2008) Speech and language processing: An introduction to natural language processing, computational linguistics and speech recognition. Prentice Hall, Lebanon, Indiana, Second edition

Krallinger M, Valencia A (2005) Text-mining and information-retrieval services for molecular biology. Genome Biol 6(7):224

Mitkov R (ed) (2003) The Oxford handbook of computational linguistics. Oxford University Press, New York

Shatkay H (2005) Hairpins in bookstacks: Information retrieval from biomedical text. Brief Bioinform 6(3):222–238

Shatkay H, Craven M (2007) Biomedical text mining, MIT Press, Cambridge

Spasic I, Ananiadou S, McNaught J, Kumar A (2005) Text mining and ontologies in biomedicine: Making sense of raw text. Brief Bioinform 6(3):239–251

Weeber M, Kors JA, Mons B (2005) Online tools to support literature-based discovery in the life sciences. Brief Bioinform 6(3):277–286

# Chapter 18
# Data and Databases

**Daniel Damian**

## 18.1 Introduction

Owing to the recent advances in technology and to the growth in the number and size of projects tasked with collecting and assembling biological and genomic information, a highly heterogeneous collection of databases have become available to the community in the last two decades. As a consequence of rapid and distributed progress throughout the field, bioinformatics databases are provided in a variety of formats and specifications. This chapter discusses the most frequently encountered data formats in bioinformatics and the tools used to access these data.

## 18.2 Fundamentals of Data Storage

In computer applications, information is stored in streams of bytes in the computer's memory – be that volatile or permanent storage. Beyond the practicalities of a storage location, accessing information requires control over two aspects. The first aspect is how the information is encoded into the stream of bytes, otherwise known as the format of the data. The second aspect is which applications can be used to perform the encoding (writing) and decoding (reading) of the data.

The two aspects are in fact closely intertwined. It is of no use to chose a complex and full-featured storage format unless a set of applications are readily available that can be used to access the data in the chosen format. Similarly, a simple and compact storage format that may be accessed by the most basic of tools becomes a limiting factor if more advanced operations cannot be performed.

D. Damian (✉)
BioWisdom Ltd., Cambridge, CB22 7GG, UK
e-mail: daniel.damian@biowisdom.com

D. Edwards et al. (eds.), *Bioinformatics: Tools and Applications*,
DOI 10.1007/978-0-387-92738-1_18, © Springer Science+Business Media, LLC 2009

This tension is the crux of bioinformatics databases today. Throughout the wealth of projects dealing with collecting and publishing bioinformatics data, a number of formats have been established. Most widely used are flat-file formats, XML or relational databases.

## 18.3 Flat-File Formats in Bioinformatics

### *18.3.1 Overview of the Flat-File Format*

Flat-file format is just another name for what is commonly understood as plain text files. Flat-file databases are formed from files that contain plain text; this text is usually formed using characters within the ASCII character set, although exceptions where the text contains characters from the extended ASCII or Unicode character set may also be considered flat files.

Data stored in flat-file format are usually structured as a collection of data entries, where an entry is a description of a specific entity of the data. For instance, in the Unified Protein Resource database, also known as UNIPROT (Uniprot 2008), an entry contains data about an individual protein sequence; it contains, among others, a list of identifiers, a description and a list of sequence features. In flat-file databases an entry is represented as a sequence of text lines, each line adhering to a well-defined format.

Entries in a flat-file database are typically listed sequentially in one or more text files. There is no established rule: in some databases all entries are stored in a single text file; in others each entry is stored in its own file. For larger databases a middle-ground approach of combining entries into several large text files is required to avoid limitations of computer file systems and associated tools.

### *18.3.2 Advantages of Flat-File Formats*

The main advantage of the flat-file format is its universality. The majority of computing machines will have readily available software capable of reading, displaying and searching text files. Writing custom applications that perform simple manipulation of text files is also relatively straightforward and does not require expert knowledge of any particular technology.

The flat-file format can thus be considered machine-independent, even though its origins can be traced to standards in UNIX operating systems, which are commonly used in data warehouses. The flat-file format is also less prone to conversion problems when transported from one machine to another, e.g., via FTP or even email, although this aspect has become less important today.

There is also, arguably, an advantage in the amount of space required for storing data in flat files. It is possible to define efficient formatting of the text that leads the space taken by a flat-file representation to be equal, if not smaller than the space required for other representations, including relational databases and especially XML formats. Compactness of the format becomes a significant problem when data are transferred across the Internet, where transfer speeds become a major bottleneck in data processing.

Following from its universality, the flat-file format is also suitable for processing through a wide number of tools. The fundamental design philosophy behind most UNIX command-line tools is line-based processing of text files. These tools are thus extensively used for efficiently processing bioinformatics flat-file data. Similarly, an extensive range of analysis tools such as BLAST (Altschul et al. 1990) and CLUSTALW (Higgins and Sharp 1988) are designed to access data in flat-file format and benefit from its performance.

### *18.3.3 Flat-File Formats: A Worked-out Example*

The example in Table 18.1 presents a cut-out of the CYC_HUMAN (human cytochrome C) flat-file entry from the UNIPROT database. The format of the entry is defined as a series of lines of text finishing with the "//" sequence on a separate line.

Each line begins with a two-character field identifier. These identifiers are used to distinguish between different parts of the entry and their meaning. In this example, the ID field identifier indicates that the specific line contains a sequence of characters that uniquely identifies the entry – in this case CYC_HUMAN. There is additional information on the first line that specifies the status of the entry and the length of the protein sequence defined by the number of amino acid residues.

The following lines provide additional information content for the entry. For example, the AC line lists previous accession numbers – these are unique identifiers by which the entry may have been referred to in the past or in other databases. The OS line indicates the organism species, which is the source of the sequence. In some cases, information may span over multiple lines: for instance, the CC lines contain multi-line comments. The sequence of lines may be repeated a number of times. For instance, each literature reference is encoded as a sequence of RN, RP, RX, RA, RT and RL lines, indicating the number, position, comment, cross-reference, authors, title and location of the reference respectively. Such repeating groups are sometimes referred to as subentries and may be considered in isolation from the main entry.

In some cases, an individual line (identified by a specific field descriptor) may also contain multiple fields of information and employs a set of specific tags to distinguish between these. For instance, the FT line defines a sequence feature; these features are described in terms of a feature key, a sequence range and further comments. Similarly, groups of CC lines use a special construction at the beginning of a comment block.

A human-readable specification of the format of entries in a particular flat-file database distribution is usually provided by the data publishers. Such specification should include detailed descriptions of both the structure and the semantic of the data contained in each line in the flat file.

**Table 18.1** Example flat-file text entry from UNIPROT (for brevity we use ellipsis within square brackets as placeholders for similar content)

```
ID   CYC_HUMAN               Reviewed;         105 AA.
AC   P99999; A4D166; P00001; Q6NUR2; Q6NX69; Q96BV4;
DT   21-JUL-1986, integrated into UniProtKB/Swiss-Prot.
DT   23-JAN-2007, sequence version 2.
DT   04-DEC-2007, entry version 56.
DE   Cytochrome c.
GN   Name=CYCS; Synonyms=CYC;
OS   Homo sapiens (Human).
OC   Eukaryota; Metazoa; Chordata; Craniata; Vertebrata; Euteleostomi;
OC   Mammalia; Eutheria; Euarchontoglires; Primates; Haplorrhini;
OC   Catarrhini; Hominidae; Homo.
OX   NCBI_TaxID=9606;
RN   [1]
RP   NUCLEOTIDE SEQUENCE [GENOMIC DNA].
RX   MEDLINE=89071748; PubMed=2849112;
RA   Evans M.J., Scarpulla R.C.;
RT   "The human somatic cytochrome c gene: two classes of processed
RT   pseudogenes demarcate a period of rapid molecular evolution.";
RL   Proc. Natl. Acad. Sci. U.S.A. 85:9625-9629(1988).
[..............]
CC   -!- FUNCTION: Electron carrier protein. The oxidized form of the
CC       cytochrome c heme group can accept an electron from the heme group
CC       of the cytochrome c1 subunit of cytochrome reductase. Cytochrome c
CC       then transfers this electron to the cytochrome oxidase complex,
CC       the final protein carrier in the mitochondrial electron-transport
CC       chain.
[..............]
CC   -----------------------------------------------------------------------
CC   Copyrighted by the UniProt Consortium, see http://www.uniprot.org/terms
CC   Distributed under the Creative Commons Attribution-NoDerivs License
CC   -----------------------------------------------------------------------
DR   EMBL; M22877; AAA35732.1; -; Genomic_DNA.
DR   EMBL; AL713681; CAD28485.1; -; mRNA.
DR   EMBL; BT006946; AAP35592.1; -; mRNA.
DR   EMBL; AC007487; AAQ96844.1; -; Genomic_DNA.
DR   EMBL; CH236948; EAL24239.1; -; Genomic_DNA.
DR   EMBL; BC005299; AAH05299.1; -; mRNA.
DR   EMBL; BC008475; AAH08475.1; -; mRNA.
[...........]
DR   PIR; A31764; CCHU.
DR   RefSeq; NP_061820.1; -.
DR   UniGene; Hs.437060; -.
DR   UniGene; Hs.617193; -.
DR   PDB; 1J3S; NMR; -; A=1-105.
DR   IntAct; P99999; -.
DR   PeptideAtlas; P99999; -.
DR   Ensembl; ENSG00000172115; Homo sapiens.
DR   GeneID; 54205; -.
[......................]
PE   1: Evidence at protein level;
KW   3D-structure; Acetylation; Apoptosis; Direct protein sequencing;
KW   Electron transport; Heme; Iron; Metal-binding; Mitochondrion;
KW   Polymorphism; Respiratory chain; Transport.
FT   INIT_MET      1      1       Removed.
FT   CHAIN         2    105       Cytochrome c.
FT                                /FTId=PRO_0000108218.
FT   METAL        19     19       Iron (heme axial ligand).
[....................]
SQ   SEQUENCE   105 AA;  11749 MW;  8EE9689E0102506B CRC64;
     MGDVEKGKKI FIMKCSQCHT VEKGGKHKTG PNLHGLFGRK TGQAPGYSYT AANKNKGIIW
     GEDTLMEYLE NPKKYIPGTK MIFVGIKKKE ERADLIAYLK KATNE
//
```

### 18.3.4 Parsing and Indexing Flat-File Data

Parsing is a computing technique that can be used to identify and to extract specific parts or syntactic components from a given text (Grune and Jacobs 1990). In practical terms, parsing is implemented as a computer algorithm that follows a set of rules (usually called syntax or grammar) to identify parts of interest in a given input text. Parsing matches an input text with the set of rules: if the match fails, the text is rejected; if the match tallies, the text is usually broken down into parts.

A simple and widely used parsing technique is based on regular expressions (Aho et al. 1986; Appel 1998; Grune and Jacobs 1990). A regular expression reflects the formal notion of finite state automata (Aho et al. 1986); informally they are finite rules describing a sequence of texts through a combination of:

- A specific sequence of characters
- A repetition of such sequences or
- An alternative of two or more sequences

A regular-expression-based parser matches the input text according to a regular expression. If a match is made, for each individual component of the regular expression the matching text is extracted.

Regular expressions are a popular concept that is accessible in most of today's programming languages. This follows naturally as regular expressions are relatively simple and easy to understand. In bioinformatics regular expressions gain even more relevance as parsers based on regular expressions are an efficient and practical solution for parsing flat-file data. Most flat-file formats use a field identifier at the beginning of the line similar to that in the UNIPROT format illustrated above. Such formats can be easily split into fields using a regular-expression matcher. UNIX shell commands or scripting languages as Perl (Wall et al. 2000) with built-in regular-expression support can be used to parse flat files and extract data.

Table 18.2 presents a small parser written in Perl. The example parses a flat file in UNIPROT format (see Table 18.1). The program reads the input line by line in the $line variable. Using Perl's regular-expression matching constructs, the program distinguishes between the ID line, AC lines and the SQ line, and prints out the first accession number (also known as primary accession) and the length of the corresponding protein sequence. The second part of the table illustrates a sample output of the program. To parse the entire data in a flat file more complex parsers have to be written, taking performance and memory usage considerations into account. See for instance BioPerl (Stajich et al 2002, http://www.bioperl.org), a wide collection of Perl bioinformatics tools.

A more complex but more powerful parsing technique is based on the notion of context-free grammars (Aho et al. 1986; Altschul et al. 1990). Context-free grammars describe text as a set of potentially mutually dependent composition rules. For instance, a context-free grammar can be used to describe a language of trees. The grammar will define a tree as a combination of two branches and a joining

**Table 18.2** Simple parsing session using regular expressions in Perl

Sample Perl parser:

```
my $newEntry = 1;
while(<STDIN>) {
   my($line) = $_;
   chomp($line);
   if($line =~ /^ID/) {
        # matched ID line
        $newEntry=1;
   }
   if($line =~ /^AC\s*(\w+)/) {
        # matched AC line
        if ($newEntry==1) {
             # print only when at first line
             print "Primary accession " . $1;
        }
        $newEntry=0;
   }
   if($line =~ /^SQ\s*\w*\s*(\w*)/) {
        # match SQ line
        print "  sequence length: " . $1 . "\n";
   }
   # ignore other lines
}
```

Sample output:

```
> perl example.pl < uniprot.dat
Primary accession Q4U9M9  sequence length: 893
Primary accession P15711  sequence length: 924
Primary accession Q43495  sequence length: 102
Primary accession P18646  sequence length: 75
Primary accession P13813  sequence length: 296
.........
```

trunk. A branch can be then defined as a leaf or another tree. Such definitions create potentially infinite structures, but they are useful as they can express more complex languages than regular expressions can (Grune and Jacobs 1990).

There are not many instances of bioinformatics flat-file databases where context-free grammars become necessary. Context-free grammars, however, may arguably be considered to provide a more practical method for defining and structuring a parser compared with regular expressions. Regular expressions tend to quickly grow in size and complexity when the text to be parsed becomes more complex; context-free grammars may provide a more compact and readable definition of the structure of the text to be parsed.

Parsers based on context-free grammars are, on the other hand, more difficult to implement. Fortunately, a number of open-source tools can be used either to parse with context-free grammar definitions or to compile context-free grammar definitions into efficient parsers. A popular choice is the open-source GNU Bison (http://www.gnu.org/software/bison/), versions of which are available in almost any programming language.

### 18.3.5 Practical Aspects of Flat-File Formats

Despite the fact that basic tools for processing text files are readily available, the sheer size and complexity of the data stored in flat-file formats demand the development of specialized software for accessing and retrieving data in this format. Most well-known flat-file bioinformatics databases have grown to be amazingly complex and simple text processing tools are no longer able to extract the information in all its complexity.

While the flat-file format solves the problem of transport and conversion across heterogeneous platforms, retrieving information from flat files remains a complex task. Some of the current flat-file databases, for instance, EMBL or GENBANK, have grown to hundreds of gigabytes of text and more than one hundred million entries at the date this chapter has been written. The set of basic text-processing tools available on most systems are not designed to cope with such sizes. Since more than a decade ago it has been unfeasible to search, locate and retrieve information stored in such flat-file databanks by using basic text-processing tools; for this purpose dedicated parsing and indexing tools have been developed.

### 18.3.6 Bioinformatics Flat-File Data Integration

A number of specialized software solutions have emerged to provide a search and retrieval service across flat-file databases. A pioneering system is Sequence Retrieval System (SRS). SRS is discussed in more detail in another section below. At present a number of commercial products provide bioinformatics flat-file data-processing features; at the same time a wide range of open-source tools developed in the academic community are available.

When specialized data integration software are not available, a dedicated flat-file data parser may be used to import the data into a data management system. Such systems are usually Relational Database Management Systems (RDBMS) that provide search and indexing facilities via SQL queries. There is a certain amount of flexibility in this approach, as the users control the import process and define which parts of the data are imported in the RDBMS and used later on. However, the import process may end up requiring a significant amount of time and resources; further effort has to be spent optimizing and tuning an RDBMS system to provide acceptable performance when operating on the largest of the bioinformatics databases currently available.

Over a period of time, bioinformatics data not only grow in terms of sheer size, but the data also frequently change format. As projects merge and changes in procedures are implemented, the precise details of a flat-file format for a specific database may change as well. Often such changes involve addition of new fields of information: new delimited lines in the format of an entry or additions of new references to other databases. Occasionally, the entire structure and semantic of the fields may change.

When such changes occur, all related flat-file parsers and associated programs have to be updated accordingly. The maintenance of dedicated parsers for each of the bioinformatics databases becomes a significant overhead over time. For these reasons other formats such as XML are increasingly adopted; we discuss these in the next sections.

### 18.3.7 Sequence Formats

A large number of bioinformatics databases contain explicit descriptions of proteins or nucleic acids in the form of a sequence. For instance, nucleic acids are described by the sequence of the constituent bases. In text form each base is represented via a single-letter code following a standard established by the International Union of Biochemistry and Molecular Biology (IUPAC 1984). The same holds for protein sequences, which are described as the sequence of amino acid building blocks of the protein.

While the alphabet of sequences has been standardized, the actual formatting of the sequence in text files differs from database to database and in between individual software applications. The format of a sequence is important, as the sequence is often extracted by users and reused across multiple applications, for analyses, display and publication. It has become customary for a number of software tools to automatically distinguish and accept sequences in different formats.

Sequence formats differ mostly in the layout and formatting of lines of sequence codes, while some formats also provide a description or meta-information line or set of lines. Table 18.3 shows examples of a sequence formatted in some of the common sequence formats.

### 18.3.8 Constructed Sequences

It is not just the annotations in a sequence database which make reference to information in other entries or databases. Some sequence databases may contain entries that do not list the actual content of a sequence, but instead give a list of instructions on how to construct the sequence from data in other entries in the database.

As an example, the EMBL nucleotide sequence database contains CON (constructed) sequence entries. These entries represent chromosomes, genomes or other

**Table 18.3** The UNIPROT/P32234 protein sequence under four different sequence formats

FASTA format:

```
>uniprot|P32234|128UP_DROME GTP-binding protein 128up.
MSTILEKISAIESEMARTQKNKATSAHLGLLKAKLAKLRRELISPKGGGGGTGEAGFEVA
KTGDARVGFVGFPSVGKSTLLSNLAGVYSEVAAYEFTTLTTVPGCIKYKGAKIQLLDLPG
IIEGAKDGKGRGRQVIAVARTCNLIFMVLDCLKPLGHKKLLEHELEGFGIRLNKKPPNIY
YKRKDKGGINLNSMVPQSELDTDLVKTILSEYKIHNADITLRYDATSDDLIDVIEGNRIY
IPCIYLLNKIDQISIEELDVIYKIPHCVPISAHHHWNFDDLLELMWEYLRLQRIYTKPKG
QLPDYNSPVVLHNERTSIEDFCNKLHRSIAKEFKYALVWGSSVKHQPQKVGIEHVLNDED
VVQIVKKV
```

Swissprot format:

```
SQ   Sequence   368 AA;
     MSTILEKISA IESEMARTQK NKATSAHLGL LKAKLAKLRR ELISPKGGGG GTGEAGFEVA
     KTGDARVGFV GFPSVGKSTL LSNLAGVYSE VAAYEFTTLT TVPGCIKYKG AKIQLLDLPG
     IIEGAKDGKG RGRQVIAVAR TCNLIFMVLD CLKPLGHKKL LEHELEGFGI RLNKKPPNIY
     YKRKDKGGIN LNSMVPQSEL DTDLVKTILS EYKIHNADIT LRYDATSDDL IDVIEGNRIY
     IPCIYLLNKI DQISIEELDV IYKIPHCVPI SAHHHWNFDD LLELMWEYLR LQRIYTKPKG
     QLPDYNSPVV LHNERTSIED FCNKLHRSIA KEFKYALVWG SSVKHQPQKV GIEHVLNDED
     VVQIVKKV
//
```

GCG format:

```
128UP_DROME  Length: 368  Check: 6459  ..

      1  MSTILEKISA IESEMARTQK NKATSAHLGL LKAKLAKLRR ELISPKGGGG GTGEAGFEVA
     61  KTGDARVGFV GFPSVGKSTL LSNLAGVYSE VAAYEFTTLT TVPGCIKYKG AKIQLLDLPG
    121  IIEGAKDGKG RGRQVIAVAR TCNLIFMVLD CLKPLGHKKL LEHELEGFGI RLNKKPPNIY
    181  YKRKDKGGIN LNSMVPQSEL DTDLVKTILS EYKIHNADIT LRYDATSDDL IDVIEGNRIY
    241  IPCIYLLNKI DQISIEELDV IYKIPHCVPI SAHHHWNFDD LLELMWEYLR LQRIYTKPKG
    301  QLPDYNSPVV LHNERTSIED FCNKLHRSIA KEFKYALVWG SSVKHQPQKV GIEHVLNDED
    361  VVQIVKKV
```

Pretty-print format:

```
1         11        21        31        41        51
MSTILEKISAIESEMARTQKNKATSAHLGLLKAKLAKLRRELISPKGGGGGTGEAGFEVA
61        71        81        91        101       111
KTGDARVGFVGFPSVGKSTLLSNLAGVYSEVAAYEFTTLTTVPGCIKYKGAKIQLLDLPG
121       131       141       151       161       171
IIEGAKDGKGRGRQVIAVARTCNLIFMVLDCLKPLGHKKLLEHELEGFGIRLNKKPPNIY
181       191       201       211       221       231
YKRKDKGGINLNSMVPQSELDTDLVKTILSEYKIHNADITLRYDATSDDLIDVIEGNRIY
241       251       261       271       281       291
IPCIYLLNKIDQISIEELDVIYKIPHCVPISAHHHWNFDDLLELMWEYLRLQRIYTKPKG
301       311       321       331       341       351
QLPDYNSPVVLHNERTSIEDFCNKLHRSIAKEFKYALVWGSSVKHQPQKVGIEHVLNDED
361
VVQIVKKV
```

long sequences constructed from sequences defined in other entries in the database. Such format makes it difficult to retrieve the consequence, as a specific algorithm is required to assemble the entire sequence from the referenced sequences.

## 18.4 The XML Format in Bioinformatics

XML (the eXtensible Markup Language) is a language for structuring data in text files. The World Wide Web consortium (http://www.w3c.com), the body which creates Internet standard recommendations, has defined the XML recommendation (XML 2006) as a generic platform-independent format for structured documents. In the last decade, the XML format has become ubiquitous in computing application as the language of choice for exchanging data between computer systems and applications.

An XML document represents a nested tree of information. Each node in the tree can contain data, a list of sub-nodes and a list of attributes, and an XML document must start with a root node. An XML document has a textual representation where the content of each node and its sub-nodes is delimited by a pair of mutually enclosing tags.

### *18.4.1 XML Format: A Worked-out Example*

A cut-down example of an XML document taken from the MEDLINE database is provided in Table 18.4. The document contains data about a publication from a life sciences journal. The root node of the document is given by the MedlineCitation tag. The children of the nodes are delimited with tags, for instance, an identifier tag (PMID), several date tags indicating creation, update and revision date, tags identifying the publication journal details, the title, abstract, authors, etc. It can be observed how the data are structured around a disciplined use of tags: for instance, all the journal data are contained inside tags within the main journal tags.

### *18.4.2 Document Type Definition (DTD)*

An XML document is structured according to a Document Type Definition (DTD) (XML 2006). A DTD defines a class of XML documents that satisfy a set of structural rules. Such rules may indicate, for instance, what types of, and how many, sub-nodes a certain type of node may have. A DTD is associated with an XML document via a Document Type Declaration, and such declaration indicates that the XML document adheres to the class of documents with the structure defined in the DTD. A DTD is itself specified via XML, and may be included directly in the type declaration of the XML document, or, more conveniently, stored and referenced as a separate file.

**Table 18.4** Example XML entry from MEDLINE (for brevity we use ellipsis within square brackets as placeholders for similar content)


```
<MedlineCitation Owner="NLM" Status="MEDLINE">
<PMID>10697468</PMID>
[... ... ... ...]
<Article PubModel="Print">
<Journal>
<ISSN IssnType="Print">0099-2399</ISSN>
<JournalIssue CitedMedium="Print">
<Volume>1</Volume>
<Issue>6</Issue>
<PubDate>
<Year>1975</Year>
<Month>Jun</Month>
</PubDate>
</JournalIssue>
<Title>Journal of endodontics</Title>
</Journal>
<ArticleTitle>Methodology and criteria in the evaluation of dental im-
plants.</ArticleTitle>
<Abstract>
<AbstractText>This study was designed to develop an inexpensive, reproduci-
ble method for studying the reaction of the tissues of the oral cavity to
the endosseous implant. Thirty-six guinea pigs, three materials (Teflon, Vi-
tallium, and Titanium 6Al-4V), two observation periods (two and 12 weeks),
and two implant designs (one with exposure to the oral cavity and one with-
out exposure to the oral cavity) were used. The inflammatory response was
significantly greater in the exposed implants than in the unexposed im-
plants. In the implants that were exposed 12 weeks, there was a strong in-
terrelationship between severe inflammation, bacteria, and epithelial in-
vagination. These factors are significant causes for failures of
implants.</AbstractText>
</Abstract>
<Affiliation>CAMM Research Institute, Wayne, NJ, USA.</Affiliation>
<AuthorList CompleteYN="Y">
<Author ValidYN="Y">
<LastName>Neuman</LastName>
<ForeName>G</ForeName>
<Initials>G</Initials>
</Author>
<Author ValidYN="Y">
<LastName>Spangberg</LastName>
<ForeName>L</ForeName>
<Initials>L</Initials>
</Author>
</AuthorList>
<Language>eng</Language>
[... ... ... ...]
</Article>
[... ... ... ...]
</MedlineCitation>
```

**Table 18.5** Document type definition (DTD) for the XML document in Table 18.4

```
<!ELEMENT MedlineCitation (PMID, Article)>
<!ELEMENT Article (Journal, ArticleTitle, Abstract?, AuthorList?)>
<!ELEMENT Journal (ISSN, JournalIssue, Title)>
<!ELEMENT JournalIssue (Volume?, Issue?, PubDate)>
<!ELEMENT Abstract (AbstractText)>
<!ELEMENT Author (LastName, ForeName, Initials?)>
<!ELEMENT AuthorList (Author+)>
<!ELEMENT PubDate (Year, Month, Day)>

<!ELEMENT AbstractText (#PCDATA)>
<!ELEMENT ArticleTitle (#PCDATA)>
<!ELEMENT Day (#PCDATA)>
<!ELEMENT ForeName (#PCDATA)>
<!ELEMENT Issue (#PCDATA)>
<!ELEMENT ISSN (#PCDATA)>
<!ELEMENT Initials (#PCDATA)>
<!ELEMENT LastName (#PCDATA)>
<!ELEMENT Month (#PCDATA)>
<!ELEMENT PMID (#PCDATA)>
<!ELEMENT Title (#PCDATA)>
<!ELEMENT Volume (#PCDATA)>
<!ELEMENT Year (#PCDATA)>

<!ATTLIST MedlineCitation
        Owner (NLM | NASA | NOTNLM) "NLM"
        Status (Completed | In-Process) #REQUIRED>
<!ATTLIST AuthorList CompleteYN (Y | N) "Y">
<!ATTLIST Article PubModel (Print | Electronic) #REQUIRED >
<!ATTLIST JournalIssue CitedMedium (Internet | Print) #REQUIRED>
<!ATTLIST Author ValidYN (Y | N) "Y">
<!ATTLIST ISSN
         IssnType  (Electronic | Print | Undetermined) #REQUIRED
>
```

A DTD for the document in Table 18.4 is presented in Table 18.5. We can follow the structure of the XML document in Table 18.4 by reading the DTD: a citation is formed from an ID and an article; in turn an article contains journal information, an optional abstract and an optional author list. The journal is defined by its ISSN number, issue and title, while the author list contains a sequence of authors identified by name, surname, etc. The DTD also specifies how tags may have attribute values and what those values may be.

### *18.4.3 Advantages of the XML Format*

Designed to become an international standard, XML has the major benefit of being a de facto lingua franca for computer applications. Almost every application programming environment in use today provides tools and libraries for reading and storing data in XML format.

### 18.4.4 Document Object Model (DOM)

The conceptual representation of an XML document is the Document Object Model (http://www.w3c.org/DOM/), a software model of the tree-like data contained in an XML document. The process of translating an XML document into a DOM object is standardized and well understood. Numerous implementations exist, varying in complexity, performance and standards support.

Similar to XML, DOM is universal, and software applications working with DOM objects are essentially processing XML data. Some of the most common software tools, for instance Internet Browsers, provide seamless XML integration and DOM-level programming support. Similarly, a whole range of document management systems, search and indexing engines are able to index, search and retrieve information from XML files in a DOM-compliant format.

### 18.4.5 Practical Aspects of Using XML

The most controversial feature of XML is, possibly, its verbosity. Most critics of XML point to the relatively poor ratio between the size of a typical XML document and the quantity of information contained within the document: XML documents tend to demand a good deal of disk space but contain a relatively smaller amount of information compared with other storage formats. The main reason for this low ratio is understood to be the use of long tag names and the amount of repetitions of these tags; similarly, the use of closing tags is considered redundant, but is required as part of the XML standard.

These drawbacks can be alleviated by using compression techniques; this, however, introduces another level of complexity and can break the universality of the format. XML was intended to be (at least partially) readable by humans as well, and therefore the redundancy provided by the liberal use of tags is helping in that respect. In bioinformatics, where data grow at a rapid pace and are often downloaded over the internet via restricted bandwidths, the size of an XML document may become a significant slowdown factor in obtaining the data rapidly. Similarly, already high storage requirements for bioinformatics databases become even higher when using the XML format.

Another potential drawback is that most software tools designed to process data in XML format expect the entire data to be contained in one XML document with a root element. This assumption creates significant problems when loading data from bioinformatics databases, which may contain tens or hundreds of gigabytes of data. In such cases specialized tools must be used to load only parts of an XML document. Alternatively, data are structured into a collection of XML documents, each adhering to the same format; even in this case specialized tools operating outside of the XML standards must be used to aggregate the data into a unified repository.

### 18.4.6 XML Formats in Bioinformatics

XML is widely used in bioinformatics as a data format. Although not specifically containing biological or genomic sequence data, a common source of information distributed in XML format are the MEDLINE and MeSH databases from the United Stated National Library of Medicine (http://www.nlm.nih.gov/).

Today, the MEDLINE database contains over 16 million literature references, complete with abstracts. The database is widely used as a preliminary source of information on studies and scientific discoveries in medical, biological, genetics and biochemistry research. Many of the existing curated genomic and proteomic sequence databases contain direct references to the entries in MEDLINE that identify relevant publications. MEDLINE is distributed as a set of XML files together with the associated DTDs via the U.S. National Library of Medicine's website. The associated Medical Subject Heading (MeSH) database (see http://www.nlm.nih.gov/mesh/) contains a hierarchical structure of terms; MeSH terms are used to index the text in MEDLINE records, and, in turn, can be used to search text articles with variable degrees of specificity.

XML is also commonly used as a communication language between software applications. Most frequently encountered is the SOAP protocol which uses XML as a format for transported data in web service-based applications. Tools for workflow management systems are just another example of a growing number of bioinformatics applications using the XML format.

### 18.4.7 Other XML Concepts

The W3C consortium has also developed a number of standards dealing with concepts used for accessing and operating with XML documents. XPath (XPath 1999) is a language used to access/identify parts of an XML documents; it is closely related to the notion of paths in a tree. XQuery (XQuery 2007) is a language used to search XML documents. Relying on these, the XSLT standard (XSLT 1999) provides a flexible and powerful means of defining XML document transformations. The use of XSLT definitions for XML documents that result in HTML code displaying information in web browsers has become increasingly common in web-enabled software applications.

## 18.5 Relational Databases

Relational databases are a form of structuring data according to a well-defined relational model.

Informally, the basic building block of a relational database is the table: a set of rows, each row containing the same number of columns, where each column adheres to a well-defined type of values. Formally (Codd 1983), the relational model organizes data in relations, defined as a set of $n$-tuples of elements, each element in the $n$-tuple belonging to a distinctive set.

Often a table will contain a designated column that holds a unique identifier (ID) for each row. Such a column is not mandatory, and it is not uncommon to obtain unique IDs by combining values contained in more than one column in a table. Tables may have attached indices to improve the performance of locating individual rows.

A relational database may contain a number of tables which are often connected via relationships and constraints. These are means to ensure consistency of the data under consideration. For instance, in a database of genes where genes are associated with organisms, an integrity constraint may demand that any genomic sequence must be connected with an organism in the database. Such a constraint will prevent the user from entering a sequence into the system without indicating to which organism it belongs (where such an organism must already exist in the database).

### *18.5.1 Relational Database Modeling*

The relational database model is designed to allow definition of complex relationships and connections between the data, to ensure that the data closely reflect the structure of the domain it records. The process of translating data within a specific domain into a relational database is called database modeling. Database modeling is a fairly complex process involving a number of trade-offs. The first step in database modeling is the definition of tables and the formats of individual columns. More complex steps involve the definition of relationships and the normalization of the data. Normalization seeks to remove redundancies in the database, namely instances where data are being duplicated, either in a single table or across several tables. While formally there are several well-defined levels of normalization, each more complete than the other, in practice the goal of normalization is to find the right trade-off between redundancy and performance. The optimal result is highly dependent of the database system being used, and the frequency or most-common-type of access that the database will be subject to.

The result of a database modeling effort is a database schema: this is a signature of the tables and relationships contained in the database. A concrete database schema defines the exact tables in the database with details for each individual column. It also defines the relationships and constraints that are established between columns and tables in the database.

### *18.5.2 Relational Databases: A Worked-out Example*

For an example of a relational database we consider a relational model of the data contained in the XML example in the previous section. The database contains a list of citation records; a visual illustration of the schema is displayed in Fig. 18.1.

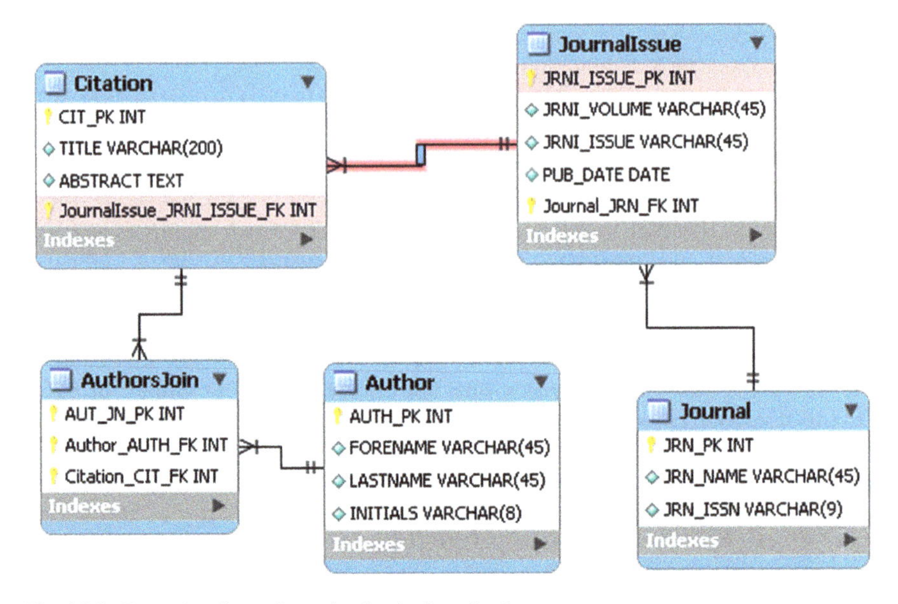


**Fig. 18.1** Example schema for a simple citations database

The data are structured in five tables, each table having a primary key identifier. The citation table contains data about a publication – the publication date, the title, abstract and a link to a journal issue record, which, in turn, contains data that identify the publication volume. The link is established via a many-to-one relationship (highlighted in Fig. 18.1) which can associate one or more publications to a journal issue. In turn, a journal issue is associated with a particular journal entity. In a slightly different approach, authors are associated with publications via a join table. This table lists associations between individual publications and author records; this way an author can be associated with more than one publication, and one publication associated with more than one author.

We visualize a sample of data from the five tables in Fig. 18.2. Three different citations are listed from three different issues of three journals, respectively. Two citations share an author ("L. Spangberg"), and this is captured by the authors join table.

Data can be extracted from the relational database using SQL queries. An example SQL query is provided in Fig. 18.3. The query selects the author name, publication title and journal name for each publication of L. Spangberg.

### 18.5.3 Relational Database Management Systems

Systems that implement the relational database model are known as RDBMS. There are several well-established software products, both commercial and open-source, that provide different levels of functionality. The standard today is in

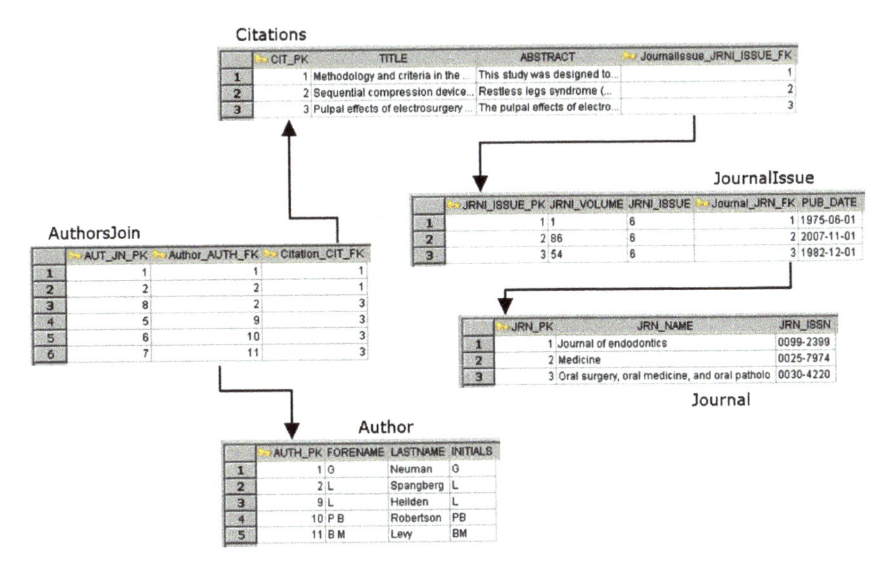


Citations

| | CIT_PK | TITLE | ABSTRACT | JournalIssue_JRNI_ISSUE_FK |
|---|---|---|---|---|
| 1 | 1 | Methodology and criteria in the ... | This study was designed to... | 1 |
| 2 | 2 | Sequential compression device... | Restless legs syndrome (... | 2 |
| 3 | 3 | Pulpal effects of electrosurgery... | The pulpal effects of electro... | 3 |

JournalIssue

| | JRNI_ISSUE_PK | JRNI_VOLUME | JRNI_ISSUE | Journal_JRN_FK | PUB_DATE |
|---|---|---|---|---|---|
| 1 | 1 | 1 | 6 | 1 | 1975-06-01 |
| 2 | 2 | 86 | 6 | 2 | 2007-11-01 |
| 3 | 3 | 54 | 6 | 3 | 1982-12-01 |

AuthorsJoin

| | AUT_JN_PK | Author_AUTH_FK | Citation_CIT_FK |
|---|---|---|---|
| 1 | 1 | 1 | 1 |
| 2 | 2 | 2 | 1 |
| 3 | 8 | 2 | 3 |
| 4 | 5 | 9 | 3 |
| 5 | 6 | 10 | 3 |
| 6 | 7 | 11 | 3 |

Journal

| | JRN_PK | JRN_NAME | JRN_ISSN |
|---|---|---|---|
| 1 | 1 | Journal of endodontics | 0099-2399 |
| 2 | 2 | Medicine | 0025-7974 |
| 3 | 3 | Oral surgery, oral medicine, and oral patholo | 0030-4220 |

Author

| | AUTH_PK | FORENAME | LASTNAME | INITIALS |
|---|---|---|---|---|
| 1 | 1 | G | Neuman | G |
| 2 | 2 | L | Spangberg | L |
| 3 | 9 | L | Heilden | L |
| 4 | 10 | P B | Robertson | PB |
| 5 | 11 | B M | Levy | BM |

**Fig. 18.2** A snapshot of data in the citation database

```
1 SELECT Author.LASTNAME, Citation.TITLE, Journal.JRN_NAME
2 FROM Author, AuthorsJoin, Citation, JournalIssue, Journal WHERE
3   Author.AUTH_PK = AuthorsJoin.Author_AUTH_FK AND
4   Citation.CIT_PK = AuthorsJoin.Citation_CIT_FK AND
5   Citation.JournalIssue_JRNI_ISSUE_FK = JournalIssue.JRNI_ISSUE_PK AND
6   JournalIssue.Journal_JRN_FK = Journal.JRN_PK AND
7   Author.LASTNAME LIKE '%Spangberg%'
```

| | LASTNAME | TITLE | JRN_NAME |
|---|---|---|---|
| 1 | Spangberg | Methodology and criteria in the evaluation of den... | Journal of endodontics |
| 2 | Spangberg | Pulpal effects of electrosurgery involving based ... | Oral surgery, oral medicine, and o... |

**Fig. 18.3** An SQL query against the citation database and its result

database servers, stand-alone software suites that serve data requests to clients over common network protocols; these store the data allowing users to add, delete or search for data in various ways via a well-defined and well-understood SQL language. Fortunately, due to the substantial effort invested in the development of relational database systems, these provide significant performance and functionality benefits. The major products (Oracle, MySQL, Microsoft SQL Server) have become ubiquitous; almost any computing department will be using a form of relational database system.

### 18.5.4 Advantages of Using Relational Databases in Bioinformatics

The major relational database systems are well known for their stability and reliability. The soundness of the model behind relational databases coupled with system reliability and well-proven record in the software industry makes an RDBMS a favourite with systems developers. In addition, any of today's major RDBMS will provide functionality for clustering and distributing data across multiple machines to cope with the increasing volume of bioinformatics data. Not least, the ability to update and add data in the system is a major improvement compared with systems that rely on static flat files on the disc.

### 18.5.5 Practical Aspects of Relational Database Systems

Any of today's RDBMS is a complex system requiring expert knowledge in configuration and deployment. Management and maintenance of large databases, especially some of the largest sequence databases, raises certain difficulties. Commercial systems may require expensive licenses to provide satisfactory performance with large volumes of data. Similarly, the formatting and modeling of data always leads to trade-off decisions: for instance, a decision on a particular length of a text field may have to be revised as more verbose data are added. It is not uncommon for restrictions on the maximum length of text fields or allowable number of binary fields to be reached with bioinformatics data.

The maintenance of a bioinformatics database within an RDBMS requires dedicated resources; when data format changes, tables may need to be recreated and indices need to be rebuilt. Depending on size, complexity and availability of hardware such operations can take significant amounts of time, during which systems may be partially unavailable. Coupled with the fast update cycle of some public databases, often maintenance of databases ends up taking a significant proportion of the available time of the total system. These problems amplify when building a system that contains a number of relational databases from disparate sources, when further translations and relationships must be custom-built.

These problems are not unique to today's RDBMSs. Indeed these problems are common to all systems that deal with large amounts of heterogeneous data that constantly change format and volume. The RDBMS is simply a tool in building such a system, while other dedicated data integration systems may have specific features that allows them to deal with the specific features of the bioinformatics domain.

Distribution-wise, relational databases are highly dependent on a particular implementation system. Due to licensing costs, relational database dumps are rarely distributed in other formats than the ones supported by open-source, freely available implementations, for instance, the MySQL Community Server.

### *18.5.6 Examples of Bioinformatics Relational Databases*

Often the data provider takes the entire effort of data modelling and management on themselves. Several projects publish bioinformatics data in relational form or directly by providing database servers access via the Internet where data can be retrieved directly.

The Ensembl database (Hubbard et al. 2007) maintained jointly by the European Bioinformatics Institute (EBI) and the Wellcome Trust Sanger Institute is a public database of annotated genome data covering a range of eukaryotic organisms, mostly concentrated on vertebrate genomes. The data are directly accessible via a MySQL server, although due to the complexities and the large number of schemas involved the Ensembl developers recommend access via a dedicated Perl API.

The Gene Ontology database (Gene Ontology 2000) is also distributed as a MySQL relational database; the database contains a database of genomic terminology structured into an ontological structure: a structure that defines a network of relationships between terms based on specific criteria. The gene ontology database is structured based on cellular component, biological process and molecular function criteria, and relationships identify, for instance, "is part of" or "is a" types of connections between terms. The Gene Ontology database contains direct information associating gene products with their functions and data existing in other public databases identifying, for instance, concrete sequences.

## 18.6 Bioinformatics Data Integration Systems

### *18.6.1 SRS*

SRS is a generic bioinformatics data integration software system. Developed initially in the early 1990s as an academic project at the European Molecular Biology Laboratory (EMBL), the system has evolved into a commercial product and is currently sold under license as a stand-alone software product.

SRS uses proprietary parsing techniques largely based on context-free grammars to parse and index flat-file data. A similar system combined with DOM-based processing rules is used to parse and index XML-formatted data. A relational database connector can be used to integrate data stored in relational database systems. SRS provides a unique common interface for accessing heterogeneous data sources and bypass complexities related to the actual format and storage mechanism for the data. SRS can exploit textual references between different databases and pull together data from disparate sources into a unified view.

SRS is designed from the ground up with extensibility and flexibility in mind, in order to cope with the ever-changing list of databases and formats in the bioinformatics world. SRS relies on a mix of database configuration via meta-definitions and hand-crafted parsers to integrate a wide range of database distributions. These meta-definitions are regularly updated and are also available for extension and modification to all users.

A number of similar commercial systems have been developed that replicate the basic functionality of SRS.

### 18.6.2 Entrez

The Entrez system is a website hosted by the National Center for Biotechnology Integration (NCBI) of the United States National Library of Medicine. The system provides a web-based interface to searching across a whole range of databases, from genomic sequence databases to literature references or taxonomical structures. The system is widely used in the academic community, especially in the United States.

### 18.6.3 Other Systems

A high level of fragmentation can be observed in the bioinformatics universe. Numerous software solutions have been built around specific databases and research projects. In terms of data analysis applications, the EMBOSS suite (Rice et al. 2000) is notable in building a comprehensive package that integrates a number of data formats and applications. The EMBL hosts a number of search and visualization interfaces for a whole range of bioinformatics data.

Recent trends focus towards semantic integration of bioinformatics databases. The goal of these projects is to provide a comprehensive integration of data using web services and semantic descriptions, establishing a common point of access towards disparate data sources. Building on ontological concepts and using Semantic Web (http://www.w3c.org/2001/sw) technologies as RDF and OWL (OWL 1999), projects as BioMoby (Wilkinson et al. 2002, http://www.biomoby.org) are on the forefront of bioinformatics data integration.

## References

Aho A, Hopcroft JE, Ullman J (1986) Compilers: Principles, techniques, and tools. Addison-Wesley, Reading

Altschul SF, Gish W, Miller W, Myers EW, Lipman DJ (1990) Basic local alignment search tool. J Mol Biol 215(3):403–410

Appel A (1998) Modern compiler implementation in C. Cambridge University Press, New York

BioPerl (http://www.bioperl.org)

Stajich et al (2002) The bioperl toolkit: Perl modules for the life sciences, Genome Res (12):1611–1618

Codd EF (1983) A relational model of data for large shared data banks. Commun ACM 26, 1 (Jan. 1983):64–69
The Document Object Model (http://www.w3c.org/DOM/)
Extensible Markup Language (XML) (2006) 1.0 (Fourth Edition). W3C Recommendation, March 2006 (http://www.w3.org/TR/REC-xml/)
The Gene Ontology Consortium (2000) Gene Ontology: Tool for the unification of biology. Nature Genet 25:25–29
GNU Bison (http://www.gnu.org/software/bison/)
Grune D, Jacobs CJH (1990) Parsing techniques – A practical guide. Ellis Horwood, Chichester, England (http://www.cs.vu.nl/~dick/PTAPG.html)
Higgins DG, Sharp PM (1988) CLUSTAL: A package for performing multiple sequence alignment on a microcomputer. Gene 73:237–244
Hubbard TJP, Aken BL, Beal K, Ballester B, Caccamo M, Chen Y et al (2007) Ensembl 2007. Nucleic Acids Res 35, Database issue:D610–D617
IUPAC-IUB Joint Commission on Biochemical Nomenclature (1984) Nomenclature and symbolism for amino acids and peptides. Recommendations. Eur J Biochem 138:9–37
OWL Web Ontology Language Reference (1999) W3C Recommendation November 1999 (http://www.w3.org/TR/owl)
Rice P, Longden I, Bleasby A (2000) EMBOSS: The European molecular biology open software suite. Trends Genet 16(6):276–277
The European Bioinformatics Institute (http://www.ebi.ac.uk/)
The UniProt Consortium (2008) The universal protein resource (UniProt). Nucleic Acids Res 36:D190–D195
The World-Wide-Web Consortium (http://www.w3c.com/)
XML (1999) Path Language (XPath). W3C Recommendation, November 1999 (http://www.w3.org/TR/xpath)
XQuery 1.0: An XML Query Language. W3C Recommendation January 2007 (http://www.w3.org/TR/xquery/)
XSL Transformations (XSLT) (1999) Version 1.0. W3C Recommendation November 1999 (http://www.w3.org/TR/xslt)
W3C Semantic Web Activity (http://www.w3.org/2001/sw)
Wall L, Christiansen T, Orwant J (2000) Programming Perl. O'Reilly & Associates, Inc.
Wilkinson M, Links M (2002) BioMOBY: An open source biological web services proposal. Brief Bioinformatics 3(4):331–341

# Chapter 19
# Programming Languages

**John Boyle**

## 19.1 Introduction

Programming and software engineering are not new disciplines. Although software engineering has under gone shifts in philosophy, the fundamental mechanism used for defining the logic within a system is still the same as it was decades ago. Advances in languages and constructs have made the development of software easier and have reduced development time. Each successive generation of programming languages has obtained this simplification by introducing further abstractions away from the complexities of generating machine specific instructions required to actually run an executable within or across operating systems. These advances are still occurring today, and we are now able to develop more complex programs more rapidly than at any time in the past.

This rapid progression has empowered the scientific developer, as modern experimental driven biological science requires the rapid development of algorithms and systems. Biology, in all its forms, is fundamentally an observational and experimental science. Whether it be ecology, neuroscience, clinical studies, or molecular biology, the high volumes of semantically rich large biological data sets require a high level of software development. This means that the software must be developed to a high standard and in a minimal amount of time. Therefore, to meet the demands of developing software to support research, the scientific developer must know about the latest tools and techniques. This chapter introduces some of these tools and techniques, in particular those that will help in the development of data intensive applications.

This chapter will introduce the facets of software engineering that are most relevant to the life sciences. The first section gives a background to programming and introduces different types of language that it is important to be familiar with. The second section describes the issues that must be considered when building applications for research driven usage. The third section introduces the relevant standards

J. Boyle (✉)
The Institute for Systems Biology, 1441 North 34th Street, Seattle, WA, 98105, USA
e-mail: jboyle@systemsbiology.org

D. Edwards et al. (eds.), *Bioinformatics: Tools and Applications*,
DOI 10.1007/978-0-387-92738-1_19, © Springer Science+Business Media, LLC 2009

and tools that are of use to life science software developers. The final sections specifically describe how to build systems to work with large volumes of heterogeneous scientific information.

## 19.2 Types of Programming Language

Even over the last decade, coding languages has evolved considerably. While the historic categorical view of programming languages (see Fig. 19.1) still exists, there is a blurring between these categories. Traditionally, languages were divided into three categories: procedural (e.g., pascal/modula 3), functional (e.g., lisp/scheme) and logical (e.g., prolog/eclipse). Because of the dominance of procedural programming, even within the life sciences, there is a tendency to also consider the style or method of programming rather than the basic underlining formalism: for example, typed or untyped, complied or interpreted, managed or unmanaged, client or server, web or desktop, application or service, and static or dynamic.

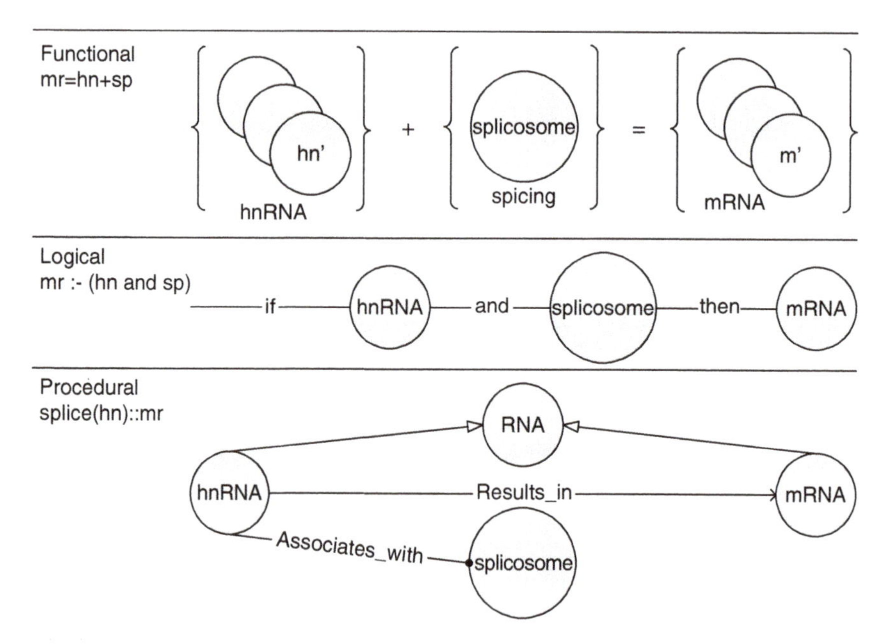


**Fig. 19.1** Historically programming languages were categorized into three, occasionally overlapping, categories. These categories broadly defined both the mode through which the software was constructed and how the software modeled both data and process. The categorizations were: functional based, which provide a framework for defining and evaluating functions; logical based, which allowed for the definition of logical rules which were typically executed in a non deterministic way; and procedural based, where all commands and sequences of operations were formally described

In common usage within the life sciences, there exist a plethora of different types of languages. Each of these languages has unique advantages and disadvantages within science, and each type of language generally has its own niche. Unfortunately, the choice of language used for a specific task generally reflects the training of a particular scientist or developer, rather than the most suitable "tool for the job." It is, therefore, important that someone working in the field have a good working knowledge of the major types of languages being used, these are compiled procedural languages (e.g., Java and C#); dynamic interpreted languages (e.g., Perl and Ruby); and mathematical function scripting environments (e.g., Matlab and R).

It is important to note that languages are still evolving, as can be seen through advances related to: dynamic querying, for example, the use of LINQ in .NET, or QL in EJB 2.0; aspects providing a means for both code injection to support migration of components and detachable "plain old java objects" (POJOs); and domain specific enhancements. New types of languages are also becoming available, which could well have direct implications in the biological sciences as they allow for a richer modeling of complex behavioral systems (e.g., using $\pi$-calculus).

The following sections introduce the types of languages that are most prevalent in bioinformatics. Further information about the use of these languages in explicit areas is given in the domain standards section.

## *19.2.1 Compiled Languages*

In bioinformatics the de facto standard for a compiled language is generally considered to be Java. Most people have an understanding of how compilation of programming languages works: a front end compiler performs checking (semantic and syntactic) and graph construction; and the back-end compiler takes this information to generate a specific optimized version of the program. Compilers are generally multi-pass and, where needed, can be just-in-time. The code that is generated can be for a specific hardware or, as is more common in bioinformatics for a virtual machine. It is this compiled behavior that gives both the advantages and disadvantages of using these languages.

Because of their mainstream nature, the evolution of compiled procedural languages can be correlated with changes in requirements (see Table 19.1). The requirements have driven the evolution of these languages, and conversely advances in computer science have enabled new requirements to be met. The actual methodology through which data can be modeled in compiled procedural languages has evolved significantly, and as we now see in frameworks such as .NET 3 or Java 6 formalized aspects are now used to simplify the development of complex coding.

Illustrative and simplified examples of how evolution of software development processes and practices have co-occurred with changes in requirements. In the 1980s, most software engineering methodologies and associated technologies were geared toward ensuring that the code was maintainable, the use of abstract data types to model generic collections and the software structure based Object

**Table 19.1** Evolution of software methodologies and technologies

| | Methodology | | Technology | |
|---|---|---|---|---|
| Requirements | Design | Process | Comms | Server |
| Code maintenance | ADT | OMT | Sockets | |
| Ease of use | Objects | UML | RPCs | X/Open DTS |
| Semantic richness | Components | RUP | Brokers | EJB 1.1 |
| Data integration | Frameworks | Agile | P(M)BV | EJB 2.0 |
| Rapid development | Aspects | MDA | SOAP | EJB 3.0 |

Modeling Technique (OMT) were common methodologies. With advances in the hardware, the requirements were for easier to develop systems, and later, the need (particularly in the scientific domain) was for the ability to model semantically rich heterogeneous data types. With the advent of distributed computing, there was a need to support data integration "as is," which required more powerful technologies. More recently, there has been a requirement for the rapid development of applications, which has meant that new methods and technologies have become available, which make it possible to write high quality code which can model and integrate semantically rich data with a substantial reduction in development time. The evolution of design methodologies has been from the use of Abstract Data Types (ADT) in the 1980s to model generic collections; to the introduction of object oriented programming ideas; OO programming evolved into component based programming, most noticeably in the use of graphical user interface components/widgets; frameworks then became popular as they provided this life cycle management for a wide range of object, and led to the current use of Aspect Oriented Programming. Software process and modeling have also evolved: OMT was naturally superseded by the complete Unified Modeling Language (UML); processes such as the rational unified process (RUP) became popular with organizations, although problems with lack of understanding or poor adoption lead to the impetus for the development of more "practical" agile processes (XP, SCRUM); more recently (in the life sciences and elsewhere), there has been a growth in the use of model driven design (MDD/MDA) which focuses on the automatic generation of code from a well defined (typically UML) model. Similarly, communication technologies have evolved from the powerful but difficult to use sockets, through to the easier to develop remote procedure calls; different CORBA versions introduced pass-by-value and other possible enhancements (e.g., component facets); difficulties with CORBA gave rise to the simpler to use Web Service architectures, which are now common place in the life sciences. The application server evolution (EJB standard) is briefly discussed in a later section.

Modern compiled procedural programming languages do offer a number of features that are essential for developing code in the fast moving area of scientific programming, they are dynamic in nature through language extensions such as LINQ and the use of dynamic scripting through systems such as Groovy (see below for information about the merging of dynamic behavior with compiled languages); a high degree of "separation of concerns" through dependency injection, so that each coded item generally performs one function and dependencies between the code and other components are injected at run time; and the use of patterns (and aspects) to prompt the development of high quality code through both reuse and easy to understand design. However, it can also be argued that some of the recent developments in software engineering are less suitable for research science, for example, "factory software" mechanisms which assume everything can be constructed using generic/reusable workflows, or MDA which assumes you can accurately describe the domain you are working in. The developments that are less useful are those that rely on the understanding that information and flow can be well modeled, which is not always the case in a research led environment.

A large number of tools are available to aid in the development of applications using these languages. With the large number of libraries, conventions, standards, and patterns available, it is imperative that an integrated development environment (IDE) is used. These environments also enable the convenient use of a number of powerful tools, including profiling, remote debugging, multi-threaded code tracing, and variable/stack watching. To date, the most suitable (free) IDE is Eclipse (www.eclipse.org) which offers support for the major computing technologies that will be used by a scientific developer.

### *19.2.2 Dynamic Interpreted Languages*

Although it is possible to compile some dynamic interpreted languages, and to interpret some "compileable" languages, the distinction between the two is a useful one. It is arguable that compiled languages are most suitable for the development of libraries and enterprise applications, while dynamic languages are best used to provide small utility applications and to glue components together. This is obviously an over simplification; however, generally this is how different types of tools are used within the life sciences. The three most popular dynamic languages are introduced below:

- *Perl* is the most popular interpreted computer language in bioinformatics and has been used in a number of large scale scientific projects (e.g., Ensembl). It is a flexible and powerful tool, which allows for the rapid development of reasonably sophisticated applications. With Perl's powerful scripting and parsing functionality, it is a good language for both "one line text processing" as well as two tier (see below) database applications. One of the advantages of Perl, over the other interpreted languages, is that its large following has built a generous repository of standardized modules which is readily available to developers. Because of space considerations, it is not possible to present the relevant features and functionality of Perl for the development of research applications. However, for information about Perl and related modules, the developer can do no better than explore the Comprehensive Perl Archive Network CPAN (www.cpan.org).
- *Ruby* is a language that has been in existence since 1993 but has recently seen growth in its audience due to the usage of its web application framework called

Rails (www.rubyonrails.org). Ruby is a modern object-oriented language, and due to its history, it has a number of advantages over other languages (e.g., dynamic class extension). While Ruby may be considered a "better" language than Perl, it does not have the same following as Perl, and so for traditional bioinformatics is unlikely to gain widespread adoption within the community. However, we can expect to see a rise in Ruby usage with the adoption of Rails for the rapid development of powerful database centric web applications. As Rails uses a series of conventions to make the development of web application faster (views communicate with controllers which update models), it is relatively easy for a new developer to adopt and use this technology. The best place to start working with Ruby is www.ruby-lang.org.

- *Python* is more similar to Ruby than Perl and is another powerful dynamic object-oriented programming language. Because of the simple and unambiguous syntax Python has been used extensively as a teaching tool. Python's popularity is also due to its ability to integrate well with other languages. The main python web site (http://www.python.org) contains more information about this language.

There exist numerous other examples of dynamic languages, although these are less important in the biological sciences. All dynamic languages share the advantages of being easy to use; highly suitable for prototype work; and flexible to allow for dynamic extensible typing. The disadvantage of these languages is that they are generally not suitable for large scale infrastructure or data intensive projects.

It is important to note that the distinction between dynamic and compiled languages is constantly narrowing. Using a script to orchestrate lower level services (or components) is becoming more common place. This means that dynamic languages are often used to provide the dynamic behavior that is often missing in compiler based applications. Compiled languages can also have dynamic behavior overlaid upon them (e.g., Groovy http://groovy.codehaus.org, JSR-274 http://www.beanshell.org), and dynamic languages can be compiled (e.g., via JRuby http://jruby.codehaus.org).

### *19.2.3 Mathematical Function Environments*

Within a life science organization, there typically exist different groups who are collectively termed "bioinformaticians." Some of these concentrate on the development of models of biological systems or undertake bespoke biological analyses. Typically, the preferred tool for such work is a mathematical or statistical functional framework such as Matlab or R. Other tools exist (e.g., S, Mathematica), but these are less prevalent in bioinformatics.

- *Matlab.* Matlab is a commercial mathematical functional scripting tool development by MathWorks (www.mathworks.com) and was originally widely used by the electrical engineering community. Since its inception, Matlab has grown as a commonly used tool across many scientific domains. For the life sciences, a number of useful "toolboxes" are provided, including one specifically for bioinformatics, as well as a statistical and simulation toolbox. Matlab is a commercial product, which has implications for many research organizations.

- *R*. R is a statistical analysis environment, which was based on S, commercial statistical package developed by AT&T Bell Laboratories. R is a freeware open source community built and maintained project (www.r-project.org). It has a rich functionality and is useful for many types of analysis. As it is not a commercial product, the level of support and documentation can cause problems for people learning to use the application (although the tutorials are generally kept up to date). R has good bioinformatics support through the bioconductor project (www.bioconductor.org), particularly in the area of genomics and microarray analysis, and a number of packages are available which can be used to support such tasks.

These functional environments offer powerful tools that provide frameworks in which analyses can be undertaken, queried, and dynamically manipulated. When undertaking a modeling or statistical analysis task, these are generally the tools of choice. The problem with these tools is that they tend to be untyped closed environments, and so integrating them with other tools is difficult. There is little problem with integrating a method locally that is to be run within one of these tools, as they both provide a means to marshal data to typed languages. However, this process is not automatic and does required the use of either standardized Java Marshalling classes (with the Matlab Java Compiler), or the use of a special (e.g., S4) extension to define the type marshalling. There are problems in using these tools in an enterprise environment, as both the tools are currently designed to work as non thread safe client based environments.

## 19.3 Scientific Programming

In the ideal world, software would be fully specified, well documented, componentized, unit tested, reliable, and robust. However, the harsh reality of working within research science frequently means that the competing requirement to deliver a timely solution often takes precedence. Thus, the developer must have a good understanding of the best practices that will enable them to rapidly develop well engineered solutions.

With any software project, there is always a chance that a project will fail. This can be avoided by understanding the reasons that are often cited for lack of project success (Booch 1996):

- *Identify the risks that are involved.* It is always a good idea to work out the risks involved in any project, and then to try and establish ways of minimizing them. Most technical risks (e.g., unknown technology, scalability, and performance issues) can be alleviated through early prototyping.

- *Do not get "blinded" by technology.* The appropriate choice of technology is important in any project, but it cannot be the sole focus of the project. If the design and end results of a piece of software becomes too focused on a specific underlying technology, then the project may ultimately fail. It is always a good idea to question why specific technologies are being used, and also to ensure that designs are not (too) restricted by your specific implementation choices.
- *Focus on what has to be built.* It is all too common in the software industry for the "wrong thing" to be built. This is largely due to the lack of communication and is probably the biggest cause of software project failure. Within science, this is a huge problem, as usually there only exists a loose idea of what is desired in the first place. The best way to avoid this issue is to ensure that release cycles are short, and that as many people as possible get to see the evolving iterations of the software. Such short iterations will allow you to more easily identify which features are appropriate.

The above points are all practical points that should always be followed in any software engineering project. Generally, within research, it is also a good idea to: first, develop a prototype to both minimize (technical) risks and to demonstrate to other scientists what you are planning on building; when actually developing the system, always consider the practicalities and cost of building on top of a legacy system versus undertaking a complete rewrite; and try to work toward short iterations, or even the adoption of automated building procedures. An important point is that you should never be afraid of a rewrite, as it generally will improve the quality and functionality of the system, and continually adding new features to an old and inappropriate codebase is a poor way to write software.

With scientific software, the following should also be considered:

- *A flexible software design is essential.* With science, it is highly probable that requirements will change within a matter of months, so a flexible and configurable design should generally always be adopted. Flexibility, and mutability to change, is an old requirement, and the same ideas about standardizing layers from over 30 years ago are still reflected in current software designs (see Figs. 19.2 and 19.3). A layered design means standardizing on components and interfaces to make maintenance of the code easier, this can be aided by adopting a software project system (e.g., Maven) and a suitable application platform (e.g., OSGi for client applications, or Spring for server based applications).
- *Do not over model "scientific information."* Because of the uncertainties and complexities of research, over specification can be a problem if the resulting software is supposed to be useful to a collection of scientists. While design is essential in a software project, over design and formalization of scientific information can cause problems as the resulting software could well have a certain model of a biological system (typically expressed as an object or data model) which may be either too simplistic or too complex for other users of the system. While any design is generally better than none, the reader should be aware of the common pitfalls when using a formal design process in a research environment. The problem with any full and unified software design endeavor is that there is

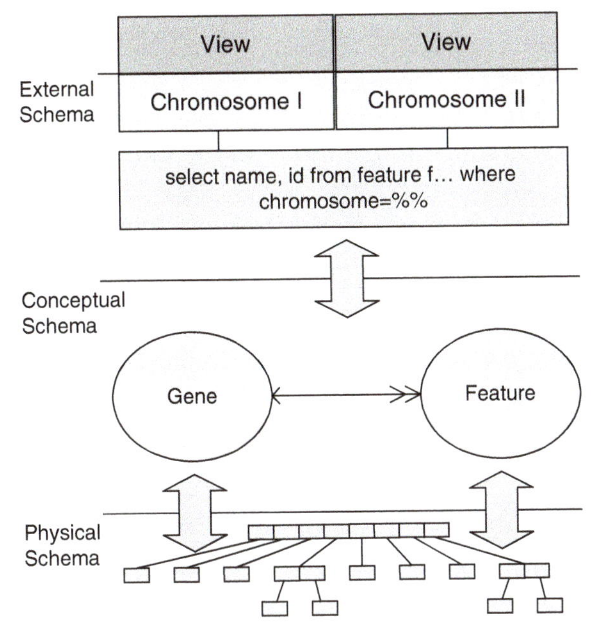


**Fig. 19.2** The ANSI-SPARC three layer database architecture was proposed in 1975 and is still used in modern RDBMS. The three proposed layers were a physical schema which defined how the data are actually stored (inode information), a conceptual schema which represented how information was related and indexed, and an external schema which represented how information was presented. The architecture was designed to provide immunity to change: the physical schema defined how the actual information was stored, and could be changed without effecting how external applications interacted with the data; and the external schema could be changed to define richer APIs, without having to change the underlying storage mechanism

no such thing as the quintessential scientist, as the only commonality that really exists between researchers is that they are all different. These differences arise as each investigator and investigation studies the unknown, which means that the rationalization, which is typically needed for the design of expedient software utilities, is missing.

- *Ensure the system is inclusive*. It is rare that any software will work in isolation, both the people using the software and the enterprise in which it runs are equally important. Although the advent of TCP (and UDP)/IP in the 1970s was heralded as the solution to interoperability, many factors (including social and developmental reasons) meant that this was not the case. Interoperability, and associated integration issues, is still a problem today. To help solve this problem over the last decade, in distributed computing, there has been a shift in philosophy from thinking in terms of the transportation of object graphs toward the retrieval of related documents. The distinction between documents and objects is subtle, albeit important: objects are for programs, whilst documents are for people. An object is by its very essence a "black box" which contains domain and platform specific information. Objects must be explicitly translated between languages,

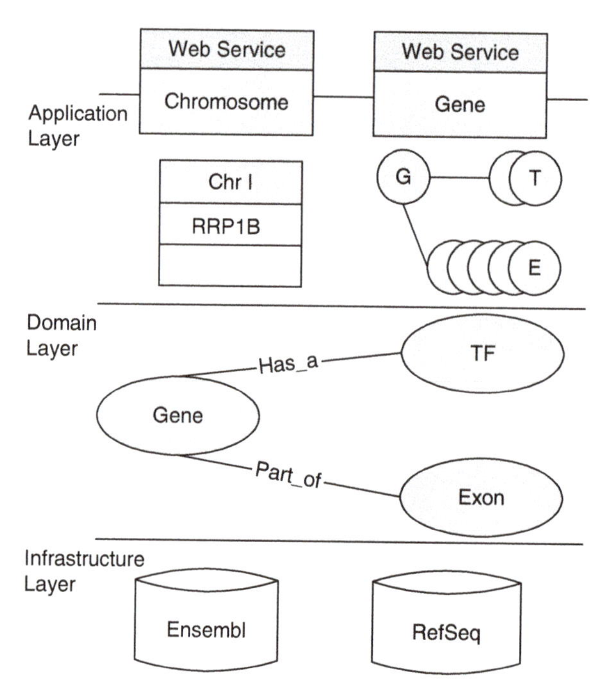


**Fig. 19.3** Design of a system within an application server typically involves three layers: an infrastructure layer which stores the data; a domain layer which is populated with vertical object models; and an application layer which models how an application can interact with the domain objects. The application server provides both resource management and service based functionality for each of these layers. In EJB terms we have: the infrastructure layer corresponding to a database, typically with O-R mappings through hibernate; the domain layer corresponding to entity beans, which basically function as an object cache; and the application layer corresponding to sessions beans or Web Services (with stateless session bean mappings)

and must be serialized (marshaled/externalised) for transmission through an object protocol (e.g., CORBA, RMI, DCOM, .NET Remoting). Documents are open and readable, and so lend themselves more easily toward the social aspects of a distributed system. With a document centric approach, the interactions are more natural and flexible: the document can be saved and retrieved from a file system using standard desktop tools; the information can be retrieved through numerous media, for example, through a web page or from an email received from a colleague; and documents can be directly browsed and their contents edited. This flexibility in delivery and interpretation means that there are multiple ways for documents to be integrated.

- *Beware the "not invented here" mentality*. The risk with any research lead computing project is that novelty can get confused with innovation. By focusing on innovating through using and extending existing solutions, the best use of the available computing technologies can be made. When building systems, it is essential to reuse the best tools from other application areas as this ensures a high return on investment. Continually developing completely new solutions is a waste of resources and funding and typically leads to a substandard and non maintained solution. By carefully considering a problem, with a good current knowledge of computer science, it is rare that a system will have to be built completely *de-novo*. Bespoke applications are required in the life sciences, but these can generally be built upon third party solutions or technologies. Project management systems, such as Maven, can help simplify this process and should always be considered if working with third party components or if working in a geographically distributed team. Basically always be wary of "reinventing the wheel," as this can lead to substandard and costly code. That stated, you should also be careful with choice of technology, as a poor decision early on will also lead to problems. There are a large number of technical and domain specific technologies which are available for use. These solutions range from class libraries which offer useful precanned solutions to full enterprise wide integration system. These standards are considered in the following section.

## 19.4 Programming Standards

Appropriate use of standards and standardizations are important as, if correctly used, they can save years of programming time and result in high quality maintainable code. The problem is that there are a large number of standards to choose from. These standards range in functionality and include: interface standards which define the method level interface that a specific implementation should expose or use; classification standards, typically defined as an ontology or a taxonomy, which represent structured knowledge and are typically used as the basis of controlled vocabularies; and reference implementations and class libraries which provide fully functional implementations with standardized (and stable) interfaces.

Standards can be categorized into one of two flavors:

- *Technology (horizontal) based*: these deal with fundamental computing issues which span more than one domain (vertical). Examples of these include communication, database, portlet, transactions, messaging, security, concurrency, and application standards. There are a number of bodies which provide these standards and reference implementations. Currently, the main bodies/processes that are relevant include the OMG, JSR, W3C, and OASIS. The most relevant of these are discussed below.
- *Domain (vertical) based*: these are standards that deal with a specific area (scientific or otherwise) and cover nearly every relevant domain (e.g., from anatomy to zoology). These standards tend to evolve from community, rather than company, based efforts. The appropriate use of these standards is harder to judge as they tend to be less regulated (and so their thoroughness, usefulness, and life span can be difficult for the non domain expert to gauge). There are a large number of these standards, and the main community based efforts are discussed below.

### 19.4.1 Technology (Horizontal) Standards

Computer science and software engineering are continually evolving fields. This evolution is pushed by changes in the demand for new types of usage for software and pulled by technical advances which open up new avenues for improvement. This continual change makes the development of solutions difficult; as today's "stunningly fabulous" technology is tomorrow's legacy system. The use of technology based standards can help alleviate this problem, as they ensure that the mechanism through which the underlying code is accessed is well understood, meaning that the developers in the future will easily be able to understand how the code works. As most standards come with reference implementation, their adoption can also save time and result in better, more maintainable applications.

Within the Java community, a number of well known open-source collections of projects are available which generally provide high quality code (e.g., apache, codehaus), as well as sites for other more specific projects. Where appropriate, this chapter gives recommendations to these projects, more information can be found in the enterprise application development section. For commercial resellers, it is worth noting that the non aligned LGPL and Apache licenses are commonly used, which enable flexible (Apache being the more flexible) redistribution of the code.

### 19.4.2 Domain (Vertical) Standards

A number of domain specific standards bodies exist, these include formal bodies (e.g., W3C, OMG) as well as academic bodies. The general standards bodies do have groups specifically dedicated to the life sciences, some of which are more active than others (e.g., HSCL from W3C (www.w3.org/2001/sw/hcls) and LSR from OMG (www.omg.org/lsr).

Standards in the life sciences cover an enormous area, they range from standards dedicated toward patient records, standards for defining clinical information and trials, through to pure scientific standards for specific domains (see Fig. 19.4). When working with high throughput experiment information, the following standards may be of interest to the bioinformatics developer:

- *Proteomics standards*. The Proteomics Standard Initiative (www.psidev.info) works with different groups to define a number of standards in a number of experiment areas (e.g., mass spectrometry, protein modifications). These standards are fairly well adopted within certain areas (e.g., use of mzData to define spectra information from MS), although the work is still progressing (e.g., unifying mzData and mzXML into mzML).
- *Microarray*. The MGED working group (www.mged.org) has a comprehensive set of standards for describing microarray experiment and results. As with other standards, this is still evolving and is largely working to simplify the use of the standard (e.g., through both MAGE-TAB and MIAMI 2).

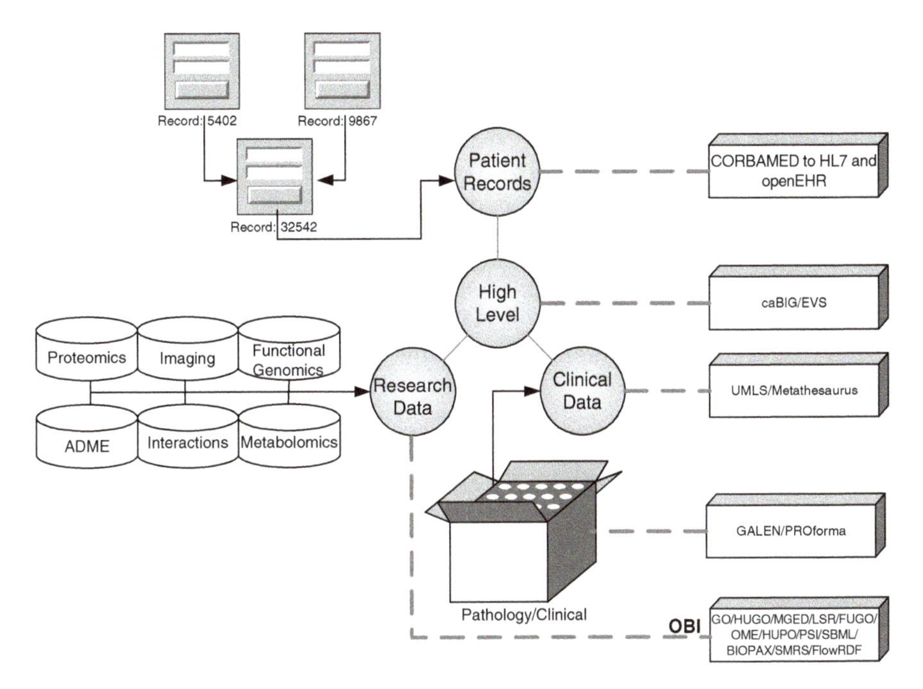


**Fig. 19.4** Overview of standards in different areas of the life sciences. Standards exist throughout the life sciences: in healthcare, there are standards for electronic patient records (EPR); in the clinical arena there are standards for defining and running clinical trials; and in R & D there are standards for numerous different research experimental areas. Standards also exist for specifying these standards, and for providing high level integration mechanisms across different domains

- *Microscopic imaging standards.* These largely mean the Open Microcopy Environment (OME www.openmicroscopy.org), which has defined a set of XML based standards for microscopy equipment (the type and setting information), information about data capture (which dimensions are measured, e.g., wavelength and position), and information about individual experiments (annotations and experimenter information). The OME go further than the standards and also provide communication specifications and file formats (e.g., OME-TIFF where meta information in OME-XML is inserted into the TIFF header), and provide two working implementations (an older Perl based solution, and a new Java rewrite).
- *Interaction standards.* BioPAX provides a standard for defining (and therefore sharing) metabolic pathway and interaction information. This information is available from Pathway Commons (www.pathwaycommons.org). This is still an active area of research, and the biopathways consortium (www.biopathways.org) is still working on standardization. There also exist standards for (generally) kinetics based models (e.g., SBML, CellML).

A number of ontology based descriptions also exist which can be used as both the basis for complex mappings and to construct a controlled vocabulary. These range in complexity from high level ontologies that describe largely abstract "experiment" concepts (e.g., OBI obi.sourceforge.net) through to detailed information about numerous domains (see obofoundry.org for more information).

As well as standards, there also exist a number of domain specific "class libraries" and "data access protocols." These can be of use when building an application in a specific area, typically their main usage is in genomics:

- *Class Libraries*. The most mature and active class library is BioPerl (Stajich et al. 2002; www.bioperl.org), although there do exist others including Biojava (Holland et al. Bioinformatics 2008), Bioruby, and Biopython. Most of these exist under the umbrella of the "open bioinformatics foundation" (www.open-bio.org). The BioPerl bioinformatics community has developed a large number of modules, largely in the area of genomics. These offer a wide range of sequence file parsers (e.g., swissprot, fasta, embl, genbank), and analysis tools (e.g., alignment operations, or data retrieval operations), which can be used to support common bioinformatics related tasks.
- *Data Access Protocols*. There exist a number of standards for accessing information from biological data sources. These range from those that provide an abstraction away from persistence information stores for specific data types (e.g., BioSQL www.biosql.org), to standardized data retrieval mechanisms from remote sites using http encoding (e.g., www.biodas.org), to systems for describing the "semantics" of specific Web Services (e.g., BioMoby (Wilkinson and Links 2002)).

## 19.5 Enterprise Application Development

As the biological sciences have advanced, so have their computational needs. With automated technologies able to reach terabytes a day throughput rates, there is a strong requirement to organize and analyze the data efficiently. Ready access to up–to-date information, and algorithms, requires a different mode of development than most computational biologists are trained for. To work with biological data, applications have to know how to access distributed data sources and know how they can function within a distributed environment. This requirement for data intensive architectures requires moving away from single machine based applications toward the development of federated infrastructure based solutions (see Fig. 19.5).

There are three issues which will be discussed in developing such applications:

- *Developing a distributed server*. There are a number of "server containers" which can be used to develop an application which is to be used in a distributed environment. These server containers have the advantage that they can both cut development time and can provide useful functionality that would otherwise be expensive to build. These containers generally allow the developer to concentrate on just defining the "business logic" of their application, rather than the complexities of more general server based operations (e.g., clustering, transactions,

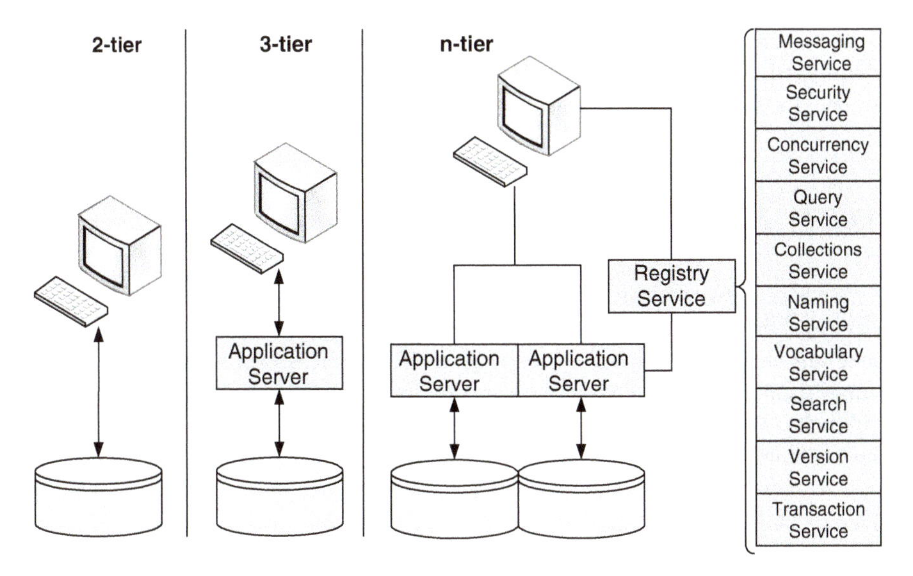


**Fig. 19.5** Distributed systems can be classified by the number of tiers that exist with the logical representation of the application. Two tier applications have a client and server component, with the client component making direct calls on the server (data storage) component. Three tier applications typically use an application server which provides a level of abstraction away from the underlying data store. N-tier applications are full distributed systems where components can be discovered (typically through a registry or directory system) and methods invoked

object-relational mappings, concurrency, security, communications). The correct choice of container depends upon the type of data, and the operations that are to be applied to the data. These issues are discussed in the "server frameworks" section.

- *Providing a service*. Once your "business logic" has been deployed within a container and is running as a server, it is important to ensure that it can be accessed by people and processes within your organization. Each of the servers can be viewed as an individual service, with some service offering "cross-cutting" functionality which can be used by other services. The term "service orientated architecture" (SOA) has been used to describe such a collection of servers within an enterprise, and has become strongly associated with the usage of Web Service and related standards. Web Services offer a simple mechanism to promote interoperability, which is the key within most life science enterprises.
- *Orchestrating multiple services*. Although SOAs offer flexible bottom-up functionality, to be useful they must be orchestrated together. Such orchestration often requires the implementation of top-down logic over the bottom-up services. These issues are discussed in the "orchestrating services" section.

The history of enterprise architectures within life sciences, including those used in both commercial entities and academic institutions, is discussed in the following section. The following three sections discuss the three main issues in developing distributed data intensive applications, by describing server frameworks, service oriented architectures (SOAs), and methods for orchestrating services.

### *19.5.1 History of Enterprise Systems within the Life Sciences*

Life science research enterprise data integration and process management systems have evolved over the last 15 years, effectively since the creation of open interoperable object based communications (e.g., CORBA). This evolution has been from single database based solutions to open, distributed, interoperable data management solutions. This has been driven by demands for rapid development, high levels of interoperability and increases in data volume and complexity.

The development of data management systems to support the life sciences has undergone a number of fundamental changes in the last decade. As in other areas, the history of enterprise systems in the life sciences is, in essence, one of a cultural change from the development of proprietary solutions, designed from the top-down, toward more flexible bottom-up architectures informed by open standards solutions (see Fig. 19.6). This change has been driven a number of factors including user requirements, where users have demanded more of systems in terms of flexibility and extensibility. This evolution of data integration and management technologies can be categorized into three stages:

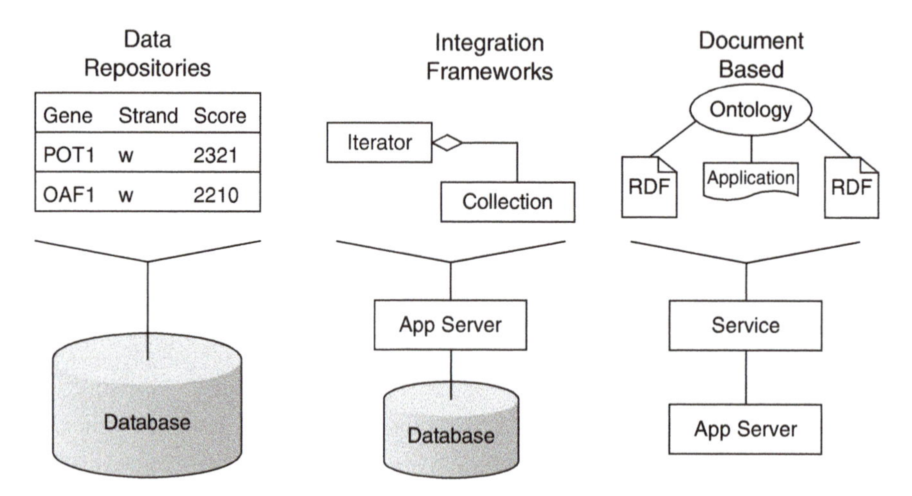


**Fig. 19.6** Evolution of enterprise architectures has occurred within the life sciences. Limitations in the flexibility of data repositories based solutions helped shape the development of integration frameworks. Integration frameworks suffered from complexity and interoperability problems, and so document based solutions are now becoming the normal model

- *Data Centric*. Initially, data repositories were developed and integrated using either external indexing services or data warehouse mechanisms. The data repositories rely on a variety of technologies, including index based (e.g., SRS (Etzold et al. 1998), DBMS (e.g., Ensembl Birney 2004), and federated database approaches (e.g., DiscoveryLink Haas et al. 2001). The development, of these data centric solutions, was driven by the availability and standardization (e.g., OleDB, ODBC) of relational database management systems and the requirement for a federated approach to data warehouse solutions. The lack of object interoperability of such data repositories gave an impetus for the development of object based top-down standards.
- *Object Centric*. With these approaches, standards bodies (e.g., LSR (http://www.omg.org/lsr), I3C, caBIO project (Covitz et al. Bioinformatics 2003; http://cabio.nci.nih.gov/)) decided on interoperable object standards. It was anticipated that these standards would be taken up by the life science industry. There has been some success with such standards, but as they had to capture all areas of a domain their complexity limited their broad-scale adoption. This object level standardization was largely driven by the maturation of object based protocols (e.g., CORBA with pass-by-value or DCOM) and associated object services (e.g., distributed registries, traders) which used interface definitions (e.g., IDL, MIDL) to formalize the static distributed interfaces. Such standards, either proprietary or open, were mainly implemented using integration frameworks built using application servers. They were introduced into a large number of pharmaceutical companies (e.g., Alliance from Synomics, GKP from Secant, Discovery Center from Netgenics). These integration frameworks suffered as their rigidly typed systems were difficult to extend and could not keep pace with evolving research requirements and new experimental technologies. Processing pipelines were also integrated within these tools. The requirements for the orchestration of analysis tools led to the growth in the number of in-house tools designed specifically for rapid development and deployment characteristics, rather than interoperability or complexity (e.g., SOAPLab (Senger et al. 2003); GenePattern (Reich et al. 2006); SBEAMS (Marzolf et al. 2006)). The integration frameworks were built upon the maturing application server products which were principally Java based (e.g., EJB 2+).
- *Document Centric*. Document based solutions (typically Web Service based) became popular as they provided a means to develop solutions that could: keep pace with the rapid advances in research; were scalable, robust, and easy to develop; and were interoperable. These are now widely used as a basis for integration (e.g., MRC's CancerGrid, NCI's caGRID). The advantages of these approaches are based on their lightweight specifications and ease of implementation. Such advantages can be seen in adoptions of related standards (e.g., DAS). Newer programming methods and frameworks simplified the development of these document centric (Web Service) systems. By using these frameworks (e.g., EJB 3.0, .NET framework), complex behaviour between Web Service can now be developed relatively simply. One of the challenges associated with Web Service architectures is their lack of semantics. In response,

ontology-based solutions were developed (S-BioMoby, Sophia from Biowisdom), although these largely depended on slowly evolving globally accepted ontologies. Designs using distributed runtime type checking and type mapping are now emerging (http://informatics.mayo.edu/LexGrid), as these provide for a means of integration and robustness that place fewer restrictions on developers. A number of eScience solutions were also associated with similar architectures, although the majority of these have converged with the mainstream Semantic Web (e.g., myGRID (Goble 2005)). A number of tools to support these service oriented solutions have been developed, including graphical tools (e.g., Taverna (Oinn et al. 2004)), schedulers (e.g., GridFlow (Cao et al. 2003)), and script translation tools (GridAnt (Amin et al. 2004)). The semantic web based solutions are being driven by both the convergence of standards for document based enterprise computing (e.g., WS-*), and the development of knowledge representations and query standards (e.g., OWL and RDF based querying solutions).

There is a natural progression with these systems, as they generally follow the traditional approaches to software designs that are prevalent at the time (e.g., MDA).

## 19.6 Architectures for Enterprise Systems

The majority of "off the shelf" scientific information integration systems are generally targeted toward the end results of research, rather than aiding in the progression of scientific understanding. The reason is that there is a dichotomy that exists between scientific understanding and scientific elucidation: understanding requires the presentation of a formalization of ideas while the elucidation requires the application of intuition and reasoning to extract meaning from a data quandary. Therefore, to support elucidation, we must also focus on understanding the evolving mechanisms through which science is undertaken and the flexibility required for scientific reasoning. This means that it is important to accept that unstructured "bottom-up" development is often required (see Figs. 19.7 and 19.8), and that scientists are often required to work in this manner if they have to carry out their research effectively. Architectural decisions are, therefore, typically driven by the need for functionality and rapid development rather than shared data models and unit tests. The traditional top-down approach to design is rarely used, with a bottom-up approach being prevalent and arguably preferred (Quackenbush et al. 2006). In the top-down approach, models are defined *a priori,* and the designer uses them as the "lingua franca." By contrast, in the bottom-up approach, models and designs arise out of ad-hoc development and are user driven.

In the future, we can expect the majority of all scientific software design to be less, not more, structured. This lack of structured design is part of a growing general trend, as the predicted rise in the number of people developing software and the simultaneous lowering of their design skills (Scaffidi et al. 2005) is being driven by a number of factors. These factors include the simplification of the development process and the development of new design tools. Aspect oriented programming,

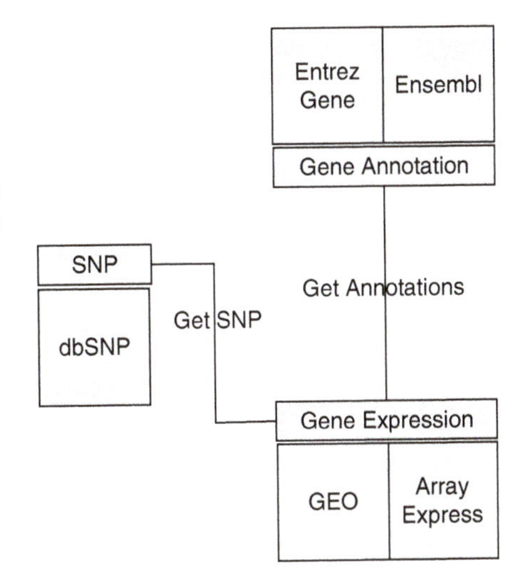


**Fig. 19.7** Top-down architectures are those where a common data model is agreed upon and shared by all the components within the system. The systems are generally more robust but are more costly to produce. The problem with top-down approaches in science is one of adoption, as it is difficult to agree on a common data model that is suitable for all

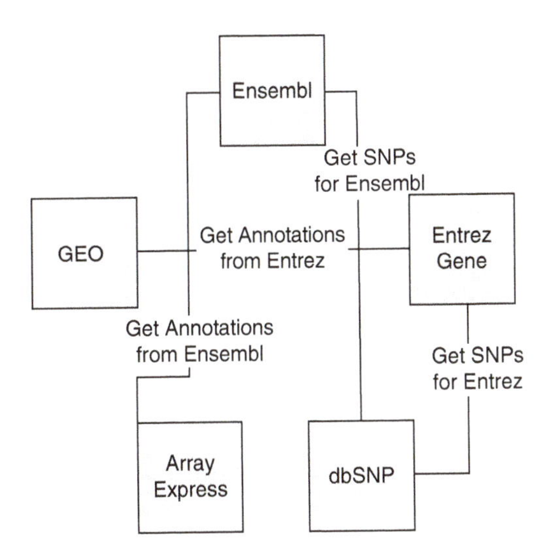


**Fig. 19.8** Bottom-up architectures are those where the design and data models arise from the data. In most cases this means that there is ad-hoc integration through the use of direct data transformations, rather than a through the use of a common data model. These systems are easy to build, but as there is no abstraction they can be fragile to change

and related ideas, is a simplifying development by providing a means to provide slices through typical software structure. These new techniques require less formal design, as they enable the incremental development of complex programs from simpler ones (Batory 2004) and are making enterprise integration more dynamic (Vambenepe et al. 2005). This trend is exasperated, when developing for research, as within the scientific community there exists a rich and varied user base, which is complex both in terms of software experience (typically including lab scientists using macro languages, statisticians and computational biologists using script based tools, and software engineers developing enterprise solutions) and the plethora of computing languages that are used.

It is not unheard of for developers, within a research environment, to integrate a new process or piece of equipment without knowing exactly how it is to be used. This means that flexibility, both in terms of design and ways of representing information are important. As discussed above, scientific enterprises can operate better through a bottom-up service oriented, rather than top-down application oriented, architecture. To achieve this flexibility, it is possible to combine a bottom-up approach with a top-down architecture with the aim of providing a system that is interoperable, so that clients and servers can be written in the scientists' language of choice; flexible, so that new analysis tools and data services can easily be integrated within the platform; and non intrusive, so that architectural decisions do not impinge on the scientists developing and using services. The use of loosely coupled service oriented systems, with no enforced data type semantics, has led to growing support for such "middle-out" development within the biological scientific community. With "middle-out" systems, top-down models are present but are not tightly bound to the individual data stores. There are a number of patterns for providing such middle-out solutions:

- *Ontology defined registry of services.* A central registry service can be used to store information about a set of individual services. This means that the formal top-down descriptions are detached from the underlying bottom-up services. Typically these descriptions are standardized through the use of an ontology and describe the analysis methods and the data that the registered services provide. This "separation of concerns" means that the services can be written easily and the definitions can be updated independently of the individual services. Such approaches do have limitations in terms of complexity of data/algorithms that can be described and do require good coordination to function. A number of specifications which follow this pattern have been proposed, the most prominent being BioMoby.
- *Unifying identity based services.* A strong unifying identity system can be used to link low level bottom-up services into a structured top-down infrastructure. The identity system must allow for the integration of a number of independent identity schemes and should, therefore, be hierarchical or namespace based (e.g., PURLs). With a unified identity scheme each item can be referenced, and so structure can be imposed through the use of additional services (e.g., relationship services to provide information about links between items); the use of a project system (e.g., to store information about collections of items and annotations); or through the association of meta data as structured information (e.g., LSIDs).
- *Semantic Web.* The Web 3.0 does provide for a means to develop middle-out solutions with each developer defining his or her services using an ontology. The problem then becomes one of how to map between the different ontologies, which can be done through the use of a higher level domain specific ontologies (e.g., OBI) or mapping ontologies. The semantic web offers a range of technologies which are useful (OWL for definitions and SPARQL for access/querying), although this work is still ongoing. The advantage of this approach is that data

and semantics can be served out using the same mechanism, although alternatives are available which are simpler (e.g., microformats).

These middle out solutions have limitations, as they attempt to overlay structure on ad-hoc and unstructured services. Generally, middle-out solutions sacrifice richness of functionality to provide flexible systems. Typically with such architectures, hybrid solutions are the most suitable (e.g., Semantic Web based services with a strong unified identity system for categorizing experiment data).

### *19.6.1 Server Frameworks*

There are numerous types of "server frameworks" which can be used to simplify the development of robust server based functionality. Typically a server provides for: operations that should be shared among a number of clients; operations that are too computationally expensive to execute on client machines; serving out shared data; a means for data integration; or a central point to access the latest version of a curated data set or a specific algorithm.

A number of different types of server frameworks exist, the main two are

- *Application Servers*. Application servers are largely designed for integrating and distributing slowly evolving data sources. As such, they have flourished in certain areas of the life sciences and are mainly used in commercial companies. Historically, they had a reasonably high development cost, although since about 2005 this is no longer the case. The purpose of application servers is to simplify the development of distributed systems, managed the life cycle of applications objects (server components), and also to provide performance and resource management.
- *Content Management Systems*. Content managements systems focus on the generic qualities of data and information and generally do not necessitate a detailed understanding of the actual data semantics. The purpose of a content management system is, as the name suggested, the management of content (or data). They typically provide versioning, locking, and searching functionality and are ideal for unstructured or continually evolving data.

#### **19.6.1.1 Application Servers**

As stated above, application servers (see Fig. 19.9) are designed to make the development of robust distributed applications easier. They provide mechanisms for controlling resources associated with applications, and also provide key functionality that would be time consuming (and wasteful) for individual developers to have to write themselves (e.g., security, messaging, leasing, concurrency). Broadly speaking, within the life sciences, three types of application server exist:

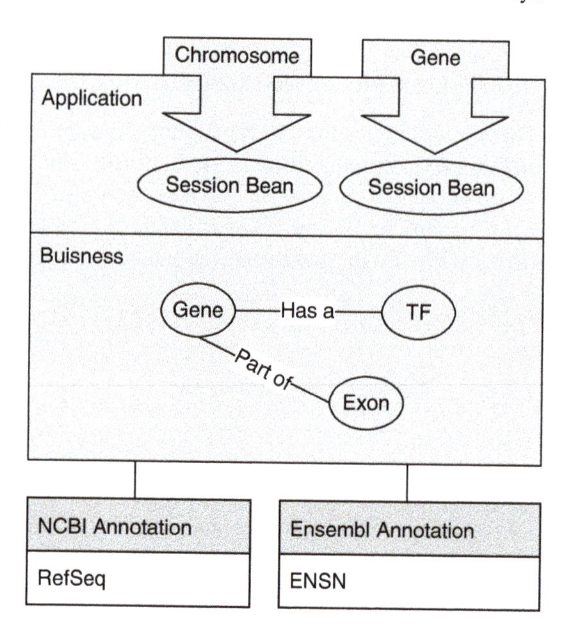


**Fig. 19.9** Application Servers provide a means to run an application within a controlling framework. Typically application servers consist of domain (or business logic) and publication (or application) layers. These are typically used to integrate information, provide common functionality and to serve out structured data

- *Enterprise Java Bean Container Application Server.* An Enterprise Java Bean (EJB) is a server side component that lives inside an EJB Container. The developer writes an EJB according to a specification which defines both how the bean will be managed and how the bean can access core services which provide useful functionality. The container controls the life cycle of the bean, facilitates access to core services, and manages server-based resources. The services that are available through a EJB Container provide: security management, including method level and ACL; transaction management; including 2 phase/distributed commits; life cycle management, including pooling of bean instances and swapping beans in/out (persisting) of memory for resource management; naming/directory management, typically through JNDI; persistence management; using ORM tools such as hibernate; remote access management, so that the bean can be accessed via RMI-IIOP/CORBA and Web Services; and a number of utility services (e.g., mailing, clustering, caching, monitoring, etc.). There are different types of EJBs, and each has a different purpose: an *entity bean* which serves as a data cache from an underlying data store, this is used for transformation and data integration logic; a *session bean* which is typically used to hold application logic which communicates with information stored within entity beans; a *stateless session bean* which typically represent simple stateless logic and generally act as end points for high level services; and a *message bean* which are used to pass messages between the other beans. The EJB standard has existed for some time and has undergone a number of revisions: the EJB 1.0 standard (1998) specified that a valid container should support bean component life cycles; the EJB 1.1 standard (1999) expanded the previous specification to mandate that entity beans with fully managed persistence (Container Managed Persistence) would be required for compatibility; the over simplicity and static nature of EJB 1 lead to the definition of the EJB 2.0 standard (2001), which defined local relationships to increase efficiency, introduced message beans to aid intra-container communication and EJB QL (query language for entity beans) to facilitate rapid application development; EJB 2.1 (2003) improved upon the standard by introducing timers and automatic stateless Web Service mappings; the EJB 3.0 standard (2006) represented a major change as the complexities of developing EJBs were inhibitive for most projects, so a simplified process of building EJBs was outlined (through the use of detachment, dependency injection, and aspects). Using EJBs is now reasonably simple, although the set up of a separate container (including all the security and configuration issues) is still an overhead. While JBOSS (www.jboss.org) is the EJB container of choice, other free solutions are available and should be considered depending upon licensing conditions (e.g., glassfish).

- *Framework application servers.* The common choice for this type of simplified application server is the Spring Framework (www.springframework.net). Although EJBs represent a powerful technology it is recommended that initially developers look at Spring, as its lightweight nature and easy to learn programming model will ease the development of application server based applications. Spring has not had such a protracted history as EJBs, mainly as it could learn from the problem with adoption of EJBs. Spring was released in 2002, with a major iteration in 2004. Since then, it has had a healthy increase in usage and has always relied heavily on inversion of control (dependency injection), to help support separation of concerns and provide flexible and portable code. The popularity of Spring is largely due to its usage of aspects to simplify coding. Aspects can be defined through annotations, property files, or separate XML configuration documents. The core of the platform allows for the development of POJOs, which are created and configured (including code injection) from the main Application Context, dependencies are then injected at run/compile/instantiation time through the use of setters or constructors. A number of core frameworks make developing applications using Spring relatively easy. The main frameworks that are of most apparent used by developers in the life sciences are the framework(s) for accessing data, these consist of a data management framework for accessing data sources and a remote access framework for accessing information from application servers (e.g., EJB) and via high level protocols (SOAP, http); and framework(s) for publishing data, Spring supports MVC to make the development easier, additionally the remote access framework can also be used to provide access to information although SOAP (typically via Apache axis). Although Spring's over reliance on aspects can add another level of confusion (particularly when debugging code), these disadvantages are offset by their easy usage and the encouragement to adopt good coding practices.
- *Light weight application servers.* A number of light weight application servers exist, which provide limited or specific functionality. The most popular of these is Tomcat from Apache. While it is a light weight application server, in terms of

services it provides, it is still a powerful product due to the technologies it supports (and in fact largely developed). These technologies include servlets, soap, jsp, axis, and portlets. It is a sign of application server maturity that it is possible to integrate other application servers with Tomcat. It is also worth noting that other application servers exist (including those that are embedded inside of operating systems and large enterprise database management systems).

In nearly all instances, where an application server is desired, it is recommended that you first consider the use of Spring. Spring can also be used to enhance other projects as it easily integrates with other systems.

#### 19.6.1.2 Content Management Systems

Content management systems (see Fig. 19.10) have evolved considerably, from static views on versionable document stores to dynamic structured service oriented data systems. A content management system provides useful functionality when dealing with data files and documents and serves a useful purpose in experimental sciences where it can be used to: share experiment data in a convenient manner (from WebDAV to more sophisticated bespoke browsing systems); provide a means to version and lock data files; allow for a means to track provenance of data; allows for different methods for searching and annotating the data files; and provide for a means to separate out data organization issues from the underlying storage mechanism.

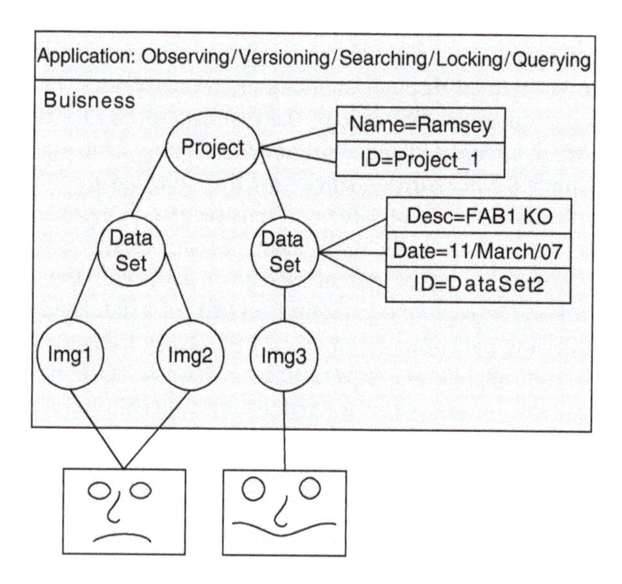


**Fig. 19.10** Content management systems are designed to handle generic data items. They are suitable for dealing with raw and processed experiment data files, which can also be versioned and annotated. With these systems, a series of higher level organizational data structures can be overlaid to make navigation (and searching) easier

In the past, the proprietary nature of content management systems meant that there was little adoption within the research community. However, over the last few years, there has been a resurgence in interest in them from other disciplines, which has meant that there now exist a number of freeware solutions which can be used.

Content management systems serve a different purpose to application servers, but can be used to serve out similar types of information using the same protocols (e.g., SOAP or similar). These systems are principally useful when unstructured experiment information is being used. Alternatively, application servers should be used where: there exists reasonably stable information which can be structured in a DBMS or similar; business logic is required to operate closely on the data; or integration logic or data transformation is required.

There are three types of content management systems which can be considered:

- *Java Content Repository (JCR) based solutions*. The definition of a content repository is that it is a system that can be structured to form a content management system. A standardized definition exists for content repositories, and the most suitable implementation is Jackrabbit from Apache (jackrabbit.apache.org). A JCR implementation offers a range of services, which are categorized by levels: *level one* provides read functionality (e.g., basic querying and structured content); *level two* provides simple write functionality (security and history); and some groups refer to a *level three* which provides advanced features (transactional control and structured searching). A JCR implementation allows for the definition of an organizational structure (as interlinked nodes of different types) to which properties (annotations and data items) can be added. Standardized services are provided for versioning and transactional behavior, as well as more advanced services for both unstructured (e.g., lucene based information retrieval) and structured querying (e.g., SQL or xpath based searching).
- *SPARQL (SPARQL Protocol and RDF Query Language) based solutions*. SPARQL is designed to allow for the querying and retrieval of documents across multiple unstructured data stores. It is designed to be a unifying communications and interface standard (and in this way shares similarities with Z39.50). The power of the system is the distributed RDF documents (or other data stores) remain unchanged, but queries can be run across them – and so it fits well with a "bottom-up" approach. If considering a SPARQL system then the reader should carefully consider the implementation that is to be used (especially in regard to both performance and licensing issues). Such a unified approach to accessing information is required to make the semantic web (Web 3.0) a reality, and there do already exist some implementations. The open source version of the Virtuoso Universal Server (www.openlinksw.com) from OpenLink is a comprehensive solution, although this is released under the GNU license which may cause problems for certain institutions.
- *Code Repository based solutions*. It should not be forgotten that simpler solutions do exist for basic concurrency control and versioning. These are generally used for simple document or code management and a plethora of tools are available that work with these standardized solutions. The two most popular solutions are CVS and Subversion, which will be readily available on any linux distro.

In nearly all instances, where a content management system is desired, it is recommended that you first consider the use of Jackrabbit.

### *19.6.2 Service Oriented Architectures*

SOAs are becoming more prevalent within most enterprises. These are architectures that involve the definition and registration of individual services that offer specific functionality. These can be considered bottom-up architectures, although the cornerstone of any such architecture is going to be a formalized registry framework with a service based taxonomy which allows for the dynamic discovery of services. The loosely coupled ad-hoc nature of these architectures makes them highly suitable to use in research. However, for a SOA to work with a research enterprise, there does need to be a high degree of "political will" to change modes of working to ensure adoption. Basically, due to their nonintrusive nature, people must be made aware of their existence and made aware of the benefits of using a formalized architecture.

Most SOAs are built around the concepts of Web Services and associated standards. Web Services are a suite of protocols that standardize the invocation of remote applications using common high level communication protocols. The naming of this set of technologies as "Web Services" is largely due to their main usage being through the passing of messages via http. The mechanisms through which Web Service operate is simple: a message is received and a reply is sent, the structure and format of the message differs depending upon the type of Web Service that is being used.

The attraction of Web Services in the biological sciences is primarily due to:

- *Simplicity*. These are relatively easy to build and have a short learning curve. A large number of tools are available which hide the complexities, of extracting message based information and marshalling (serializing) to languages specific types, from the application developer. The advent of AOP has also simplified the development and deployment process (see Table 19.2).
- *Interoperability*. As messages are encoded into a XML document based intermediate format, these services have a high degree of interoperability. With care, Web Services can relatively easily interoperate between the major programming languages being used in bioinformatics. This interoperability comes at a cost in terms of the level of complexity that Web Services can offer, although this rarely becomes an issue.
- *Reliability*. Web Services are generally implemented in a stateless manner and use high level protocols for communications, which means they rarely go wrong. Their simplicity is their strength, they are reliable and will stay up with little (if any) administration overhead. Additionally, they are relatively easy to debug, especially compared to binary communication mechanisms.

Although all Web Services operate in the same manner, the high level protocol is used to send a message to a specific endpoint, there do exist three main types:

**Table 19.2** Using annotations to define POJOs for Web Service endpoint

```
  @WebService
  @SOAPBinding(style          =          SOAPBinding.Style.DOCUMENT,use=
SOAPBinding.Use.LITERAL)
  @Stateless
  Public class HelloWorldService implements HelloWorld {
      @WebMethod
      public String getString () {
          return "Hello World";
      }
}
```

Java snippet showing how annotations can be used to specify a SOAP Web Service within any standardized Java Application Server (e.g. Spring, JBOSS). The annotations state that this POJO should be mapped to a Web Service, and the encoding will be document/literal. The HelloWorld method is to exposed through this Web Service. The annotations also specify that this POJO is a Stateless EJB, and so if this class exists within JBOSS then this POJO will be instantiated as a stateless bean and this will be mapped to the Web Service instance.

```
  [WebService]
  [WebServiceBinding(ConformsTo=WsiProfiles.BasicProfile1_1)]
  public class HelloWorldService:WebService
  {
      [WebMethod]
      public string HelloWorld() {
          return "Hello World";
      }
}
```

C# snippet showing how annotations can be used to specify a SOAP Web Service within IIS. The annotations specify that this class will provide the implementation of the Web Service, and the encoding should follow the WSDL 1.1 basic profile. The HelloWorld method is exposed through this Web Service.

- *SOAP*. Historically SOAP was an acronym for Simple Object Access Protocol, but as it has evolved this definition is becoming less meaningful. SOAP is a means to encapsulate requests for information to an application and to receive their response (see Fig. 19.11). SOAP has been around for about a decade, but its aim to produce a light weight interoperable protocol had problems due to incompatible implementations and multiple methods for structuring information. Most of the problems arose as there were different ways of defining the target for a call (e.g., *RPC* methods which tightly coupled the call to a specific method, or *document* methods which provided a "loosely" coupled mechanism) and the way in which data were structured within the message (e.g., *encoded* using external complex language specific types, or *literal* where all data items were defined). The WS profiles provide the required standardization, and SOAP implementations are now interoperable (although problems still exist). SOAP achieves this high level of interoperability due to the fact that the interface for a specific service offers is formally defined in a WSDL (Web Service Description Language) document. This WSDL document can be automatically downloaded, and tools can use it to generate convenience classes for specific languages, so that no XML parsing code needs to be written by the developer. When writing a WSDL, the use of

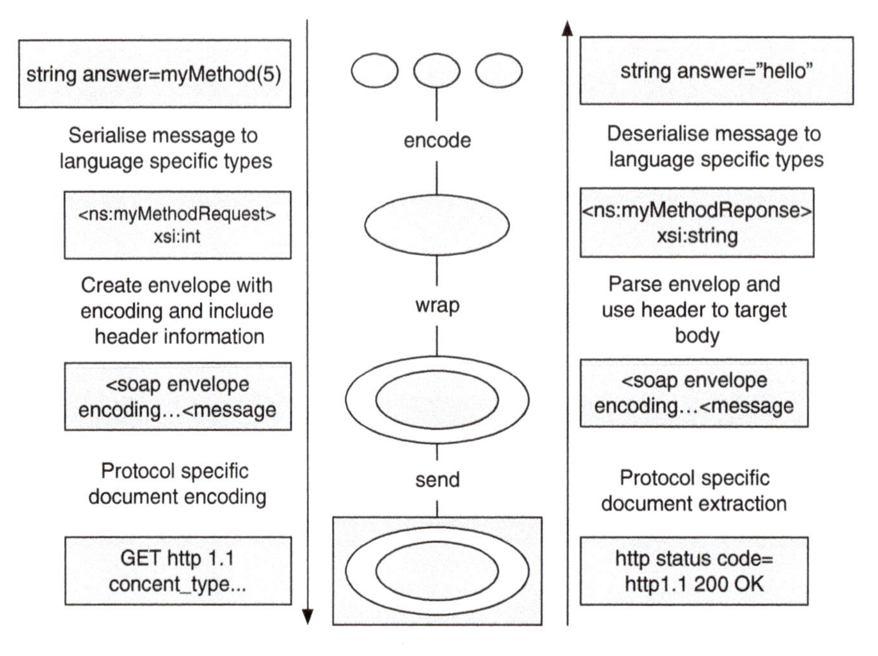


**Fig. 19.11** The protocol for transporting messages using SOAP is described as a stack, with different levels of information being available at each level of the stack. When a method is invoked by a client it is first encoded as XML, which includes information about the types and values of the arguments, as well as the remote method information; second a header is added which contains extra information which is required, with the adoption of WS-* standards the complexity of this header will increase over time; third the envelop is added which contains encoding information relevant to the encapsulated document; and last information related to the transport protocol is added (e.g., http or SMTP headers). The response message is returned and must be decoded using the reverse of the encoding stack: the transport protocol information must be used to extract the SOAP envelope; envelope information is used to decode the header and the body; the header information is used for specific targeting and other operations; and the body is then decoded and the results returned to the calling method

profiles with literal/document "styles" ensures a high level of interoperability. Associated with SOAP are a large number of standards which provide useful additional functionality, these are the WS-* standards. More information about these standards can be found at OASIS (www.oasis-open.org). The most important one of these, to the life science developer, is the WS-RF (resource frameworks) which allow for a means to provide stateful web services.

- *REST*. Representational State Transfer (REST) has gained popularity, as it retains some of the power of SOAP but is considerably easier to implement. The idea is to use preexisting technologies as the basis for the protocol (e.g., "the web is the platform"). There exists some confusion about what represents a Restful service, rather than just an HTTP encoded request for an XML document. True REST is based upon the verb/noun/type based calls, where you apply an operation (verb e.g., POST, GET, PUT and DELETE) to a URI (noun) with a certain view (type).

- *Definition based extension.* A number of extensions and new "Web Service like" protocols have been developed by the life science community. The majority of these were defined due to deficiencies in the broader community standards. These standards largely define how data are served out from web services and how they are described, and thus, serve to improve integration (e.g., BioMoby, caDSR within caGRID/caBIG).

Generally a REST based service should be considered first, however if complex functionality (e.g. stateful calls, true interoperability, distributed transactions) is required then it is recommended that Web Services are constructed using SOAP. The advantages of SOAP are the plethora of tools that are available, due mainly to the high level of standardization (and commercial backing). Most SOAP tools will automatically generate REST based interfaces as well.

When developing SOAP interfaces you should be aware of:

The disadvantage of SOAP is that it is largely static and difficult to implement.

- *Debugging issues.* The parsing of SOAP messages typically occurs in a chain, with specific information being added or removed between elements of the chain. This means that actual complexity involved in debugging differs enormously depending on the programming environment that is used. The use of proxy tools (e.g., ProxyTrace from www.pocketsoap.com) can also aid in debugging, as the actual message documents can be captured and examined. Additionally, there are frameworks available that can be used for testing SOAP services (e.g., SOAPUI from www.soapui.org).
- *Different implementations exist for different languages.* For JAVA, Apache offers Axis, and a wide range of implementations for a number of the different WS-* standards. Alternatively, the Globus Alliance (www.globus.org) also offers implementations. Unfortunately, when working with Perl, the simplest (and best) way is to directly access the document messages using XPATH.
- *Interface definitions.* When building an interface for a service two mechanisms are available: class driven, where the WSDL is automatically generated from an underlying implementation; or interface driven, where the WSDL is created by the developer and the language specific stubs are generated by a tool. To promote interoperability, it is always better to start with a WSDL, and generate the language specific bindings (e.g., top-down).

### 19.6.3 Orchestrating Services

Control and data flow across distributed services typically require a system of message passing through the use of a high level control systems (see Fig. 19.12), for example, an enterprise bus or similar. An enterprise bus is a powerful architecture that allows for the high level orchestration of distributed services. These typically operate in an asynchronous message based manner, as a message queue or scheduling system is required for their operation. Asynchronous messaging has shown itself to be a stable and versatile method for developing enterprise systems,

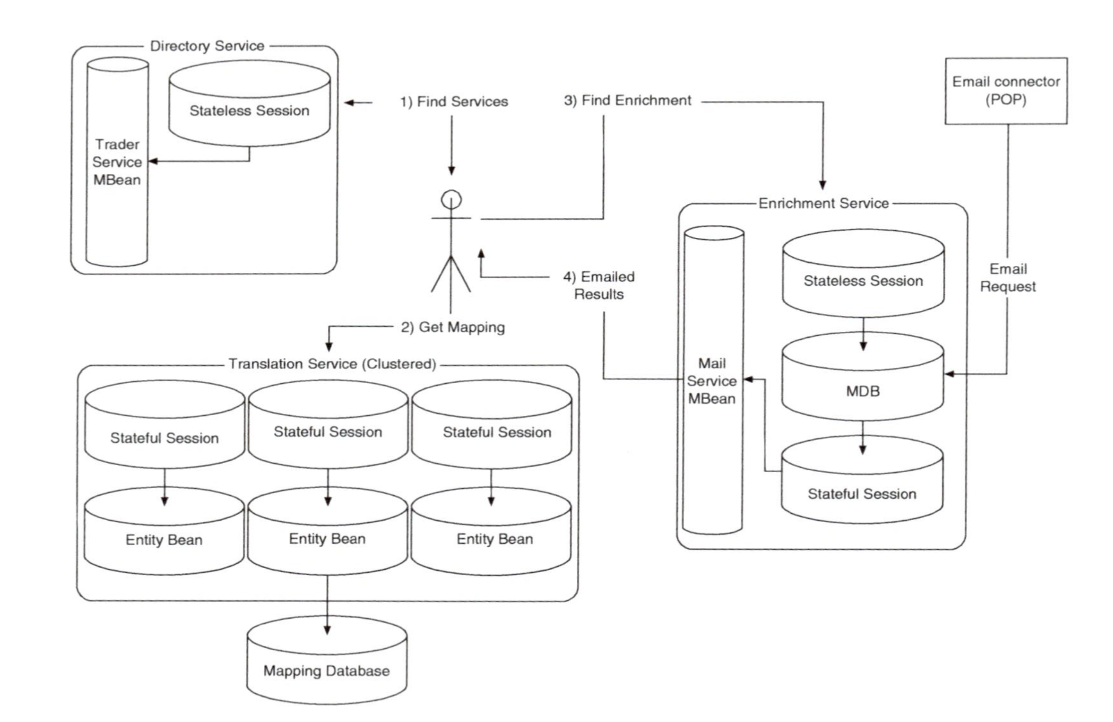


**Fig. 19.12** Example of an enterprise architecture based around services. The services themselves typically hide an underlying application server or similar. To be able to coordinate data flow between these low-level services a top-down definition can be imposed. This orchestration on the services is typically done through a high level message passing architecture. The example above is a simplified version of an analysis, where the orchestration allows for: (1) the discovery of the required services; (2) the retrieval of relevant (chromosome) mapping information to retrieve relevant sets of genes; (3) an analysis on the sets of genes (in this case, looking to see if they have similar characteristics); (4) use of a synchronous message system to perform a notification when the analysis is complete

because it has a high tolerance for failure and can behave in a failsafe manner (so jobs and state are not lost during a serious error).

Typically, within an enterprise bus, a workflow is established and then executed. This workflow is of the form "execute service A, extract information and pass to service B." Depending on how they are coordinated, these workflows can differ enormously in complexity.

If such distributed coordination or distributed processing is required, then two architectures are worth considering:

- Channel based solutions, for example, Mule ESB. Mule is a comprehensive enterprise service bus, which will handle the message passing between distributed services. Mule functions as both a middleware (object broker) and a message system. Mule is simple to use, and orchestration between UMOs (Universal Message Objects, which are simply services) is defined through configuration files. The UMOs are the only code that the developer needs to provide, and as mule adopts a "separation of concerns"/POJO policy the UMOs are concerned solely with the business logic rather than parsing or transportation code. The UMOs can themselves be Web Services. Mule provides standardized components for working with different protocols, these are categorized as: message receivers (and connectors) which understand how to receive communications from different communication protocols, for example, http; routers which target and filter specific messages/events to and from specific UMOs; and transformers which understand specific formats (e.g., SOAP envelopes) and can transform them into alternative representations require by UMOs. Mule is an established technology and should be considered for any distributed processing project as it is easy to use; robust; removes most of the "grunt" work of working with different protocols; reasonably easy to reconfigure; and used by a large number of groups and has a stable development team. More information can be found on the Mule web site (www.mulesource.org).
- *BPEL based solutions.* Web Services do have a high level orchestration language, called the Business Process Orchestration Language (BPEL). This defines a number of high level primitives for controlling: invocation of methods, requesting/receiving data from services; flow control across a number of machines; as well as a series of basic control structures (see Fig. 19.13).

There are also some specialized life science high level orchestration tools that have been developed, examples include those from both commercial and academic groups (e.g., Taverna).

Enterprise buses have their natural place but can be overkill if all that is needed is a simple batch file style of operation. In a large number of cases, bioinformatics "workflows" can be best executed as a toolset, where each tool is executed in a serial fashion with the output of one tool being piped into the next tool in the chain. Toolsets are notoriously prone to failure and difficult to debug; however, they are simple to build. If a toolset is required, then there exist a number of toolset builders and execution environments which will be of use. One such toolset environment is GenePattern (www.broad.mit.edu/cancer/software/genepattern).

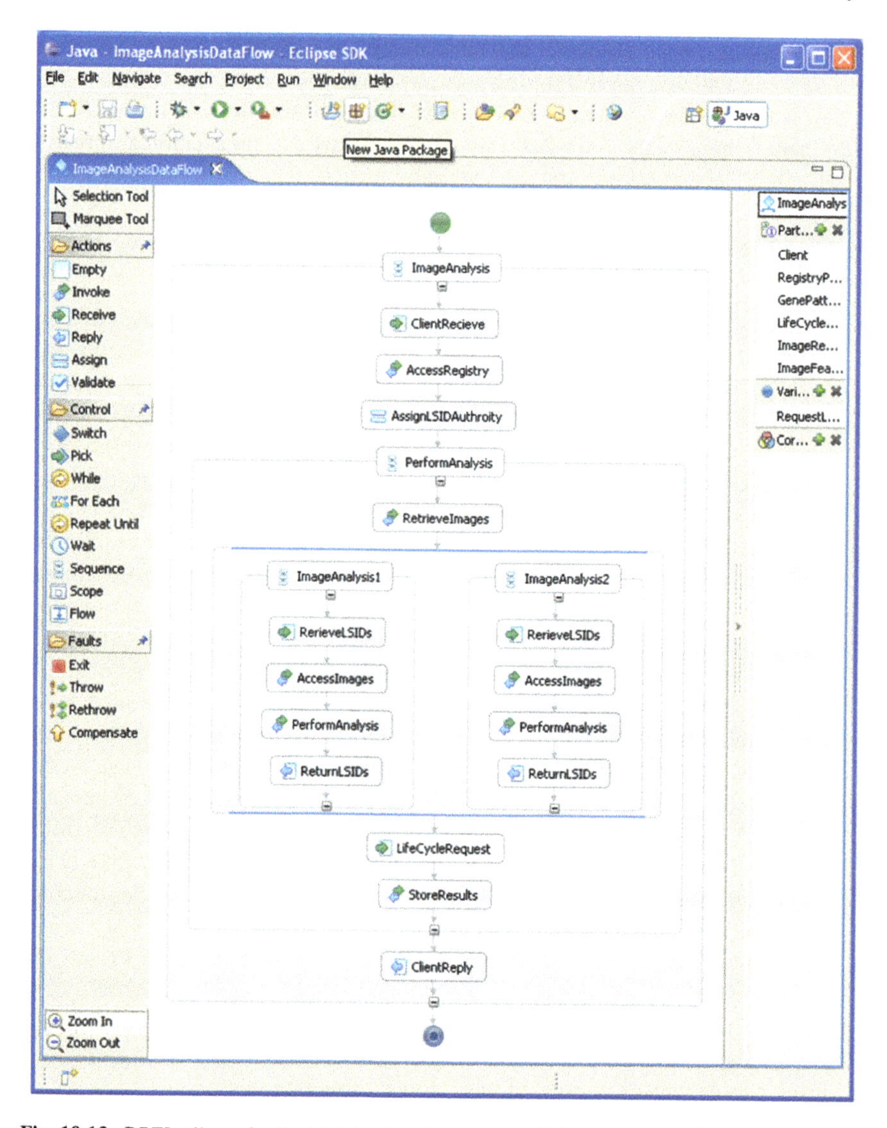


**Fig. 19.13** BPEL allows for the high level orchestration of Web Services. BPEL can be visually specified using a variety of systems. The above example uses the Eclipse plugin, and specifies orchestration of a set of service to perform image analysis

## 19.7 Conclusion

This chapter gives an introduction to programming, and programming languages, for the life sciences. This chapter can only serve as a guide, as the only real way of learning the required skills is to actually undertake software projects.

Within the life sciences, we have been seeing a steady increase in the volumes and complexity of the data being generated by experiments. High-throughput experimentation (e.g., using high throughput sequencing, imaging or proteomics) with output measured in terms of "CD a minute" are already exasperating this trend. This trend is placing further demands on the teams of software engineers working within the life science arena. The demand for rapid development of high quality code is being driven by: the growth in both magnitude and complexity of data; the constant introduction of new automated experiment technologies; and the growing requirements for integration and interoperability from fields such as systems biology.

Fortunately, modern programming languages are now approaching a level of sophistication where we can rapidly develop reliable distributed interoperable applications. Such rapid development can be achieved through a good understanding of how technologies can be adapted and reused, allowing the programmer to concentrate their efforts mainly on the development of the business logic of the software. Therefore, to build well architected and maintainable software, knowledge of both the life sciences and current computing technologies are required. Armed with this knowledge you will be able to meet the challenges of building systems to support biological research.

**Acknowledgments** This work was supported by Grant Number P50GMO76547 from the National Institute of General Medical Sciences. The content is solely the responsibility of the author and does not necessarily represent the official views of the NIGMS or the NIH.

## Glossary

**ADT** Abstract data types pre-date the adoption of object-oriented programming. They provided a means to reuse storage and retrieval structures, and are similar to "generics" (e.g., lists, tables, queues).

**AOP** Aspect Oriented Programming is used to easily apply cross-cutting functionality (e.g. logging) to programs. A programmer typically defines a method as having a particular aspect, and a separate framework will be responsible for ensuring that the correct behavior occurs (e.g., when, how and where the code injection occurs). This cross-cutting can be injected into method calls at a variety of times in an objects life cycle.

**BPEL** The business process execution language is a specification designed to support the high level orchestration of web services. The heart of the BPEL specification is the scripting language which defines how services and data produced by them are linked together. This specification is rich enough to allow for most workflows and defines both how method invocations and data are linked and the how web services should be coordinated (e.g., concurrency, choices, sequential operations). The specification also defines extensions to the WSDL which can be used to specify links between services.

**CORBA** The Common Object Request Broker Architecture supports interoperability between distributed processes (applications). Central to the architecture is an ORB (object request broker) which both marshals data and controls compartmentalization (to allow for invocation on specific remote threads etc.) of the different processes. The specification was defined by the OMG, and ORBS are available for most platforms.

**DCOM** Provides for a means to make distributed calls between COM (Component Object Model) objects. Thread compartmentalization and marshalling (using low level XML interchange) are handled automatically. For application developers, this has largely been superseded by .NET Remoting.*EJB*. Enterprise Java Beans provides a means for writing application servers in Java. A container manages a number of enterprise beans and provides access to common functionality and imposes control over the beans (e.g., life cycle, resource management, transactions and security). The EJB specification has evolved considerably since its first release and is now a feature rich framework which can be used to easily develop complex server side functionality.

**I3C** The I3C was a short lived commercially led organization established to standardize aspects of life science informatics. The organization was led by Oracle, Sun and IBM. The I3C did promote the use of LSIDs, which have been adopted by the OMG.

**IDL** The Interface Definition Language formalizes the remote interfaces that can be accessed through CORBA. IDL has evolved considerably, with the advent of pass-by-value and components (facets). A WSDL serves the same type of purpose for Web Services.

**IIOP** The Internet Inter-Orb Protocol is the means through which Object Request Brokers (ORBS) communicate. This allows for discovery, life cycle and compartmentalization of object requests.

**JCR** The Java Content Repository is a specific Java Standard (JSR-170) for defining the interface to a content repository. A content repository is a flexible system that is typically customized for a specific usage, when customized, it is referred to as a Content Management System (CMS).

**JNDI** The Java Naming and Directory Interface are the specification for the directory and naming system using within Java. The underlying system can use a variety of systems (e.g., RMI Registry) and provides a means to discovery and query resources.

**JSR** Java Specification Requests is the process through which community standards are achieved for Java. The requests are diverse and have led to a number of useful reference implementations.

**LINQ** This is a .NET project that extends the platform to allow for general resource querying from within code. Resources that are queried can then be accessed as objects within the framework.

**LSID** The Life Science Identifier standard provides a concentrate definition and implementation of a URN. The LSID specification outlines how the URN is resolved to two locations (the data and the metadata) through the use of "an authority." In this way, the authority acts as a registry. The documents that are retrieved are returned as objects and an associated RDF data file which encodes the metadata. The standard also encompasses many aspects of using URNS and includes specifications for associated services (e.g., assignment).

**LSR** The Life Science Research group of the OMG defines standard in the "vertical" life science domain. The body has defined and adopted a number of standards. These standards cover a wide range of areas (including the "sequence" and "literature").

**Maven** Maven is a build and artifact management tool available from Apache. Its primary use is for Java.

**MDA** A Model Driven Architecture is one where the model underlying the system is defined in a language independent way, and the corresponding services/classes are automatically pushed out from that model. Typically, the model is defined in UML, and then XMI is used to automatically generate stubs/skeletons which can be used to provide implementations of the model.

**MIDL** The Microsoft Interface Definition Language serves a similar purpose to IDL but is generally based on specifying the remote procedure call interface which is used between COM components.

**MVC** Model view controller pattern is commonly used in both web application frameworks and GUI frameworks. Commands are managed by the controller, which directs changes to an underlying model, and (multiple) views provide representations of the model.

**OASIS** The Organization for the Advancement of Structured Information Standards is a standard body made up of members from a large number of organizations. They have been particular effective in driving forward standards for Web Service extensions.

**ODBC/OLEDB** The Open Database Base Connectivity is a definition of the interface presented by a DBMS. The ODBC specification is well established and bridges with other technologies (including JDBC). The OLEDB is an extension to the ODBC offer richer functionality.

**OMG** The Object Management Group is an open not for profit standardization body. The OMG have produced a number of horizontal (e.g., Trader service, Naming service, Event Service) and vertical (see LSR) standards for use with CORBA.

**OMT** The Object Modeling Technique is a predecessor to UML and provides a formal representation of the design of software.

**ORM** Object Relation Mapping provides a means to map object onto relational databases, and to map relational databases into objects. A number of ORM solutions are available with hibernate being the most prevalent.

**OSGi** OSGi is a standards organization which provides a framework for building applications. The framework provides for both a means for components within an application to be discovered, and also an updating mechanism.

**OWL** The Web Ontology Language is an RDF description of an underlying data resource. The ontology describes the data items produced through a web service as well as the relationships between them.

**P(M)BV** Pass (or Marshall) By Value in distributed systems allows for objects to be moved between nodes, rather than using remote references.

**POJO** A Plain Old Java Object is one that uses "separation of concerns," so that only business logic (and not, for example, server logic) is implemented. Any required dependencies and services are injected after the code has been written.

**PURL** A Persistence URL is one that points to a resolution service, which ensures that the underlying resource can always be located with the PURL.

**QL** In the EJB 2.0 standard, a query language was introduced, this was originally to standardizes the "finder" logic in the now obsolete EJB Homes.

**RDBMS** A relational database management system is the environment in which relational database instances exists. A RDBMS provides a unified framework which can be used to control the physical (tablespaces), conceptual (logical schemas), and external (views) of databases.

**REST** Representational State Transfer (REST) can be considered an alternative to SOAP, although it is considerably easier to implement. REST uses pre-existing technologies as the basis for the protocol (e.g., "the web is the platform"). There exists some confusion about what represents a Restful service, rather than just an HTTP encoded request for an XML document. True REST is based upon the verb/noun/type based calls, where you apply an operation (verb e.g., POST, GET, PUT and DELETE) to a URI (noun) with a certain view (type).

**RMI** Remote Method Innovation is a Java-to-Java solution for communication between distributed Java threads/applications. RMI uses a number of abstraction layers (remote reference layer/RRL and transport layer), this has a number of advantages including the fact that different underlying protocols can be used to actually provide the communication (e.g., IIOP). Marshalling is done through serialization, leasing is available, and distributed GC is supported. RMI is a convenient, but not interoperable, protocol.

**RUP** The rational unified process was a software development process that was popularized through the release of Rational Rose and associated tools. It centered on UML and provided a means to gather use cases, match them to features, and track feature development and defects. Its popularity has decreased significantly over the last decade.

**SOA** A Service Oriented Architecture is one which consists of loosely coupled federated services. There is typically little linkage between these services, and they are generally discovered dynamically using a registry system or similar. SOAs have grown in popularity within many enterprises, as they provide a practical and convenient for disparate groups to share information/processes.

**SOAP** SOAP is a protocol for making requests on remote services to return structured data. It is designed to use any high level protocol that supports the sending of information and is primarily used with http. Much like CORBA, interoperability is the big draw of SOAP, and (unlike CORBA) SOAP has the advantage of being simple to develop and test. The original stateless nature of SOAP limited its usage; however, with the advent of WS-RF (and other standards) SOAP is maturing into a general purpose object protocol.

**SPARQL** The SPARQL Protocol and RDF Query Language are designed to allow for the querying and retrieval of documents across multiple unstructured data stores. The power of the system is the distributed RDF documents (or other data stores) remain unchanged, but queries can be run across them – and so it fits well with a "bottom-up" approach. Such a unified approach to accessing information is required to make the semantic web (Web 3.0) a reality, and there do already exist some implementations.

**UML** The Unified Modeling Language formalizes visual representations for most aspects of software design. This formalization encompasses uses cases, class structure, state transitions, sequence of method calls, and deployment scenarios.

**URN** A Uniform Resource Name is a type of URI (Uniform Resource Identifier). It is the logical counterpart to a URL, in that it provides the name of a resource rather than the exact location of a resource. A number of URN implementations are available, including LSIDs.

**WS-*** The WS-* are a series of specifications for adding functionality to SOAP. These extensions provide new functionality such as security, messaging, binary object attachment and state. These extensions generally involve the addition of information to the SOAP message (within the envelope). State information can be maintained between SOAP calls through the use of resource frameworks (e.g., WS-RF).

**WSDL** The Web Service Description Language provides a means to specify the interface exposed by a SOAP Web Service. The WSDL document can be automatically retrieved, and tools can be use to generate convenience classes for specific languages, so that no XML parsing code needs to be written by the developer. When writing a WSDL a number of standards (e.g., WS-I) are available to ensure interoperability, typically though the use of profiles with literal/document "styles."

**XP** Extreme Programming was largely a reaction to the "over specification" that was proposed though the RUP. XP advocated a number of approaches that were designed to ensure more responsive (Agile) well written code could be developed.

## References

Booch G (1996) Object Solutions: Managing the Object-Oriented Project
Wilkinson M, Links M (2002) BioMOBY: An open source biological web services proposal. Brief Bioinform 3(4):331–341
Etzold T, Ulyanov A, Argos P (1998) SRS: Information retrieval system for molecular biology data banks. Meth Enzymol 266:114–128
Birney E (2004) An overview of Ensembl. Genome Research 14(5):925–928
Haas LM, Schwarz PM, Kodali P, Kotlar E, Rice JE, Swope WC (2001) Discovery link: A system for integrated access to life sciences data sources. IBM Syst J 40(2):489–511
LSR. (cited: Available from: http://www.omg.org/lsr)
caBIO. (cited: Available from: http://cabio.nci.nih.gov/)
Senger M, Rice P, Oinn T (2003) SOAPLab – a unified Sesame door to analysis tools. In: UK e-Science All Hands Meeting, Nottingham, UK
Reich M, Liefeld T, Gould J, Lerner J, Tamayo P, Mesirov JP (2006) GenePattern 2.0. Nat Genet 38:500–501
Marzolf B, Deutsch EW, Moss P, Campbell D, Johnson MH, Galitski T (2006) SBEAMS-Microarray: Database software supporting genomic expression analyses for systems biology. BMC Bioinformatics 7:286–291
LexGrid. (cited: Available from http://informatics.mayo.edu/LexGrid/)
Goble C (2005) Putting semantics into e-science and grids in proceedings E-science. In: 1st IEEE international conference on e-science and grid technologies, Melbourne, Australia
Oinn T, Addis M, Ferris J, Marvin D, Senger M, Greenwood M, Carver T, Glover K, Pocock M, Wipat A, Li P (2004) Taverna: A tool for the composition and enactment of bioinformatics workflows. Bioinformatics 20(17):3045–3054
Cao J et al (2003) GridFlow: Workflow management for grid computing. In: 3rd international symposium on cluster computing and the grid. IEEE
Covitz P, Hartel F, Schaefer C, De Coronado S, Fragoso G, Sahri H, Gustafson S and Buetow K (2003) "caCORE: A common infrastructure for cancer informatics" Bioinformatics, 19, 18, pp 2404–2412
Amin K et al (2004) GridAnt: A client-controllable grid workflow system. In: 7th international conference on system sciences. IEEE: Hawaii
Quackenbush J et al (2006) Top-down standards will not serve systems biology. Nature 440(7080):24
Scaffidi C, Shaw M, Myers B (2005) Estimating the numbers of end users and end user programmers. In: Proceedings of 2005 IEEE symposium on visual languages and human-centric computing, Dallas, Texas
Batory D (2004) Feature-oriented programming and the AHEAD tool suite. In: international conference on software engineering, Edinburgh, UK
Stajich et al (2002) The bioperl toolkit: Perl modules for the life sciences. Genome Res (12):1611–1618
Vambenepe WT, Thompson C, Talwar V, Rafaeli S, Murray B, Milojicic D, Iyer S, Farkas K, Arlitt M (2005) Dealing with scale and adaptation of global web services management. In: IEEE International Conference on Web Services (*KWS 2005*), ISBN 0-7695-2409-5

# Index

**R**

www.ingramcontent.com/pod-product-compliance
Ingram Content Group UK Ltd.
Pitfield, Milton Keynes, MK11 3LW, UK
UKHW020812210125
453990UK00001B/6